Introductory Algebra

Introductory Algebra

Dennis Bila
Washtenaw Community College

Ralph Bottorff
Washtenaw Community College

Paul Merritt
Highland Park Community College

Donald Ross
Washtenaw Community College

WORTH PUBLISHERS, INC.

Introductory Algebra

Copyright © 1975 by Worth Publishers, Inc.

Printed in the United States of America

Library of Congress Catalog Card No. 74-84641
ISBN: 0-87901-037-1

Second printing, August, 1977

WORTH PUBLISHERS, INC.
444 Park Avenue South
New York, New York 10016

To our students

Preface

For many years we have struggled with the problem of providing well-written programmed materials for those students who wish to learn mathematics. Many of these students have had unrewarding prior experiences with math and are reluctant to take further courses. Our attempts to help them have been hindered by the lack of adequate text material. As a result, we began writing semiprogrammed booklets, each covering a single concept, to supplement the textbook we were using. We were encouraged by the positive reactions of our students to these booklets and decided to develop a series of semiprogrammed texts written specifically for students similar to those in our own classes. *Introductory Algebra* is the second in a series which also includes *Core Mathematics* and *Intermediate Algebra*. We sincerely hope that your experiences with these texts will be as rewarding as ours have been.

Introductory Algebra is intended to be used in a one-semester course for which students need only a background in fundamental operations with whole numbers. Students who require a working knowledge of algebra for their later study of intermediate algebra or other courses are provided here with a rigorously class-tested worktext which allows them to proceed at their own pace, and which reinforces them at each carefully graded step along the way.

Chapter 1 introduces the symbols and vocabulary of sets that are used throughout the text. In this chapter the basic properties of whole numbers are reviewed.

Chapters 2 and 3 review operations with integers and rational numbers, and additional properties of integers and rational numbers are included where they are appropriate.

By Chapter 4, it is expected that students will have acquired sufficient recognition and recall of the basic properties to be able to use them in the simplification of algebraic expressions. This chapter provides an opportunity to develop skills in the use of the basic properties.

Chapter 5 develops the methods of solving equations by use of the principles of equality involving addition, subtraction, multiplication, and division.

Chapters 6 and 7 carefully build confidence and present basic strategies having to do with the difficult area of solving verbal problems. The student is provided with techniques which simplify analysis and permit him to set up appropriate equations.

Chapter 8 presents the solutions to linear inequalities in one variable. Special emphasis is given to the interpretation and use of "and" and "or" in mathematical sentences.

Chapter 9 introduces the notion of a relation as a set of ordered pairs and develops the graphs of linear equations and inequalities in two variables.

In Chapter 10, systems of equations in two variables are solved, first by graphing. Then algebraic methods of addition and substitution are presented. Verbal problems are provided so that students will have the opportunity to practice these techniques and compare them with procedures learned in Chapter 7.

Operations with polynomials are studied in Chapter 11. Special emphasis is given to the use of the distributive properties in the process of obtaining products. Factoring of polynomials is performed in an algebraic manner, again with emphasis on the distributive properties.

Chapter 12 presents operations with rational algebraic expressions. Because students often have difficulty with fractions, many practice problems are provided.

Operations with radical expressions are presented in Chapter 13. The set of irrational numbers is introduced. The real numbers are then defined to be the union of the set of rational numbers and the set of irrational numbers.

In Chapter 14, the solution of quadratic equations by factoring is studied. Then the quadratic formula is developed and applied to the solution of quadratic equations with real roots.

Features of *Introductory Algebra*

Format

Each unit of material (a single concept or a few closely related concepts) is presented in a section. Within a section, short, boxed, numbered frames contain all instructional material, including sample problems and sample solutions. Frames are followed by practice problems, with work space provided. Answers immediately follow, with numerous solutions and supplementary comments.

Exercises

Exercises at the end of each section are quite traditional in nature. However, they are generally shorter than those found in most texts because the student has already done numerous problems and answered many questions during the study of the section. Hence, the exercises serve as a review of the content of the section and as a guide that will help students to recognize whether they have mastered the material. Answers to the problems are provided immediately, permitting the student to advance through the text without having to turn to the back to check answers. Word problems are used throughout.

Sample Chapter Tests

At the end of each chapter we have provided a Sample Chapter Test. It is keyed to each section with the answers provided immediately for student convenience. It may also serve as a pretest for each chapter. However, if an instructor wishes to pretest without the availability of the answers (to the student), one form of the post-test for the chapter (provided in the *Test Manual*) could be used for this purpose.

Upon finishing the chapter, the student should complete the Chapter Sample Test, and the results should be shown to the instructor and discussed with the student. If the instructor and student are confident about mastery of the content of the chapter, a post-test can then be administered. Used in this manner, outstanding results can be anticipated on the first attempt for each post-test.

Objectives

The Sample Chapter Tests serve as objectives for both student and instructor. The instructor can readily ascertain the objectives of any chapter by examining the sample test at its end, and the student is in a good position to see what is expected of him by examining problems and questions that are going to be asked at the completion of the chapter. We believe that objectives stated verbally are of less benefit to the student than the statement of a problem that must be solved.

Glossary

The glossary at the back of the text provides the student with the pronunciation and the definition of all mathematical words and phrases used in the text. This is particularly important wherever the text is used in a laboratory situation where the student will not hear the words used in class discussions.

Acknowledgments

We would like to thank the reviewers of the manuscript for their many useful comments and suggestions: Mel Avery, Tarrant County Junior College; William E. Coppage, Wright State University; C. A. Powers, Lansing Community College; Richard Spangler, Tacoma Community College; and Martha M. Wood, Clayton Junior College.

Our thanks go also to Janet Hastings for her help in developing the supplemental testing programs. Student reviewers and readers provided us with many valuable suggestions; in particular we would like to thank Donald Bolen, John Dowding, William McCracken, James Schultz, Alan Turner, and James Weston. Thanks also to our typists, Vonda Bottorff, Phyllis Bostwick, Marilyn Myers, Kathy Stripp, Lillian Thurston, and Carolyn Williams.

Dennis Bila
Ralph Bottorff
Paul Merritt
Donald Ross

February, 1975

To the Student

You are about to brush up on some skills you may already have learned in a previous course; but they may be a bit rusty. You will learn some new techniques, too. This book is designed to help pinpoint those areas for which you need to sharpen your skills and to permit you to move quickly through those topics you know well.

If you follow the suggestions below, you will make best use of your time, proceeding through the course as quickly as possible, and you will have mastered all the material in this book.

Chapter Sample Tests

The Chapter Sample Test at the end of each chapter will help you to determine whether you can skip certain sections of the chapter. It is a cumulative review of the entire chapter, and questions are keyed to the individual sections of the chapter. To use this test to best advantage, we suggest this procedure:

1. Work the solutions to all problems neatly on your own paper to the best of your ability. You want to determine what you know about the material without help from someone else. If you receive help in completing this self-test, the result will show another person's knowledge rather than your own.
2. Once you have done as much as you can, check the test with the answers provided. On the basis of errors made, determine the most appropriate course of study. You may need to study all sections in the chapter or you may be able to skip one or two sections.
3. When you have completed the chapter, rework all problems originally missed on the Chapter Sample Test and review the entire test in preparation for the post-test that will cover the entire chapter.

Instructional Sections

To use each section most effectively:

1. Study the boxed frames carefully.
2. Following each frame are questions based on the information presented in the frame. When completing these questions, do the work in the spaces provided.
3. Use a blank piece of paper as a mask to cover the answers below the stop rule. For example:

Q1 How many eggs can 21 chickens lay in 1 hour, if 1 chicken can lay 7 eggs in $\frac{1}{2}$ hour?

(Workspace)

STOP • **STOP** • **STOP** • **STOP** • **STOP** • **STOP** • **STOP** • **STOP** • **STOP**

Paper Mask

4. After you have written your calculations and response in the workspace, slide the paper mask down, uncovering the correct solution, and check your response with the correct answer given. In this way you can check your progress as you go without accidentally seeing the response before you have completed the necessary thinking or work.

Q1 How many eggs can 21 chickens lay in 1 hour, if 1 chicken can lay 7 eggs in $\frac{1}{2}$ hour?

$7 \times 2 = 14$

$$
\begin{array}{r}
14 \\
\times 21 \\
\hline
14 \\
28 \\
\hline
294
\end{array}
$$

STOP • **STOP** • **STOP** • **STOP** • **STOP** • **STOP** • **STOP** • **STOP** • **STOP**

A1 294: because 1 chicken can lay 14 in 1 hour, 21 chickens can lay 21(14) in 1 hour.

Paper Mask

Notice that the answer is often followed by a colon (:). The information following the colon is one of the following:
a. The complete solution,
b. A partial solution, possibly a key step frequently missed by many students,
c. A remark to remind you of an important point, or
d. A comment about the solution.
If your answer is correct, advance immediately to the next step of instruction.

5. Space is provided for you to work in the text. However, you may prefer to use the paper mask to work out your solutions. If you do, be sure to show complete solutions to all problems, clearly numbered, on the paper mask. When both sides of the paper are full, file it in a notebook for future reference.

6. Make all necessary corrections before continuing to the next frame or problem.

7. If something is not clear, talk to your instructor immediately.

8. If difficulty arises when you are studying outside of class, be sure to note the difficulty so that you can ask about it at your earliest convenience.

9. When preparing for a chapter post-test, the frames of each section serve as an excellent review of the chapter.

Section Exercises

The exercises at the end of each section are provided for additional practice on the content of the section. For your convenience, answers are given immediately following the exercises. However, detailed solutions to these problems are not shown.

You should:

1. Work all problems neatly on your own paper.
2. Check your responses against the answers given.
3. Rework any problem with which you disagree. If you cannot verify the response given, discuss the result with your instructor.
4. In the process of completing the exercise, use the instructional material in the section for review when necessary.
5. Problems marked with an asterisk (*) are considered more difficult than examples in the instructional section. They are intended to challenge the interested student. Problems of this difficulty will not appear on the Chapter Sample Test or the post-test.

Contents

Introductory Algebra

Chapter 1

Sets and Properties of Whole Numbers

Sections 1.1 and 1.2 are designed to provide the student with the necessary background for understanding the role of sets when developing number ideas throughout the text.

1.1   Set Ideas

| 1 | A *set* is a well-defined collection of things. The things that make up a set are called *elements*. The collection of letters in the English alphabet is a set of 26 letters. The letter b is an element of this set. The collection of months in a year is a set of 12 months. July is an element of this set. |

Q1   The collection of days in a week is a _____ of seven days.

STOP • STOP • STOP • STOP • STOP • STOP • STOP • STOP • STOP

A1   set

Q2   Name an element of the set in Q1._____

STOP • STOP • STOP • STOP • STOP • STOP • STOP • STOP • STOP

A2   Any day of the week: For example, Wednesday.

| 2 | A set may contain elements that have a common property, such as the presidents of the United States or the residents of a particular city. A set may also consist of unrelated objects, such as a chair, a pencil, and a book. However, a set must be *well defined*. This means that membership in the set is clear. The set of men who have been President of the United States is a well-defined collection because it is clear exactly which men are included in this set. A collection of any three letters of the English alphabet is not a well-defined collection because many possibilities exist and you would not be sure which is intended. |

Q3   Choose the sets from the following collections:
   a. The collection of states in the United States bordered by the Great Lakes
   b. The collection of three consecutive warm months
   c. The collection of beautiful people
   d. The collection of women who have been President of the United States

   _____

STOP • STOP • STOP • STOP • STOP • STOP • STOP • STOP • STOP

A3   a and d: Part b is not well defined because temperature is dependent on location and the meaning of "warm" is debatable. Part c is not well defined because there would not be agreement as to who is beautiful.

---

3   Using braces, { }, to indicate a set is called *set notation*. The elements of the set are enclosed in braces and are separated by commas. For example, the set of one-digit numbers* between 1 and 7 is denoted as {2, 3, 4, 5, 6}. The braces are read "the set containing," so {1, 2, 5} is read "the set containing 1, 2, and 5."

*The "digits" are often referred to as 0, 1, 2, 3, 4, 5, 6, 7, 8, and 9.

Q4   Use set notation to indicate the set of digits between 2 and 9._____

STOP • **STOP** • **STOP** • **STOP** • **STOP** • **STOP** • **STOP** • **STOP** • **STOP**

A4   {3, 4, 5, 6, 7, 8}

---

4   The use of set notation is not always convenient. When listing the elements would be difficult or impractical, a description is used. Such an example would be the set of U.S. Congressmen.

Q5   Where convenient, use set notation to indicate the set:

a. The set of vowels in the English alphabet   _____

b. The set of different kinds of flowers found in Ontario   _____

c. The set of digits in the numeral 100,537,375   _____

STOP • **STOP** • **STOP** • **STOP** • **STOP** • **STOP** • **STOP** • **STOP** • **STOP**

A5   a. {a, e, i, o, u} (*Note:* It is common to arrange letter elements of a set alphabetically.)
b. Although this information could be listed, it would not be convenient to do so in set notation.
c. {0, 1, 3, 5, 7} (*Note:* The order in which the elements are arranged does not affect the set. That is, {0, 1, 3, 5, 7} = {1, 0, 5, 3, 7}. Also, duplicate elements need be listed only once.)

---

5   The set that contains no elements is called the *empty set*. It is sometimes written using braces with no elements, { }. The symbol ∅ also denotes the empty set. The set of one-digit numbers between 4 and 5 is an example of the empty set.

Q6   The set of living 200-year-old people represents an _____ set.

STOP • **STOP** • **STOP** • **STOP** • **STOP** • **STOP** • **STOP** • **STOP** • **STOP**

A6   empty

Q7   Use set notation to indicate the set of two-digit numbers greater than 99._____

STOP • **STOP** • **STOP** • **STOP** • **STOP** • **STOP** • **STOP** • **STOP** • **STOP**

A7   { } or ∅

---

6   How many men had walked on the moon by 1960? The answer to this question is zero. That is, the set of men who had walked on the moon by 1960 is empty. You should observe that the number of elements in the empty set is zero; that is, there are no elements in the empty set.

Q8   **a.**   How many elements are there in $\{0\}$? _____

b.   How many elements are there in $\{\ \ \}$? _____

c.   Is $\{0\} = \{\ \ \}$? _____

STOP • STOP • STOP • STOP • STOP • STOP • STOP • STOP • STOP

A8   **a.** one, the element 0   **b.** zero   **c.** no

---

7   If each element of one set is also a member of a second set, the first set is a *subset* of the second. The set of dogs is a subset of the set of animals.

Q9   The set of New England States is a _____ of the set of states in the United States.

STOP • STOP • STOP • STOP • STOP • STOP • STOP • STOP • STOP

A9   subset

Q10   Determine whether the first set is a subset of the second set:

a. $\{\text{Jane, Mary}\}$ $\{\text{June, Judy, Jane}\}$ _____

b. $\{1, 2, 5\}$ $\{0, 1, 2, 3, 4, 5\}$ _____

c. $\{2, 4, 6\}$ $\{1, 2, 3, 4, 5\}$ _____

d. $\{1, 2, 3\}$ $\{1, 2, 3\}$ _____

e. set of vowels   set of all letters in the English alphabet _____

STOP • STOP • STOP • STOP • STOP • STOP • STOP • STOP • STOP

A10   **a.** no   **b.** yes   **c.** no   **d.** yes   **e.** yes

---

8   The symbol $\in$ is used to abbreviate "is an element of." $x \in \{x, y, z\}$ is read "x is an element of the set containing x, y, and z." The slant bar, $/$, is used as a negation symbol. $\notin$ means "is *not* an element of." $1 \notin \{0, 2, 4, 6\}$ is read "1 is not an element of the set containing 0, 2, 4, and 6."

Q11   Insert the correct symbol, $\in$ or $\notin$, in the following blanks:

a. r _____ $\{a, e, i, o, u\}$

b. 5 _____ $\{1, 3, 5, 7, 9\}$

STOP • STOP • STOP • STOP • STOP • STOP • STOP • STOP • STOP

A11   **a.** $\notin$ **b.** $\in$

---

9   Since the set of letters of the English alphabet are numerous, this set may be indicated as $\{a, b, c, \ldots, z\}$. The three dots, read "and so on," indicate that not all elements of the set have been listed and that additional elements of the set can be found by following the pattern established in the elements preceding the dots.

Q12   Insert the correct symbol, $\in$ or $\notin$, in the following blanks:

    a. 17 _____ $\{1, 2, 3, \ldots, 20\}$

    b. 12 _____ $\{1, 3, 5, \ldots, 19\}$

    c. 10 _____ $\{2, 4, 6, \ldots, 20\}$

STOP • **STOP** • **STOP** • **STOP** • **STOP** • **STOP** • **STOP** • **STOP** • **STOP**

A12   **a.** $\in$ **b.** $\notin$ **c.** $\in$

10	If the elements of a set can be counted and the count has a last number, the set is a *finite* set. A set whose count is unending is an *infinite* set. For example, $\{a, b, c, d\}$ is a finite set, whereas $\{1, 2, 3, 4, \ldots\}$ is an infinite set.

Q13   Describe each of the following sets as finite or infinite:

    a. The set of letters in the word "infinite" _____

    b. $\{1, 2, 3, 4, \ldots, 15\}$ _____

    c. $\{5, 10, 15, \ldots\}$ _____

    d. The set of provincial parks in Canada _____

    e. $\varnothing$ _____

STOP • **STOP** • **STOP** • **STOP** • **STOP** • **STOP** • **STOP** • **STOP** • **STOP**

A13   **a.** finite   **b.** finite   **c.** infinite   **d.** finite
    e. finite (the empty set has a count of zero)

This completes the instruction for this section.

1.1   Exercises

1. Give the word or phrase that best describes each of the following:
   a. well-defined collection
   b. set that contains no elements
   c. set in which the elements can be counted and the count has a last number
   d. set whose count is unending
2. Specify the members of each of the following sets:
   a. states in the United States bordered by the Pacific Ocean
   b. last three letters of the English alphabet
   c. $\{1, 2, 3\}$
3. Which of the following are not sets:
   a. $\{1, 3, 5, 7, 9, 11, \ldots\}$
   b. collection of ex-Presidents of the United States
   c. collection of consonants in the English alphabet
   d. collection of large numbers
4. Where convenient, use set notation to indicate the following sets:
   a. letters in the word "college"
   b. digits in the number 10,274
   c. five-letter words beginning with C
   d. days of the week beginning with R
   e. digits between 0 and 9

5. True or false:
   a. {a, c, d} is a subset of {a, b, c, d}.
   b. The set of cities that are provincial capitals in Canada is a subset of the set of all cities in Canada.
   c. {0} is a subset of {1, 2, 3, ...}.
   d. {Jane, Mary, Sue, Ann} is a subset of {Mary, Sue, Jane, Ann}.
6. Insert the correct symbol, $\in$ or $\notin$, in each of the following:
   a. h _____ {a, e, i, o, u}
   b. 0 _____ {1, 2, 3, ..., 12}
   c. 32 _____ {0, 2, 4, 6, ...}
   d. 2,000 _____ {0, 1, 2, 3, ...}

7. Describe each of the following sets as finite or infinite:
   a. consonants in the English alphabet
   b. {0, 1, 2, 3, 4, ...}
   c. digits between 5 and 6
   d. {0, 2, 4, 6, ..., 20}

1.1 Exercise Answers

1. a. set   b. empty set   c. finite set   d. infinite set
2. a. Alaska, California, Hawaii, Oregon, and Washington
   b. x, y, and z   c. 1, 2, and 3
3. d
4. a. {c, e, g, l, o}   b. {0, 1, 2, 4, 7}   c. not convenient   d. { } or ∅
   e. {1, 2, 3, 4, 5, 6, 7, 8} or {1, 2, 3, ..., 8}
5. a. true   b. true   c. false   d. true
6. a. $\notin$ b. $\notin$ c. $\in$ d. $\in$
7. a. finite   b. infinite   c. finite   d. finite

1.2 Set Intersection and Union

1 | The *intersection* of two sets is a third set which contains those elements, and only those elements, that belong to *both* (one and the other) of the original sets. Let

$$A = \{a, e, i, o, u\}$$
$$B = \{a, r, i, s, t, o, c, e\}$$

(*Note:* For convenience it is common to label sets with capital letters.) The intersection of sets A and B is {a, e, i, o}.

Q1 | Let $C = \{1, 2, 3, 4\}$ and $D = \{2, 4, 6\}$. Determine the intersection of sets C and D.

STOP • STOP • STOP • STOP • STOP • STOP • STOP • STOP • STOP

A1 | {2, 4}

2 | The symbol ∩ is used to express the operation of intersection. For example, let

$$A = \{1, 2, 3, 5, 7\}$$
$$B = \{2, 3, 7, 8, 9\}$$

The intersection of A and B may be shown:

$$A \cap B = \{2, 3, 7\}$$

Q2   Given $A = \{a, b, c\}$ and $B = \{b, c, d\}$, determine $A \cap B$. _____

STOP • STOP • STOP • STOP • STOP • STOP • STOP • STOP • STOP

A2 $\{b, c\}$

---

3   Two sets are *disjoint sets* when they have no common elements. $\{1, 3, 5\}$ and $\{0, 2, 4\}$ are disjoint sets. These sets have no elements in common; hence, the intersection of these sets has no elements and is, therefore, the empty set. Letting

$$C = \{1, 3, 5\}$$
$$D = \{0, 2, 4\}$$
$$C \cap D = \varnothing$$

If the intersection of two sets is empty, the sets are disjoint.

Q3 $A = \{2, 4, 6, 8, 10, 12\}$ $B = \{1, 3, 5, 7, 9, 11\}$ $C = \{3, 6, 9, 12\}$
           $D = \{4, 8, 12\}$ $E = \varnothing$

     a. $A \cap B =$ _____   **b.** $B \cap C =$ _____

     c. $A \cap C =$ _____   **d.** $A \cap D =$ _____

     e. $B \cap D =$ _____   **f.** $C \cap D =$ _____

     g. $B \cap E =$ _____

STOP • STOP • STOP • STOP • STOP • STOP • STOP • STOP • STOP

A3   **a.** $\varnothing$ **b.** $\{3, 9\}$ **c.** $\{6, 12\}$
         d. $\{4, 8, 12\}$ **e.** $\varnothing$ **f.** $\{12\}$
         g. $\varnothing$

Q4   Consider the sets A, B, C, D, and E in Q3 and indicate the pairs of sets that are disjoint. _____

STOP • STOP • STOP • STOP • STOP • STOP • STOP • STOP • STOP

A4 A and B, A and E, B and D, B and E, C and E, D and E

---

4   The *union* of two sets is a third set that contains all those elements that belong to *either* (one or the other) of the original sets. Let

$$A = \{a, c, t\}$$
$$B = \{a, e, r, t\}$$

The union of sets A and B is $\{a, c, e, r, t\}$.

Q5   Let $C = \{1, 2, 3, 4\}$ and $D = \{2, 4, 6\}$. Determine the union of sets C and D.
_____

STOP • STOP • STOP • STOP • STOP • STOP • STOP • STOP • STOP

A5 $\{1, 2, 3, 4, 6\}$

---

5   The symbol $\cup$ is used to express the operation of union. For example, let

$A = \{1, 2, 3, 5, 7\}$
$B = \{2, 3, 7, 8, 9\}$

The union of A and B may be shown:

$A \cup B = \{1, 2, 3, 5, 7, 8, 9\}$

Q6   Given $A = \{a, b, c\}$ and $B = \{b, c, d\}$, determine $A \cup B$. _____

STOP • **STOP** • **STOP** • **STOP** • **STOP** • **STOP** • **STOP** • **STOP** • **STOP**

A6 $\{a, b, c, d\}$

Q7 $A = \{2, 4, 6, 8, 10, 12\}$ $B = \{1, 3, 5, 7, 9, 11\}$ $C = \{3, 6, 9, 12\}$
       $D = \{4, 8, 12\}$ $E = \varnothing$

   a. $A \cup B = $ _____   **b.** $B \cup C = $ _____

   c. $A \cup C = $ _____   **d.** $A \cup D = $ _____

   e. $B \cup D = $ _____   **f.** $B \cup E = $ _____

STOP • **STOP** • **STOP** • **STOP** • **STOP** • **STOP** • **STOP** • **STOP** • **STOP**

A7   **a.** $\{1, 2, 3, \ldots, 12\}$ **b.** $\{1, 3, 5, 6, 7, 9, 11, 12\}$
       c. $\{2, 3, 4, 6, 8, 9, 10, 12\}$ **d.** $\{2, 4, 6, 8, 10, 12\}$
       e. $\{1, 3, 4, 5, 7, 8, 9, 11, 12\}$ **f.** $\{1, 3, 5, 7, 9, 11\}$

This completes the instruction for this section.

1.2   Exercises

1. Write the correct symbol for:
   a. union   **b.** intersection
2. Given the sets below, determine:
   $A = \{1, 3, 5, 7, \ldots\}$ $B = \{0, 2, 4, 6, \ldots\}$ $C = \{3, 7, 8, 9\}$
   $D = \{2, 5, 7, 8\}$ $E = \varnothing$

   a. $B \cap C$ **b.** $C \cup D$ **c.** $A \cup B$
   d. $A \cap C$ **e.** $B \cap D$ **f.** $C \cap D$
   g. $B \cup E$ **h.** $A \cap E$ **i.** $A \cap B$
   *j. $(A \cup C) \cap D$

3. Which pairs of sets below are disjoint?
   $A = \{1, 3, 5, 7, 9, \ldots\}$
   $B = \{0, 2, 4, 6, \ldots\}$
   $C = \{3, 7, 8, 12\}$
   $D = \varnothing$

1.2   Exercise Answers

1. **a.** $\cup$ **b.** $\cap$
2. **a.** $\{8\}$ **b.** $\{2, 3, 5, 7, 8, 9\}$ **c.** $\{0, 1, 2, 3, \ldots\}$
   d. $\{3, 7, 9\}$ **e.** $\{2, 8\}$ **f.** $\{7, 8\}$

    g. $\{0, 2, 4, 6, \ldots\}$ **h.** $\varnothing$ **i.** $\varnothing$

    *j. $\{5, 7, 8\}$

3. A and B, A and D, B and D, C and D

1.3   Evaluating Open Expressions

1

The evaluation of numerical expressions such as $432 - 178 + 92$, $3 + 5 \cdot 7$, $(13 - 2)(15 + 8)$, $12 \cdot 9 - 5 \cdot 6$, and $5(9 - 4)$ require rules for order of operations:

Step 1: Perform operations within grouping symbols first.

Step 2: Next, perform all multiplications and divisions in the order in which they appear from left to right.

Step 3: Finally, perform all additions and subtractions in the order in which they appear from left to right.

The raised dot, $\cdot$, indicates multiplication. Also, parentheses side by side or a number preceding parentheses indicates multiplication. That is,

$$(5)(7) = 5 \cdot 7 \quad \text{or} \quad 5(7) = 5 \cdot 7$$

The form of the evaluation of a numerical expression may vary. Two acceptable forms are:

$$5 + 4 \cdot 7 = 5 + 28 \quad \text{and} \quad 5 + 4 \cdot 7$$
$$= 33 \qquad\qquad 5 + 28$$
$$33$$

When steps are arranged vertically, as in the second example, each line is assumed to be equal to the previous line.

Q1   Evaluate the following numerical expressions (show your work):

    a. $4 + 5 \cdot 6$ **b.** $3 \cdot 5 - 8$

    c. $12(13 - 5)$ **d.** $15 \cdot 3 - 10 \cdot 2$

STOP • **STOP** • **STOP** • **STOP** • **STOP** • **STOP** • **STOP** • **STOP** • **STOP**

A1
    a. 34: $4 + 5 \cdot 6 = 4 + 30$ **b.** 7: $3 \cdot 5 - 8 = 15 - 8$
$$= 34 \qquad\qquad\qquad\qquad = 7$$
    c. 96: $12(13 - 5)$ **d.** 25: $15 \cdot 3 - 10 \cdot 2$
$$12(8) \qquad\qquad\qquad\qquad 45 \quad - \quad 20$$
$$96 \qquad\qquad\qquad\qquad\qquad 25$$

Q2   Evaluate the following numerical expressions (show your work):

    a. $2 + 4 \cdot 5 - 3$ **b.** $35 - 3(5 + 12 \div 3)$

  c. $(32 - 13)(12 + 9)$ **d.** $45 - 15 \div 5$

STOP • **STOP** • **STOP** • **STOP** • **STOP** • **STOP** • **STOP** • **STOP** • **STOP**

A2   **a.** 19: $2 + 20 - 3$ **b.** 8: $35 - 3(5 + 4)$
       $22 - 3$ $35 - 3 \cdot 9$
       19 $35 - 27$
                 8
   c. 399: $19 \cdot 21$ **d.** 42: $45 - 3$
        399 42

---

2    $\square + 7$ is called an *open expression*. Other examples of open expressions are:

  $15 + \square$ $2 + 3 \cdot \square - 4$
  $5(3 \cdot \square + 7)$ $45 - 18 - \square$

  Open expressions may be evaluated when the $\square$ is replaced by some number.

  Example: Evaluate $15 + \square$ when $\square$ is replaced by 7.

  Solution

  $15 + \square$ or $15 + \square = 15 + 7$
  $15 \mid 7$ $= 22$
  22

Q3   Evaluate each of the following open expressions when the $\square$ is replaced by the given value (show your work):
   a. $(\square + 3) \cdot 5$ when $\square$ is 7

   b. $(\square - 4)(\square + 8)$ when $\square$ is $12\frac{1}{2}$

   c. $3(\square \div 3 - 2)$ when $\square$ is 9

    d. $5(3 \cdot \square + 4 \cdot \square)$ when $\square$ is $\frac{1}{3}$

    e. $\square \cdot (\square + 5 - 2 \cdot \square)$ when $\square$ is 5

STOP • **STOP** • **STOP** • **STOP** • **STOP** • **STOP** • **STOP** • **STOP** • **STOP**

A3   **a.** 50: $(7 + 3) \cdot 5$
              $10 \cdot 5$
              50

        b. $174\frac{1}{4}$: $\left(12\frac{1}{2} - 4\right)\left(12\frac{1}{2} + 8\right)$

                     $8\frac{1}{2} \cdot 20\frac{1}{2}$

                     $\frac{17}{2} \cdot \frac{41}{2}$

                     $\frac{697}{4} = 174\frac{1}{4}$

    c. 3: $3(9 \div 3 - 2)$
              $3(3 - 2)$
              $3 \cdot 1$
              3

        d. $11\frac{2}{3}$: $5\left(3 \cdot \frac{1}{3} + 4 \cdot \frac{1}{3}\right)$

                     $5\left(1 + 1\frac{1}{3}\right)$

                     $5 \cdot 2\frac{1}{3}$

                     $5 \cdot \frac{7}{3}$

    e. 0: $5 \cdot (5 + 5 - 2 \cdot 5)$
             $5 \cdot (5 + 5 - 10)$
             $5 \cdot 0$
             0

                     $\frac{35}{3}$

                     $11\frac{2}{3}$

| 3 | Any letter of the alphabet may be used in place of the $\square$ in an open expression. Using x, $\square + 5$ may be written as $x + 5$. $x + 5$ is also called an *open expression*. Other examples of open expressions are $a - 3$, $5 \cdot y + 2 \cdot y$, $5(b + 7)$, $(x - 2)(x + 3)$, and $16 + 5 \cdot z$. |

Q4 $2(y - 6)$ is an _____ expression.

STOP • **STOP** • **STOP** • **STOP** • **STOP** • **STOP** • **STOP** • **STOP** • **STOP**

A4   open

| 4 | When a letter may be replaced by any one of a set of many numbers, the letter is called a *variable*. The set of permissible values of the variable is referred to as its *replacement set*. |

When variables are replaced by values from their replacement set, the resulting numerical expression may then be evaluated. (In this section, the replacement set will be whole numbers, common fractions, and decimals.)

Example: Evaluate $5(b + 7)$ when b is replaced by 4.

Solution

$5(b + 7)$ or $5(b + 7) = 5(4 + 7)$
$5(4 + 7)$ $= 5 \cdot 11$
$5 \cdot 11$ $= 55$
55

Q5   Evaluate:
          a. $5x$ when x is 7 (*Note:* $5x = 5 \cdot x$.)   **b.** $3y - 12$ when y is 10

          c. $7(t - 8)$ when t is $9\frac{1}{2}$ **d.** $2a + 3a$ when a is 5

          e. $b(b + 6)$ when b is 2

STOP • **STOP** • **STOP** • **STOP** • **STOP** • **STOP** • **STOP** • **STOP** • **STOP**

A5   **a.** 35: $5x$ **b.** 18: $3y - 12$
                  $5 \cdot 7$ $3 \cdot 10 - 12$
                  35 $30 - 12$
                                                           18

          c. $10\frac{1}{2}$: $7(t - 8)$ **d.** 25: $2a + 3a$
                              $7\left(9\frac{1}{2} - 8\right)$ $2 \cdot 5 + 3 \cdot 5$
                               $7 \cdot 1\frac{1}{2}$ $10 + 15$
                               $10\frac{1}{2}$ 25

          e. 16: $b(b + 6)$
                    $2(2 + 6)$
                    $2 \cdot 8$
                    16

---

5   Open expressions may contain more than one variable.

Example: Evaluate $3x + y - 2z$ when $x = 2$, $y = 5$, and $z = 4$.

Solution

$3x + y - 2z$
$3 \cdot 2 + 5 - 2 \cdot 4$
$6 + 5 - 8$
$11 - 8$
3

Q6   Evaluate (show your work):

a. $x + 7y$ when $x = 5$ and $y = 0.2$

b. $(5 + 2a) + 3b$ when $a = 4$ and $b = 5$

c. $0.5x + (3y - 2)$ when $x = 10$ and $y = 4$

d. $2(a - 2b) + 3a$ when $a = 12$ and $b = \dfrac{1}{2}$

e. $x(y + z)$ when $x = 2$, $y = 3$, and $z = 1.5$

f. $cd - 5c$ when $c = 7$ and $d = 5$ (*Note: cd = c \cdot d.*)

g. $\dfrac{x}{y}$ when $x = \dfrac{2}{3}$ and $y = \dfrac{4}{5}$

STOP • **STOP** • **STOP** • **STOP** • **STOP** • **STOP** • **STOP** • **STOP** • **STOP**

A6

a. 6.4: $x + 7y$
$\quad\quad 5 + 7 \cdot 0.2$
$\quad\quad 5 + 1.4$
$\quad\quad 6.4$

b. 28: $(5 + 2a) + 3b$
$\quad\quad (5 + 2 \cdot 4) + 3 \cdot 5$
$\quad\quad (5 + 8) + 15$
$\quad\quad 13 + 15$
$\quad\quad 28$

c. 15: $0.5x + (3y - 2)$
$\quad\quad 0.5 \cdot 10 + (3 \cdot 4 - 2)$
$\quad\quad 5 + (12 - 2)$
$\quad\quad 5 + 10$
$\quad\quad 15$

d. 58: $2(a - 2b) + 3a$
$\quad\quad 2\left(12 - 2 \cdot \dfrac{1}{2}\right) + 3 \cdot 12$
$\quad\quad 2(12 - 1) + 36$
$\quad\quad 2 \cdot 11 + 36$
$\quad\quad 22 + 36$
$\quad\quad 58$

e. 9: $x(y + z)$
$\quad\quad 2(3 + 1.5)$
$\quad\quad 2(4.5)$
$\quad\quad 9$

f. 0: $cd - 5c$
$\quad\quad 7 \cdot 5 - 5 \cdot 7$
$\quad\quad 35 - 35$
$\quad\quad 0$

g. $\dfrac{5}{6}$: $\dfrac{x}{y}$

$\quad\quad \dfrac{\frac{2}{3}}{\frac{4}{5}} = \dfrac{2}{3} \cdot \dfrac{5}{4} = \dfrac{5}{6}$

---

6 | The product of 3 and 4 is 12. 3 and 4 are said to be *factors* of 12. In such products as $2x$. $3y$, cd, and $7xyz$, the parts of the product are also called factors. For example, the factors of $2x$ are 2 and x. The factors of cd are c and d. The factors of $7xyz$ are 7, x, y, and z.

Q7 | Indicate the factors of each open expression:

a. $5y$ _____

b. $3ab$ _____

c. rst _____

d. $0.6x$ _____

STOP • **STOP** • **STOP** • **STOP** • **STOP** • **STOP** • **STOP** • **STOP** • **STOP**

A7

a. 5 and y

b. 3, a, and b

c. r, s, and t

d. 0.6 and x

---

7 | In an open expression such as $5x$, 5 is called the number factor and x is called the letter factor. Mathematicians often use *numerical coefficient* for number factor and *literal coefficient* for letter factor.

Q8 | **a.** The numerical coefficient of $5x$ is _____.

b. The literal coefficient of $5x$ is _____.

c. The numerical coefficient of $3ab$ is _____.

d. The literal coefficients of $3ab$ are _____.

STOP • STOP • STOP • STOP • STOP • STOP • STOP • STOP • STOP

A8   **a.** 5   **b.** x **c.** 3   **d.** a and b

8	Parentheses are used for grouping expressions. $5 + (a + 3)$ is read "5 plus the *quantity* a plus 3." $2(x - 1)$ is read "2 times the *quantity* x minus 1." The word "quantity" indicates the expression that has been placed within the parentheses.

Q9   **a.** $2x - (3x + 5)$ is read "$2x$ minus the _____ $3x$ plus 5."

b. $h(h + 8)$ is read "h times the _____ h plus 8."

STOP • STOP • STOP • STOP • STOP • STOP • STOP • STOP • STOP

A9   **a.** quantity   **b.** quantity

This completes the instruction for this section.

1.3   Exercises

1. Evaluate each open expression for the value of the variable given:
   a. $x - 5$ when $x = 32.6$
   b. $3a + 7$ when $a = 4$
   c. $4(y + 3)$ when $y = \dfrac{1}{2}$
   d. $12 + 2(2x - 9)$ when $x = 5$
   e. $3b(b - 8)$ when $b = 12$
   f. $2x + 3y$ when $x = 9$ and $y = \dfrac{2}{3}$
   g. rst when $r = 0.2$, $s = 20$, and $t = 0.6$
   h. $\dfrac{2a}{b} + 6$ when $a = 10$ and $b = 5$
   i. $7x - 4x$ when $x = 4.3$
   j. $(2y - 3)(y + 6)$ when $y = 7$

2. Use the following words or phrases to correctly answer each of the following questions. If more than one applies, you need only select one. (variable, open expression, numerical coefficient, literal coefficient, quantity, factor)
   a. What is each of the following called? $x - 5$, $3a + 7$, $3b(b - 8)$, and $2x + 3y$
   b. What is x called in $x - 5$, xy, and $2x + 3y$?
   c. What is $b - 8$ called in $3b(b - 8)$?
   d. What is 2 called in $2x$ and $2(3 + x)$?
   e. What is x called in $2x$?
   f. What is $4x$ called in $(7x)(4x)$?
   g. What is $3b$ called in $3b(b - 8)$?
   h. What is $x + 2$ called in $5 + (x + 2)$?

1.3   Exercise Answers

1. a. 27.6   **b.** 19   **c.** 14   **d.** 14   **e.** 144   **f.** 20   **g.** 2.4
   h. 10   **i.** 12.9   **j.** 143
2. a. open expression
   c. quantity or factor
   e. variable, factor, or literal coefficient
   g. factor

   b. variable
   d. factor or numerical coefficient
   f. factor or quantity
   h. quantity

1.4   Properties of Whole Numbers

---

1   The *stated equality* of two *numerical* expressions is a *mathematical statement.* $5 + 3 = 8$ is a mathematical statement.

Q1   **a.** $4 \cdot 3 - 2$ and $12 - 2$ are _____ expressions.

    b. $4 \cdot 3 - 2 = 12 - 2$ is a _____ statement.

STOP • **STOP** • **STOP** • **STOP** • **STOP** • **STOP** • **STOP** • **STOP** • **STOP**

A1   **a.** numerical   **b.** mathematical

---

2   Mathematical statements may be judged as being true or false. $2 + (5 + 3) = 2 + 8$ is true. $(15 - 3) - 8 = 11 - 8$ is false.

Q2   Answer true or false:

    a. $5 \cdot 4 = 4 \cdot 5$ _____

    b. $2 + 3 = 3 + 2$ _____

    c. $\dfrac{12}{3} = \dfrac{3}{12}$ _____

STOP • **STOP** • **STOP** • **STOP** • **STOP** • **STOP** • **STOP** • **STOP** • **STOP**

A2   **a.** true   **b.** true   **c.** false

---

3   When two numerical expressions have the same evaluation, they are said to be *equivalent.* $3 + 8$ and $8 + 3$ are equivalent. $5 - 2$ and $1 + 2$ are equivalent. $12 - 2$ and $2 - 12$ are not equivalent.

Q3   **a.**   Are $12 + 5$ and $5 + 12$ equivalent?   _____

    b.   Are $12 - 5$ and $13 - 7$ equivalent?   _____

STOP • **STOP** • **STOP** • **STOP** • **STOP** • **STOP** • **STOP** • **STOP** • **STOP**

A3   **a.** yes   **b.** no

---

4   If two numerical expressions are equivalent, their stated equality is always true. The sums $12 + 5$ and $5 + 12$ are equivalent, so $12 + 5 = 5 + 12$ is true.

Answer yes or no to the following questions:

Q4   **a.** Are $5 + (4 + 1)$ and $(5 + 4) + 1$ equivalent?   _____

     b. Is $5 + (4 + 1) = (5 + 4) + 1$ a true statement?   _____

STOP • **STOP** • **STOP** • **STOP** • **STOP** • **STOP** • **STOP** • **STOP** • **STOP**

A4   **a.** yes: because both $5 + (4 + 1)$ and $(5 + 4) + 1$ equal 10

     b. yes

Q5   **a.** Are $\frac{2}{5} \cdot 1$ and $1 \cdot \frac{2}{5}$ equivalent?   _____

     b. Is $\frac{2}{5} \cdot 1 = 1 \cdot \frac{2}{5}$ a true statement?   _____

STOP • **STOP** • **STOP** • **STOP** • **STOP** • **STOP** • **STOP** • **STOP** • **STOP**

A5   **a.** yes   **b.** yes

Q6   **a.** Are $17 \cdot 0$ and 0 equivalent?   _____

     b. Is $17 \cdot 0 = 0$ a true statement?   _____

STOP • **STOP** • **STOP** • **STOP** • **STOP** • **STOP** • **STOP** • **STOP** • **STOP**

A6   **a.** yes   **b.** yes

Q7   **a.** Are $\frac{7}{8} + 0$ and $\frac{7}{8}$ equivalent?   _____

     b. Is $\frac{7}{8} + 0 = \frac{7}{8}$ a true statement?   _____

STOP • **STOP** • **STOP** • **STOP** • **STOP** • **STOP** • **STOP** • **STOP** • **STOP**

A7   **a.** yes   **b.** yes

Q8   **a.** Are $2(3 \cdot 4)$ and $(2 \cdot 3)4$ equivalent? [*Note:* $(2 \cdot 3)4 = (2 \cdot 3) \cdot 4.$]   _____

     b. Is $2(3 \cdot 4) = (2 \cdot 3)4$ a true statement?   _____

STOP • **STOP** • **STOP** • **STOP** • **STOP** • **STOP** • **STOP** • **STOP** • **STOP**

A8   **a.** yes   **b.** yes

Q9   **a.** Are $20 - (12 - 3)$ and $(20 - 12) - 3$ equivalent?   _____

     b. Is $20 - (12 - 3) = (20 - 12) - 3$ a true statement?   _____

STOP • **STOP** • **STOP** • **STOP** • **STOP** • **STOP** • **STOP** • **STOP** • **STOP**

A9   **a.** no: because $20 - (12 - 3) = 11$, whereas $(20 - 12) - 3 = 5$

     b. no

Q10   **a.** Verify that $2(5 - 2)$ and $2 \cdot 5 - 2 \cdot 2$ are equivalent.

     b. Is $2(5 - 2) = 2 \cdot 5 - 2 \cdot 2$ a true statement?   _____

STOP • **STOP** • **STOP** • **STOP** • **STOP** • **STOP** • **STOP** • **STOP** • **STOP**

A10   **a.** $2(5 - 2) = 2 \cdot 3 = 6$
$2 \cdot 5 - 2 \cdot 2 = 10 - 4 = 6$

b. yes

---

5      $\{1, 2, 3, 4, \ldots\}$ is called the set of *counting numbers*. This set is also called the set of *natural numbers*. It is common to denote this set by N:

$$N = \{1, 2, 3, 4, \ldots\}$$

The set of *whole numbers* is the set of natural numbers together with zero. It is common to denote this set by W:

$$W = \{0, 1, 2, 3, 4, \ldots\}$$

Q11   Answer true or false:

a. The set of natural numbers is a subset of the set of whole numbers.   _____

b. $0 \in W$ but $0 \notin N$.   _____

c. $5 \in N$ and $5 \in W$.   _____

d. The set of whole numbers is an infinite set.   _____

e. $\{0, 1, 2, 3, \ldots\}$ is a well-defined collection.   _____

STOP • **STOP** • **STOP** • **STOP** • **STOP** • **STOP** • **STOP** • **STOP** • **STOP**

A11   **a.** true: Because each element of the set of natural numbers is also an element of the set of whole numbers.

b. true: Zero belongs to the set of whole numbers, but zero does not belong to the set of natural numbers.

c. true: 5 belongs to both sets; hence, 5 is both a natural number and a whole number. Every natural number is also a whole number.

d. true: Because the count is unending.

e. true: Membership is clear.

(Note: In the remaining portion of this section, the replacement set for all variables will be the set of whole numbers.)

---

6      $x + 2$ is called an *open expression*. $x + 2 = 5$ is called an *open sentence*. An open sentence cannot be judged as being true or false until a replacement is made for the variable. When x is replaced by 3, $x + 2 = 5$ becomes the true statement $3 + 2 = 5$. When x is replaced by 4, $x + 2 = 5$ becomes the false statement $4 + 2 = 5$.

Q12   In each open sentence, when the variable is replaced by the given value, determine whether the resulting statement is true or false:

a. $3x - 4 = 17$; $x = 7$ **b.** $x + 5 = 5$; $x = 0$

c. $x + 2 = 2 + x$; $x = 7$ **d.** $5x = x \cdot 5$; $x = 3$

  e. $x + 2 = x;\ x = 0$

STOP • STOP • STOP • STOP • STOP • STOP • STOP • STOP • STOP

A12   **a.** true: because $3 \cdot 7 - 4 = 17$ **b.** true: because $0 + 5 = 5$
          c. true: because $7 + 2 = 2 + 7$ **d.** true: because $5 \cdot 3 = 3 \cdot 5$
          e. false: because $0 + 2 \neq 0$

---

7   The expressions $x + 7$ and $7 + x$ are open expressions. They are equivalent because they have the same evaluation for all replacements of the variable. Hence, $x + 7 = 7 + x$ is true for all whole-number replacements of x. For example,

$$0 + 7 = 7 + 0$$
$$1 + 7 = 7 + 1$$
$$2 + 7 = 7 + 2$$
$$3 + 7 = 7 + 3$$
     etc.

(*Note:* Although $x + 7 = 7 + x$ is an open sentence, it is true for all whole-number replacements of x.)

Q13   Answer yes or no:
       a. Are $a + b$ and $b + a$ equivalent for all whole-number replacements of a and b? _____
       (Try to find a pair of whole numbers to replace a and b that do not give the same evaluation for both expressions.)
       b. Is $a + b = b + a$ true for all whole-number replacements of a and b? _____

STOP • STOP • STOP • STOP • STOP • STOP • STOP • STOP • STOP

A13   **a.** yes   (it is reasonable to assume that they are equivalent)
          b. yes

---

8   The assumption that $a + b = b + a$ is true for all whole-number replacements of a and b is called the *commutative property of addition*. It is concerned with the *order* in which two numbers are added. Addition is called a commutative operation because the order in which two numbers are added does not affect the sum. Examples of this property are:

$$0 + 5 = 5 + 0$$
$$2 + 7 = 7 + 2$$
$$52 + 1 = 1 + 52$$

Q14 $15 + 7 = 7 + 15$ is an example of the _____ property of addition.

STOP • STOP • STOP • STOP • STOP • STOP • STOP • STOP • STOP

A14   commutative

Q15   Use the commutative property of addition to complete a true statement:

       a. $17 + 8 =$ _____   **b.** $0 + 9 =$ _____

       c. $1 + 12 =$ _____   **d.** $92 + 18 =$ _____

STOP • STOP • STOP • STOP • STOP • STOP • STOP • STOP • STOP

A15	**a.** $8 + 17$	**b.** $9 + 0$	**c.** $12 + 1$	**d.** $18 + 92$

Q16   The commutative property of addition indicates that the _____ of two numbers in an addition problem may be reversed without affecting the sum.

STOP • **STOP** • **STOP** • **STOP** • **STOP** • **STOP** • **STOP** • **STOP** • **STOP**

A16   order

---

9   The commutative property of addition includes an infinite number of specific cases. The statement that $a + b = b + a$ is true for all possible whole-number replacements of a and b illustrates the power of making one statement that covers an infinite number of cases. A similar property is true for multiplication. You know that $3 \cdot 5 = 5 \cdot 3$. You realize that the *order* of the factors in a multiplication does not affect the product. The *commutative property of multiplication* assumes that $ab = ba$ is true for all whole-number replacements of a and b. Other examples of this property are:

$2 \cdot 17 = 17 \cdot 2$
$93 \cdot 6 = 6 \cdot 93$
$0 \cdot 15 = 15 \cdot 0$

Q17   Use the commutative property of multiplication to complete a true statement:

a. $1 \cdot 15 = $ _____   **b.** $107 \cdot 49 = $ _____

c. $7 \cdot 12 = $ _____   **d.** $50 \cdot 75 = $ _____

STOP • **STOP** • **STOP** • **STOP** • **STOP** • **STOP** • **STOP** • **STOP** • **STOP**

A17   **a.** $15 \cdot 1$ **b.** $49 \cdot 107$ **c.** $12 \cdot 7$ **d.** $75 \cdot 50$

Q18   **a.** Use the commutative property of addition to change $2x + 5$ into an equivalent expression. _____
      b. Use the commutative property of multiplication to change $3x \cdot 5$ into an equivalent expression. _____

STOP • **STOP** • **STOP** • **STOP** • **STOP** • **STOP** • **STOP** • **STOP** • **STOP**

A18   **a.** $5 + 2x$ **b.** $5 \cdot 3x$

Q19   Verify that $1 + (2 + 3) = (1 + 2) + 3$ is a true statement by showing that $1 + (2 + 3)$ and $(1 + 2) + 3$ are equivalent expressions.

STOP • **STOP** • **STOP** • **STOP** • **STOP** • **STOP** • **STOP** • **STOP** • **STOP**

A19 $1 + (2 + 3) = 1 + 5 = 6$
      $(1 + 2) + 3 = 3 + 3 = 6$

Q20   **a.** Are $2 + (a + 3)$ and $(2 + a) + 3$ equivalent for all whole-number replacements of a? _____
      b. Is $2 + (a + 3) = (2 + a) + 3$ true for all whole-number replacements of a? _____

STOP • **STOP** • **STOP** • **STOP** • **STOP** • **STOP** • **STOP** • **STOP** • **STOP**

A20   **a.** yes   **b.** yes

Q21 **a.** Are $a + (b + c)$ and $(a + b) + c$ equivalent for all whole-number replacements of a, b, and c?_____

b. Is $a + (b + c) = (a + b) + c$ true for all whole-number replacements of a, b, and c?_____

STOP • **STOP** • **STOP** • **STOP** • **STOP** • **STOP** • **STOP** • **STOP** • **STOP**

A21 **a.** yes　　**b.** yes

10　The assumption that $a + (b + c) = (a + b) + c$ is true for all whole-number replacements of a, b, and c is called the *associative property of addition*. It indicates that in addition, the way three numbers are *grouped* does not affect the sum. Other examples of this property are:

$$(5 + 6) + 2 = 5 + (6 + 2)$$
$$12 + (5 + 0) = (12 + 5) + 0$$

Q22 **a.** $15 + (5 + 12) = (15 + 5) + 12$ is an example of the _____ property of addition.

b. The associative property of addition indicates that the _____ of three numbers in an addition problem may be changed without affecting the sum.

STOP • **STOP** • **STOP** • **STOP** • **STOP** • **STOP** • **STOP** • **STOP** • **STOP**

A22 **a.** associative　　**b.** grouping

Q23 Use the associative property of addition to complete the following in the form of true statements:

a. $7 + (103 + 28) =$ _____

b. $(9 + 12) + 8 =$ _____

c. $(8 + 4) + 6 =$ _____

d. $9 + (31 + 12) =$ _____

STOP • **STOP** • **STOP** • **STOP** • **STOP** • **STOP** • **STOP** • **STOP** • **STOP**

A23 **a.** $(7 + 103) + 28$　**b.** $9 + (12 + 8)$　**c.** $8 + (4 + 6)$　**d.** $(9 + 31) + 12$

11　A similar property is true for multiplication. In multiplication, the way three *factors* are grouped does not affect the product. This assumption is the *associative property of multiplication*, which states that $a(bc) = (ab)c$ is true for all whole-number replacements of a, b, and c. Examples of this property are:

$$7(3 \cdot 2) = (7 \cdot 3)2$$
$$(3 \cdot 4)5 = 3(4 \cdot 5)$$

Q24 The associative property of multiplication indicates that the grouping of three _____ in a multiplication problem may be changed without affecting the product.

STOP • **STOP** • **STOP** • **STOP** • **STOP** • **STOP** • **STOP** • **STOP** • **STOP**

A24 factors

Q25 Use the associative property of multiplication to complete the following in the form of true statements:

a. $(72 \cdot 5)2 =$ _____　　**b.** $4(25 \cdot 7) =$ _____

c. $125(8 \cdot 9) =$ _____   d. $(19 \cdot 8)5 =$ _____

STOP • STOP • STOP • STOP • STOP • STOP • STOP • STOP • STOP

A25   **a.** $72(5 \cdot 2)$ **b.** $(4 \cdot 25)7$ **c.** $(125 \cdot 8)9$ **d.** $19(8 \cdot 5)$

Q26   **a.** Use the associative property of addition to change $(x + 3) + 2$ into an equivalent expression. _____
         b. Use the associative property of multiplication to change $2(3x)$ into an equivalent expression. _____

STOP • STOP • STOP • STOP • STOP • STOP • STOP • STOP • STOP

A26   **a.** $x + (3 + 2)$ **b.** $(2 \cdot 3)x$

| 12 | You should note that subtraction and division are not associative operations. |

Q27   Verify the preceding true statement by showing that each of the following are false statements:
         a. $(15 - 7) - 3 = 15 - (7 - 3)$ **b.** $60 \div (15 \div 3) = (60 \div 15) \div 3$

STOP • STOP • STOP • STOP • STOP • STOP • STOP • STOP • STOP

A27   **a.** $8 - 3 \neq 15 - 4$ **b.** $60 \div 5 \neq 4 \div 3$

Q28   True or false:

a. $1x = x \cdot 1$ _____

b. $2 + (x + 5) = 2 + (5 + x)$ _____

c. $(x - 2)5 = 5(x - 2)$ _____

d. $(3 - x) - 4 = 3 - (x - 4)$ _____

e. $(3x + 7) + 9 = 3x + (7 + 9)$ _____

f. $x + 0 = 0 + x$ _____

g. $24 \div (x \div 4) = (24 \div x) \div 4$ _____

h. $2(5x) = (2 \cdot 5)x$ _____

STOP • STOP • STOP • STOP • STOP • STOP • STOP • STOP • STOP

A28   **a.** true: commutative property of multiplication
         b. true: commutative property of addition
         c. true: commutative property of multiplication
         d. false: (subtraction is not an associative operation)
         e. true: associative property of addition
         f. true: commutative property of addition
         g. false: (division is not an associative operation)
         h. true: associative property of multiplication

| 13 | The following properties are much easier to recognize and remember. The *addition property of zero* (additive identity property) states that $a + 0 = 0 + a = a$ is true for all whole-number replacements of a. If zero is added to any whole number, the result is always the identical whole number: |

$$5 + 0 = 0 + 5 = 5$$
$$17 + 0 = 0 + 17 = 17$$

The *multiplication property of zero* states that $a \cdot 0 = 0a = 0$ is true for all whole-number replacements of a. If zero is multiplied by any whole number, the result is always zero:

$$5 \cdot 0 = 0 \cdot 5 = 0$$
$$32 \cdot 0 = 0 \cdot 32 = 0$$

The *multiplication property of one* (multiplicative identity property) states that $a \cdot 1 = 1a = a$ is true for all whole-number replacements of a. If 1 is multiplied by any whole number, the result is always the identical whole number:

$$5 \cdot 1 = 1 \cdot 5 = 5$$
$$108 \cdot 1 = 1 \cdot 108 = 108$$

The *division property of one* states that $\frac{a}{1} = a$ is true for all whole-number replacements of a. If any whole number is divided by 1, the result is the same whole number:

$$\frac{5}{1} = 5$$

$$\frac{297}{1} = 297$$

$$\frac{0}{1} = 0$$

Q29   True or false:

a. $\dfrac{5x}{1} = 5x$ _____

b. $(2x + 1)1 = 2x + 1$ _____

c. $(5x + 0) + 3 = 5x + 3$ _____

d. $\dfrac{2}{3}x \cdot 0 = 0$ _____

STOP • STOP • STOP • STOP • STOP • STOP • STOP • STOP • STOP

A29   a. true: division property of one
b. true: multiplication property of one (multiplicative identity property)
c. true: addition property of zero (additive identity property)
d. true: multiplication property of zero

This completes the instruction for this section.

1.4   Exercises

1. Insert $=$ or $\neq$ in the blank to form a true statement (the replacement set for all variables is the set of whole numbers):

a. $(xy)4$ _____ $4(xy)$

b. $7a + 0$ _____ $7a$

c. $x - 8$ _____ $8 - x, x \neq 8$

d. $c \div 5$ _____ $5 \div c, c \neq 5$

e. $1(x + y)$ _____ $x + y$

f. $(2a + b) + 3b$ _____ $2a + (b + 3b)$

g. $(5 + b) + 3$ _____ $(b + 5) + 3$ **h.** $4(5x)$ _____ $(4 \cdot 5)x$

i. $5x \cdot 0$ _____ 0 **j.** $\dfrac{2x - 1}{1}$ _____ $2x - 1$

k. $14 \div (7 \div x)$ _____ $(14 \div 7) \div x, \, x \neq 1$
l. $(x - 13) - 6$ _____ $x - (13 - 6)$

2. Indicate the number of the general statement that corresponds to its name. (Assume that general statements are true for all whole-number replacements of x, y, and z.)
   a. commutative property of addition   **1.** $x \cdot 1 = 1x = x$
   b. commutative property of multiplication   **2.** $x(yz) = (xy)z$
   c. associative property of addition   **3.** $x \cdot 0 = 0x = 0$
   d. associative property of multiplication   **4.** $x + y = y + x$

   e. addition property of zero   **5.** $\dfrac{x}{1} = x$

   f. multiplication property of zero   **6.** $(x + y) + z = x + (y + z)$
   g. multiplication property of one   **7.** $xy = yx$
   h. division property of one   **8.** $x + 0 = 0 + x = x$

3. Give one numerical example to illustrate that
   a. Subtraction of whole numbers is not a commutative operation.
   b. Division of whole numbers is not a commutative operation.
   c. Subtraction of whole numbers is not an associative operation.
   d. Division of whole numbers is not an associative operation.
4. Use the property given to write a true statement:
   a. By the associative property of addition, $2 + (3 + 4) =$ _____.
   b. By the multiplication property of one, $32 \cdot 1 =$ _____.
   c. By the commutative property of multiplication, $1 \cdot 15 =$ _____.
   d. By the addition property of zero, $(72 - 5) + 0 =$ _____.
   e. By the multiplication property of zero, $0(15 \cdot 3) =$ _____.
   f. By the associative property of multiplication, $1(4 \cdot 7) =$ _____.

   g. By the division property of one, $\dfrac{42}{1} =$ _____.

   h. By the commutative property of addition, $2(3 + 4) =$ _____.

1.4   Exercise Answers

1. **a.** $=$: commutative property of multiplication
   b. $=$: addition property of zero (additive identity property)
   c. $\neq$: (subtraction is not a commutative operation)
   d. $\neq$: (division is not a commutative operation)
   e. $=$: multiplication property of one (multiplicative identity property)
   f. $=$: associative property of addition
   g. $=$: commutative property of addition
   h. $=$: associative property of multiplication
   i. $=$: multiplication property of zero
   j. $=$: division property of one
   k. $\neq$: (division is not an associative operation)
   l. $\neq$: (subtraction is not an associative operation)
2. **a.** 4   **b.** 7   **c.** 6   **d.** 2   **e.** 8   **f.** 3   **g.** 1
   h. 5

3. **a.** $5 - 1 \neq 1 - 5$ (one of many examples)
    b. $6 \div 2 \neq 2 \div 6$ (one of many examples)
    c. $10 - (5 - 2) \neq (10 - 5) - 2$ (one of many examples)
    d. $20 \div (10 \div 2) \neq (20 \div 10) \div 2$ (one of many examples)
4. **a.** $(2 + 3) + 4$ **b.** 32   **c.** $15 \cdot 1$ **d.** $72 - 5$
    e. 0   **f.** $(1 \cdot 4)7$ **g.** 42   **h.** $2(4 + 3)$

1.5   Distributive Properties

1

Consider the following evaluations:

$2(3 + 4)$ and $2 \cdot 3 + 2 \cdot 4$
$2 \cdot 7$ $6 + 8$
14   14

$2(3 + 4)$ and $2 \cdot 3 + 2 \cdot 4$ are equivalent expressions because both expressions have the same evaluation. Therefore, $2(3 + 4) = 2 \cdot 3 + 2 \cdot 4$ is a true statement.

Q1 Verify that $5 \cdot 6 + 5 \cdot 2 = 5(6 + 2)$ is a true statement by showing that $5 \cdot 6 + 5 \cdot 2$ and $5(6 + 2)$ are equivalent expressions.

STOP • **STOP** • **STOP** • **STOP** • **STOP** • **STOP** • **STOP** • **STOP** • **STOP**

A1 $5 \cdot 6 + 5 \cdot 2$ and $5(6 + 2)$
     $30 + 10$ $5 \cdot 8$
     40   40

Q2 Complete the following in order to form true statements:
a. $3(2 + 5)$ **b.** $7(8 + 6)$
     $3 \cdot 2 + 3 \cdot$ _____ $7 \cdot$ _____ $+ 7 \cdot 6$

STOP • **STOP** • **STOP** • **STOP** • **STOP** • **STOP** • **STOP** • **STOP** • **STOP**

A2 **a.** $3 \cdot 2 + 3 \cdot \underline{5}$ **b.** $7 \cdot \underline{8} + 7 \cdot 6$

2

The true statement $9(10 + 3) = 9 \cdot 10 + 9 \cdot 3$ is an example of the *left distributive property of multiplication over addition*. You should observe that $9(10 + 3)$ is a *product* of 9 times 13, whereas $9 \cdot 10 + 9 \cdot 3$ is a *sum* of 90 and 27. This property changes a product to a sum. It also changes a sum to a product. For example,

$2 \cdot 6 + 2 \cdot 4 = 2(6 + 4)$

Q3 Use the left distributive property of multiplication over addition to change each sum to a product or each product to a sum (parts a and b are given as examples):

a. $5 \cdot 3 + 5 \cdot 7 =$ $5(3 + 7)$ **b.** $6(4 + 5) =$ $6 \cdot 4 + 6 \cdot 5$

c. $9(2 + 1) =$ _____   **d.** $17(5 + 6) =$ _____

e. $10 \cdot 2 + 10 \cdot 3 =$ _____   **f.** $8 \cdot 12 + 8 \cdot 8 =$ _____

STOP • **STOP** • **STOP** • **STOP** • **STOP** • **STOP** • **STOP** • **STOP** • **STOP**

A3   c. $9 \cdot 2 + 9 \cdot 1$ d. $17 \cdot 5 + 17 \cdot 6$ e. $10(2 + 3)$ f. $8(12 + 8)$

---

3   In general, the *left distributive property of multiplication over addition* states that $a(b + c) = ab + ac$ is true for all whole-number replacements of a, b, and c.

Q4   Use the left distributive property of multiplication over addition to complete the evaluation of each numerical expression:

   a. $4 \cdot 7 + 4 \cdot 3$ **b.** $15(10 + 2)$

      $4(\underline{\qquad} + \underline{\qquad})$ $15 \cdot \underline{\qquad} + 15 \cdot \underline{\qquad}$

      $4 \cdot \underline{\qquad}$ $\underline{\qquad} + \underline{\qquad}$

      $\underline{\qquad}$ $\underline{\qquad}$

STOP • **STOP** • **STOP** • **STOP** • **STOP** • **STOP** • **STOP** • **STOP** • **STOP**

A4   **a.** $4(\underline{7} + \underline{3})$ **b.** $15 \cdot \underline{10} + 15 \cdot \underline{2}$

         $4 \cdot \underline{10}$ $\underline{150} + \underline{30}$

         $\underline{40}$ $\underline{180}$

---

4   Consider these evaluations:

   $(3 + 4)5$ and $3 \cdot 5 + 4 \cdot 5$
   $7 \cdot 5$ $15 + 20$
   35 35

   Hence, $(3 + 4)5 = 3 \cdot 5 + 4 \cdot 5$. This is an example of the *right distributive property of multiplication over addition*. This property also changes products to sums or sums to products.

Q5   Use the right distributive property of multiplication over addition to change each sum to a product or each product to a sum:

   a. $(4 + 2)7 =$ $\underline{\qquad\qquad}$ **b.** $2 \cdot 3 + 21 \cdot 3 =$ $\underline{\qquad\qquad}$

   c. $18 \cdot 12 + 2 \cdot 12 =$ $\underline{\qquad\qquad}$ **d.** $(13 + 19)10 =$ $\underline{\qquad\qquad}$

STOP • **STOP** • **STOP** • **STOP** • **STOP** • **STOP** • **STOP** • **STOP** • **STOP**

A5   **a.** $4 \cdot 7 + 2 \cdot 7$ **b.** $(2 + 21)3$
   c. $(18 + 2) \cdot 12$ **d.** $13 \cdot 10 + 19 \cdot 10$

---

5   In general, the *right distributive property of multiplication over addition* states that $(a + b)c = ac + bc$ is true for all whole-number replacements of a, b, and c.

Q6   Complete the following evaluations:
   a. $14 \cdot 9 + 6 \cdot 9$ **b.** $(17 + 8)12$

      $(\underline{\qquad} + \underline{\qquad}) \cdot 9$ $17 \cdot \underline{\qquad} + 8 \cdot \underline{\qquad}$

      $\underline{\qquad} \cdot 9$ $\underline{\qquad} + \underline{\qquad}$

      $\underline{\qquad}$ $\underline{\qquad}$

STOP • **STOP** • **STOP** • **STOP** • **STOP** • **STOP** • **STOP** • **STOP** • **STOP**

A6   **a.** $(\underline{14} + \underline{6}) \cdot 9$ **b.** $17 \cdot \underline{12} + 8 \cdot \underline{12}$

         $\underline{20} \cdot 9$ $\underline{204} + \underline{96}$

         $\underline{180}$ $\underline{300}$

6

Notice that:

$7(5 - 2)$ and $7 \cdot 5 - 7 \cdot 2$
$7 \cdot 3$ $35 - 14$
21 21

Hence, $7(5 - 2) = 7 \cdot 5 - 7 \cdot 2$. This is an example of the *left distributive property of multiplication over subtraction*. Observe that $7(5 - 2)$ is a product, whereas $7 \cdot 5 - 7 \cdot 2$ is a difference. This property also changes a difference to a product. For example,

$7 \cdot 5 - 7 \cdot 2 = 7(5 - 2)$

Q7

Use the left distributive property of multiplication over subtraction to change each product to a difference or each difference to a product (parts a and b are given as examples):

   a. $5 \cdot 7 - 5 \cdot 3 = $ $5(7 - 3)$ b. $10(19 - 7) = $ $10 \cdot 19 - 10 \cdot 7$

   c. $12(13 - 6) = $ _____   d. $8 \cdot 7 - 8 \cdot 3 = $ _____

   e. $24 \cdot 12 - 24 \cdot 2 = $ _____   f. $3(36 - 27) = $ _____

STOP   •   STOP   •   STOP   •   STOP   •   STOP   •   STOP   •   STOP   •   STOP   •   STOP

A7

   c. $12 \cdot 13 - 12 \cdot 6$ d. $8(7 - 3)$
   e. $24(12 - 2)$ f. $3 \cdot 36 - 3 \cdot 27$

7

In general, the left distributive property of multiplication over subtraction states that $a(b - c) = ab - ac$ is true for all whole-number replacements of a, b, and c.

Q8

Complete the following evaluations:
   a. $24(12 - 2)$ b. $7 \cdot 23 - 7 \cdot 3$

       $24 \cdot \underline{\hspace{1cm}} - 24 \cdot \underline{\hspace{1cm}}$ $7(\underline{\hspace{1cm}} - \underline{\hspace{1cm}})$

       $\underline{\hspace{1cm}} - \underline{\hspace{1cm}}$ $7 \cdot \underline{\hspace{1cm}}$

       $\underline{\hspace{1cm}}$ $\underline{\hspace{1cm}}$

STOP   •   STOP   •   STOP   •   STOP   •   STOP   •   STOP   •   STOP   •   STOP   •   STOP

A8

   a. $24 \cdot \underline{12} - 24 \cdot \underline{2}$ b. $7(\underline{23} - \underline{3})$
       $\underline{288} - \underline{48}$ $7 \cdot \underline{20}$
       $\underline{240}$ $\underline{140}$

8

The true statement $(12 - 9)2 = 12 \cdot 2 - 9 \cdot 2$ is an example of the *right distributive property of multiplication over subtraction*. This property also changes products to differences or differences to products.

Q9

Use the right distributive property of multiplication over subtraction to change each product to a difference or each difference to a product:

   a. $(12 - 2)6 = $ _____

   b. $7 \cdot 4 - 5 \cdot 4 = $ _____

STOP   •   STOP   •   STOP   •   STOP   •   STOP   •   STOP   •   STOP   •   STOP   •   STOP

A9

   a. $12 \cdot 6 - 2 \cdot 6$ b. $(7 - 5)4$

9

In general, the right distributive property of multiplication over subtraction states that $(a - b)c = ac - bc$ is true for all whole-number replacements of a, b, and c.

Q10   Complete the following evaluations:

    a. $(45 - 7)100$

        _____ $\cdot$ 100 $-$ _____ $\cdot$ 100

        _____ $-$ _____

        _____

    b. $35 \cdot 9 - 5 \cdot 9$

        (_____ $-$ _____)9

        _____ $\cdot$ 9

        _____

STOP • STOP • STOP • STOP • STOP • STOP • STOP • STOP • STOP

A10   **a.** $\underline{45} \cdot 100 - \underline{7} \cdot 100$

        $\underline{4{,}500} - \underline{700}$

        $\underline{3{,}800}$

    b. $(\underline{35} - \underline{5})9$

        $\underline{30} \cdot 9$

        $\underline{270}$

---

10   Consider the following examples of the distributive properties:

$$5(7 + 9) = 5 \cdot 7 + 5 \cdot 9$$
$$(12 - 5)6 = 12 \cdot 6 - 5 \cdot 6$$
$$10(15 - 7) = 10 \cdot 15 - 10 \cdot 7$$
$$(32 + 11)8 = 32 \cdot 8 + 11 \cdot 8$$

In each example, a product was changed to a sum or difference. When the distribution properties are used in this manner, it is usually said that "the parentheses have been removed."

Q11   Use a distributive property to remove the parentheses from each of the following numerical expressions and complete the evaluation (part a is given as an example):

    a. $2(7 - 3) =$ $2 \cdot 7 - 2 \cdot 3 = 14 - 6 = 8$

    b. $18(5 + 3) =$ _____

    c. $(12 - 8)6 =$ _____

    d. $20(9 + 7) =$ _____

    e. $35(2 - 1) =$ _____

STOP • STOP • STOP • STOP • STOP • STOP • STOP • STOP • STOP

A11   **b.** $18 \cdot 5 + 18 \cdot 3 = 90 + 54 = 144$

    c. $12 \cdot 6 - 8 \cdot 6 = 72 - 48 = 24$

    d. $20 \cdot 9 + 20 \cdot 7 = 180 + 140 = 320$

    e. $35 \cdot 2 - 35 \cdot 1 = 70 - 35 = 35$

This completes the instruction for this section.

1.5   Exercises

1. Change each product to a sum or each sum to a product:

    a. $9(6 + 7)$ **b.** $2 \cdot 4 + 2 \cdot 7$ **c.** $(15 + 9)2$ **d.** $6 \cdot 8 + 10 \cdot 8$

2. Change each product to a difference or each difference to a product:

    a. $7 \cdot 9 - 7 \cdot 4$ **b.** $3(12 - 8)$ **c.** $(17 - 2)5$ **d.** $27 \cdot 8 - 7 \cdot 8$

3. Use one of the distributive properties to evaluate each expression (remove the parentheses):

    a. $2(31 - 6)$ **b.** $(15 + 7)3$ **c.** $(18 - 3)4$ **d.** $12(10 - 8)$

4. True or false:

    a. $3(x + 5) = 3x + 3 \cdot 5$ **b.** $4a - 1a = (4 - 1)a$

    c. $5y - 5 \cdot 3 = 5(y - 3)$ **d.** $7(b - 2) = 7b - 7 \cdot 2$

    e. $5x + 3x = (5 + 3)x$ **f.** $12(x + 5) = 12x + 5$

5. Complete the following general statements assumed true for all whole-number replacements of x, y, and z:

    a. $x(y - z) =$ _____   **b.** $(x + y)z =$ _____

    c. $(x - y)z =$ _____   **d.** $x(y + z) =$ _____

1.5   Exercise Answers

1. **a.** $9 \cdot 6 + 9 \cdot 7$ **b.** $2(4 + 7)$ **c.** $15 \cdot 2 + 9 \cdot 2$ **d.** $(6 + 10)8$

2. **a.** $7(9 - 4)$ **b.** $3 \cdot 12 - 3 \cdot 8$ **c.** $17 \cdot 5 - 2 \cdot 5$ **d.** $(27 - 7)8$

3. **a.** 50   **b.** 66   **c.** 60   **d.** 24

4. **a.** true: left distributive property of multiplication over addition

    b. true: right distributive property of multiplication over subtraction

    c. true: left distributive property of multiplication over subtraction

    d. true: left distributive property of multiplication over subtraction

    e. true: right distributive property of multiplication over addition

    f. false: 12 has not been multiplied times 5

5. **a.** $xy - xz$ **b.** $xz + yz$ **c.** $xz - yz$ **d.** $xy + xz$

Chapter 1 Sample Test

At the completion of Chapter 1 it is expected that you will be able to work the following problems.

1.1   Set Ideas

1. What word best describes each phrase?

    a. well-defined collection   **b.** things that belong to a set

    c. set whose count has a last number   **d.** set whose count is unending

2. Explain briefly why the set of people in the room in which you are located is a well-defined collection.

3. Use set notation where convenient to indicate the following sets:

    a. books in your city's public library

    b. digits in the number 1,041,257

4. Insert $\in$ or $\notin$ in the blank to form a true statement:

    a. 15 _____ $\{1, 3, 5, 7, 9, \ldots\}$ **b.** 12 _____ $\{1, 2, 3, 4, \ldots, 20\}$

5. Is the set on the left a subset of the set on the right?

    a. $\{1, 3, 6\}$ $\{1, 2, 3, \ldots\}$ **b.** $\{2, 10, 15\}$ $\{2, 4, 6, \ldots\}$

    c. $\{0, 7, 12\}$ $\{0, 7, 12\}$

6. **a.** Write two symbols that represent a set with no elements.

    b. What is the name of such a set?

    c. Explain briefly why $\{0\} \neq \{\ \ \}$.

1.2   Set Intersection and Union

7. Let $A = \{2, 4, 6, 8, 10, 12\}$, $B = \{1, 3, 5, 7, 9, 11\}$, $C = \{3, 6, 9, 12\}$, $D = \{2, 3, 4, 5, 6, 7\}$, and $E = \{\ \ \}$. Determine:

a. $A \cup B$ **b.** $A \cap B$ **c.** $B \cap D$ **d.** $C \cup D$
e. $D \cup E$ **f.** $C \cap E$

8. What does it mean if two sets are disjoint?

1.3   **Evaluating Open Expressions**

9. Evaluate each open expression for the value of the variable given:
   a. $7(x - 3)$ when $x = 14$ **b.** $5a + 9$ when $a = 13$
   c. $15 - 3(y + 2)$ when $y = 3$ **d.** $7x - 9y$ when $x = 5$ and $y = 2$
   e. $(3b + 1)(b - 7)$ when $b = 9$

10. Use the following words or phrases to complete each open sentence: variable, open expression, numerical coefficient, literal coefficient, factor(s), quantity.
    a. In $5x$, 5 is called the _____.
    b. A _____ is a letter that may be replaced by any one of a set of many numbers.
    c. $2x + 3$, $2(x - 5)$, and $3x - 2y$ are each examples of a(n) _____.
    d. In $7x$ the letter factor x is called a _____ (two words).
    e. In $2(x - 2)$, the _____ $x - 2$ is a factor.
    f. The _____ of abc are a, b, and c.
    g. In $2 + (x + 2)$, the open expression $x + 2$ may be called a _____.

1.4   **Properties of Whole Numbers**

11. **a.** Use set notation to indicate the set of whole numbers.
    b. Complete: Two open expressions are _____ if they have the same evaluation for all replacements of the variable.
    c. The stated equality of two equivalent expressions forms a _____.
    d. $x + 2 = 5$ is an example of an _____ sentence.

12. The open sentence on the left is an application of which property listed on the right? Write the number associated with the property. (The replacement set is the set of whole numbers.)
    a. $(15y) \cdot 6 = 6(15y)$ 1. multiplication property of one
    b. $a \cdot 1 = a$ 2. associative property of addition
    c. $5(3z) = (5 \cdot 3)z$ 3. commutative property of addition
    d. $2x + 0 = 2x$ 4. associative property of multiplication
    e. $3 + 5c = 5c + 3$ 5. commutative property of multiplication

    f. $\dfrac{7d}{1} = 7d$ 6. multiplication property of zero

    g. $3 + (8 + 2m) = (3 + 8) + 2m$ 7. addition property of zero
    h. $0 \cdot 17t = 0$ 8. division property of one

1.5   **Distributive Properties**

13. The open sentence on the left is an application of which property on the right? Write the number associated with the property. (The replacement set is the set of whole numbers.)
    a. $(2x - 3)7 = 2x \cdot 7 - 3 \cdot 7$ 1. right distributive property of multiplication over addition

b. $2(6y + 5) = 2 \cdot 6y + 2 \cdot 5$
c. $5a + 3a = (5 + 3)a$
d. $4(3z - 2) = 4 \cdot 3z - 4 \cdot 2$

2. left distributive property of multiplication over addition

3. right distributive property of multiplication over subtraction

4. left distributive property of multiplication over subtraction

14. Change each product to a sum or difference. Change each sum or difference to a product:
    a. $2(3 + 7)$ **b.** $4 \cdot 5 - 2 \cdot 5$ **c.** $(12 - 8) \cdot 3$ **d.** $4 \cdot 7 + 4 \cdot 8$
    e. $12 \cdot 7 - 12 \cdot 2$

15. Use a distributive property to remove the parentheses (do not evaluate):
    a. $5(13 - 9)$ **b.** $(12 + 7)2$ **c.** $9(2 + 8)$ **d.** $(20 - 13)8$

Chapter 1 Sample Test Answers

1. **a.** set   **b.** elements   **c.** finite   **d.** infinite
2. membership is clear
3. **a.** not convenient **b.** $\{0, 1, 2, 4, 5, 7\}$
4. **a.** $\in$ **b.** $\in$
5. **a.** yes   **b.** no   **c.** yes
6. **a.** $\{\ \}, \varnothing$ **b.** empty set
   c. $\{0\}$ has one element; $\{\ \}$ has zero elements
7. **a.** $\{1, 2, 3, 4, 5, 6, 7, 8, 9, 10, 11, 12\}$ or $\{1, 2, 3, \ldots, 12\}$
   b. $\{\ \}$ or $\varnothing$
   c. $\{3, 5, 7\}$
   d. $\{2, 3, 4, 5, 6, 7, 9, 12\}$
   e. $\{2, 3, 4, 5, 6, 7\}$
   f. $\{\ \}$ or $\varnothing$
8. Their intersection is empty.
9. **a.** 77   **b.** 74   **c.** 0   **d.** 17   **e.** 56
10. **a.** numerical coefficient   **b.** variable   **c.** open expression
    d. literal coefficient   **e.** quantity   **f.** factors
    g. quantity
11. **a.** $\{0, 1, 2, 3, \ldots\}$ **b.** equivalent   **c.** true statement   **d.** open
12. **a.** 5   **b.** 1   **c.** 4   **d.** 7
    e. 3   **f.** 8   **g.** 2   **h.** 6
13. **a.** 3   **b.** 2   **c.** 1   **d.** 4
14. **a.** $2 \cdot 3 + 2 \cdot 7$ **b.** $(4 - 2)5$ **c.** $12 \cdot 3 - 8 \cdot 3$ **d.** $4(7 + 8)$
    e. $12(7 - 2)$
15. **a.** $5 \cdot 13 - 5 \cdot 9$ **b.** $12 \cdot 2 + 7 \cdot 2$ **c.** $9 \cdot 2 + 9 \cdot 8$ **d.** $20 \cdot 8 - 13 \cdot 8$

Chapter 2

The Set of Integers

2.1   Introduction

---

1 | In Chapter 1 we discussed the set of whole numbers and their properties. The natural numbers, $N = \{1, 2, 3, \ldots\}$, were presented as a subset of the set of whole numbers, $W = \{0, 1, 2, \ldots\}$. The set of whole numbers, W, can be thought of as an extension of the set of natural numbers, N.

Q1 | Label each of the numbers below as a natural number, a whole number, or both a natural number and a whole number:

a. 5 _____

b. 0 _____

c. 1 _____

STOP • STOP • STOP • STOP • STOP • STOP • STOP • STOP • STOP

A1 | **a.** both a natural and a whole number
b. whole number
c. both a natural and a whole number

---

2 | The set of whole numbers can be pictured using what is called a *number line* as follows:

Each dot on the number line indicates the point where a specific number occurs and is called the *graph* of the number. The number that corresponds with each dot is called the *coordinate* of the point. The arrow on the end of the number line and the three dots that follow the last number shown both serve to indicate that the set of whole numbers is an infinite set.

Q2 | Graph the following whole numbers by placing dots at the appropriate points on the number line:

a. 7   **b.** 2   **c.** 0

STOP • STOP • STOP • STOP • STOP • STOP • STOP • STOP • STOP

31

A2

3

The number line of Frame 2 pictures the whole numbers spaced an equal distance apart, with each number indicating the *number of units that it is to the right of zero*. That is, the number 2 indicates the point that is 2 units to the right of zero. A new set of numbers will now be defined by extending the number line to the *left* of zero and again marking off points an equal distance apart. It is called the set of *negative integers* and is written $\{^-1, ^-2, ^-3, \ldots\}$. (Read "negative one," "negative two," "negative three," etc.) In the same way that 2 represents 2 units to the *right of zero*, the negative integer $^-2$ (negative two) will represent a distance of 2 units to the *left of zero*. To emphasize the difference in direction between the negative integers $\{^-1, ^-2, ^-3, \ldots\}$ and the natural numbers $\{1, 2, 3, \ldots\}$, the set of natural numbers is sometimes written $\{^+1, ^+2, ^+3, \ldots\}$. (Read "positive one," "positive two," "positive three," etc.) This set is also correctly referred to as the set of *positive integers*. Thus any positive integer can be written either with or without the positive sign. The positive integer $^+7$, for example, is also correctly represented as 7.

Q3

Graph each of the following numbers by placing dots at the appropriate points on the number line:

a. $^-5$ **b.** $^+3$ **c.** 0   **d.** $^-1$ **e.** 6

STOP • STOP • STOP • STOP • STOP • STOP • STOP • STOP • STOP

A3

4

The set that contains the set of negative integers, zero, and the set of positive integers is called the set of *integers* and is represented by the letter *I*. That is,

$$I = \{\ldots, ^-3, ^-2, ^-1, 0, ^+1, ^+2, ^+3, \ldots\}$$

Each of the integers can be thought of as having two parts, *distance* and *direction*. The integer $^+5$, for example, represents a distance of 5 units to the *right* of zero. Similarly, the integer $^-9$ represents a distance of 9 units to the *left* of zero.

Q4

Write the distance and direction represented by each of the following integers:

		Distance	Direction
a.	$^-6$	6 units	left of zero
b.	$^+42$	_____	_____
c.	$^-42$	_____	_____
d.	0	_____	_____

STOP • STOP • STOP • STOP • STOP • STOP • STOP • STOP • STOP

A4

b. 42 units, right of zero
c. 42 units, left of zero
d. 0 units, neither direction

| 5 | The integers, ... , $^-3, ^-2, ^-1, 0, ^+1, ^+2, ^+3, ...$, are also referred to as *signed numbers* because each one (except zero) has a direction designated by either a "$-$" or a "$+$" sign. Zero is an integer but is considered neither positive nor negative since its coordinate is neither to the right nor to the left of the zero point. |

Q5   What is the sign of each of the following integers?

    a. $^+7$ _____   **b.** 6 _____

    c. $^-3$ _____   **d.** 0 _____

STOP • STOP • STOP • STOP • STOP • STOP • STOP • STOP • STOP

A5   **a.** positive   **b.** positive   **c.** negative
    d. none: zero has no sign

| 6 | As a result of the definition of the set of integers, each integer can be paired with a second integer that is the same distance from zero but in a different direction. The paired integers are called *opposites* and are shown in the following drawing: |

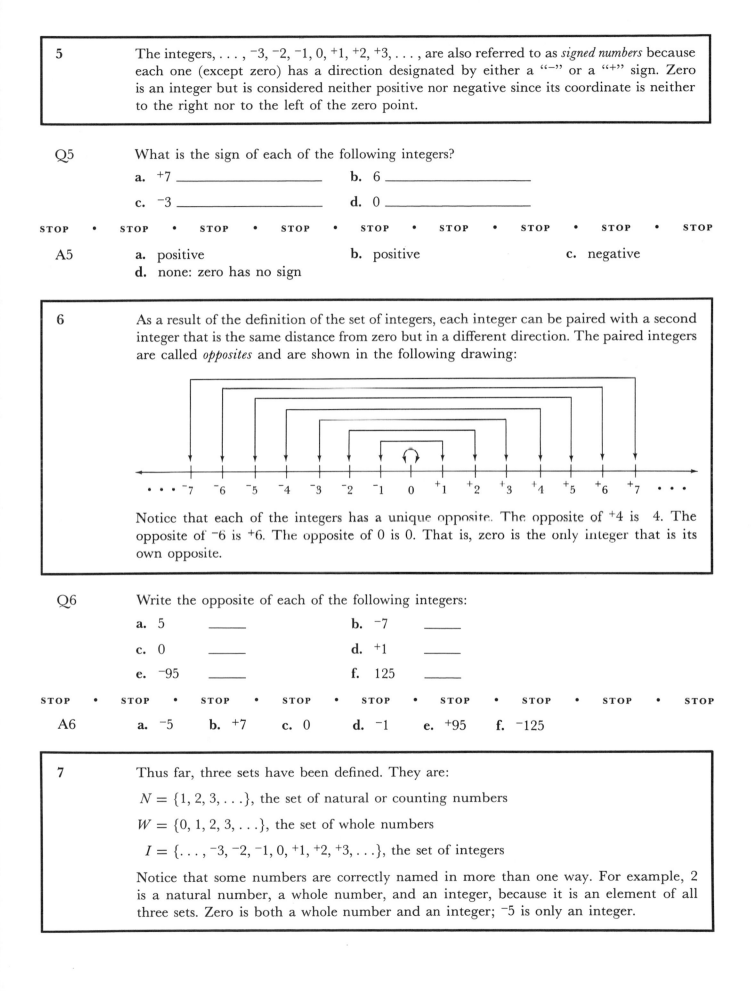

Notice that each of the integers has a unique opposite. The opposite of $^+4$ is 4. The opposite of $^-6$ is $^+6$. The opposite of 0 is 0. That is, zero is the only integer that is its own opposite.

Q6   Write the opposite of each of the following integers:

    a. 5 _____   **b.** $^-7$ _____

    c. 0 _____   **d.** $^+1$ _____

    e. $^-95$ _____   **f.** 125 _____

STOP • STOP • STOP • STOP • STOP • STOP • STOP • STOP • STOP

A6   **a.** $^-5$ **b.** $^+7$ **c.** 0   **d.** $^-1$ **e.** $^+95$ **f.** $^-125$

| 7 | Thus far, three sets have been defined. They are: |

$N = \{1, 2, 3, ...\}$, the set of natural or counting numbers

$W = \{0, 1, 2, 3, ...\}$, the set of whole numbers

$I = \{..., ^-3, ^-2, ^-1, 0, ^+1, ^+2, ^+3, ...\}$, the set of integers

Notice that some numbers are correctly named in more than one way. For example, 2 is a natural number, a whole number, and an integer, because it is an element of all three sets. Zero is both a whole number and an integer; $^-5$ is only an integer.

Q7   Insert $\in$ or $\notin$ to form true statements:

a.	5	___ N	b.	5	___ W	c.	5	___ I
d.	0	___ N	e.	0	___ W	f.	0	___ I
g.	⁻12	___ N	h.	⁻12	___ W	i.	⁻12	___ I
j.	100	___ I	k.	⁻101	___ N	l.	⁻101	___ W

STOP • STOP • STOP • STOP • STOP • STOP • STOP • STOP • STOP

A7   a. $\in$ b. $\in$ c. $\in$ d. $\notin$ e. $\in$ f. $\in$ g. $\notin$
          h. $\notin$ i. $\in$ j. $\in$ k. $\notin$ l. $\notin$

_____

This completes the instruction for this section.

2.1   Exercises

1. The set $\{\ldots, ^-2, ^-1, 0, ^+1, ^+2, \ldots\}$ is called the set of _____.
2. Each integer is thought of as having what two parts?
3. Write the distance and direction represented by each of the following integers:
   a. 15   b. ⁻3   c. 0   d. ⁺7
4. What is the sign of the integer 0?
5. If N, W, and I are defined as in Frame 7, insert $\in$ or $\notin$ to make each of the following a true statement:
   a. ⁻4 ___ N   b. 5 ___ I
   c. 0 ___ W   d. ⁺6 ___ I
6. Graph each of the following integers:
   a. ⁻2   b. 4   c. 0   d. ⁺3

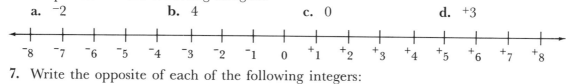

7. Write the opposite of each of the following integers:
   a. ⁺2   b. ⁻3   c. 11   d. 0

2.1   Exercise Answers

1. integers
2. direction, distance
3. a. right of zero   15 units
   b. left of zero   3 units
   c. neither right nor left   0 units
   d. right of zero   7 units
4. Zero has no sign, because it is neither right nor left of zero on the number line.
5. a. $\notin$ b. $\in$ c. $\in$ d. $\in$
6.

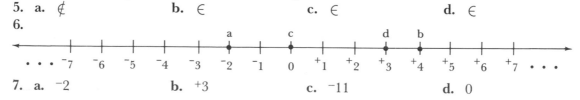

7. a. ⁻2   b. ⁺3   c. ⁻11   d. 0

2.2   Addition

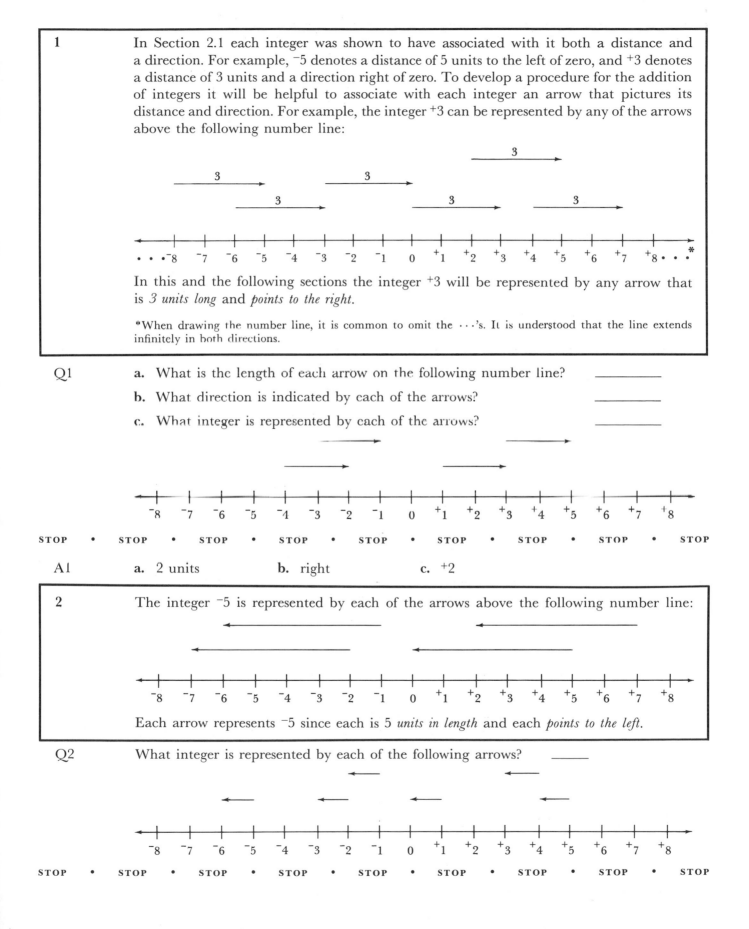

1   In Section 2.1 each integer was shown to have associated with it both a distance and a direction. For example, $^-5$ denotes a distance of 5 units to the left of zero, and $^+3$ denotes a distance of 3 units and a direction right of zero. To develop a procedure for the addition of integers it will be helpful to associate with each integer an arrow that pictures its distance and direction. For example, the integer $^+3$ can be represented by any of the arrows above the following number line:

In this and the following sections the integer $^+3$ will be represented by any arrow that is *3 units long* and *points to the right*.

*When drawing the number line, it is common to omit the $\cdots$'s. It is understood that the line extends infinitely in both directions.

Q1   **a.** What is the length of each arrow on the following number line?   _____

b. What direction is indicated by each of the arrows?   _____

c. What integer is represented by each of the arrows?   _____

STOP • **STOP** • **STOP** • **STOP** • **STOP** • **STOP** • **STOP** • **STOP** • **STOP**

A1   **a.** 2 units   **b.** right   **c.** $^+2$

2   The integer $^-5$ is represented by each of the arrows above the following number line:

Each arrow represents $^-5$ since each is 5 *units in length* and each *points to the left*.

Q2   What integer is represented by each of the following arrows?   _____

STOP • **STOP** • **STOP** • **STOP** • **STOP** • **STOP** • **STOP** • **STOP** • **STOP**

A2 $^-1$: each is 1 unit long and each points to the left.

Q3   Draw three arrows which each represent the integer $^-4$.

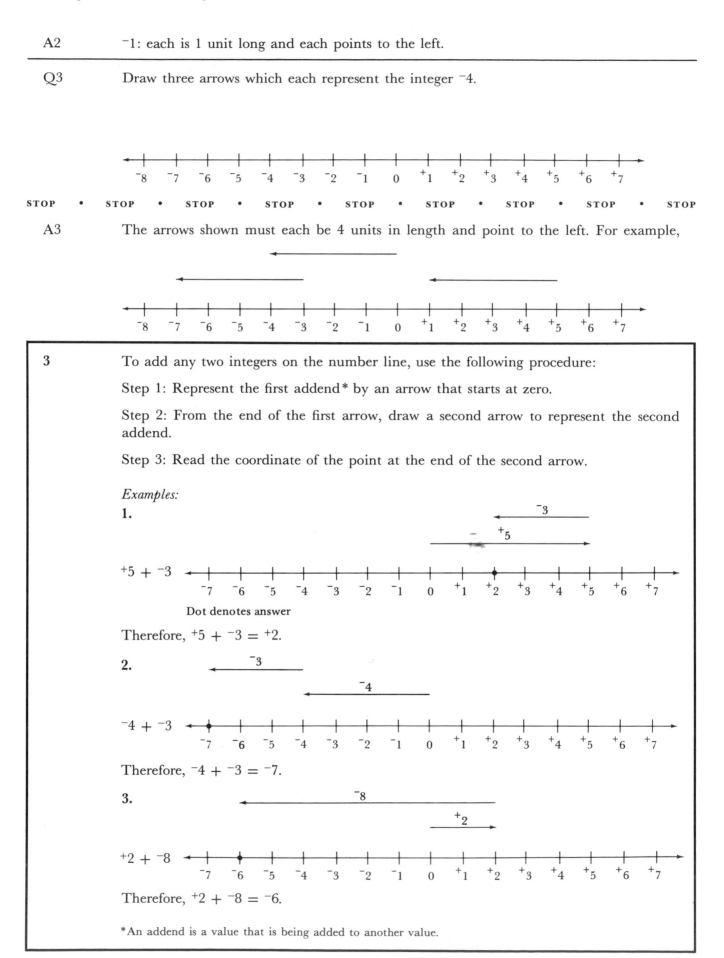

STOP • STOP • STOP • STOP • STOP • STOP • STOP • STOP • STOP

A3   The arrows shown must each be 4 units in length and point to the left. For example,

3   To add any two integers on the number line, use the following procedure:

Step 1: Represent the first addend* by an arrow that starts at zero.

Step 2: From the end of the first arrow, draw a second arrow to represent the second addend.

Step 3: Read the coordinate of the point at the end of the second arrow.

Examples:
1.

$^+5 + {}^-3$

Dot denotes answer

Therefore, $^+5 + {}^-3 = {}^+2$.

2.

$^-4 + {}^-3$

Therefore, $^-4 + {}^-3 = {}^-7$.

3.

$^+2 + {}^-8$

Therefore, $^+2 + {}^-8 = {}^-6$.

*An addend is a value that is being added to another value.

Q4   Find the sum ⁻3 + ⁺4 by use of arrows for each of the addends on the following number line:

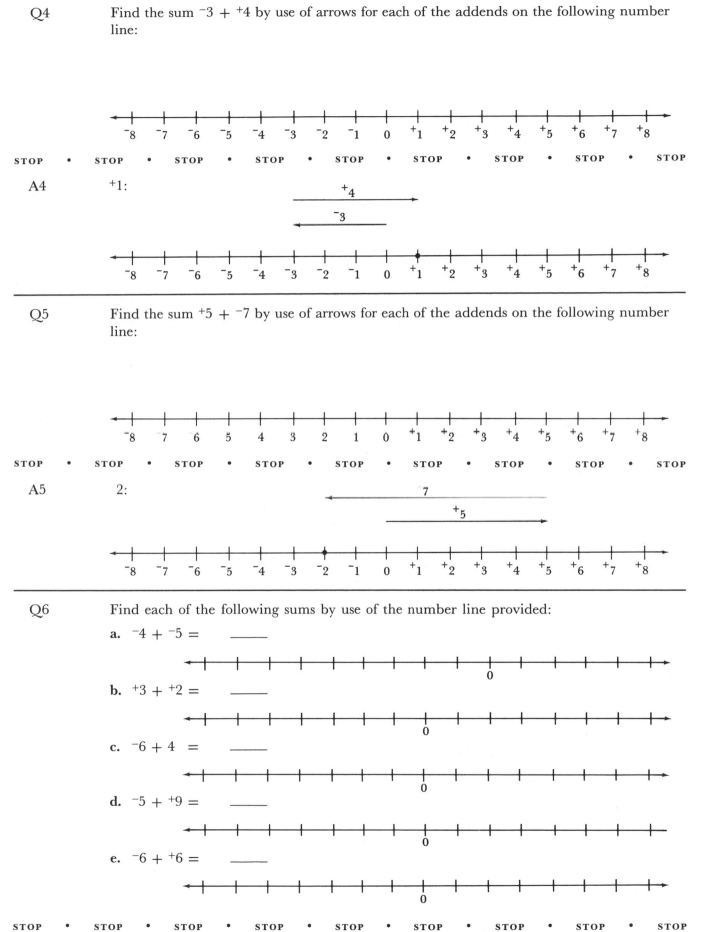

A6   **a.** $^-9$ **b.** $^+5$ **c.** $^-2$ **d.** $^+4$ **e.** 0

4

It is sometimes helpful to notice certain facts about the sum of two integers. Study each of the following three example sets to see if you can discover the three facts demonstrated.

Sum of two positives:	Sum of two negatives:	Sum of a positive and a negative:
$^+5 + {}^+3 = {}^+8$	$^-5 + {}^-3 = {}^-8$	$^-5 + {}^+3 = {}^-2$
$^+7 + {}^+6 = {}^+13$	$^-7 + {}^-6 = {}^-13$	$^+7 + {}^-6 = {}^+1$
$^+2 + {}^+9 = {}^+11$	$^-2 + {}^-9 = {}^-11$	$^+2 + {}^-2 = 0$

The three facts that correspond to the three preceding examples are:

1. The sum of two positive integers is a positive integer.
2. The sum of two negative integers is a negative integer.
3. The sum of a positive integer and a negative integer is sometimes positive, sometimes negative, and sometimes zero.

Q7

Use the facts of Frame 4 or a number line to find each of the following sums:

a. $^-3 + {}^-6 =$ _____   **b.** $^+5 + {}^+8 =$ _____

c. $^+4 + {}^-1 =$ _____   **d.** $^+2 + {}^-7 =$ _____

e. $^+3 + {}^-3 =$ _____   **f.** $^-5 + {}^+5 =$ _____

STOP   •   STOP   •   STOP   •   STOP   •   STOP   •   STOP   •   STOP   •   STOP   •   STOP

A7   **a.** $^-9$ (fact 2)   **b.** $^+13$ (fact 1)   **c.** $^+3$ (fact 3)   **d.** $^-5$
          e. 0 (fact 3)   **f.** 0

5

Recall from Section 2.1 that every integer has a unique opposite, and that opposites represent the same distance but different directions. Consider the sum of $^-7$ and its opposite, $^+7$, shown on the following number line:

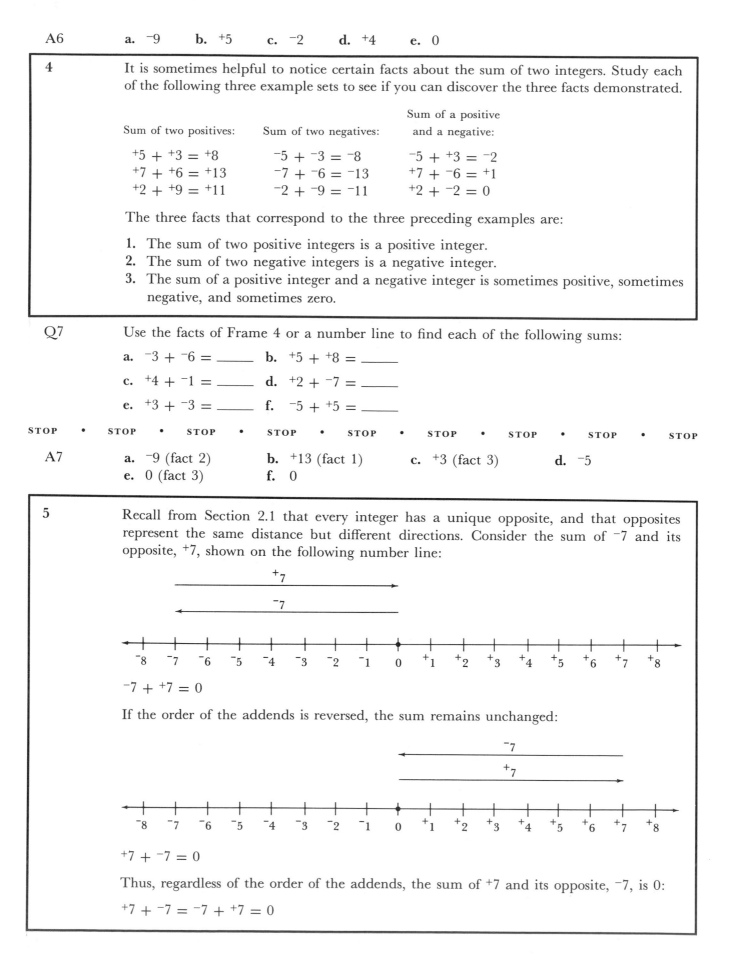

$^-7 + {}^+7 = 0$

If the order of the addends is reversed, the sum remains unchanged:

$^+7 + {}^-7 = 0$

Thus, regardless of the order of the addends, the sum of $^+7$ and its opposite, $^-7$, is 0:

$$^+7 + {}^-7 = {}^-7 + {}^+7 = 0$$

Q8   **a.** Find the sum $^-5 + {}^+5$.

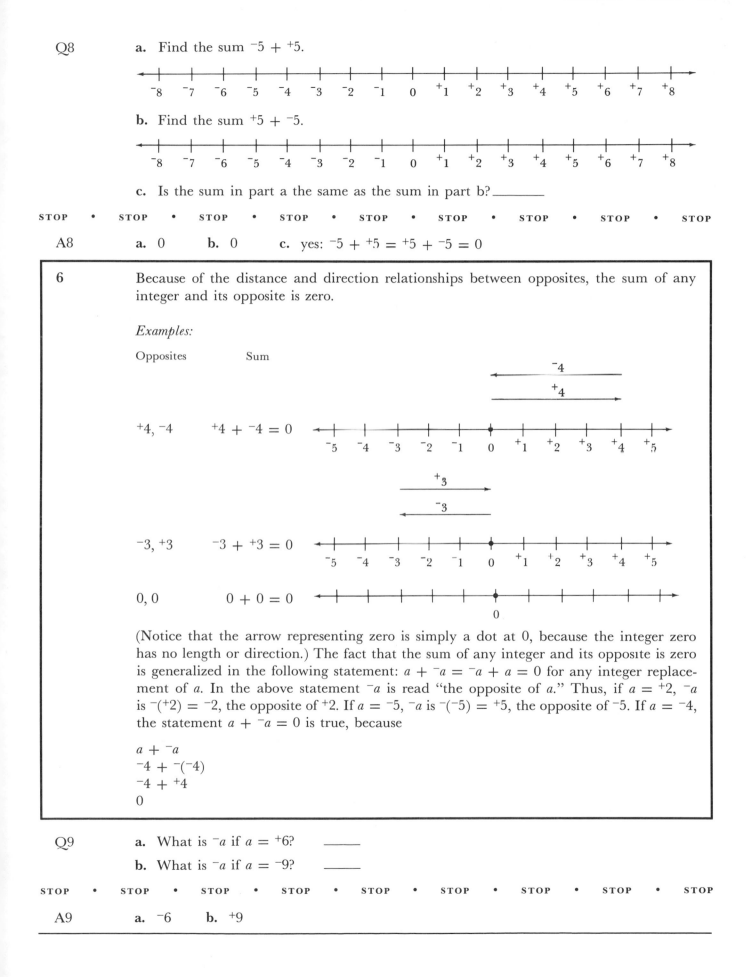

b. Find the sum $^+5 + {}^-5$.

c. Is the sum in part a the same as the sum in part b? _____

STOP • STOP • STOP • STOP • STOP • STOP • STOP • STOP • STOP

A8   **a.** 0   **b.** 0   **c.** yes: $^-5 + {}^+5 = {}^+5 + {}^-5 = 0$

---

6   Because of the distance and direction relationships between opposites, the sum of any integer and its opposite is zero.

Examples:

Opposites   Sum

$^+4, {}^-4$ $^+4 + {}^-4 = 0$

$^-3, {}^+3$ $^-3 + {}^+3 = 0$

$0, 0$ $0 + 0 = 0$

(Notice that the arrow representing zero is simply a dot at 0, because the integer zero has no length or direction.) The fact that the sum of any integer and its opposite is zero is generalized in the following statement: $a + {}^-a = {}^-a + a = 0$ for any integer replacement of a. In the above statement ^-a is read "the opposite of a." Thus, if $a = {}^+2$, ^-a is $^-(^+2) = {}^-2$, the opposite of $^+2$. If $a = {}^-5$, ^-a is $^-(^-5) = {}^+5$, the opposite of $^-5$. If $a = {}^-4$, the statement $a + {}^-a = 0$ is true, because

$a + {}^-a$
$^-4 + {}^-(^-4)$
$^-4 + {}^+4$
0

Q9   **a.** What is ^-a if $a = {}^+6$?   _____

b. What is ^-a if $a = {}^-9$?   _____

STOP • STOP • STOP • STOP • STOP • STOP • STOP • STOP • STOP

A9   **a.** $^-6$ **b.** $^+9$

Q10   Show that $a + {}^-a = 0$ is true for $a = {}^-7$.

STOP • *STOP* • *STOP* • *STOP* • *STOP* • *STOP* • *STOP* • *STOP* • *STOP*

A10 $a + {}^-a$
$${}^-7 + {}^-({}^-7)$$
$${}^-7 + {}^+7$$
$$0$$

Q11   Find each of the following sums:

      a. ${}^-3 + {}^-3 = $ _____   **b.** ${}^+5 + {}^-5 = $ _____

      c. ${}^+7 + {}^+7 = $ _____   **d.** ${}^-6 + {}^+6 = $ _____

      e. ${}^-8 + {}^+8 = $ _____   **f.** $0 + 0 = $ _____

STOP • *STOP* • *STOP* • *STOP* • *STOP* • *STOP* • *STOP* • *STOP* • *STOP*

A11   **a.** ${}^-6$: the sum of two negatives is a negative.
      b. 0: the sum of two opposites is zero.
      c. ${}^+14$: the sum of two positives is a positive.
      d. 0 **e.** 0 **f.** 0

Q12   Find the replacement for x that converts each of the following open sentences to a true statement:

      a. ${}^+2 + {}^-2 = x$ _____   **b.** $x + {}^+7 = 0$ _____

      c. ${}^-15 + x = 0$ _____   **d.** ${}^+13 + {}^-13 = x$ _____

STOP • *STOP* • *STOP* • *STOP* • *STOP* • *STOP* • *STOP* • *STOP* • *STOP*

A12   **a.** 0 **b.** ${}^-7$ **c.** ${}^+15$ **d.** 0

---

7   The sum of any natural number and zero is the natural number: for example, $0 + 2 = 2$ and $8 + 0 = 8$. The same is true of any integer and zero: for example,

$${}^-3 + 0 = {}^-3$$
$$0 + {}^+12 = {}^+12$$
$${}^-9 + 0 = {}^-9$$

This fact is called the *addition property of zero* and is generalized $a + 0 = 0 + a = a$ for any integer a.

Q13   Find each of the following sums:

      a. $0 + {}^-5 = $ _____   **b.** ${}^-4 + {}^-9 = $ _____

      c. ${}^+4 + {}^-4 = $ _____   **d.** ${}^+11 + 0 = $ _____

      e. ${}^+5 + {}^+7 = $ _____   **f.** ${}^-7 + {}^+3 = $ _____

STOP • *STOP* • *STOP* • *STOP* • *STOP* • *STOP* • *STOP* • *STOP* • *STOP*

A13   **a.** ${}^-5$ **b.** ${}^-13$ **c.** 0 **d.** ${}^+11$ **e.** ${}^+12$ **f.** ${}^-4$

---

8   To find the sum of more than two integers, use the methods presented earlier to add the integers two at a time. For example, the sum $^-3 + {}^+7 + {}^-6$ is found:

$$^-3 + {}^+7 + {}^-6 = (^-3 + {}^+7) + {}^-6$$
$$= {}^+4 + {}^-6$$
$$= {}^-2$$

Notice that each sum of two integers can be found by starting at zero and using the number-line procedure if necessary.

Q14   Find the sum $^+3 + {}^-2 + {}^+5$.

STOP • **STOP** • **STOP** • **STOP** • **STOP** • **STOP** • **STOP** • **STOP** • **STOP**

A14 $^+6$: $^+3 + {}^-2 + {}^+5 = (^+3 + {}^-2) + {}^+5$
$$= {}^+1 + {}^+5$$
$$= {}^+6$$

Q15   Find the sum $^-4 + {}^-7 + {}^+3$.

STOP • **STOP** • **STOP** • **STOP** • **STOP** • **STOP** • **STOP** • **STOP** • **STOP**

A15 $^-8$: $^-4 + {}^-7 + {}^+3 = (^-4 + {}^-7) + {}^+3$
$$= {}^-11 + {}^+3$$
$$= {}^-8$$

---

9   The sum $^-3 + {}^+5 + {}^-6 + {}^-2$ is found as follows:

$$^-3 + {}^+5 + {}^-6 + {}^-2 = (^-3 + {}^+5) + {}^-6 + {}^-2$$
$$= {}^+2 + {}^-6 + {}^-2$$
$$= (^+2 + {}^-6) + {}^-2$$
$$= {}^-4 + {}^-2$$
$$= {}^-6$$

Q16   Find each of the following sums:
a. $^-3 + {}^-5 + {}^+8$ b. $^-4 + {}^+5 + {}^-1 + {}^-3$ c. $^-2 + {}^-3 + {}^-6$

d. $^+1 + {}^+5 + {}^+7$

STOP • **STOP** • **STOP** • **STOP** • **STOP** • **STOP** • **STOP** • **STOP** • **STOP**

A16 **a.** 0 **b.** $^-3$ **c.** $^-11$ **d.** $^+13$

---

10 In Chapter 1 it was established that $a + b = b + a$ is true for all *whole-number replacements* of a and b (commutative property of addition). It is similarly true that $a + b = b + a$ for all *integer* replacements of a and b. An infinite number of examples can be used to demonstrate this fact. A few are:

$^-1 + {}^+2 = {}^+2 + {}^-1$ (both sums are $^+1$)
$^-7 + {}^-4 = {}^-4 + {}^-7$ (both sums are $^-11$)
$^+9 + {}^-2 = {}^-2 + {}^+9$ (both sums are $^+7$)

Q17 Verify that $a + b = b + a$ is true for $a = {}^-5$ and $b = {}^+8$.

STOP • **STOP** • **STOP** • **STOP** • **STOP** • **STOP** • **STOP** • **STOP** • **STOP**

A17 $^-5 + {}^+8 = {}^+8 + {}^-5$ (both sums are $^+3$)

---

11 In Chapter 1 it was also established that $(a + b) + c = a + (b + c)$ is true for all *whole-number* replacements of a, b, and c (associative property of addition). The associative property of addition is also true for all *integer* replacements of a, b, and c. For example,

1. $(^-3 + {}^+4) + {}^-5 = {}^-3 + (^+4 + {}^-5)$
$^+1 + {}^-5 = {}^-3 + {}^-1$
$^-4 = {}^-4$

2. $(^+7 + {}^+5) + {}^-10 = {}^+7 + (^+5 + {}^-10)$
$^+12 + {}^-10 = {}^+7 + {}^-5$
$^+2 = {}^+2$

Q18 Verify that $(a + b) + c = a + (b + c)$ is true for $a = {}^-2$, $b = {}^-7$, and $c = {}^+3$.

STOP • **STOP** • **STOP** • **STOP** • **STOP** • **STOP** • **STOP** • **STOP** • **STOP**

A18 $(^-2 + {}^-7) + {}^+3 = {}^-2 + (^-7 + {}^+3)$
$^-9 + {}^+3 = {}^-2 + {}^-4$
$^-6 = {}^-6$

This completes the instruction for this section.

2.2   Exercises

1. Find each of the following sums:

a. $^-3 + {}^-5$ **b.** $^+5 + {}^+9$
c. $^+7 + {}^-6$ **d.** $^-6 + {}^+6$
e. $^-9 + 0$ **f.** $^-3 + {}^+4 + {}^-3$
g. $^-7 + {}^+6 + {}^+1$ **h.** $0 + {}^+4$
i. $^-4 + {}^-3 + {}^+7 + {}^+4$ **j.** $^-1 + {}^+6 + {}^-6 + {}^+1$

2. Find the replacement for x that converts each of the following open sentences to a true mathematical statement:

    a. $x + {}^{-}2 = 0$ **b.** ${}^{-}8 + {}^{+}8 = x$

    c. ${}^{-}12 + x = {}^{-}12$ **d.** ${}^{-}12 + x = 0$

3. Is the statement $a + b = b + a$ true for all integer replacements of a and b?

4. Match each of the following with the number of the statement that demonstrates it correctly.

    a. associative property of addition   **1.** ${}^{-}5 + {}^{+}3 = {}^{+}3 + {}^{-}5$

    b. addition property of zero   **2.** ${}^{-}2 + 0 = 0 + {}^{-}2 = {}^{-}2$

    c. the sum of two opposites is zero   **3.** ${}^{-}9 + {}^{+}9 = 0$

    d. commutative property of addition   **4.** $({}^{-}3 + {}^{+}7) + {}^{-}2 = {}^{-}3 + ({}^{+}7 + {}^{-}2)$

2.2 Exercise Answers

1. a. ${}^{-}8$ **b.** ${}^{+}14$ **c.** ${}^{+}1$ **d.** 0 **e.** ${}^{-}9$ **f.** ${}^{-}2$ **g.** 0

    h. ${}^{+}4$ **i.** ${}^{+}4$ **j.** 0

2. a. ${}^{+}2$ **b.** 0 **c.** 0 **d.** ${}^{+}12$

3. yes: commutative property of addition

4. a. 4 **b.** 2 **c.** 3 **d.** 1

2.3 Subtraction

1

Subtraction and addition can be thought of as opposite operations.* Consider, for example, the effect on any number x of first adding 5 and then performing the opposite operation of subtracting 5.

x

$x + 5$ add 5

$x + 5 - 5$ subtract 5

x

Notice that the operation of subtraction undoes what the operation of addition does, and the result is again the number x. The operation of addition also undoes an equal subtraction.

x

$x - 5$ subtract 5

$x - 5 + 5$ add 5

x

Thus, regardless of the order in which they are done, the operations of addition and subtraction are opposites. One operation undoes the other operation.

*Some mathematicians refer to addition and subtraction as "inverse operations."

Q1   **a.** What is the opposite operation of addition?   _____

         b. What is the opposite of adding 7 to any number?   _____

STOP • **STOP** • **STOP** • **STOP** • **STOP** • **STOP** • **STOP** • **STOP** • **STOP**

A1   **a.** subtraction   **b.** subtracting 7

Q2   **a.** What is the opposite operation of subtraction?   _____

           b. What is the opposite of subtracting 3 from any number?   _____

STOP • STOP • STOP • STOP • STOP • STOP • STOP • STOP • STOP

A2   **a.** addition   **b.** adding 3

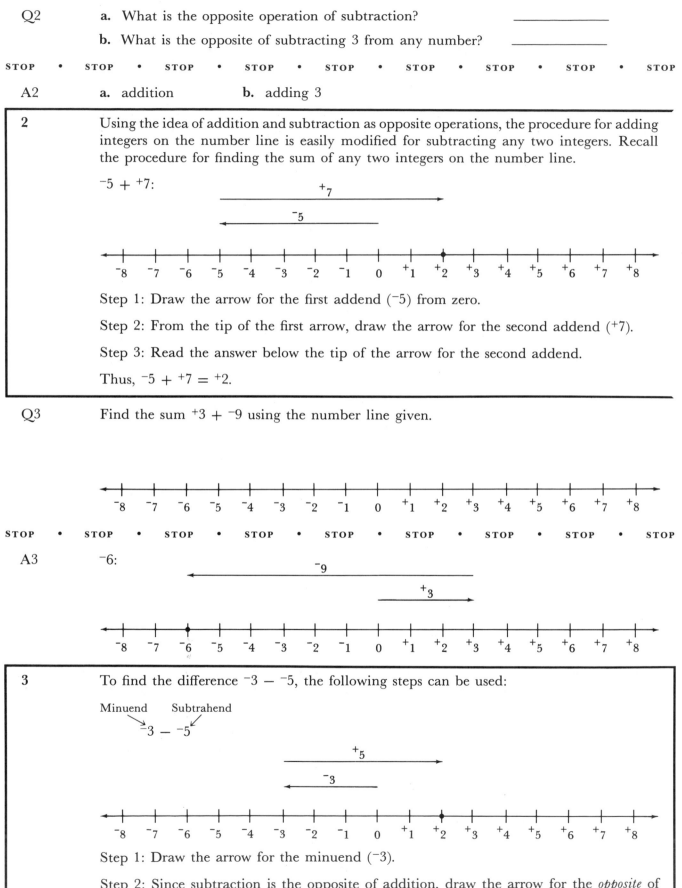

2   Using the idea of addition and subtraction as opposite operations, the procedure for adding integers on the number line is easily modified for subtracting any two integers. Recall the procedure for finding the sum of any two integers on the number line.

$^-5 + {}^+7$:

Step 1: Draw the arrow for the first addend ($^-5$) from zero.

Step 2: From the tip of the first arrow, draw the arrow for the second addend ($^+7$).

Step 3: Read the answer below the tip of the arrow for the second addend.

Thus, $^-5 + {}^+7 = {}^+2$.

Q3   Find the sum $^+3 + {}^-9$ using the number line given.

STOP • STOP • STOP • STOP • STOP • STOP • STOP • STOP • STOP

A3 $^-6$:

3   To find the difference $^-3 - {}^-5$, the following steps can be used:

Minuend   Subtrahend

$^-3 - {}^-5$

Step 1: Draw the arrow for the minuend ($^-3$).

Step 2: Since subtraction is the opposite of addition, draw the arrow for the *opposite* of the subtrahend ($^+5$) from the tip of the first arrow.

Step 3: Read the answer below the tip of the second arrow.

Thus, $^-3 - {}^-5 = {}^+2$.

Q4   Use the following number line to find the difference $^-7 - {}^-4$:

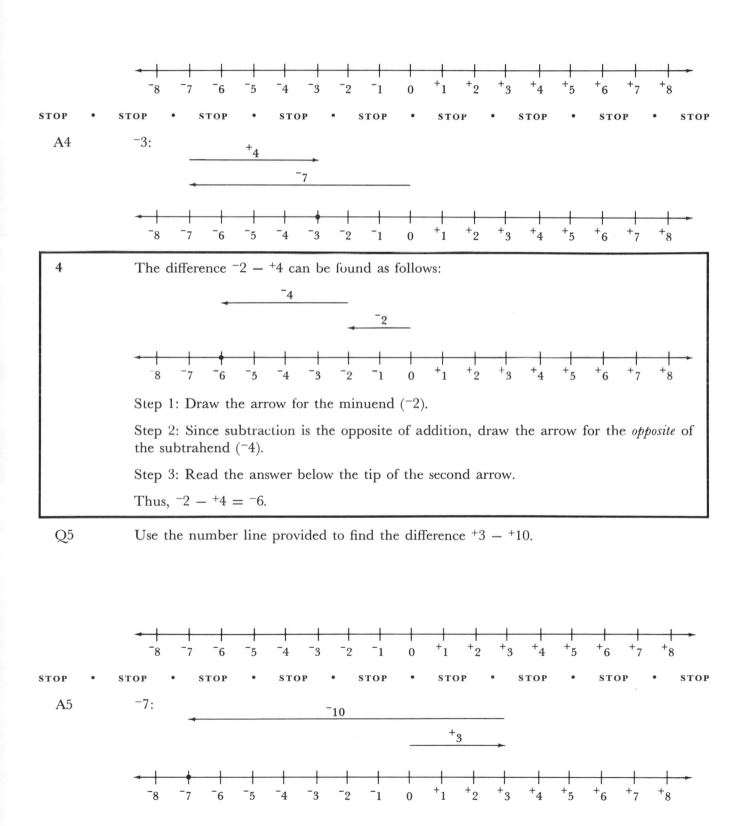

A4 $^-3$:

4   The difference $^-2 - {}^+4$ can be found as follows:

Step 1: Draw the arrow for the minuend ($^-2$).

Step 2: Since subtraction is the opposite of addition, draw the arrow for the *opposite* of the subtrahend ($^-4$).

Step 3: Read the answer below the tip of the second arrow.

Thus, $^-2 - {}^+4 = {}^-6$.

Q5   Use the number line provided to find the difference $^+3 - {}^+10$.

A5 $^-7$:

5 We repeat here the examples of Frames 3 and 4:

Minuend Subtrahend Difference

$$^-3 \;-\; ^-5 \;=\; ^+2$$
$$^-2 \;-\; ^+4 \;=\; ^-6$$

In each case the difference was found by first drawing the arrow for the minuend and then drawing the arrow for the *opposite* of the subtrahend.

This procedure is the basis for defining subtraction in terms of addition: *To find the difference of two integers, add the minuend to the opposite of the subtrahend.*

Examples:

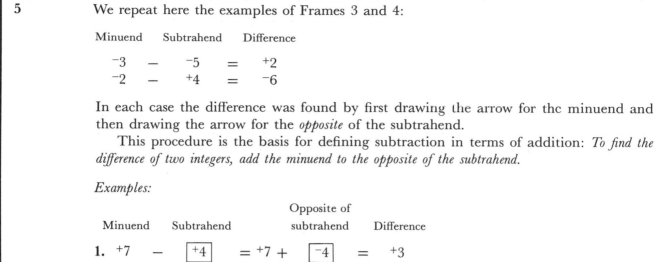

	Minuend	Subtrahend		Opposite of subtrahend		Difference
1.	$^+7$	$-$	$\boxed{^+4}$	$= \;^+7 \;+$	$\boxed{^-4}$	$= \quad ^+3$
2.	$^-3$	$-$	$\boxed{^+5}$	$= \;^-3 \;+$	$\boxed{^-5}$	$= \quad ^-8$
3.	$^-4$	$-$	$\boxed{^-6}$	$= \;^-4 \;+$	$\boxed{^+6}$	$= \quad ^+2$
4.	$^+2$	$-$	$\boxed{^+6}$	$= \;^+2 \;+$	$\boxed{^-6}$	$= \quad ^-4$

Notice that in each example, two changes are involved: The operation sign for subtraction is changed to addition, and the subtrahend is changed to its opposite.

Q6 Find the difference $^-3 - {}^+1$ using the number line provided.

STOP • STOP • STOP • STOP • STOP • STOP • STOP • STOP • STOP

A6 $^-4$: $^-3 - {}^+1 = {}^-3 + {}^-1$

Q7 Find the difference by rewriting as a sum.

$$^-4 - {}^+7 = {}^-4 + \underline{\hphantom{xxx}} = \underline{\hphantom{xxx}}$$

STOP • STOP • STOP • STOP • STOP • STOP • STOP • STOP • STOP

A7 $^-7, ^-11$

Q8 Find each of the following differences:
a. $^-2 - {}^+5$ **b.** $^+7 - {}^-5$ **c.** $^+1 - {}^+9$ **d.** $^-5 - {}^-3$

STOP • STOP • STOP • STOP • STOP • STOP • STOP • STOP • STOP

A8
  a. $^-7$: $^-2 - {}^+5 = {}^-2 + {}^-5 = {}^-7$
  b. $^+12$: $^+7 - {}^-5 = {}^+7 + {}^+5 = {}^+12$
  c. $^-8$: $^+1 - {}^+9 = {}^+1 + {}^-9 = {}^-8$
  d. $^-2$: $^-5 - {}^-3 = {}^-5 + {}^+3 = {}^-2$

---

6   It is important to realize that the procedure of "adding the opposite" is done only with *subtraction*.

1. To find a *sum*, follow the procedure for adding integers directly.
2. To find a *difference, rewrite the problem as a sum* (by adding the opposite of the subtrahend to the minuend) and follow the procedure for adding integers.

Examples are:

$^-4 - {}^+6 = {}^-4 + {}^-6 = {}^-10$
$^+5 + {}^-6 = {}^-1$
$^-7 + {}^-4 = {}^-11$
$^-4 - {}^-3 = {}^-4 + {}^+3 = {}^-1$

Q9   Which of the following problems must be rewritten?
  a. $^-4 - \ 5$ b. $^-2 + {}^-1$
  c. $^+6 + {}^-9$ d. $^+6 - {}^+15$ _____

STOP • STOP • STOP • STOP • STOP • STOP • STOP • STOP • STOP

A9   a and d: subtraction problems must be rewritten (b and c are addition problems and thus can be answered directly).

Q10   Complete the problems of Q9.

STOP • STOP • STOP • STOP • STOP • STOP • STOP • STOP • STOP

A10   a. $^-4 - {}^-5 = {}^-4 + {}^+5 = {}^+1$ b. $^-3$ c. $^-3$
  d. $^+6 - {}^+15 = {}^+6 + {}^-15 = {}^-9$

Q11   Complete each of the following as a sum or difference as indicated:
  a. $^-4 + {}^-8$ b. $^+7 - {}^+4$ c. $^-6 - {}^+4$ d. $^-6 + {}^+7$

  e. $^+2 - {}^+5$ f. $^-4 - {}^+8$ g. $^+4 + {}^+7$ h. $^-1 - {}^+1$

  i. $^+11 + {}^-9$ j. $0 - {}^-5$ k. $^-4 + 0$ l. $^+1 - {}^+1$

m. $^+5 + ^-5$ **n.** $^-2 + ^-3$ **o.** $^-3 - ^-3$ **p.** $^+5 + ^-7$

q. $^+5 - ^+9$ **r.** $^-1 + ^+7$ **s.** $^-2 - 0$ **t.** $0 - ^+1$

STOP • STOP • STOP • STOP • STOP • STOP • STOP • STOP • STOP

A11   **a.** $^-12$ **b.** $^+3$ **c.** $^-10$ **d.** $^+1$
e. $^-3$ **f.** $^-12$ **g.** $^+11$ **h.** $^-2$
i. $^+2$ **j.** $^+5$ **k.** $^-4$ **l.** 0
m. 0 **n.** $^-5$ **o.** 0 **p.** $^-2$
q. $^-4$ **r.** $^+6$ **s.** $^-2$ **t.** $^-1$

---

7   When evaluating number sentences involving a combination of sums and differences, rewrite each of the differences as a sum and proceed as in Section 2.2, Frame 8. For example, the expression

$$^+3 + ^-7 - ^+5$$

(sum / difference)

involves a sum and a difference. First rewrite the difference as

$$^+3 + ^-7 + ^-5$$

The problem is now completed:

$$^-4 + ^-5$$
$$^-9$$

Study the following examples:

1. $^-4 - ^-2 + ^+3$ **2.** $^-3 - ^+4 - ^+6 + ^+7$
$^-4 + ^+2 + ^+3$ $^-3 + ^-4 + ^-6 + 7$
$^-2 + ^+3$ $^-13 + 7$
$^+1$ $^-6$

Q12   Evaluate $^-3 + ^+5 - ^+7$.

STOP • STOP • STOP • STOP • STOP • STOP • STOP • STOP • STOP

A12 $^-5$: $^-3 + ^+5 - 7$
$^-3 + 5 + ^-7$
$^+2 + ^-7$
$^-5$

Q13   Evaluate $^+1 - ^+5 - ^+6$.

STOP • STOP • STOP • STOP • STOP • STOP • STOP • STOP • STOP

A13   ⁻10:   ⁺1 − ⁺5 − ⁺6
                    ⁺1 + ⁻5 + ⁻6
                    ⁻4 + ⁻6
                    ⁻10

Q14   Evaluate ⁺4 + ⁻3 − ⁺7 − ⁻6.

STOP • **STOP** • **STOP** • **STOP** • **STOP** • **STOP** • **STOP** • **STOP** • **STOP**

A14   0:   ⁺4 + ⁻3 − ⁺7 − ⁻6
                  4 + ⁻3 + ⁻7 + ⁺6
                  ⁺1 + ⁻7 + ⁺6
                  ⁻6 + ⁺6
                  0

Q15   Evaluate:

    a. ⁻2 − ⁺5 + ⁺3   **b.** ⁺3 + ⁻5 + ⁻4

    c. ⁺4 − ⁻6 + ⁺7   **d.** ⁻6 − ⁺2 + ⁻7

    e. ⁻2 − ⁺3 − ⁺4   **f.** ⁺7 + ⁻2 − ⁺8 + ⁺3

    g. ⁺4 − 0 + ⁻4   **h.** 0 − ⁺4 − ⁺3 + 2

    i. ⁺6 − ⁺3 − ⁺5   **j.** ⁻6 + ⁻3 − ⁺10

STOP • **STOP** • **STOP** • **STOP** • **STOP** • **STOP** • **STOP** • **STOP** • **STOP**

A15   **a.** ⁻4: ⁻2 − ⁺5 + 3 = ⁻2 + ⁻5 + ⁺3 = ⁻7 + ⁺3 = ⁻4
        b. ⁻6
        c. ⁺17
        d. ⁻15: ⁻6 − ⁺2 + ⁻7 = ⁻6 + ⁻2 + ⁻7 = ⁻15
        e. ⁻9
        f. 0: ⁺7 + ⁻2 − ⁺8 + ⁺3 = 7 + ⁻2 + ⁻8 + ⁺3 = 0
        g. 0
        h. ⁻5: 0 − ⁺4 − ⁺3 + 2 = 0 + ⁻4 + ⁻3 + ⁺2 = ⁻7 + 2 = ⁻5
        i. ⁻2: ⁺6 − ⁺3 − ⁺5 = ⁺6 + ⁻3 + ⁻5 = ⁺3 + ⁻5 = ⁻2
        j. ⁻19

---

8   Consider the following examples:

$^+3 - {}^+4 = {}^+3 + {}^-4 = {}^-1$
$^+4 - {}^+3 = {}^+1$

The expressions $^+3 - {}^+4$ and $^+4 - {}^+3$ are not equivalent because they have different evaluations. Thus, $^+3 - {}^+4 = {}^+4 - {}^+3$ is a false statement. $^+3 - {}^+4 \neq {}^+4 - {}^+3$ is a true statement.

Q16   **a.** Are $^-2 - {}^+3$ and $^+3 - {}^-2$ equivalent expressions?   _____

b. Is $^-2 - {}^+3 = {}^+3 - {}^-2$ a true statement?   _____

STOP • STOP • STOP • STOP • STOP • STOP • STOP • STOP • STOP

A16   **a.** no: $^-2 - {}^+3 = {}^-5$, whereas $^+3 - {}^-2 = {}^+5$
b. no

Q17   Verify that $a - b$ and $b - a$ are not equivalent expressions for $a = {}^-5$ and $b = {}^-2$ by evaluating both expressions.

STOP • STOP • STOP • STOP • STOP • STOP • STOP • STOP • STOP

A17 $a - b = {}^-5 - {}^-2 = {}^-5 + {}^+2 = {}^-3$
$b - a = {}^-2 - {}^-5 = {}^-2 + {}^+5 = {}^+3$

---

9 $a - b$ and $b - a$ are not equivalent expressions since they do not have the same evaluation for all integer replacements of a and b.

Q18   For most integer replacements of a and b, $a - b = b - a$ is a _____ statement.
                                                                                        true/false

STOP • STOP • STOP • STOP • STOP • STOP • STOP • STOP • STOP

A18   false: $a - b = b - a$ is true when $a = b$

---

10   The fact that $a - b$ and $b - a$ are not equivalent expressions for all replacements of a and b demonstrates that subtraction is *not* a commutative operation.

Q19   Verify that $(^+3 - {}^+5) - {}^+7$ and $^+3 - (^+5 - {}^+7)$ do not have the same evaluation.

STOP • STOP • STOP • STOP • STOP • STOP • STOP • STOP • STOP

A19 $(^+3 - {}^+5) - {}^+7 = (^+3 + {}^-5) + {}^-7 = {}^-2 + {}^-7 = {}^-9$
$^+3 - (^+5 - {}^+7) = {}^+3 - (^+5 + {}^-7) = {}^+3 - (^-2) = {}^+3 + {}^+2 = {}^+5$

Q20 $(^+3 - {}^+5) - {}^+7 = {}^+3 - (^+5 - {}^+7)$ is a _____ statement.
                                                            true/false

STOP • STOP • STOP • STOP • STOP • STOP • STOP • STOP • STOP

A20	false

Q21   Verify that $(a - b) - c$ and $a - (b - c)$ are not equivalent by evaluating both expressions for $a = {}^+3$, $b = {}^-5$, and $c = {}^-4$.

STOP • **STOP** • **STOP** • **STOP** • **STOP** • **STOP** • **STOP** • **STOP** • **STOP**

A21
$(a - b) - c$ $a - (b - c)$
$({}^+3 - {}^-5) - {}^-4$ ${}^+3 - ({}^-5 - {}^-4)$
$({}^+3 + {}^+5) + {}^+4$ ${}^+3 - ({}^-5 + {}^+4)$
${}^+8 + {}^+4$ ${}^+3 - ({}^-1)$
${}^+12$ ${}^+3 + {}^+1$
                  ${}^+4$

---

11 $(a - b) - c$ and $a - (b - c)$ are not equivalent expressions, because they do not have the same evaluation for all integer replacements of a, b, and c. This fact demonstrates that subtraction is *not* an associative operation.

Q22   Answer true or false:

     a. $a + b = b + a$ is true for all integer replacements of a and b.   _____

     b. $a - b = b - a$ is true for all integer replacements of a and b.   _____

     c. $a + (b + c) = (a + b) + c$ is true for all integer replacements of a, b, and c.

                                                           _____

     d. $a - (b - c) = (a - b) - c$ is true for all integer replacements of a, b, and c.

                                                            _____

STOP • **STOP** • **STOP** • **STOP** • **STOP** • **STOP** • **STOP** • **STOP** • **STOP**

A22   **a.** true   **b.** false   **c.** true   **d.** false

---

This completes the instruction for this section.

2.3   Exercises

1. Find the following sums:

   a. ${}^+3 + {}^-7$ **b.** ${}^+12 + {}^-9$ **c.** ${}^-5 + {}^-6$ **d.** ${}^-9 + {}^+12$

   e. ${}^-4 + {}^+11$ **f.** ${}^+6 + {}^-6$ **g.** ${}^-2 + {}^+2$ **h.** ${}^+4 + 0$

   i. $0 + {}^-7$ **j.** ${}^-3 + {}^-4$

2. Write the opposite for each of the following integers:

   a. ${}^-5$ **b.** 0 **c.** ${}^+6$ **d.** ${}^-4$ **e.** 2 **f.** ${}^-1$

3. Find the following differences:

   a. ${}^-2 - 4$ **b.** ${}^-5 - {}^-2$ **c.** ${}^-7 - {}^-5$ **d.** ${}^-3 - 0$

   e. ${}^+6 - {}^+9$ **f.** ${}^-4 - {}^-4$ **g.** ${}^+3 - {}^+2$ **h.** $0 - {}^-1$

   i. $0 - {}^+3$ **j.** ${}^-2 - {}^-8$

4. Complete each of the following:
   a. $^+7 + ^-3$ b. $^-3 + 0$ c. $^-6 - ^+4$ d. $^+7 - ^+5$
   e. $^-5 + ^-9$ f. $^+7 + ^-7$ g. $0 + ^-4$ h. $^-6 - ^-7$
   i. $^-1 - ^+1$ j. $0 + ^+3$ k. $^-3 - ^+4$ l. $^-4 + ^+9$
   m. $^-5 - 0$ n. $0 - ^-5$ o. $^+6 + ^-9$ p. $^+4 - ^+5$
   q. $^-3 + ^+12$ r. $0 + ^-4$ s. $^-5 + ^+5$ t. $0 - ^-2$
5. Complete each of the following:
   a. $^+4 - ^+3 + ^-7$ b. $^-2 - ^-5 - ^-6$ c. $^-6 + ^+7 + ^-3$
   d. $^-4 + ^+6 - ^-4 - ^+6$ e. $^+2 + ^-2 - 0$ f. $^+3 - ^+2 - ^+4$
   g. $^+4 + ^-7 - ^+3$ h. $0 - ^+4 + ^+7$ i. $^-2 - ^+5 - ^+6$
   j. $^+6 - 0 + ^-3$
6. Answer true or false:
   a. $^-3 + ^+7 = ^+7 + ^-3$
   b. $^+2 - ^+1 = ^+1 - ^+2$
   c. $(^-3 + ^+5) + ^-7 = ^-3 + (^+5 + ^-7)$
   d. $^-5 + 0 = 0 + ^-5 = ^-5$
   e. $^+6 - (^+3 - ^-2) = (^+6 - ^+3) - ^-2$

2.3   Exercise Answers

1. a. $^-4$ b. $^+3$ c. $^-11$ d. $^+3$ e. $^+7$ f. 0 g. 0
   h. $^+4$ i. $^-7$ j. $^-7$
2. a. $^+5$ b. 0 c. $^-6$ d. $^+4$ e. $^-2$ f. $^+1$
3. a. $^-6$ b. $^-3$ c. $^-2$ d. $^-3$ e. $^-3$ f. 0 g. $^+1$
   h. $^+1$ i. $^-3$ j. $^+6$
4. a. $^+4$ b. $^-3$ c. $^-10$ d. $^+2$ e. $^-14$ f. 0 g. $^-4$
   h. $^+1$ i. $^-2$ j. $^+3$ k. $^-7$ l. $^+5$ m. $^-5$ n. $^+5$
   o. $^-3$ p. $^-1$ q. $^+9$ r. $^-4$ s. 0 t. $^+2$
5. a. $^-6$ b. $^+9$ c. $^-2$ d. 0 e. 0 f. $^-3$ g. $^-6$
   h. $^+3$ i. $^-13$ j. $^+3$
6. a. true: commutative property of addition
   b. false
   c. true: associative property of addition
   d. true: addition property of zero (additive identity property)
   e. false

2.4   Multiplication

1   The operation of multiplication was developed as a shortcut procedure for addition. For example, the product $2 \cdot 3$ can be represented either as the sum of 2 threes or as the sum of 3 twos:

$$2 \cdot 3 = \underbrace{3 + 3}_{\text{2 addends of 3}} = 6$$

or

$$3 \cdot 2 = \underbrace{2 + 2 + 2}_{\text{3 addends of 2}} = 6$$

Similarly, the product $7 \cdot 4$ can be represented as either 7 fours or 4 sevens.

$$7 \cdot 4 = 4 + 4 + 4 + 4 + 4 + 4 + 4 = 28$$

or

$$4 \cdot 7 = 7 + 7 + 7 + 7 = 28$$

Q1   Write $5 \cdot 6$ as a sum in two ways.

STOP • **STOP** • **STOP** • **STOP** • **STOP** • **STOP** • **STOP** • **STOP** • **STOP**

A1 $$5 \cdot 6 = 6 + 6 + 6 + 6 + 6 = 30$$

or

$$6 \cdot 5 = 5 + 5 + 5 + 5 + 5 + 5 = 30$$

Q2   Write $3 \cdot 9$ as a sum in two ways.

STOP • **STOP** • **STOP** • **STOP** • **STOP** • **STOP** • **STOP** • **STOP** • **STOP**

A2 $$3 \cdot 9 = 9 + 9 + 9 = 27$$

or

$$9 \cdot 3 = 3 + 3 + 3 + 3 + 3 + 3 + 3 + 3 + 3 = 27$$

2   The procedure of writing a product as a sum can also be used to find the product of two integers. For example, the product $^+4 \cdot {}^-2$ can be written as the sum of 4 negative twos:

$$^+4 \cdot {}^-2 = \underbrace{{}^-2 + {}^-2 + {}^-2 + {}^-2}_{\text{4 addends of } {}^-2}$$

Since the sum on the right is equal to $^-8$, the product $^+4 \cdot {}^-2$ is $^-8$:

$$^+4 \cdot {}^-2 = {}^-8$$

Similarly, the product $^+3 \cdot {}^-7$ is $^-21$, because

$$\begin{aligned} ^+3 \cdot {}^-7 &= {}^-7 + {}^-7 + {}^-7 \\ &= {}^-21 \end{aligned}$$

Q3   Write $^+2 \cdot {}^-6$ as a sum.   _____

STOP • **STOP** • **STOP** • **STOP** • **STOP** • **STOP** • **STOP** • **STOP** • **STOP**

A3 $$^+2 \cdot {}^-6 = {}^-6 + {}^-6$$

Q4   Find the product $^+2 \cdot {}^-6$.

STOP • **STOP** • **STOP** • **STOP** • **STOP** • **STOP** • **STOP** • **STOP** • **STOP**

A4 $^-12$

Q5   Find the product $^+3 \cdot {}^-5$ by writing it as a sum.

STOP • **STOP** • **STOP** • **STOP** • **STOP** • **STOP** • **STOP** • **STOP** • **STOP**

A5 $^-15: {}^+3 \cdot {}^-5 = {}^-5 + {}^-5 + {}^-5$
$$= {}^-15$$

Q6   Find the product $^+7 \cdot {}^-1$ by writing it as a sum.

STOP • **STOP** • **STOP** • **STOP** • **STOP** • **STOP** • **STOP** • **STOP** • **STOP**

A6 $^-7: 7 \cdot {}^-1 = {}^-1 + {}^-1 + {}^-1 + {}^-1 + {}^-1 + {}^-1 + {}^-1$
$$= {}^-7$$

Q7   Find each of the following products:

         a. $^+5 \cdot {}^-2 =$ _____   **b.** $^+6 \cdot {}^-9 =$ _____

         c. $^+1 \cdot {}^-4 =$ _____   **d.** $0 \cdot {}^+4 =$ _____

         e. $^+3 \cdot {}^+3 =$ _____   **f.** $^+2 \cdot {}^-8 =$ _____

STOP • **STOP** • **STOP** • **STOP** • **STOP** • **STOP** • **STOP** • **STOP** • **STOP**

A7   **a.** $^-10$ **b.** $^-54$ **c.** $^-4$ **d.** 0 **e.** $^+9$ **f.** $^-16$

---

3   The product $^-5 \cdot {}^+4$ can be found by computing the sum of 4 negative fives:

$$^-5 \cdot {}^+4 = \underbrace{{}^-5 + {}^-5 + {}^-5 + {}^-5}_{4 \text{ addends of } {}^-5}$$
$$= {}^-20$$

Q8   Find the product $^-7 \cdot {}^+5$ by writing it as a sum.

STOP • **STOP** • **STOP** • **STOP** • **STOP** • **STOP** • **STOP** • **STOP** • **STOP**

A8 $^-35: {}^-7 \cdot {}^+5 = {}^-7 + {}^-7 + {}^-7 + {}^-7 + {}^-7$
$$= {}^-35$$

Q9   Find the product $^+5 \cdot {}^-7$

STOP • **STOP** • **STOP** • **STOP** • **STOP** • **STOP** • **STOP** • **STOP** • **STOP**

A9 $^-35: {}^+5 \cdot {}^-7 = {}^-7 + {}^-7 + {}^-7 + {}^-7 + {}^-7$
$$= {}^-35$$

Q10   Find the product $^-4 \cdot {}^+6$.

STOP   •   *STOP*   •   *STOP*   •   *STOP*   •   *STOP*   •   *STOP*   •   *STOP*   •   *STOP*   •   *STOP*

A10 $^-24$: $^-4 \cdot {}^+6 = {}^-4 + {}^-4 + {}^-4 + {}^-4 + {}^-4 + {}^-4$
                    $= {}^-24$

Q11   Find the product $^-7 \cdot {}^+9$.

STOP   •   *STOP*   •   *STOP*   •   *STOP*   •   *STOP*   •   *STOP*   •   *STOP*   •   *STOP*   •   *STOP*

A11 $^-63$

---

4 | In each of the preceding products where the two factors had different signs (one positive and one negative), the product was negative. For example,

$$\underbrace{{}^+5 \cdot {}^-9}_{} = {}^-45 \qquad \underbrace{{}^-7 \cdot {}^+8}_{} = {}^-56$$

different   negative   different   negative
signs   product   signs   product

These examples demonstrate the following rule for multiplying integers with different signs: *The product of two integers with different signs (one positive and one negative) is a negative integer.*

To make the multiplication of integers consistent with the multiplication of natural numbers, the rule for multiplying two positive integers is as follows: *The product of two positive integers is a positive integer.* For example,

$$^+2 \cdot {}^+7 = {}^+14 \qquad 5 \cdot 8 = 40$$

The "+" sign is frequently omitted from positive numbers such as in the second example above. A number written without a sign is assumed to be positive.

Q12   Find the product in each of the following:

   a. $^-2 \cdot {}^+3 =$ _____   **b.** $^+6 \cdot {}^-3 =$ _____   **c.** $4 \cdot 7 =$ _____

   d. $^+1 \cdot {}^-5 =$ _____   **e.** $^+10 \cdot {}^-8 =$ _____   **f.** $^+5 \cdot {}^+3 =$ _____

   g. $^-7 \cdot {}^+7 =$ _____   **h.** $^-8 \cdot {}^+6 =$ _____

STOP   •   *STOP*   •   *STOP*   •   *STOP*   •   *STOP*   •   *STOP*   •   *STOP*   •   *STOP*   •   *STOP*

A12   **a.** $^-6$ **b.** $^-18$ **c.** 28 **d.** $^-5$ **e.** $^-80$ **f.** 15 **g.** $^-49$
           h. $^-48$

---

5 | In Section 1.3 the multiplication property of zero was stated as being true for all whole-number replacements of the variable. This property is also true for all integer replacements of the variable; that is, $a \cdot 0 = 0a = 0$ for all integers a. Thus, any integer times zero is equal to zero. For example,

$$^-5 \cdot 0 = 0 \qquad 0 \cdot 3 = 0$$

Q13 Find each of the following products:

    a. $^-4 \cdot 0 =$ _____   **b.** $^+8 \cdot ^-1 =$ _____

    c. $^-3 \cdot ^+4 =$ _____   **d.** $0 \cdot ^+5 =$ _____

    e. $6 \cdot 7 =$ _____   **f.** $^+9 \cdot ^-9 =$ _____

STOP • STOP • STOP • STOP • STOP • STOP • STOP • STOP • STOP

A13   **a.** 0   **b.** $^-8$ **c.** $^-12$ **d.** 0
    e. 42   **f.** $^-81$

---

6 The product of two integers with *different signs* is *negative*. The product of *two positive integers* is a *positive* integer. To discover the product of *two negative integers*, study the following series of products and notice the pattern that is present in the answers on the right.

$$^-2 \cdot ^+4 = ^-8$$
$$^-2 \cdot ^+3 = ^-6$$
$$^-2 \cdot ^+2 = ^-4$$
$$^-2 \cdot ^+1 = ^-2$$
$$^-2 \cdot 0 = 0$$
$$^-2 \cdot ^-1 = ?$$
$$^-2 \cdot ^-2 = ?$$

The pattern in the products on the right is that each answer increases by 2. Hence, to complete the pattern, the products are:

$$^-2 \cdot ^-1 = ^+2$$
$$^-2 \cdot ^-2 = ^+4$$

It is, thus, appropriate to state the rule for the product of two negative integers as follows: *The product of two negative integers is a positive integer.* For example,

$$^-7 \cdot ^-4 = ^+28 \qquad ^-11 \cdot ^-5 = ^+55$$

Q14 Find the product $^-4 \cdot ^-6$.

STOP • STOP • STOP • STOP • STOP • STOP • STOP • STOP • STOP

A14 $^+24$ (or simply 24)

Q15 $^-3 \cdot 5 =$ _____

STOP • STOP • STOP • STOP • STOP • STOP • STOP • STOP • STOP

A15 $^-15$

Q16 $^-9 \cdot ^-6 =$ _____

STOP • STOP • STOP • STOP • STOP • STOP • STOP • STOP • STOP

A16   54

Q17 Find the following products:

    a. $^-3 \cdot 4 =$ _____   **b.** $^-4 \cdot ^-9 =$ _____

    c. $^-2 \cdot ^-5 =$ _____   **d.** $9 \cdot ^-8 =$ _____

e. $0 \cdot {}^-6 =$ _____   **f.** ${}^-12 \cdot 0 =$ _____

g. $5 \cdot {}^-1 =$ _____   **h.** ${}^-7 \cdot {}^-5 =$ _____

STOP • STOP • STOP • STOP • STOP • STOP • STOP • STOP • STOP

A17   **a.** ${}^-12$ **b.** 36   **c.** 10   **d.** ${}^-72$
       e. 0   **f.** 0   **g.** ${}^-5$ **h.** 35

---

7   Study the effect of multiplying any integer by ${}^-1$ in the following examples:

${}^-1 \cdot 4 = {}^-4$ ${}^-1 \cdot {}^-5 = {}^+5$

In the first example, multiplying 4 by negative one changes it to ${}^-4$, its opposite. In the second example, the product of ${}^-5$ and negative one changes ${}^-5$ to ${}^+5$, its opposite. The fact that negative one times a number is the opposite of the number is often called the multiplication property of ${}^-1$. *The multiplication property of ${}^-1$ states that ${}^-1 \cdot a = {}^-a$ is true for all integer replacements of a.* (*Note:* ${}^-a$ is read "the *opposite* of a.")

Q18   Find each of the following products:

a. ${}^-1 \cdot {}^-7 =$ _____   **b.** $1 \cdot {}^-7 =$ _____

c. ${}^-1 \cdot 9 =$ _____   **d.** ${}^-1 \cdot 0 =$ _____

STOP • STOP • STOP • STOP • STOP • STOP • STOP • STOP • STOP

A18   **a.** 7   **b.** ${}^-7$ **c.** ${}^-9$ **d.** 0

---

8   It is important for later work that the student be able to read the multiplication property of ${}^-1$ both from left to right and from right to left. Reading from left to right it says that multiplying by ${}^-1$ gives the opposite of the integer involved:

The   Its
integer   opposite
${}^-1 \cdot {}^+6 = {}^-6$
${}^-1 \cdot {}^-9 = {}^+9$

Reading from right to left it says that any integer can be expressed as a product of ${}^-1$ and its opposite.

The   Its
integer   opposite
${}^-5 = {}^-1 \cdot {}^+5$
${}^-7 = {}^-1 \cdot {}^+7$
${}^+6 = {}^-1 \cdot {}^-6$
${}^+3 = {}^-1 \cdot {}^-3$

Q19   Complete the following statements using the multiplication property of ${}^-1$:

a. ${}^-1 \cdot {}^+7 =$ _____   **b.** ${}^-1 \cdot$ _____ $= {}^+5$

c. ${}^-3 = {}^-1 \cdot$ _____   **d.** _____ $\cdot {}^-8 = {}^+8$

STOP • STOP • STOP • STOP • STOP • STOP • STOP • STOP • STOP

A19   **a.** ${}^-7$ **b.** ${}^-5$ **c.** ${}^+3$ **d.** ${}^-1$

Q20   Find the product:

a. $^-3 \cdot {}^+2 = $ _____   b. $^+2 \cdot {}^-3 = $ _____

STOP • STOP • STOP • STOP • STOP • STOP • STOP • STOP • STOP

A20   a. $^-6$ b. $^-6$

Q21   Evaluate the expressions ab and ba for $a = {}^-7$ and $b = {}^-8$.

STOP • STOP • STOP • STOP • STOP • STOP • STOP • STOP • STOP

A21 $ab = {}^-7 \cdot {}^-8$ $ba = {}^-8 \cdot {}^-7$
$\quad\;\; = {}^+56 \qquad\qquad\;\; = {}^+56$

Q22   Do you think the statement $ab = ba$ is true for all integer replacements of a and b? _____

STOP • STOP • STOP • STOP • STOP • STOP • STOP • STOP • STOP

A22   yes: the order of the factors when finding the product of two integers does not affect the product.

---

9 | The set of integers is commutative with respect to the operation of multiplication. That is, the statement

$$ab = ba$$

is true for all integer replacements of a and b. For example,

1. $^-3 \cdot {}^+9 = {}^+9 \cdot {}^-3$ 2. $^-12 \cdot {}^-5 = {}^-5 \cdot {}^-12$
$\quad\;\; ^-27 = {}^-27 \qquad\qquad\qquad\quad\; ^+60 = {}^+60$

Q23   Is $^-253 \cdot {}^+479 = {}^+479 \cdot {}^-253$ a true statement? _____

STOP • STOP • STOP • STOP • STOP • STOP • STOP • STOP • STOP

A23   yes: the commutative property of multiplication is true for all integers.

---

10 | Frame 9 makes it possible to state the following procedure for finding a product of more than two numbers. To find the product of more than two numbers:

Step 1: Do all work within parentheses first.

Step 2: Find the product of two numbers at a time in *any order desired*.

Examples:

1. $^-2 \cdot {}^+4 \cdot {}^-5$
$^+10 \cdot {}^+4$ (since there are no parentheses, follow step 2)
40

2. $^+3(^-6 \cdot {}^-5) \cdot {}^+9$
$^+3(^+30)^+9$ (since parentheses are involved,
$^+90 \cdot {}^+9$ use step 1 and then step 2)
810

Q24   Find the product $^-2 \cdot {}^-3 \cdot {}^+5$.

STOP • **STOP** • **STOP** • **STOP** • **STOP** • **STOP** • **STOP** • **STOP** • **STOP**

A24   30

Q25   Find the product $(^-1 \cdot {}^+3)(^-4 \cdot {}^-7)$.

STOP • **STOP** • **STOP** • **STOP** • **STOP** • **STOP** • **STOP** • **STOP** • **STOP**

A25 $^-84$: $(^-1 \cdot {}^+3)(^-4 \cdot {}^-7)$
             $^-3 \cdot {}^+28$
             $^-84$

Q26   Find each of the following products:
             a. $^-2 \cdot {}^-4 \cdot 0$ **b.** $(^-3 \cdot {}^+4) \cdot {}^-5$

             c. $^-6 \cdot {}^+11 \cdot {}^+10$ **d.** $^-1(^+3 \cdot {}^-5)$

             c. $(^+4 \cdot 0)(^-7 \cdot {}^-6)$ **f.** $(^-2 \cdot {}^+3)(^-4 \cdot {}^-6)$

STOP • **STOP** • **STOP** • **STOP** • **STOP** • **STOP** • **STOP** • **STOP** • **STOP**

A26   **a.** 0   **b.** 60   **c.** $^-660$ **d.** 15   **e.** 0   **f.** $^-144$

---

11   In Chapter 1 it was established that the set of whole numbers for the operation of multiplication was associative, commutative, and distributive with respect to addition and subtraction. These properties are likewise true for the set of integers. Thus, the following statements are true for all integer replacements of a, b, and c:

            1. Associative property of multiplication

               $a(bc) = (ab)c$

            2. Commutative property of multiplication

               $ab = ba$

            3. Distributive property of multiplication over addition

               $a(b + c) = ab + ac$ (left)
               $(a + b)c = ac + bc$ (right)

            4. Distributive property of multiplication over subtraction

               $a(b - c) = ab - ac$ (left)
               $(a - b)c = ac - bc$ (right)

Q27   Verify that $a(bc) = (ab)c$ is true when $a = {}^-3$, $b = {}^+5$, and $c = {}^+7$.

$a(bc) =$ $(ab)c =$

STOP • **STOP** • **STOP** • **STOP** • **STOP** • **STOP** • **STOP** • **STOP** • **STOP**

A27
$$
\begin{aligned}
a(bc) &= {}^-3({}^+5 \cdot {}^+7) \\
&= {}^-3({}^+35) \\
&= {}^-105
\end{aligned}
\qquad
\begin{aligned}
(ab)c &= ({}^-3 \cdot {}^+5) \cdot {}^+7 \\
&= ({}^-15) \cdot {}^+7 \\
&= {}^-105
\end{aligned}
$$

Q28   Verify that $a(b - c) = ab - ac$ is true when $a = {}^+5$, $b = {}^-1$, and $c = {}^-4$.

$a(b - c) =$ $ab - ac =$

STOP • **STOP** • **STOP** • **STOP** • **STOP** • **STOP** • **STOP** • **STOP** • **STOP**

A28
$$
\begin{aligned}
a(b - c) &= {}^+5({}^-1 - {}^-4) \\
&= {}^+5({}^-1 + {}^+4) \\
&= {}^+5({}^+3) \\
&= {}^+15
\end{aligned}
\qquad
\begin{aligned}
ab - ac &= {}^+5 \cdot {}^-1 - {}^+5 \cdot {}^-4 \\
&= {}^-5 - {}^-20 \\
&= {}^-5 + {}^+20 \\
&= {}^+15
\end{aligned}
$$

This completes the instruction for this section.

2.4   Exercises

1. The product of two integers with different signs is a _____ integer.
2. The product of any integer and zero is _____.
3. The product of any two integers with the same sign is a _____ integer.
4. Fill in the blanks so that a true statement results:
   a. _____ $\cdot 5 = 0$ b. ${}^-1 \cdot 7 =$ _____   c. ${}^+8 =$ _____ $\cdot {}^-8$
   d. ${}^-1 \cdot {}^-5 =$ _____

5. Find each of the following products:
   a. ${}^+4 \cdot {}^-9$ b. ${}^-3 \cdot {}^-4$
   c. ${}^-2 \cdot 0$ d. ${}^-1 \cdot {}^+9$
   e. ${}^-1 \cdot {}^+7$ f. $5 \cdot {}^-7$
   g. $0 \cdot {}^-9$ h. ${}^-6 \cdot {}^-9$
   i. $11 \cdot {}^-5$ j. $7 \cdot {}^-8$
6. Find each of the following products:
   a. ${}^+2 \cdot {}^-3 \cdot {}^-4$ b. $({}^+7 \cdot {}^-1) \cdot {}^-7$
   c. ${}^-1 \cdot {}^-7 \cdot {}^-8$ d. ${}^-1 \cdot {}^-1 \cdot 0$
   e. ${}^-5 \cdot 0 \cdot {}^-6 \cdot {}^+3$ f. $4 \cdot 3 \cdot {}^-5$
   g. $({}^-4 \cdot {}^-4)({}^-2 \cdot {}^+2)$ h. ${}^-7(9 \cdot {}^-3) \cdot {}^+2$
   i. ${}^+6({}^-3 \cdot {}^+2)$ j. $({}^-2 \cdot 4)(3 \cdot {}^-3)$
   k. $({}^+6 \cdot {}^-3) \cdot {}^+2$ l. ${}^-4 \cdot 5({}^-6 \cdot 0)$
7. Verify that $a(b - c) = ab - ac$ when $a = {}^-2$, $b = {}^-3$, and $c = {}^+5$.

8. True or false:
   a. $ab = ba$ for all integers a and b.
   b. $a - b = b - a$ for all integers a and b.
   c. $(ab)c = a(bc)$ for all integers a, b, and c.

2.4   Exercise Answers

1. negative
2. zero
3. positive
4. a. 0   b. $^-7$ c. $^-1$ d. 5
5. a. $^-36$ b. 12   c. 0   d. $^-9$ e. $^-7$
   f. $^-35$ g. 0   h. 54   i. $^-55$ j. $^-56$
6. a. 24   b. 49   c. $^-56$ d. 0   e. 0   f. $^-60$
   g. $^-64$ h. 378   i. $^-36$ j. 72   k. $^-36$ l. 0
7. $\quad a(b - c) = ab - ac$
   $^-2(^-3 - {}^+5) = {}^-2 \cdot {}^-3 - {}^-2 \cdot {}^+5$
   $^-2(^-3 + {}^-5) = {}^+6 - {}^-10$
   $\quad\quad ^-2(^-8) = {}^+6 + {}^+10$
   $\quad\quad\quad\quad 16 = 16$
8. a. true   b. false   c. true

2.5   Division

1	The operation of division is closely related to that of multiplication. A division problem is often checked using the multiplication operation. As was the case with addition and subtraction, multiplication and division are also opposite or inverse operations; that is, one undoes the effect of the other. For example,

$$10 \div 2 = 5 \quad \text{because} \quad 5 \cdot 2 = 10 \quad \text{or} \quad 8 \overline{)56}^{\,7} \quad \text{because} \quad 7 \cdot 8 = 56$$

Q1 $12 \div 3 = 4$, because _____ $\cdot$ _____ = _____.

STOP • STOP • STOP • STOP • STOP • STOP • STOP • STOP • STOP

A1 $4 \cdot 3 = 12$

Q2 $7 \overline{)63}^{\,9}$, because _____ $\cdot$ _____ = _____.

STOP • STOP • STOP • STOP • STOP • STOP • STOP • STOP • STOP

A2 $9 \cdot 7 = 63$

Q3   Use multiplication to find the quotient and write the check: $45 \div 9 =$ _____, because
_____.

STOP • STOP • STOP • STOP • STOP • STOP • STOP • STOP • STOP

A3 $45 \div 9 = 5$, because $5 \cdot 9 = 45$

Q4   Use multiplication to find the quotient and check the result: $82 \div 2 =$ _____, because
_____.

STOP • STOP • STOP • STOP • STOP • STOP • STOP • STOP • STOP

A4 $82 \div 2 = 41$, because $41 \cdot 2 = 82$

---

2   Consider the following quotient and check:

$^-27 \div {}^+9 = \square$ because $\square \cdot {}^+9 = {}^-27$

Since $^-3$ makes the product statement true, it also satisfies the quotient statement:

$^-27 \div {}^+9 = {}^-3$ because $^-3 \cdot {}^+9 = {}^-27$

Q5   Place an integer in the $\square$ to form a true statement: $^+12 \div {}^-3 = \square$, because
$\square \cdot {}^-3 = {}^+12$

STOP • STOP • STOP • STOP • STOP • STOP • STOP • STOP • STOP

A5 $^+12 \div {}^-3 = {}^-4$, because $^-4 \cdot {}^-3 = {}^+12$

Q6   Place an integer in the $\square$ to form a true statement: $^-16 \div {}^+2 = \square$, because
$\square \cdot {}^+2 = {}^-16$

STOP • STOP • STOP • STOP • STOP • STOP • STOP • STOP • STOP

A6 $^-16 \div {}^+2 = {}^-8$, because $^-8 \cdot {}^+2 = {}^-16$

Q7   Find the quotient and write the check as a multiplication problem: $^+56 \div {}^-8 =$
_____, because _____ $\cdot {}^-8 = {}^+56$

STOP • STOP • STOP • STOP • STOP • STOP • STOP • STOP • STOP

A7 $^+56 \div {}^-8 = {}^-7$, because $^-7 \cdot {}^-8 = {}^+56$

Q8   Find the quotient and write the check as a multiplication problem: $^-30 \div {}^+10 =$ _____,
because _____

STOP • STOP • STOP • STOP • STOP • STOP • STOP • STOP • STOP

A8 $^-3$, because $^-3 \cdot {}^+10 = {}^-30$

Q9 $^+15 \div {}^-3 =$ _____

STOP • STOP • STOP • STOP • STOP • STOP • STOP • STOP • STOP

A9 $^-5$

---

3   Each of the problems in Q5 through Q9 involved the quotient of two integers with different
signs. Study the problems and their answers below:

$^+12 \div {}^-3 = {}^-4$
$^-16 \div {}^+2 = {}^-8$
$^+56 \div {}^-8 = {}^-7$
$^-30 \div {}^+10 = {}^-3$
$^+15 \div {}^-3 = {}^-5$

Q10   What is true of all the answers in Frame 3?

_____

STOP • STOP • STOP • STOP • STOP • STOP • STOP • STOP • STOP

A10   Each answer is a negative integer.

---

4   The examples of Frame 3 demonstrate the following definition for the division of two integers with different signs: *The quotient of two integers with different signs is a negative integer.*
           A division problem is also correctly written as a fraction. Thus, $20 \div 4 = 5$ can also be written $\frac{20}{4} = 5$. Study the following examples:

$$\frac{^-36}{^+4} = {^-9}$$

$$48 \div {^-6} = {^-8}$$

$${^+27} \div {^+9} = {^+3}$$

$$\frac{^+14}{^-2} = {^-7}$$

Q11   Find the following quotients:

   a. $\dfrac{^-25}{^+5} = $ _____   **b.** $^-25 \div {^+5} = $ _____

   c. $\dfrac{^-90}{^+10} = $ _____   **d.** $^+12 : {^-2} = $ _____

   e. $^+1 \div {^-1} = $ _____   **f.** $\dfrac{^+45}{^-9} = $ _____

STOP • STOP • STOP • STOP • STOP • STOP • STOP • STOP • STOP

A11   **a.** $^-5$ **b.** $^-5$ **c.** $^-9$ **d.** $^-6$ **e.** $^-1$ **f.** $^-5$

---

5   To determine the sign of the *quotient* of two *integers* with the *same sign*, consider the integer that converts the following open sentence into a true statement.

           $^-20 \div {^-5} = \square$ because $\square \cdot {^-5} = {^-20}$

           The correct integer replacement is $^+4$. That is, $^-20 \div {^-5} = {^+4}$, because $^+4 \cdot {^-5} = {^-20}$.

Q12   Place an integer in the $\square$ to form a true statement: $^-15 \div {^-3} = \square$, because $\square \cdot {^-3} = {^-15}$

STOP • STOP • STOP • STOP • STOP • STOP • STOP • STOP • STOP

A12 $^+5$: $^-15 \div {^-3} = \boxed{^+5}$, because $\boxed{^+5} \cdot {^-3} = {^-15}$.

Q13   Find the quotient and write the check as a multiplication problem: $^-63 \div {^-7} = $ _____, because _____ $\cdot {^-7} = {^-63}$

STOP • STOP • STOP • STOP • STOP • STOP • STOP • STOP • STOP

A13 $^+9$: $^-63 \div {^-7} = \underline{^+9}$, because $\underline{^+9} \cdot {^-7} = {^-63}$

Q14 $^-81 \div ^-9 =$ _____

STOP   •   STOP   •   STOP   •   STOP   •   STOP   •   STOP   •   STOP   •   STOP   •   STOP

A14 $^+9$ (or simply 9)

---

6   The quotients of Q12 through Q14 involved integers with the same sign. The answer in each case was a positive integer. The rule suggested for quotients of this type is as follows: *The quotient of two integers with the same sign is a positive integer.* Some examples of this rule are:

$$\frac{-36}{-4} = {}^+9 \qquad ^-26 \div ^-2 = {}^+13$$

$$^+42 \div {}^+21 = {}^+2 \qquad \frac{18}{6} = 3$$

Q15   Find the quotient $^-24 \div ^-6$.

STOP   •   STOP   •   STOP   •   STOP   •   STOP   •   STOP   •   STOP   •   STOP   •   STOP

A15   4

Q16 $\dfrac{-56}{-7} =$ _____

STOP   •   STOP   •   STOP   •   STOP   •   STOP   •   STOP   •   STOP   •   STOP   •   STOP

A16   8

Q17   Find each of the following quotients:

    a. $\dfrac{-12}{-3} =$ _____   **b.** $^+72 \div ^-6 =$ _____

    c. $^+4 \div ^-4 =$ _____   **d.** $^-93 \div ^-3 =$ _____

    e. $\dfrac{-19}{-19} =$ _____   **f.** $14 \div 7 =$ _____

    g. $\dfrac{+24}{-6} =$ _____   **h.** $\dfrac{-24}{+6} =$ _____

    i. $^-28 \div ^-7 =$ _____   **j.** $^-1 \div ^+1 =$ _____

    k. $^-1 \div ^-1 =$ _____   **l.** $\dfrac{-2}{-2} =$ _____

STOP   •   STOP   •   STOP   •   STOP   •   STOP   •   STOP   •   STOP   •   STOP   •   STOP

A17   **a.** 4   **b.** $^-12$ **c.** $^-1$ **d.** 31   **e.** 1   **f.** 2   **g.** $^-4$

       h. $^-4$ **i.** 4   **j.** $^-1$ **k.** 1   **l.** 1

---

7   The *quotient* of two integers with *different signs* is a *negative* integer: for example, $^-12 \div {}^+6 = {}^-2$. The *product* of two integers with *different signs* is also a *negative* integer: for example, $^-4 \cdot {}^+6 = {}^-24$.

Q18   The quotient of two integers with the *same sign* is a _____ integer.

STOP   •   STOP   •   STOP   •   STOP   •   STOP   •   STOP   •   STOP   •   STOP   •   STOP

A18   positive

Q19   The product of two integers with the *same sign* is a _____ integer.

STOP • STOP • STOP • STOP • STOP • STOP • STOP • STOP • STOP

A19   positive

Q20   **a.** The quotient of two integers with *different signs* is a _____ integer.

b. The product of two integers with *different signs* is a _____ integer.

STOP • STOP • STOP • STOP • STOP • STOP • STOP • STOP • STOP

A20   **a.** negative   **b.** negative

---

8   The rules for multiplication and division are the same. They may be summarized:

1. *The product or quotient of two integers with the same sign is positive.*
2. *The product or quotient of two integers with different signs is negative.*

(*Note:* The above rules are sometimes abbreviated as follows.)
In multiplication or division of integers:

1. Same sign—positive.
2. Different signs—negative.

This gives a quick and easy means to remember the signs in a multiplication or division problem.

Q21   Find the following products and quotients:

a. $^-15 \div {}^-3 =$ _____   **b.** $\dfrac{^+18}{^-3} =$ _____

c. $^-4 \cdot {}^-8 =$ _____   **d.** $\dfrac{^-14}{^-7} =$ _____

e. $^+7 \cdot {}^-3 =$ _____   **f.** $^-5 \cdot {}^-9 =$ _____

g. $\dfrac{^-12}{^+3} =$ _____   **h.** $^+7 \cdot {}^-6 =$ _____

i. $^-8 \cdot 0 =$ _____   **j.** $\dfrac{^+27}{^-9} =$ _____

k. $^+48 \div {}^-12 =$ _____   **l.** $^-32 \div {}^-4 =$ _____

STOP • STOP • STOP • STOP • STOP • STOP • STOP • STOP • STOP

A21   **a.** 5   **b.** $^-6$ **c.** 32   **d.** 2   **e.** $^-21$ **f.** 45   **g.** $^-4$
h. $^-42$ **i.** 0   **j.** $^-3$ **k.** $^-4$ **l.** 8

---

9   In Section 2.4 the multiplication property of $^-1$ was stated:

$^-1 \cdot a = {}^-a$ is true for all integer replacements of a

According to this property, *negative one times a number is the opposite of the number.* For example, $^-1 \cdot {}^+5 = {}^-5$ and $^-1 \cdot {}^-3 = {}^+3$. Consider the result of dividing an integer by $^-1$ in the following examples:

$$\frac{^+3}{^-1} = {}^-3 \qquad \frac{^-5}{^-1} = {}^+5$$

Q22   When $^+3$ is divided by $^-1$, the quotient is _____.

STOP • **STOP** • **STOP** • **STOP** • **STOP** • **STOP** • **STOP** • **STOP** • **STOP**

A22 $^-3$: the opposite of $^+3$

Q23   When $^-5$ is divided by $^-1$, the quotient is _____.

STOP • **STOP** • **STOP** • **STOP** • **STOP** • **STOP** • **STOP** • **STOP** • **STOP**

A23 $^+5$: the opposite of $^-5$

---

10   The preceding examples demonstrate that the result of dividing an integer by $^-1$ is the same as when multiplying an integer by $^-1$. In each case, the answer is the opposite of the integer being multiplied or divided. *The division property of $^-1$ states that $\frac{a}{^-1} = {}^-a$ is true for all integer replacements of a.*

Q24   Complete each of the following:

    a. $^-1 \cdot {}^+7 =$ _____   **b.** $\frac{^+7}{^-1} =$ _____

    c. $^-1 \cdot {}^-5 =$ _____   **d.** $\frac{^-5}{^-1} =$ _____

    e. $^-x = {}^-1 \cdot$ _____   **f.** $\frac{x}{^-1} =$ _____

STOP • **STOP** • **STOP** • **STOP** • **STOP** • **STOP** • **STOP** • **STOP** • **STOP**

A24   **a.** $^-7$ **b.** $^-7$ **c.** 5   **d.** 5

    e. $^-x = {}^-1 \cdot x$ **f.** $\frac{x}{^-1} = {}^-x$

---

11   The number zero is frequently confusing when involved as a divisor or dividend in a division problem. Consider the open sentence

$0 \div 5 = \square$ because $\square \cdot 5 = 0$

The integer that converts the open sentence to a true statement is zero:

$0 \div 5 = 0$ because $0 \cdot 5 = 0$

Q25   What integer converts the open sentence to a true statement? $0 \div {}^-9 = \square$, because $\square \cdot {}^-9 = 0$

STOP • **STOP** • **STOP** • **STOP** • **STOP** • **STOP** • **STOP** • **STOP** • **STOP**

A25 $0: 0 \div {}^-9 = 0$, because $0 \cdot {}^-9 = 0$

Q26 $0 \div {}^+2 =$ _____, because _____

STOP • **STOP** • **STOP** • **STOP** • **STOP** • **STOP** • **STOP** • **STOP** • **STOP**

A26 0, because $0 \cdot {}^+2 = 0$

Q27 $0 \div {}^-8 =$ _____

STOP • STOP • STOP • STOP • STOP • STOP • STOP • STOP • STOP

A27   0

---

12   When zero is the divisor, the quotient is said to be *undefined*. To understand why, consider the open sentence

$$4 \div 0 = \square \quad \text{because} \quad \square \cdot 0 = 4$$

There is no answer to the product $\square \cdot 0 = 4$, so there is no answer to the corresponding quotient $4 \div 0 = \square$. That is, $4 \div 0$ is undefined, because *no number* $\cdot \, 0 = 4$.

Q28 $7 \div 0 =$ _____, because _____

STOP • STOP • STOP • STOP • STOP • STOP • STOP • STOP • STOP

A28   *undefined,* because *no number* $\cdot \, 0 = 7$

Q29 ${}^-3 \div 0 =$ _____

STOP • STOP • STOP • STOP • STOP • STOP • STOP • STOP • STOP

A29   undefined

---

13   *When zero is the divisor* in a division problem, *the quotient is* said to be *undefined*. That is,

$\dfrac{x}{0} =$ undefined. For example,

$$\dfrac{{}^-3}{0} = \text{undefined} \qquad {}^+12 \div 0 = \text{undefined}$$

When zero is divided by any nonzero integer, the quotient is zero. That is, $\dfrac{0}{x} = 0$ for $x \in I$, $x \neq 0$.

For example,

$$\dfrac{0}{{}^-7} = 0 \qquad 0 \div {}^+6 = 0$$

Q30   Complete each of the following quotients:

    a. $0 \div {}^-2 =$ _____   **b.** ${}^-2 \div 0 =$ _____

    c. $\dfrac{0}{{}^+7} =$ _____   **d.** $\dfrac{7}{0} =$ _____

    e. ${}^-3 \div 0 =$ _____   **f.** $10 \div {}^-1 =$ _____

    g. $\dfrac{{}^-4}{{}^-2} =$ _____   **h.** $0 \div {}^-4 =$ _____

    i. $\dfrac{42}{{}^-6} =$ _____   **j.** $\dfrac{8}{0} =$ _____

STOP • STOP • STOP • STOP • STOP • STOP • STOP • STOP • STOP

A30   **a.** 0   **b.** undefined   **c.** 0
        d. undefined   **e.** undefined   **f.** ${}^-10$
        g. 2   **h.** 0   **i.** ${}^-7$
        j. undefined

Q31   Evaluate:

     a. $^{+}4 \div {}^{-}2 =$ _____   **b.** $^{-}2 \div 4 =$ _____

STOP • *STOP* • *STOP* • *STOP* • *STOP* • *STOP* • *STOP* • *STOP* • *STOP*

A31   **a.** $^{-}2$ **b.** $\dfrac{^{-}2}{4} = \dfrac{^{-}1}{2}$

Q32   Is $^{+}4 \div {}^{-}2 = {}^{-}2 \div {}^{+}4$ a true statement? _____

STOP • *STOP* • *STOP* • *STOP* • *STOP* • *STOP* • *STOP* • *STOP* • *STOP*

A32   no

14	The statement $a \div b = b \div a$ is false for most integer replacements of a and b. Therefore, the set of integers is not commutative with respect to the operation of division.

Q33   Evaluate:
     a. $(^{-}20 \div {}^{-}10) \div {}^{+}2$ **b.** $^{-}20 \div ({}^{-}10 \div {}^{+}2)$

STOP • *STOP* • *STOP* • *STOP* • *STOP* • *STOP* • *STOP* • *STOP* • *STOP*

A33   **a.** 1: $(^{-}20 \div {}^{-}10) \div {}^{+}2 = {}^{+}2 \div {}^{+}2$
                                $= 1$
     b. 4: $^{-}20 \div ({}^{-}10 \div {}^{+}2) = {}^{-}20 \div {}^{-}5$
                                 $= 4$

Q34   Is $(^{-}20 \div {}^{-}10) \div {}^{+}2 = {}^{-}20 \div ({}^{-}10 \div {}^{+}2)$ a true statement? _____

STOP • *STOP* • *STOP* • *STOP* • *STOP* • *STOP* • *STOP* • *STOP* • *STOP*

A34   no

15	The statement $(a \div b) \div c = a \div (b \div c)$ is false for most integer replacements of a, b, and c. Therefore, the set of integers is not associative with respect to the operation of division.

Q35   Complete the following:

     a. The set of integers _____ commutative with respect to multiplication.
                              is/is not

     b. The set of integers _____ commutative with respect to division.
                              is/is not

     c. $(^{-}16 \div {}^{+}4) \div {}^{+}2 = {}^{-}16 \div ({}^{+}4 \div {}^{+}2)$ _____ a true statement.
                                        is/is not

STOP • *STOP* • *STOP* • *STOP* • *STOP* • *STOP* • *STOP* • *STOP* • *STOP*

A35   **a.** is   **b.** is not   **c.** is not

This completes the instruction for this section.

2.5   Exercises

1. Find each of the following quotients:

   a. $72 \div {}^-9$ b. $\dfrac{-18}{2}$ c. $^-15 \div {}^-5$ d. $^-5 \div {}^-5$

   e. $0 \div {}^-7$ f. $\dfrac{-24}{-6}$ g. $\dfrac{10}{0}$ h. $\dfrac{0}{-2}$

   i. $^-13 \div {}^-1$ j. $\dfrac{-8}{-8}$

2. Find each of the following products and quotients:

   a. $\dfrac{-51}{3}$ b. $\dfrac{0}{-12}$ c. $^-12 \cdot {}^-5$ d. $\dfrac{-45}{-3}$

   e. $\dfrac{5}{-1}$ f. $\dfrac{4}{0}$ g. $^-5 \cdot 0$ h. $^-4 \cdot 0$

   i. $0 \div 3$ j. $^-1 \cdot 3$ k. $\dfrac{0}{-5}$ l. $^-6 \cdot {}^-7$

   m. $\dfrac{-75}{-15}$ n. $7 \cdot {}^-9$ o. $^-3 \cdot 0$ p. $\dfrac{-8}{-2}$

   q. $^-1 \cdot 6$ r. $^-32 \div 4$ s. $^-8 \cdot {}^-1$ t. $7 \cdot 0$

3. The product or quotient of two integers with the same signs is a _____ integer.
4. The product or quotient of two integers with different signs is a _____ integer.
5. Answer true or false: The set of integers is
   a. associative for division
   b. commutative for multiplication
   c. commutative for division
   d. associative for multiplication

2.5   Exercise Answers

1. a. $^-8$ b. $^-9$ c. 3   d. 1
   e. 0   f. 4   g. undefined   h. 0
   i. 13   j. 1
2. a. $^-17$ b. 0   c. 60   d. 15
   e. $^-5$ f. undefined   g. 0   h. 0
   i. 0   j. $^-3$ k. 0   l. 42
   m. 5   n. $^-63$ o. 0   p. 4
   q. $^-6$ r. $^-8$ s. 8   t. 0
3. positive
4. negative
5. a. false   b. true   c. false   d. true

Chapter 2 Sample Test

At the completion of Chapter 2 it is expected that you will be able to work the following problems.

2.1 Introduction

1. Label each of the following sets as the set of integers, whole numbers, or natural numbers:
   a. $\{1, 2, 3, \ldots\}$
   b. $\{0, 1, 2, 3, \ldots\}$
   c. $\{\ldots, {}^-2, {}^-1, 0, 1, 2, \ldots\}$
2. Each integer is thought of as having what two parts?
3. Graph each of the following integers on the number line:
   a. 2   b. $^-7$ c. 5   d. 0

$$\begin{array}{cccccccccccccccccc} & {}^-8 & {}^-7 & {}^-6 & {}^-5 & {}^-4 & {}^-3 & {}^-2 & {}^-1 & 0 & {}^+1 & {}^+2 & {}^+3 & {}^+4 & {}^+5 & {}^+6 & {}^+7 & {}^+8 \end{array}$$

4. Write the opposite of each of the following integers:
   a. $^-5$ b. 3   c. 0   d. 7

2.2 Addition

5. Find the sum:
   a. $^-3 + {}^-6$ b. $^-5 + 8$ c. $3 + {}^-11$ d. $0 + {}^-2$
   e. $5 + {}^-5$ f. $^-2 + 7 + {}^-5$ g. $^-3 + {}^-5 + 6$ h. $9 + {}^-3 + 6$
6. The following properties are true for all integer replacements of a, b, and c. Match the property that each demonstrates with its name.
   a. $a + b = b + a$
   b. $a + 0 = 0 + a = a$
   c. $(a + b) + c = a + (b + c)$

   1. associative property of addition
   2. commutative property of addition
   3. addition property of zero

2.3 Subtraction

7. Find the difference:
   a. $^-2 - 5$ b. $3 - 7$ c. $6 - {}^-3$ d. $^-1 - {}^-2$
   e. $11 - 9$ f. $5 - 7 - {}^-3$
8. Complete each of the following:
   a. $^-5 + 7$ b. $^-5 - 7$ c. $0 - 5$ d. $^-3 + 0$
   e. $^-4 + {}^-9$ f. $6 - {}^-9$ g. $^-4 + 7 - 6$ h. $2 - 1 - {}^-1$

2.4 Multiplication

9. Find the product:
   a. $^-2 \cdot 5$ b. $^-4 \cdot {}^-6$ c. $^-3 \cdot 0$ d. $5 \cdot {}^-9$
   e. $^-1 \cdot 13$ f. $0 \cdot {}^-1$ g. $^-2 \cdot 4 \cdot {}^-5$ h. $(6 \cdot {}^-2)({}^-3 \cdot {}^-2)$
   i. $^-3(4 + {}^-7)$ j. $2({}^-4 + 4)$
10. The following properties are true for all integer replacements of a, b, and c. Match the property that each demonstrates with its name.
    a. $a \cdot 0 = 0a = 0$
    b. $1a = a \cdot 1 = a$
    c. $a(bc) = (ab)c$
    d. $ab = ba$
    e. $a(b + c) = ab + ac$

    1. associative property of multiplication
    2. commutative property of multiplication
    3. multiplication property of one
    4. left distributive property of multiplication over addition
    5. multiplication property of zero

2.5   **Division**

11. Find the quotient:

a. $^-36 \div 9$

b. $^-18 \div ^-6$

c. $\dfrac{^-12}{4}$

d. $\dfrac{-25}{-5}$

e. $10 \div 0$

f. $\dfrac{0}{^-6}$

g. $0 \div ^-2$

h. $\dfrac{4}{0}$

12. Answer true or false:

a. $a \div b = b \div a$ is true for all integers a and b.

b. $ab = ba$ is true for all integers a and b.

c. $a \div (b \div c) = (a \div b) \div c$ is true for all integers a, b, and c.

d. $a(bc) = (ab)c$ is true for all integers a, b, and c.

Chapter 2 Sample Test Answers

1. a. the set of natural numbers
    b. the set of whole numbers
    c. the set of integers
2. distance and direction
3.

4. a. 5   b. $^-3$ c. 0   d. $^-7$
5. a. $^-9$ b. 3   c. $^-8$ d. $^-2$ e. 0   f. 0   g. $^-2$
    h. 12
6. a. 2, commutative property of addition
    b. 3, addition property of zero
    c. 1, associative property of addition
7. a. $^-7$ b. $^-4$ c. 9   d. 1   e. 2   f. 1
8. a. 2   b. $^-12$ c. $^-5$ d. $^-3$ e. $^-13$ f. 15   g. $^-3$
    h. 2
9. a. $^-10$ b. 24   c. 0   d. $^-45$ e. $^-13$ f. 0   g. 40
    h. $^-72$ i. 9   j. 0
10. a. 5, multiplication property of zero
    b. 3, multiplication property of one
    c. 1, associative property of multiplication
    d. 2, commutative property of multiplication
    e. 4, left distributive property of multiplication over addition
11. a. $^-4$ b. 3   c. $^-3$ d. 5
    e. undefined   f. 0   g. 0   h. undefined
12. a. false   b. true   c. false   d. true

Chapter 3

The Set of Rational Numbers

The purpose of this chapter is to develop skill in performing the fundamental operations with rational numbers. The following number sets have been studied previously:

$N = \{1, 2, 3, \ldots\}$ natural numbers
$W = \{0, 1, 2, \ldots\}$ whole numbers
$I = \{\ldots, {}^{-}2, {}^{-}1, 0, 1, 2, \ldots\}$ integers

3.1   Introduction

---

1   The set of rational numbers is an extension of the set of integers developed in Chapter 2. Because it is impossible to list the set of rational numbers, a verbal description is usually given as a means of defining the set. A rational number is defined as follows:

A rational number is any number that can be written in the form $\dfrac{p}{q}$, where p and q are integers and q $\neq$ 0. Some examples of rational numbers are:

$$\frac{2}{3} \qquad \frac{-5}{7} \qquad 4 \qquad {}^{-}2 \qquad 0 \qquad 3\frac{1}{7}$$

$\dfrac{2}{3}$ and $\dfrac{-5}{7}$ are rational numbers because they are written in the form $\dfrac{p}{q}$ and their denominators (q) are not zero.

4, $^{-}$2, and 0 are rational numbers because they can be written in the form $\dfrac{p}{q}$ by the use of 1 as each denominator. That is, $4 = \dfrac{4}{1}$, $^{-}2 = \dfrac{^{-}2}{1}$, and $0 = \dfrac{0}{1}$.

$3\dfrac{1}{7}$ is a rational number because it can be written in the form $\dfrac{p}{q}$ by use of its improper-fraction equivalent. That is, $3\dfrac{1}{7} = \dfrac{3 \cdot 7 + 1}{7} = \dfrac{22}{7}$.

Q1   Which of the following are rational numbers?

$^{-}5, \dfrac{3}{0}, \dfrac{1}{2}, 2\dfrac{1}{6}$ _____

A1 $^-5$, $\frac{1}{2}$, and $2\frac{1}{6}$: $\frac{3}{0}$ is not a rational number, because its denominator is zero and its value is thus undefined.

Q2   Write $^-5$, $\frac{1}{2}$, and $2\frac{1}{6}$ in the $\frac{p}{q}$ form:

$$^-5 = \underline{\qquad} \qquad\qquad \frac{1}{2} = \underline{\qquad} \qquad\qquad 2\frac{1}{6} = \underline{\qquad}$$

STOP • **STOP** • **STOP** • **STOP** • **STOP** • **STOP** • **STOP** • **STOP** • **STOP**

A2 $\dfrac{^-5}{1}, \dfrac{1}{2}, \dfrac{13}{6}$

---

2   By its definition, the set of rational numbers includes many different types of numbers. It includes positive and negative *fractions,* both proper and improper, because they are already in the required $\frac{p}{q}$ form. It includes positive and negative *mixed numbers,* because any mixed number can be written as an improper fraction that is in the required $\frac{p}{q}$ form. Finally, the set of rational numbers includes the natural numbers, the whole numbers, and the integers, because each of the elements of these sets can be written in the $\frac{p}{q}$ form by the use of the number 1 as the denominator q. As was true of the sets N, W, and I, the set of rational numbers is an infinite set.

Q3   Write each of the following rational numbers in the $\frac{p}{q}$ form:

a. $^-3 = \underline{\qquad}$ **b.** $4\frac{2}{7} = \underline{\qquad}$ **c.** $0 = \underline{\qquad}$

STOP • **STOP** • **STOP** • **STOP** • **STOP** • **STOP** • **STOP** • **STOP** • **STOP**

A3   **a.** $\dfrac{^-3}{1}$ **b.** $\dfrac{30}{7}$ **c.** $\dfrac{0}{1}$

---

3   A negative mixed number can be written in $\frac{p}{q}$ form as follows:

$$^-3\frac{1}{2} = {}^-\left(3\frac{1}{2}\right)$$

$$= {}^-\left(\frac{3 \cdot 2 + 1}{2}\right)$$

$$= {}^-\left(\frac{7}{2}\right) \qquad \text{(read ``the opposite of seven halves'')}$$

$$= \frac{^-7}{2} \qquad \text{(read ``negative seven halves'')}$$

Q4   Write each of the following negative mixed numbers in $\frac{p}{q}$ form:

a. $-4\frac{1}{7}$ b. $-2\frac{2}{9}$

STOP • STOP • STOP • STOP • STOP • STOP • STOP • STOP • STOP

A4   a. $\dfrac{-29}{7}$: $-4\dfrac{1}{7} = -\left(4\dfrac{1}{7}\right)$ b. $\dfrac{-20}{9}$

$$= -\left(\frac{4\cdot 7 + 1}{7}\right)$$

$$= -\left(\frac{29}{7}\right)$$

$$= \frac{-29}{7}$$

Q5   Write each of the following rational numbers in $\dfrac{p}{q}$ form:

a. $7 =$ _____   b. $-5\dfrac{1}{4} =$ _____   c. $\dfrac{2}{3} =$ _____

d. $-4 =$ _____   e. $4\dfrac{1}{2} =$ _____   f. $-2\dfrac{4}{5} =$ _____

STOP • STOP • STOP • STOP • STOP • STOP • STOP • STOP • STOP

A5   a. $\dfrac{7}{1}$ b. $\dfrac{-21}{4}$ c. $\dfrac{2}{3}$ d. $\dfrac{-4}{1}$

e. $\dfrac{9}{2}$ f. $\dfrac{-14}{5}$

Q6   If Q stands for the set of rational numbers, answer true or false for each of the following (*Note:* Q is used to denote the set of rational numbers because in future work R is used to denote the set of real numbers):

a. N is a subset of Q. _____

b. W is a subset of Q. _____

c. I is a subset of Q. _____

STOP • STOP • STOP • STOP • STOP • STOP • STOP • STOP • STOP

A6   a. true   b. true   c. true
Each of the sets (natural numbers, whole numbers, and integers) is a subset of the set of rational numbers.

---

4   The number line is used to graph rational numbers in a manner similar to that used with integers. The rational numbers $\dfrac{3}{4}$ and $-4\dfrac{2}{3}$ and their opposites are graphed on the following number line:

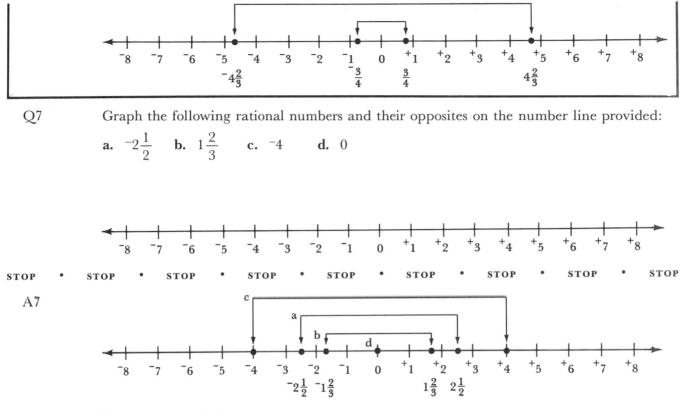

Q7   Graph the following rational numbers and their opposites on the number line provided:

a. $-2\frac{1}{2}$ b. $1\frac{2}{3}$ c. -4 d. 0

STOP • STOP • STOP • STOP • STOP • STOP • STOP • STOP • STOP

A7

Note that zero is its own opposite.

---

5   It is also possible to describe a rational number in terms of its decimal representation: *A rational number is any number whose decimal representation is either a terminating or an infinite repeating decimal.* For example,

$\dfrac{1}{2} = 0.5$ terminating

$\dfrac{1}{3} = 0.\overline{3}\ (0.3333\cdots)*$ infinite repeating

$\dfrac{-5}{8} = -0.625$ terminating

$\dfrac{2}{7} = 0.\overline{285714}\ (0.285714285714\cdots)$ infinite repeating

$-4 = -4.0$ terminating

The decimal representation of a rational number in $\dfrac{p}{q}$ form is found by dividing the numerator (p) by the denominator (q). For example, $\dfrac{3}{4}$ has decimal representation 0.75 because

```
       0.75
   4)3.00
     2 8
     ───
       20
       20
       ──
        0
```

*Bar indicates the digit or block of digits that repeat endlessly.

The decimal 0.75 is terminating because its division comes out evenly (0 remainder). $\frac{5}{33}$ has decimal representation $0.\overline{15}$ because

$$
\begin{array}{r}
0.1515\cdots \quad \text{(repeating 15s)} \\
33\overline{)5.0000} \\
\underline{3\,3} \\
1\,70 \\
\underline{1\,65} \\
50 \\
\underline{33} \\
170 \\
\underline{165} \\
5 \quad \text{etc.}
\end{array}
$$

The decimal $0.\overline{15}$ is *infinite* because its division does not come out evenly, and *repeating* because it repeats itself in blocks of the digits 15. That is, $0.\overline{15} = 0.151515\cdots$ (repeating 15s).

Q8   **a.** Find the decimal representation for the rational number $\frac{3}{8}$.

        b. Is it terminating or infinite repeating? _____

STOP • **STOP** • **STOP** • **STOP** • **STOP** • **STOP** • **STOP** • **STOP** • **STOP**

A8   **a.** 0.375: because
$$
\begin{array}{r}
0.375 \\
8\overline{)3.000} \\
\underline{2\,4} \\
60 \\
\underline{56} \\
40 \\
\underline{40} \\
0
\end{array}
$$
        b. terminating

Q9   **a.** Find the decimal representation for the rational number $\frac{5}{11}$.

   b. Is it terminating or infinite repeating? _____

STOP • **STOP** • **STOP** • **STOP** • **STOP** • **STOP** • **STOP** • **STOP** • **STOP**

A9   **a.** $0.\overline{45}$: because $11\overline{)5.0000}$ with quotient $0.4545\cdots$ (repeating 45s)

$$
\begin{array}{r}
0.4545\cdots \\
11\,\overline{)5.0000} \\
4\,4 \\ \hline
60 \\
55 \\ \hline
50 \\
44 \\ \hline
60 \\
55 \\ \hline
5
\end{array}
$$

   b. infinite repeating

---

6   The decimal representation of $-7\dfrac{1}{3}$ is found by performing the division of the $\dfrac{p}{q}$ form of $-7\dfrac{1}{3}$. Thus, since $-7\dfrac{1}{3} = \dfrac{-22}{3}$, the decimal representation is found as follows:

$$
\begin{array}{r}
7.33\cdots \\
3\,\overline{)22.00} \\
21 \\ \hline
10 \\
9 \\ \hline
10 \\
9 \\ \hline
10
\end{array}
$$

(repeating 3s) (notice that the negative is not used in the division, but merely applied to the quotient)

Therefore, $-7\dfrac{1}{3} = -7.\overline{3}$.

Q10   Find the decimal representation of the rational number $\dfrac{-1}{8}$.

STOP • **STOP** • **STOP** • **STOP** • **STOP** • **STOP** • **STOP** • **STOP** • **STOP**

A10 -0.125

Q11   Find the decimal representation for $^-1\dfrac{5}{9}$.

STOP • **STOP** • **STOP** • **STOP** • **STOP** • **STOP** • **STOP** • **STOP** • **STOP**

A11 $^-1.\overline{5}$ $(^-1.555\cdots)$

> **7**   Some decimals are neither terminating nor infinite repeating and thus are *not* rational numbers. An example is $5.01001000100001\cdots$. This number is an infinite nonrepeating decimal. Infinite nonrepeating decimals are called *irrational* numbers and will be studied further in Chapter 13.

Q12   Choose the rational numbers in the following list: $^-5$, 0, $2.\overline{3}$, $5.10110111011110\cdots$, $2\dfrac{1}{3}$, $1.\overline{513}$, $^-7.8$. _____

STOP • **STOP** • **STOP** • **STOP** • **STOP** • **STOP** • **STOP** • **STOP** • **STOP**

A12 $^-5$, 0, $2.\overline{3}$, $2\dfrac{1}{3}$, $1.\overline{513}$, $^-7.8$

Q13   If W, I, and Q denote the whole numbers, integers, and rational numbers, respectively, write $\in$ or $\notin$ in each of the following blanks to form a true statement:

  a. 2.75 _____ W **b.** 2.75 _____ I **c.** 2.75 _____ Q

  d. $^-3.42$ _____ W **e.** $^-3.\overline{42}$ _____ I **f.** $^-3.\overline{42}$ _____ Q

  g. 7.0 _____ W **h.** 7.0 _____ I **i.** 7.0 _____ Q

STOP • **STOP** • **STOP** • **STOP** • **STOP** • **STOP** • **STOP** • **STOP** • **STOP**

A13   **a.** $\notin$ **b.** $\notin$ **c.** $\in$ **d.** $\notin$ **e.** $\notin$ **f.** $\in$ **g.** $\in$
  h. $\in$ **i.** $\in$

This completes the instruction for this section.

3.1   Exercises

  1. Choose the whole numbers in the following list: 2.5, 1, $\dfrac{2}{3}$, 6, $^-5$, 0, $-2\dfrac{1}{2}$, $3.\overline{24}$, $4\dfrac{1}{3}$.

  2. Choose the integers in the following list: 2.5, 1, $\dfrac{2}{3}$, 6, $^-5$, 0, $^-2\dfrac{1}{2}$, $3.\overline{24}$, $4\dfrac{1}{3}$.

3. Choose the rational numbers in the following list: 2.5, 1, $\frac{2}{3}$, 6, ⁻5, 0, ⁻2$\frac{1}{2}$, 3.$\overline{24}$, 4$\frac{1}{3}$.

4. Write each of the following rational numbers in $\frac{p}{q}$ form:

    a. ⁻3   **b.** 0   **c.** $\frac{2}{3}$ **d.** 8$\frac{1}{6}$

    e. ⁻5$\frac{1}{4}$ **f.** 7   **g.** $\frac{-5}{9}$ **h.** ⁻1$\frac{1}{3}$

5. Graph the following rational numbers and their opposites on the number line provided:

    a. ⁻3$\frac{1}{3}$ **b.** 1$\frac{1}{2}$ **c.** ⁻5   **d.** $\frac{2}{3}$

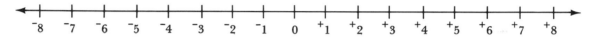

6. Find the decimal representation for each of the following rational numbers and state whether each is terminating or infinite repeating:

    a. $\frac{-2}{3}$ **b.** 2$\frac{1}{5}$ **c.** 5   **d.** $\frac{-7}{9}$

    e. ⁻3$\frac{1}{7}$ **f.** $\frac{5}{8}$ **g.** $\frac{-7}{11}$

3.1   Exercise Answers

1. 1, 6, 0

2. 1, 6, ⁻5, 0

3. 2.5, 1, $\frac{2}{3}$, 6, ⁻5, 0, ⁻2$\frac{1}{2}$, 3.$\overline{24}$, 4$\frac{1}{3}$

4. a. $\frac{-3}{1}$ **b.** $\frac{0}{1}$ **c.** $\frac{2}{3}$ **d.** $\frac{49}{6}$

    e. $\frac{-21}{4}$ **f.** $\frac{7}{1}$ **g.** $\frac{-5}{9}$ **h.** $\frac{-4}{3}$

5.

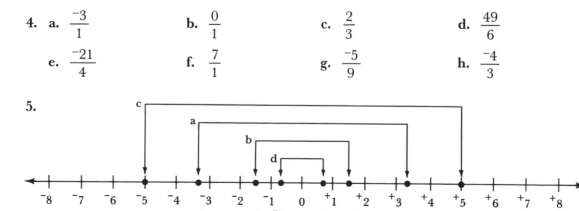

6. **a.** $^-0.\overline{6}$ infinite repeating
   b. 2.2   terminating
   c. 5.0   terminating
   d. $^-0.\overline{7}$ infinite repeating
   e. $^-3.\overline{142857}$ infinite repeating
   f. 0.625   terminating
   g. $^-0.\overline{63}$ infinite repeating

3.2 Addition

1

Since the set of fractions is the set of nonnegative rationals, the procedure used in the addition of rational numbers is much the same as that used for the addition of fractions. To add two rational numbers with a common (same) denominator:

Step 1: Write the sum of the numerators over the common denominator.

Step 2: Apply the rules for the addition of integers.

Step 3: Reduce the rational number to lowest terms, if possible.

Examples:

1. $\dfrac{^-3}{5} + \dfrac{4}{5} = \dfrac{-3+4}{5}$ step 1

              $= \dfrac{1}{5}$ step 2

2. $\dfrac{3}{32} + \dfrac{^-11}{32} = \dfrac{3 + \,^-11}{32}$ step 1

              $= \dfrac{^-8}{32}$ step 2

              $= \dfrac{^-1}{4}$ step 3

3. $\dfrac{^-5}{8} + \dfrac{^-7}{8} = \dfrac{-5 + \,^-7}{8}$ step 1

              $= \dfrac{^-12}{8}$ step 2

              $= \dfrac{^-3}{2}$ step 3

In algebra, rational numbers are usually left in $\dfrac{p}{q}$ form.

Q1   Evaluate:

     a. $\dfrac{^-5}{18} + \dfrac{7}{18}$ **b.** $\dfrac{^-7}{12} + \dfrac{^-2}{12}$

STOP • STOP • STOP • STOP • STOP • STOP • STOP • STOP • STOP

A1 **a.** $\frac{1}{9}$: $\frac{-5}{18} + \frac{7}{18} = \frac{-5 + 7}{18}$ **b.** $\frac{-3}{4}$: $\frac{-7}{12} + \frac{-2}{12} = \frac{-7 + -2}{12}$

$$= \frac{2}{18}$$ $$= \frac{-9}{12}$$

$$= \frac{1}{9}$$ $$= \frac{-3}{4}$$

Q2   Evaluate:

a. $\frac{17}{25} + \frac{-22}{25}$ **b.** $\frac{-5}{6} + \frac{-9}{6}$

STOP • **STOP** • **STOP** • **STOP** • **STOP** • **STOP** • **STOP** • **STOP** • **STOP**

A2   **a.** $\frac{-1}{5}$: $\frac{17}{25} + \frac{-22}{25} = \frac{17 + -22}{25}$ **b.** $\frac{-7}{3}$: $\frac{-5}{6} + \frac{-9}{6} = \frac{-5 + -9}{6}$

$$= \frac{-5}{25}$$ $$= \frac{-14}{6}$$

$$= \frac{-1}{5}$$ $$= \frac{-7}{3}$$

Q3   Find each of the following sums:

a. $\frac{16}{17} + \frac{-4}{17}$ **b.** $\frac{-7}{40} + \frac{2}{40}$ **c.** $\frac{-8}{15} + \frac{-4}{15}$ **d.** $\frac{17}{35} + \frac{-10}{35}$

e. $\frac{-9}{11} + \frac{-10}{11}$ **f.** $\frac{-18}{13} + \frac{5}{13}$ **g.** $\frac{4}{14} + \frac{-11}{14}$ **h.** $\frac{-2}{5} + \frac{-11}{5}$

i. $\frac{-5}{24} + \frac{8}{24}$ **j.** $\frac{-12}{36} + \frac{12}{36}$

STOP • **STOP** • **STOP** • **STOP** • **STOP** • **STOP** • **STOP** • **STOP** • **STOP**

A3   **a.** $\dfrac{12}{17}$ **b.** $\dfrac{-1}{8}$ **c.** $\dfrac{-4}{5}$ **d.** $\dfrac{1}{5}$

   e. $\dfrac{-19}{11}$ **f.** -1 **g.** $\dfrac{-1}{2}$ **h.** $\dfrac{-13}{5}$

   i. $\dfrac{1}{8}$ **j.** 0

---

2   To add any two rational numbers with different denominators it is first necessary to express them as rational numbers with a common denominator. A common denominator for a set of rational numbers is any number that all the denominators will divide into evenly. The computation is simplified when adding rational numbers if the smallest or least common denominator (LCD) is used. Some common denominators for the rational numbers $\dfrac{5}{6}$ and $\dfrac{3}{10}$ are 30, 60, 90, and 120. The LCD is 30.

Q4   Guess the LCD for the rational numbers $\dfrac{2}{3}$ and $\dfrac{4}{5}$. _____

STOP • **STOP** • **STOP** • **STOP** • **STOP** • **STOP** • **STOP** • **STOP** • **STOP**

A4   15

Q5   Guess the LCD for $\dfrac{-5}{9}$ and $\dfrac{8}{27}$. _____

STOP • **STOP** • **STOP** • **STOP** • **STOP** • **STOP** • **STOP** • **STOP** • **STOP**

A5   27

---

3   If the LCD cannot be guessed, the following procedure is helpful in determining the least common denominator:

Step 1: Write the multiples of the largest denominator.

Step 2: Choose the smallest multiple that the remaining denominators will divide into evenly.

Examples:

1. $\dfrac{1}{3}, \dfrac{5}{7}$ 7 is the largest denominator

The multiples of 7 are 7, 14, 21, 28, 35, ...
The smallest multiple that 3 will divide evenly is 21. Thus, the LCD is 21.

2. $\dfrac{2}{5}, \dfrac{1}{3}, \dfrac{5}{6}$ 6 is the largest denominator

The multiples of 6 are 6, 12, 18, 24, 30, 36, ...
The smallest multiple that 3 and 5 will divide evenly is 30. Thus, the LCD is 30.

Q6   Use the multiple method to find the LCD for the rational numbers $\dfrac{5}{9}$ and $\dfrac{7}{15}$.

STOP • **STOP** • **STOP** • **STOP** • **STOP** • **STOP** • **STOP** • **STOP** • **STOP**

A6   45: the multiples of the largest denominator are 15, 30, 45, 60, 75, ... The smallest
            multiple that 9 will divide evenly is 45.

Q7   Find the LCD for the rational numbers $\dfrac{7}{12}$, $\dfrac{3}{10}$, and $\dfrac{5}{6}$.

STOP • **STOP** • **STOP** • **STOP** • **STOP** • **STOP** • **STOP** • **STOP** • **STOP**

A7   60: the multiples of 12 are 12, 24, 36, 48, 60, 72, ... The smallest multiple that 10
            and 6 divide evenly is 60.

Q8   Find the LCD for each of the following:

   a. $\dfrac{2}{7}, \dfrac{3}{6}$ **b.** $\dfrac{5}{6}, \dfrac{9}{24}$ **c.** $\dfrac{1}{2}, \dfrac{3}{5}, \dfrac{4}{7}$ **d.** $\dfrac{2}{15}, \dfrac{3}{10}$

   e. $\dfrac{3}{8}, \dfrac{2}{3}, \dfrac{1}{4}$ **f.** $\dfrac{2}{3}, \dfrac{5}{7}, \dfrac{1}{9}$

STOP • **STOP** • **STOP** • **STOP** • **STOP** • **STOP** • **STOP** • **STOP** • **STOP**

A8   **a.** 42   **b.** 24   **c.** 70   **d.** 30
            e. 24   **f.** 63

---

4   To find the sum of rational numbers with different denominators:

Step 1: Find the LCD.

Step 2: Use equivalent fractions to express each of the rational numbers over the LCD.

Step 3: Follow the rules of Frame 1 for adding rational numbers with a common denominator.

(Recall that an equivalent-fraction form of any rational number is found by either multiplying or dividing *both* numerator and denominator by the *same* natural number.) For example,

$$\frac{7}{15} + \frac{-1}{6}$$

The LCD is 30. To find equivalent fractions, the open expressions $\dfrac{7}{15} = \dfrac{?}{30}$ and

$\dfrac{-1}{6} = \dfrac{?}{30}$ must be completed. This is done as follows:

$$\frac{7}{15} = \frac{14}{30} \qquad \left(\text{because } \frac{7 \times 2}{15 \times 2} = \frac{14}{30} \right)$$

$$\frac{-1}{6} = \frac{-5}{30} \qquad \left(\text{because } \frac{-1 \times 5}{6 \times 5} = \frac{-5}{30} \right)$$

Therefore,

$$\frac{7}{15} + \frac{-1}{6} = \frac{14}{30} + \frac{-5}{30}$$

$$= \frac{14 + {}^-5}{30}$$

$$= \frac{9}{30}$$

$$= \frac{3}{10}$$

Q9   **a.** What is the LCD for $\frac{1}{8} + \frac{-5}{12}$? _____

    b. Complete the equivalent fractions over their LCD:

$$\frac{1}{8} = \text{_____} \qquad \frac{-5}{12} = \text{_____}$$

    c. Find the sum $\frac{1}{8} + \frac{-5}{12}$.

STOP • STOP • STOP • STOP • STOP • STOP • STOP • STOP • STOP

A9   **a.** 24   **b.** $\frac{1}{8} = \frac{3}{24}, \frac{-5}{12} = \frac{-10}{24}$ **c.** $\frac{-7}{24}$: $\frac{1}{8} + \frac{-5}{12} = \frac{3}{24} + \frac{-10}{24}$

$$= \frac{-7}{24}$$

Q10   Find the sum:

    a. $\frac{-1}{6} + \frac{4}{5}$ **b.** $\frac{-3}{7} + \frac{-5}{8}$

STOP • STOP • STOP • STOP • STOP • STOP • STOP • STOP • STOP

A10   **a.** $\frac{19}{30}$: $\frac{-1}{6} + \frac{4}{5} = \frac{-5}{30} + \frac{24}{30}$ **b.** $\frac{-59}{56}$: $\frac{-3}{7} + \frac{-5}{8} = \frac{-24}{56} + \frac{-35}{56}$

$$= \frac{19}{30} \qquad\qquad\qquad\qquad\qquad = \frac{-59}{56}$$

Q11   Find each of the following sums:

    a. $\frac{-2}{3} + \frac{-5}{9}$ **b.** $\frac{5}{6} + \frac{-7}{8}$ **c.** $\frac{-2}{15} + \frac{19}{20}$ **d.** $\frac{7}{12} + \frac{4}{7}$

    e. $\frac{5}{14} + \frac{-3}{4}$ **f.** $\frac{-7}{16} + \frac{-1}{4}$ **g.** $\frac{7}{9} + \frac{-4}{12}$ **h.** $\frac{-1}{8} + \frac{-4}{15}$

STOP • STOP • STOP • STOP • STOP • STOP • STOP • STOP • STOP

A11   a. $\dfrac{-11}{9}$ b. $\dfrac{-1}{24}$ c. $\dfrac{49}{60}$ d. $\dfrac{97}{84}$

     e. $\dfrac{-11}{28}$ f. $\dfrac{-11}{16}$ g. $\dfrac{4}{9}$ h. $\dfrac{-47}{120}$

---

5   The sum $\dfrac{-3}{4} + \dfrac{7}{12} + \dfrac{-5}{9}$ is found as follows:

$$\frac{-3}{4} + \frac{7}{12} + \frac{-5}{9} = \frac{-27}{36} + \frac{21}{36} + \frac{-20}{36}$$

$$= \frac{-27 + 21 + {}^-20}{36}$$

$$= \frac{-26}{36}$$

$$= \frac{-13}{18}$$

Q12   Find the sum $\dfrac{1}{2} + \dfrac{3}{4} + \dfrac{-5}{8}$.

STOP • **STOP** • **STOP** • **STOP** • **STOP** • **STOP** • **STOP** • **STOP** • **STOP**

A12 $\dfrac{5}{8}$: $\dfrac{1}{2} + \dfrac{3}{4} + \dfrac{-5}{8} = \dfrac{4}{8} + \dfrac{6}{8} + \dfrac{-5}{8}$

$$= \frac{4 + 6 + {}^-5}{8}$$

$$= \frac{5}{8}$$

Q13   Find the sum $\dfrac{-1}{6} + \dfrac{-4}{5} + \dfrac{7}{15}$.

STOP • **STOP** • **STOP** • **STOP** • **STOP** • **STOP** • **STOP** • **STOP** • **STOP**

A13 $\dfrac{-1}{2}$: $\dfrac{-1}{6} + \dfrac{-4}{5} + \dfrac{7}{15} = \dfrac{-5}{30} + \dfrac{-24}{30} + \dfrac{14}{30}$

$$= \frac{-5 + {}^-24 + 14}{30}$$

$$= \frac{-15}{30}$$

$$= \frac{-1}{2}$$

---

6   When an integer is present in a sum of rational numbers, write the integer in $\dfrac{p}{q}$ form with a denominator of 1 and proceed as before.

Examples:

$$-5 + \frac{2}{7} = \frac{-5}{1} + \frac{2}{7} \qquad 3 + \frac{-5}{8} = \frac{3}{1} + \frac{-5}{8}$$

$$= \frac{-35}{7} + \frac{2}{7} \qquad\qquad = \frac{24}{8} + \frac{-5}{8}$$

$$= \frac{-33}{7} \qquad\qquad\qquad = \frac{19}{8}$$

Q14   Find the sum:

     a. $-3 + \dfrac{6}{7}$ **b.** $\dfrac{-3}{11} + 2$

STOP • **STOP** • **STOP** • **STOP** • **STOP** • **STOP** • **STOP** • **STOP** • **STOP**

A14   **a.** $\dfrac{-15}{7}$: $-3 + \dfrac{6}{7} = \dfrac{-3}{1} + \dfrac{6}{7}$ **b.** $\dfrac{19}{11}$: $\dfrac{-3}{11} + 2 = \dfrac{-3}{11} + \dfrac{2}{1}$

$$= \frac{-21}{7} + \frac{6}{7} \qquad\qquad\qquad\qquad\qquad = \frac{-3}{11} + \frac{22}{11}$$

$$= \frac{-15}{7} \qquad\qquad\qquad\qquad\qquad\qquad = \frac{19}{11}$$

---

7   A mixed rational number, or simply a mixed number, is one that has an integer part and a fraction (positive or negative) part. Thus $-3\dfrac{1}{2}$, $5\dfrac{4}{7}$, and $12\dfrac{2}{9}$ are all mixed rational numbers. Recall that any mixed number can be written in $\dfrac{p}{q}$ form.

Examples:

$$5\frac{4}{7} = \frac{7(5) + 4}{7} \qquad\qquad -5\frac{2}{3} = {}^{-}\!\left(\frac{5 \cdot 3 + 2}{3}\right)$$

$$= \frac{35 + 4}{7} \qquad\qquad\qquad = {}^{-}\!\left(\frac{17}{3}\right)$$

$$= \frac{39}{7} \qquad\qquad\qquad\qquad = \frac{-17}{3}$$

Q15   Write $2\dfrac{3}{8}$ in $\dfrac{p}{q}$ form.

STOP • **STOP** • **STOP** • **STOP** • **STOP** • **STOP** • **STOP** • **STOP** • **STOP**

A15 $\frac{19}{8}$: $2\frac{5}{8} = \frac{8(2) + 3}{8}$

$= \frac{19}{8}$

Q16 Write $^-5\frac{6}{7}$ in $\frac{p}{q}$ form.

STOP • STOP • STOP • STOP • STOP • STOP • STOP • STOP • STOP

A16 $\frac{-41}{7}$: $^-5\frac{6}{7} = ^-\left(\frac{5 \cdot 7 + 6}{7}\right)$

$= ^-\left(\frac{41}{7}\right)$

$= \frac{-41}{7}$

---

8 To find a sum involving one or more mixed numbers:

Step 1: Write the mixed number(s) in $\frac{p}{q}$ form.

Step 2: Follow the rules for adding rational numbers as previously stated.

Examples:

$^-1\frac{1}{5} + 3\frac{2}{5} = \frac{-6}{5} + \frac{17}{5}$

$= \frac{-6 + 17}{5}$

$= \frac{11}{5}$

$^-5\frac{2}{7} + \frac{3}{5} = \frac{-37}{7} + \frac{3}{5}$

$= \frac{-185}{35} + \frac{21}{35}$

$= \frac{-185 + 21}{35}$

$= \frac{-164}{35}$

Q17 Write the mixed numbers $2\frac{1}{3}$ and $^-3\frac{1}{4}$ in $\frac{p}{q}$ form:

$2\frac{1}{3} = $ _____ $^-3\frac{1}{4} = $ _____

STOP • STOP • STOP • STOP • STOP • STOP • STOP • STOP • STOP

A17 $\frac{7}{3}, \frac{-13}{4}$

Q18 Write the rational numbers of A17 over their LCD.

$$\frac{7}{3} = \underline{\hspace{1cm}} \qquad\qquad \frac{^-13}{4} = \underline{\hspace{1cm}}$$

STOP • STOP • STOP • STOP • STOP • STOP • STOP • STOP • STOP

A18 The LCD is 12. $\frac{7}{3} = \frac{28}{12}$, $\frac{^-13}{4} = \frac{^-39}{12}$

Q19 Use the equivalent fractions of A18 to find the sum $2\frac{1}{3} + {}^-3\frac{1}{4}$.

STOP • STOP • STOP • STOP • STOP • STOP • STOP • STOP • STOP

A19 $\frac{^-11}{12}$: $2\frac{1}{3} + {}^-3\frac{1}{4} = \frac{28}{12} + \frac{^-39}{12}$

$$= \frac{^-11}{12}$$

Q20 Find the sum:

a. $^-1\frac{3}{5} + 4\frac{1}{6}$

b. $^-9\frac{2}{3} + {}^-7\frac{2}{3}$

STOP • STOP • STOP • STOP • STOP • STOP • STOP • STOP • STOP

A20 a. $\frac{77}{30}$: $^-1\frac{3}{5} + 4\frac{1}{6} = \frac{^-8}{5} + \frac{25}{6}$

$$= \frac{^-48}{30} + \frac{125}{30}$$

$$= \frac{^-48 + 125}{30}$$

$$= \frac{77}{30}$$

b. $\frac{^-52}{3}$: $^-9\frac{2}{3} + {}^-7\frac{2}{3} = \frac{^-29}{3} + \frac{^-23}{3}$

$$= \frac{^-29 + {}^-23}{3}$$

$$= \frac{^-52}{3}$$

Q21 Find each of the following sums:

a. $2\frac{7}{8} + {}^-6\frac{3}{8}$

b. $^-7\frac{3}{5} + 5\frac{3}{8}$

c. $^-6\frac{2}{9} + 4\frac{5}{9}$

d. $6\frac{1}{5} + {}^-3\frac{4}{7}$

e. $^-1\dfrac{2}{3} + ^-4\dfrac{5}{6}$ f. $^-1\dfrac{4}{7} + 5\dfrac{3}{8}$ g. $1\dfrac{4}{5} + \dfrac{^-3}{8}$ h. $\dfrac{5}{9} + ^-3\dfrac{1}{2}$

STOP • STOP • STOP • STOP • STOP • STOP • STOP • STOP • STOP

A21   a. $\dfrac{^-7}{2}$ b. $\dfrac{^-89}{40}$ c. $\dfrac{^-5}{3}$ d. $\dfrac{92}{35}$

e. $\dfrac{^-13}{2}$ f. $\dfrac{213}{56}$ g. $\dfrac{57}{40}$ h. $\dfrac{^-53}{18}$

---

9   To find the sum of a mixed number and an integer, write each in $\dfrac{p}{q}$ form and follow the procedures previously developed. For example,

$$^-2\dfrac{3}{5} + 7 = \dfrac{^-13}{5} + \dfrac{7}{1}$$

$$= \dfrac{^-13}{5} + \dfrac{35}{5}$$

$$= \dfrac{^-13 + 35}{5}$$

$$= \dfrac{22}{5}$$

Q22   Find the sum:

a. $^-2\dfrac{3}{5} + 8$ b. $6 + ^-5\dfrac{3}{8}$

STOP • STOP • STOP • STOP • STOP • STOP • STOP • STOP • STOP

A22   a. $\dfrac{27}{5}$: $^-2\dfrac{3}{5} + 8 = \dfrac{^-13}{5} + \dfrac{8}{1}$ b. $\dfrac{5}{8}$: $6 + ^-5\dfrac{3}{8} = \dfrac{6}{1} + \dfrac{^-43}{8}$

$$= \dfrac{^-13}{5} + \dfrac{40}{5}$$ $$= \dfrac{48}{8} + \dfrac{^-43}{8}$$

$$= \dfrac{27}{5}$$ $$= \dfrac{48 + ^-43}{8}$$

$$= \dfrac{5}{8}$$

Q23 Find each of the following sums:

a. $\dfrac{-5}{7} + {}^{-}3$ b. $\dfrac{1}{6} + {}^{-}2\dfrac{4}{6}$ c. $^{-}4\dfrac{5}{6} + \dfrac{2}{3}$

d. $^{-}8 + \dfrac{3}{11}$ e. $\dfrac{8}{13} + {}^{-}6$ f. $^{-}3\dfrac{1}{7} + {}^{-}2\dfrac{5}{8}$

STOP • STOP • STOP • STOP • STOP • STOP • STOP • STOP • STOP

A23   a. $\dfrac{-26}{7}$ b. $\dfrac{-5}{2}$ c. $\dfrac{-25}{6}$ d. $\dfrac{-85}{11}$ e. $\dfrac{-70}{13}$ f. $\dfrac{-323}{56}$

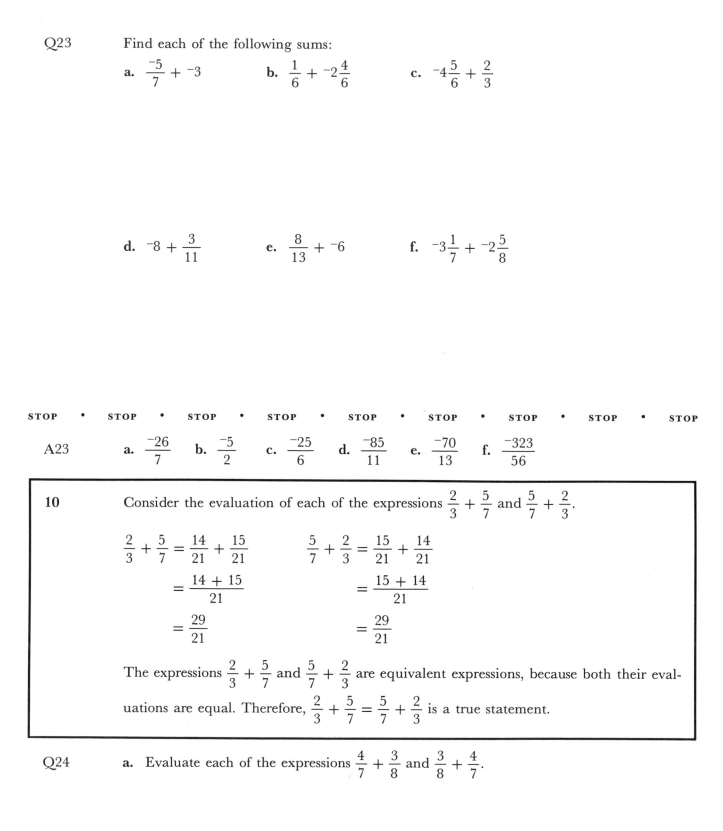

10   Consider the evaluation of each of the expressions $\dfrac{2}{3} + \dfrac{5}{7}$ and $\dfrac{5}{7} + \dfrac{2}{3}$.

$$\dfrac{2}{3} + \dfrac{5}{7} = \dfrac{14}{21} + \dfrac{15}{21} \qquad \dfrac{5}{7} + \dfrac{2}{3} = \dfrac{15}{21} + \dfrac{14}{21}$$

$$= \dfrac{14 + 15}{21} \qquad\qquad = \dfrac{15 + 14}{21}$$

$$= \dfrac{29}{21} \qquad\qquad = \dfrac{29}{21}$$

The expressions $\dfrac{2}{3} + \dfrac{5}{7}$ and $\dfrac{5}{7} + \dfrac{2}{3}$ are equivalent expressions, because both their evaluations are equal. Therefore, $\dfrac{2}{3} + \dfrac{5}{7} = \dfrac{5}{7} + \dfrac{2}{3}$ is a true statement.

Q24   a. Evaluate each of the expressions $\dfrac{4}{7} + \dfrac{3}{8}$ and $\dfrac{3}{8} + \dfrac{4}{7}$.

b. Are $\dfrac{4}{7} + \dfrac{3}{8}$ and $\dfrac{3}{8} + \dfrac{4}{7}$ equivalent expressions?   _____

c. Is $\dfrac{4}{7} + \dfrac{3}{8} = \dfrac{3}{8} + \dfrac{4}{7}$ a true statement? _____

STOP • STOP • STOP • STOP • STOP • STOP • STOP • STOP • STOP

A24 a. $\dfrac{4}{7} + \dfrac{3}{8} = \dfrac{32}{56} + \dfrac{21}{56}$, $\dfrac{3}{8} + \dfrac{4}{7} = \dfrac{21}{56} + \dfrac{32}{56}$

$\qquad\qquad\quad = \dfrac{53}{56} \qquad\qquad\qquad\qquad = \dfrac{53}{56}$

b. yes: both equal the same number

c. yes

Q25 Verify that $a + b$ and $b + a$ are equivalent expressions for $a = \dfrac{1}{2}$ and $b = \dfrac{3}{4}$.

STOP • STOP • STOP • STOP • STOP • STOP • STOP • STOP • STOP

A25 $a + b = \dfrac{1}{2} + \dfrac{3}{4}$, $b + a = \dfrac{3}{4} + \dfrac{1}{2}$

$\qquad\qquad = \dfrac{2}{4} + \dfrac{3}{4} \qquad\qquad = \dfrac{3}{4} + \dfrac{2}{4}$

$\qquad\qquad = \dfrac{5}{4} \qquad\qquad\qquad = \dfrac{5}{4}$

> **11** The expressions $a + b$ and $b + a$ are equivalent for all rational-number replacements of a and b. Thus, $a + b = b + a$ is a true statement for all rational numbers a and b. This fact demonstrates that just as addition was commutative for the set of whole numbers and integers, addition is also commutative for the set of rational numbers. That is, a change in the order of addition does not affect the sum.

Q26 a. Is $a + b = b + a$ a true statement for all rational numbers a and b? _____

b. Is the addition operation commutative for the set of rational numbers? _____

STOP • STOP • STOP • STOP • STOP • STOP • STOP • STOP • STOP

A26 a. yes   b. yes

> **12** Consider the evaluation of each of the expressions $\left(\dfrac{3}{4} + \dfrac{1}{2}\right) + \dfrac{1}{3}$ and $\dfrac{3}{4} + \left(\dfrac{1}{2} + \dfrac{1}{3}\right)$.
>
> $\left(\dfrac{3}{4} + \dfrac{1}{2}\right) + \dfrac{1}{3} = \left(\dfrac{3}{4} + \dfrac{2}{4}\right) + \dfrac{1}{3}$ $\dfrac{3}{4} + \left(\dfrac{1}{2} + \dfrac{1}{3}\right) = \dfrac{3}{4} + \left(\dfrac{3}{6} + \dfrac{2}{6}\right)$
>
> $\qquad\qquad\qquad\quad = \dfrac{5}{4} + \dfrac{1}{3} \qquad\qquad\qquad\qquad = \dfrac{3}{4} + \dfrac{5}{6}$
>
> $\qquad\qquad\qquad\quad = \dfrac{15}{12} + \dfrac{4}{12} \qquad\qquad\qquad\quad = \dfrac{9}{12} + \dfrac{10}{12}$
>
> $\qquad\qquad\qquad\quad = \dfrac{19}{12} \qquad\qquad\qquad\qquad\quad = \dfrac{19}{12}$

> The expressions $\left(\frac{3}{4} + \frac{1}{2}\right) + \frac{1}{3}$ and $\frac{3}{4} + \left(\frac{1}{2} + \frac{1}{3}\right)$ are equivalent expressions because the evaluation of both expressions are equal. Thus, $\left(\frac{3}{4} + \frac{1}{2}\right) + \frac{1}{3} = \frac{3}{4} + \left(\frac{1}{2} + \frac{1}{3}\right)$ is a true mathematical statement.

Q27   Verify that $(a + b) + c$ and $a + (b + c)$ are equivalent expressions when $a = \frac{3}{5}$, $b = \frac{7}{10}$, and $c = \frac{3}{20}$.

STOP • STOP • STOP • STOP • STOP • STOP • STOP • STOP • STOP

A27

$$(a + b) + c = \left(\frac{3}{5} + \frac{7}{10}\right) + \frac{3}{20}, \qquad a + (b + c) = \frac{3}{5} + \left(\frac{7}{10} + \frac{3}{20}\right)$$

$$= \left(\frac{6}{10} + \frac{7}{10}\right) + \frac{3}{20} \qquad\qquad = \frac{3}{5} + \left(\frac{14}{20} + \frac{3}{20}\right)$$

$$= \frac{13}{10} + \frac{3}{20} \qquad\qquad = \frac{3}{5} + \frac{17}{20}$$

$$= \frac{26}{20} + \frac{3}{20} \qquad\qquad = \frac{12}{20} + \frac{17}{20}$$

$$= \frac{29}{20} \qquad\qquad = \frac{29}{20}$$

Q28   Is $(a + b) + c = a + (b + c)$ a true statement for $a = \frac{3}{5}$, $b = \frac{7}{10}$, and $c = \frac{3}{20}$? _____

STOP • STOP • STOP • STOP • STOP • STOP • STOP • STOP • STOP

A28   yes

---

13 $(a + b) + c = a + (b + c)$ is a true statement for any rational number a, b, and c. Thus, just as addition was associative for the sets of whole numbers and the set of integers, addition is associative for the set of rational numbers. That is, a change in the grouping when doing a series of additions does not affect the sum.

Q29   **a.** Is $(a + b) + c = a + (b + c)$ a true statement for all rational numbers a, b, and c? _____

   b. Is addition an associative operation for the set of rational numbers? _____

STOP • STOP • STOP • STOP • STOP • STOP • STOP • STOP • STOP

A29   **a.** yes   **b.** yes

This completes the instruction for this section.

3.2   Exercises

1. Find each of the following sums:

   a. $\dfrac{11}{13} + \dfrac{^-7}{13}$ b. $\dfrac{^-4}{5} + \dfrac{^-2}{5}$ c. $\dfrac{14}{15} + \dfrac{^-4}{15}$ d. $\dfrac{2}{3} + \dfrac{^-5}{3}$

   e. $\dfrac{^-6}{7} + \dfrac{6}{7}$ f. $\dfrac{1}{4} + \dfrac{3}{5}$ g. $\dfrac{^-4}{5} + \dfrac{3}{8}$ h. $\dfrac{^-1}{6} + \dfrac{^-4}{18}$

2. Find each of the following sums:

   a. $^-3\dfrac{3}{4} + 1\dfrac{1}{4}$ b. $^-4\dfrac{3}{7} + ^-4\dfrac{5}{7}$ c. $1\dfrac{1}{3} + ^-7\dfrac{2}{3}$ d. $6\dfrac{2}{9} + ^-3\dfrac{5}{9}$

   e. $^-12\dfrac{7}{8} + 15\dfrac{3}{8}$ f. $45\dfrac{1}{5} + ^-2\dfrac{4}{5}$ g. $^-3\dfrac{2}{5} + 7$ h. $15 + ^-2\dfrac{4}{11}$

   i. $\dfrac{^-3}{19} + 2$ j. $^-7 + \dfrac{4}{5}$

3. Find each of the following sums:

   a. $\dfrac{3}{10} + \dfrac{^-7}{6} + \dfrac{5}{12}$ b. $\dfrac{^-5}{24} + \dfrac{7}{8} + \dfrac{^-7}{12}$ c. $12\dfrac{3}{4} + ^-5\dfrac{1}{4}$

   d. $^-2\dfrac{3}{5} + 1\dfrac{4}{5}$ e. $3\dfrac{2}{9} + ^-4\dfrac{5}{12}$ f. $^-13\dfrac{1}{5} + 6\dfrac{4}{7}$

   g. $6\dfrac{1}{3} + \dfrac{^-4}{11}$ h. $^-1\dfrac{2}{3} + ^-7\dfrac{3}{5}$

4. True or false:
   The set of rational numbers is:
   a. associative for the operation addition.
   b. commutative for the operation addition.

5. True or false:

   a. $\dfrac{1}{3} + \dfrac{^-7}{16} = \dfrac{^-7}{16} + \dfrac{1}{3}$ b. $\left(\dfrac{^-3}{5} + \dfrac{2}{9}\right) + \dfrac{1}{7} \neq \dfrac{^-3}{5} + \left(\dfrac{2}{9} + \dfrac{1}{7}\right)$

3.2   Exercise Answers

1. a. $\dfrac{4}{13}$ b. $\dfrac{^-6}{5}$ c. $\dfrac{2}{3}$ d. $^-1$ e. 0 f. $\dfrac{17}{20}$ g. $\dfrac{^-17}{40}$

   h. $\dfrac{^-7}{18}$

2. a. $\dfrac{^-5}{2}$ b. $\dfrac{^-64}{7}$ c. $\dfrac{^-19}{3}$ d. $\dfrac{8}{3}$ e. $\dfrac{5}{2}$ f. $\dfrac{212}{5}$ g. $\dfrac{18}{5}$

   h. $\dfrac{139}{11}$ i. $\dfrac{35}{19}$ j. $\dfrac{^-31}{5}$

3. a. $\dfrac{^-9}{20}$ b. $\dfrac{1}{12}$ c. $\dfrac{15}{2}$ d. $\dfrac{^-4}{5}$ e. $\dfrac{^-43}{36}$ f. $\dfrac{^-232}{35}$ g. $\dfrac{197}{33}$

   h. $\dfrac{^-139}{15}$

    4. a. true   **b.** true

    5. a. true   **b.** false

3.3   Subtraction

---

1   The definition of subtraction for integers as defined in Chapter 2 is also true for the set of rational numbers. That is,

$$x - y = x + \ ^-y \qquad \text{for all rational-number replacements of } x \text{ and } y$$

To subtract any two rational numbers:

Step 1: Use the definition of subtraction to rewrite the difference as a sum.

Step 2: Apply the procedures of Section 3.2 to find the sum.

Q1   Use the definition of subtraction to rewrite $\dfrac{4}{5} - \dfrac{7}{12}$ as a sum. _____

STOP • STOP • STOP • STOP • STOP • STOP • STOP • STOP • STOP

A1 $\dfrac{4}{5} + \dfrac{^-7}{12}$

Q2   Find the sum $\dfrac{4}{5} + \dfrac{^-7}{12}$.

STOP • STOP • STOP • STOP • STOP • STOP • STOP • STOP • STOP

A2 $\dfrac{13}{60}$: $\dfrac{4}{5} + \dfrac{^-7}{12} = \dfrac{48}{60} + \dfrac{^-35}{60}$

$$= \dfrac{13}{60}$$

Recall that two rational numbers with different denominators are combined by first finding the least common denominator (LCD).

Q3   Rewrite $\dfrac{^-4}{7} - \dfrac{^-5}{6}$ as a sum. _____

STOP • STOP • STOP • STOP • STOP • STOP • STOP • STOP • STOP

A3 $\dfrac{^-4}{7} + \dfrac{5}{6}$

Q4   Find the sum $\dfrac{^-4}{7} + \dfrac{5}{6}$.

STOP • STOP • STOP • STOP • STOP • STOP • STOP • STOP • STOP

A4 $\dfrac{11}{42}$: $\dfrac{^-4}{7} + \dfrac{5}{6} = \dfrac{^-24}{42} + \dfrac{35}{42}$

$\qquad\qquad\qquad = \dfrac{11}{42}$

Q5   Find the difference by using the definition of subtraction to rewrite the problem as a sum: $\dfrac{^-7}{12} - \dfrac{5}{8}$.

STOP • STOP • STOP • STOP • STOP • STOP • STOP • STOP • STOP

A5 $\dfrac{^-29}{24}$: $\dfrac{^-7}{12} - \dfrac{5}{8} = \dfrac{^-7}{12} + \dfrac{^-5}{8}$

$\qquad\qquad\qquad = \dfrac{^-14}{24} + \dfrac{^-15}{24}$

$\qquad\qquad\qquad = \dfrac{^-29}{24}$

Q6   Find the difference $\dfrac{^-1}{4} - \dfrac{^-5}{6}$.

STOP • STOP • STOP • STOP • STOP • STOP • STOP • STOP • STOP

A6 $\dfrac{7}{12}$: $\dfrac{^-1}{4} - \dfrac{^-5}{6} = \dfrac{^-1}{4} + \dfrac{5}{6}$

$\qquad\qquad\qquad = \dfrac{^-6}{24} + \dfrac{20}{24}$

$\qquad\qquad\qquad = \dfrac{14}{24}$

$\qquad\qquad\qquad = \dfrac{7}{12}$

Q7   Find each of the following differences:

a. $\dfrac{4}{5} - \dfrac{3}{4}$ b. $\dfrac{^-2}{7} - \dfrac{5}{11}$ c. $\dfrac{16}{19} - \dfrac{^-3}{19}$ d. $\dfrac{^-5}{24} - \dfrac{7}{24}$

e. $\dfrac{2}{3} - \dfrac{5}{7}$ f. $\dfrac{^-7}{21} - \dfrac{^-5}{6}$

STOP • **STOP** • **STOP** • **STOP** • **STOP** • **STOP** • **STOP** • **STOP** • **STOP**

A7   a. $\dfrac{1}{20}$ b. $\dfrac{^-57}{77}$ c. 1 d. $\dfrac{^-1}{2}$

e. $\dfrac{^-1}{21}$ f. $\dfrac{1}{2}$

---

2   To find the difference $3\dfrac{2}{7} - 4\dfrac{1}{2}$, use the steps of Frame 1 as follows:

$$3\dfrac{2}{7} - 4\dfrac{1}{2} = 3\dfrac{2}{7} + {}^-4\dfrac{1}{2}$$
$$= \dfrac{23}{7} + \dfrac{^-9}{2}$$
$$= \dfrac{46}{14} + \dfrac{^-63}{14}$$
$$= \dfrac{^-17}{14}$$

Q8   Find the difference ${}^-4\dfrac{2}{3} - 1\dfrac{2}{5}$.

STOP • **STOP** • **STOP** • **STOP** • **STOP** • **STOP** • **STOP** • **STOP** • **STOP**

A8 $\dfrac{^-91}{15}$: ${}^-4\dfrac{2}{3} - 1\dfrac{2}{5} = {}^-4\dfrac{2}{3} + {}^-1\dfrac{2}{5}$

$$= \dfrac{^-14}{3} + \dfrac{^-7}{5}$$
$$= \dfrac{^-70}{15} + \dfrac{^-21}{15}$$
$$= \dfrac{^-91}{15}$$

Q9   Find the difference $12\frac{1}{6} - 5\frac{3}{8}$.

STOP • **STOP** • **STOP** • **STOP** • **STOP** • **STOP** • **STOP** • **STOP** • **STOP**

A9 $\frac{163}{24}$: $12\frac{1}{6} - 5\frac{3}{8} = 12\frac{1}{6} + {}^{-}5\frac{3}{8}$

$$= \frac{73}{6} + \frac{{}^{-}43}{8}$$

$$= \frac{292}{24} + \frac{{}^{-}129}{24}$$

$$= \frac{163}{24}$$

Q10   Find the difference ${}^{-}7\frac{1}{3} - 2$.

STOP • **STOP** • **STOP** • **STOP** • **STOP** • **STOP** • **STOP** • **STOP** • **STOP**

A10 $\frac{{}^{-}28}{3}$: ${}^{-}7\frac{1}{3} - 2 = {}^{-}7\frac{1}{3} + {}^{-}2$ or ${}^{-}7\frac{1}{3} - 2 = {}^{-}7\frac{1}{3} + {}^{-}2$

$$= \frac{{}^{-}22}{3} + \frac{{}^{-}2}{1} \qquad\qquad = {}^{-}9\frac{1}{3}$$

$$= \frac{{}^{-}22}{3} + \frac{{}^{-}6}{3} \qquad\qquad = \frac{{}^{-}28}{3}$$

$$= \frac{{}^{-}28}{3}$$

Q11   Find the difference for each of the following:

a. $2\frac{1}{4} - 3\frac{3}{4}$ **b.** ${}^{-}4\frac{3}{7} - 6\frac{1}{7}$ **c.** $5\frac{3}{8} - 7\frac{4}{9}$ **d.** ${}^{-}2\frac{1}{9} - 3\frac{2}{5}$

e. $18 - \dfrac{5}{17}$ f. $\dfrac{^{-}3}{16} - 4$ g. $2\dfrac{1}{3} - \dfrac{9}{11}$ h. $\dfrac{^{-}6}{13} - 4\dfrac{1}{2}$

i. $16\dfrac{3}{10} - 5\dfrac{7}{8}$ j. $^{-}15\dfrac{1}{6} - {}^{-}2\dfrac{7}{15}$

STOP   •   **STOP**   •   **STOP**   •   **STOP**   •   **STOP**   •   **STOP**   •   **STOP**   •   **STOP**   •   **STOP**

A11   a. $\dfrac{^{-}3}{2}$ b. $\dfrac{^{-}74}{7}$ c. $\dfrac{^{-}149}{72}$ d. $\dfrac{^{-}248}{45}$ e. $\dfrac{301}{17}$ f. $\dfrac{^{-}67}{16}$ g. $\dfrac{50}{33}$

h. $\dfrac{^{-}129}{26}$ i. $\dfrac{417}{40}$ j. $\dfrac{^{-}127}{10}$

---

3   Consider the evaluation of the expressions $\dfrac{2}{3} - \dfrac{5}{8}$ and $\dfrac{5}{8} - \dfrac{2}{3}$.

$\dfrac{2}{3} - \dfrac{5}{8}$ $\dfrac{5}{8} - \dfrac{2}{3}$

$\dfrac{2}{3} + \dfrac{^{-}5}{8}$ $\dfrac{5}{8} + \dfrac{^{-}2}{3}$

$\dfrac{16}{24} + \dfrac{^{-}15}{24}$ $\dfrac{15}{24} + \dfrac{^{-}16}{24}$

$\dfrac{1}{24}$ $\dfrac{^{-}1}{24}$

The expressions $\dfrac{2}{3} - \dfrac{5}{8}$ and $\dfrac{5}{8} - \dfrac{2}{3}$ are not equivalent, because they have different evaluations. Thus, $\dfrac{2}{3} - \dfrac{5}{8} = \dfrac{5}{8} - \dfrac{2}{3}$ is a false statement. $\dfrac{2}{3} - \dfrac{5}{8} \neq \dfrac{5}{8} - \dfrac{2}{3}$ is a true statement.

Q12   a. Evaluate the expressions $\dfrac{3}{7} - \dfrac{4}{5}$ and $\dfrac{4}{5} - \dfrac{3}{7}$.

b. Are $\dfrac{3}{7} - \dfrac{4}{5}$ and $\dfrac{4}{5} - \dfrac{3}{7}$ equivalent expressions?   _____

c. Is $\dfrac{3}{7} - \dfrac{4}{5} = \dfrac{4}{5} - \dfrac{3}{7}$ a true statement?   _____

STOP • **STOP** • **STOP** • **STOP** • **STOP** • **STOP** • **STOP** • **STOP** • **STOP**

A12   **a.** $\dfrac{3}{7} - \dfrac{4}{5} = \dfrac{^-13}{35}, \dfrac{4}{5} - \dfrac{3}{7} = \dfrac{13}{35}$

b. no   **c.** no

---

4   The open sentence $a - b = b - a$ is false for most whole numbers a and b. Since the whole numbers are a subset of the rational numbers, the statement $a - b = b - a$ is likewise false for most rational numbers a and b. Thus, the rational numbers are not commutative for subtraction.

Q13   **a.** Is addition a commutative operation for the set of rational numbers? _____
           b. Is subtraction a commutative operation for the set of rational numbers? _____

STOP • **STOP** • **STOP** • **STOP** • **STOP** • **STOP** • **STOP** • **STOP** • **STOP**

A13   **a.** yes   **b.** no

Q14   Evaluate each of the expressions $\left(\dfrac{1}{2} - \dfrac{2}{3}\right) - \dfrac{3}{5}$ and $\dfrac{1}{2} - \left(\dfrac{2}{3} - \dfrac{3}{5}\right).$

STOP • **STOP** • **STOP** • **STOP** • **STOP** • **STOP** • **STOP** • **STOP** • **STOP**

A14 $\left(\dfrac{1}{2} - \dfrac{2}{3}\right) - \dfrac{3}{5}$ $\qquad$ $\dfrac{1}{2} - \left(\dfrac{2}{3} - \dfrac{3}{5}\right)$

$\left(\dfrac{3}{6} + \dfrac{^-4}{6}\right) - \dfrac{3}{5}$ $\qquad$ $\dfrac{1}{2} - \left(\dfrac{10}{15} - \dfrac{9}{15}\right)$

$\dfrac{^-1}{6} - \dfrac{3}{5}$ $\qquad$ $\dfrac{1}{2} - \dfrac{1}{15}$

$\dfrac{^-5}{30} + \dfrac{^-18}{30}$ $\qquad$ $\dfrac{15}{30} + \dfrac{^-2}{30}$

$\dfrac{^-23}{30}$ $\qquad$ $\dfrac{13}{30}$

Q15   **a.** Are the expressions of Q14 equivalent? _____

b. Is $\left(\dfrac{1}{2} - \dfrac{2}{3}\right) - \dfrac{3}{5} = \dfrac{1}{2} - \left(\dfrac{2}{3} - \dfrac{3}{5}\right)$ a true statement? _____

STOP • **STOP** • **STOP** • **STOP** • **STOP** • **STOP** • **STOP** • **STOP** • **STOP**

A15   **a.** no   **b.** no

5	The open sentence $(a - b) - c = a - (b - c)$ is false for most whole numbers a, b, and c. Since the whole numbers are a subset of the rational numbers, the statement $(a - b) - c = a - (b - c)$ is likewise false for most rational numbers a, b, and c. Thus, the rational numbers are not associative for subtraction.

Q16   **a.** Is addition an associative operation for the set of rational numbers?_____

                 b. Is subtraction an associative operation for the set of rational numbers?_____

STOP • **STOP** • **STOP** • **STOP** • **STOP** • **STOP** • **STOP** • **STOP** • **STOP**

A16   **a.** yes   **b.** no

This completes the instruction for this section.

3.3   Exercises

1. Find the difference for each of the following:

    a. $\dfrac{2}{3} - \dfrac{5}{3}$ **b.** $\dfrac{3}{7} - \dfrac{7}{9}$ **c.** $4 - 2\dfrac{4}{7}$ **d.** $2\dfrac{1}{3} - 5\dfrac{1}{4}$

    e. $2\dfrac{1}{6} - 1$ **f.** $\dfrac{5}{12} - \dfrac{3}{4}$ **g.** $^{-}3\dfrac{2}{5} - 4\dfrac{3}{5}$ **h.** $\dfrac{^{-}5}{6} - \dfrac{^{-}10}{12}$

    i. $^{-}3 - 2\dfrac{3}{7}$ **j.** $3\dfrac{2}{5} - 4\dfrac{1}{7}$

2. Complete each of the following:

    a. $1\dfrac{3}{7} + ^{-}2\dfrac{4}{7}$ **b.** $\dfrac{^{-}5}{8} - \dfrac{1}{8}$ **c.** $\dfrac{4}{5} - 1$ **d.** $^{-}2\dfrac{1}{3} + 5$

    e. $3\dfrac{1}{2} + ^{-}7\dfrac{5}{8}$ **f.** $^{-}6\dfrac{1}{5} - 1\dfrac{3}{7}$ **g.** $2\dfrac{1}{3} + \dfrac{^{-}5}{6} + \dfrac{2}{9}$ **h.** $1\dfrac{2}{11} + 3\dfrac{15}{11}$

    i. $2\dfrac{1}{3} - 4 + 1\dfrac{2}{5}$ **j.** $4\dfrac{1}{5} - 2\dfrac{2}{5} + ^{-}1\dfrac{4}{5}$

3. Is addition an associative operation in the set of rational numbers?

4. Is subtraction an associative operation in the set of rational numbers?

5. Is addition a commutative operation in the set of rational numbers?

6. Is subtraction a commutative operation in the set of rational numbers?

3.3   Exercise Answers

    1. a. $^{-}1$ **b.** $\dfrac{^{-}22}{63}$ **c.** $\dfrac{10}{7}$ **d.** $\dfrac{^{-}35}{12}$ **e.** $\dfrac{7}{6}$ **f.** $\dfrac{^{-}1}{3}$ **g.** $^{-}8$

       h. 0 **i.** $\dfrac{^{-}38}{7}$ **j.** $\dfrac{^{-}26}{35}$

2. a. $\dfrac{-8}{7}$ **b.** $\dfrac{-3}{4}$ **c.** $\dfrac{-1}{5}$ **d.** $\dfrac{8}{3}$ **e.** $\dfrac{-33}{8}$ **f.** $\dfrac{-267}{35}$ **g.** $\dfrac{31}{18}$

   h. $\dfrac{61}{11}$ **i.** $\dfrac{-4}{15}$ **j.** 0

3. yes

4. no

5. yes

6. no

3.4   Multiplication

---

1   Let $\dfrac{a}{b}$ and $\dfrac{c}{d}$ stand for any two rational numbers. The product of $\dfrac{a}{b}$ and $\dfrac{c}{d}$ is defined as:

$$\frac{a}{b} \cdot \frac{c}{d} = \frac{ac}{bd}$$

where ac is the product of the numerators and bd is the product of the denominators.

Examples:

$$\frac{-3}{5} \cdot \frac{6}{7} = \frac{-18}{35} \qquad \frac{-6}{11} \cdot \frac{-5}{7} = \frac{30}{77}$$

As is the case with the sum and difference of two rational numbers, the product of two rational numbers is always reduced to lowest terms.

Q1   Find the product $\dfrac{-3}{5} \cdot \dfrac{4}{7}$.

STOP • STOP • STOP • STOP • STOP • STOP • STOP • STOP • STOP

A1 $\dfrac{-12}{35}$: $\dfrac{-3}{5} \cdot \dfrac{4}{7} = \dfrac{-12}{35}$

Q2   Find the product:

   a. $\dfrac{-6}{5} \cdot \dfrac{-2}{3}$ **b.** $\dfrac{3}{8} \cdot \dfrac{4}{15}$

STOP • STOP • STOP • STOP • STOP • STOP • STOP • STOP • STOP

A2   **a.** $\dfrac{4}{5}$: $\dfrac{-6}{5} \cdot \dfrac{-2}{3} = \dfrac{12}{15}$ **b.** $\dfrac{1}{10}$

$$= \dfrac{4}{5}$$

---

2   A common procedure when multiplying rational numbers is to divide any numerator and any denominator by a common factor. This procedure is called *reducing* (sometimes referred to as "canceling").

Examples:

$$\dfrac{3}{\cancel{5}_{1}} \cdot \dfrac{\cancel{10}^{2}}{13} = \dfrac{6}{13}$$ (common factor of 5)

$$\dfrac{\cancel{4}^{-1}}{\cancel{7}_{1}} \cdot \dfrac{\cancel{-14}^{-2}}{\cancel{20}_{5}} = \dfrac{2}{5}$$ (common factors of 4 and 7)

$$\dfrac{\cancel{-9}^{-3}}{\cancel{12}_{4}} \cdot \dfrac{5}{7} = \dfrac{-15}{28}$$ (common factor of 3)

Q3   Find the product $\dfrac{-4}{8} \cdot \dfrac{3}{7}$ by first dividing out common factors.

STOP • **STOP** • **STOP** • **STOP** • **STOP** • **STOP** • **STOP** • **STOP** • **STOP**

A3 $\dfrac{-3}{14}$: $\dfrac{\cancel{4}^{-1}}{\cancel{8}_{2}} \cdot \dfrac{3}{7} = \dfrac{-3}{14}$

Q4   Find the product $\dfrac{-15}{39} \cdot \dfrac{-13}{10}$.

STOP • **STOP** • **STOP** • **STOP** • **STOP** • **STOP** • **STOP** • **STOP** • **STOP**

A4 $\dfrac{1}{2}$: $\dfrac{\cancel{-15}^{-3}}{\cancel{39}_{3}} \cdot \dfrac{\cancel{-13}^{-1}}{\cancel{10}_{2}} = \dfrac{1}{2}$

Reducing order
Step 1: $^{-}15$ and 10
Step 2: 39 and $^{-}13$
Step 3: $^{-}3$ and 3

Notice that the problem could also be completed using other orders of reducing. One other order is the following:

$$\frac{\overset{-1}{\cancel{\overset{-5}{\cancel{-15}}}}}{\underset{1}{\cancel{\underset{2}{39}}}} \cdot \frac{\overset{-1}{\cancel{-13}}}{\underset{2}{\cancel{10}}} = \frac{1}{2}$$

Reducing order

Step 1: ⁻15 and 39

Step 2: ⁻5 and 10

Step 3: 13 and ⁻13

However, regardless of the order used, the final product is always the same.

---

3 One of the quotient rules from Chapter 2 states that when dividing integers with unlike signs, the answer is negative. For example,

$$\frac{-10}{5} = {}^-2 \qquad \text{and} \qquad \frac{10}{-5} = {}^-2$$

Q5 **a.** What is the sign of the quotient $\frac{-2}{3}$? _____

b. What is the sign of the quotient $\frac{2}{-3}$? _____

STOP • STOP • STOP • STOP • STOP • STOP • STOP • STOP • STOP

A5 **a.** negative **b.** negative

---

4 The numbers $\frac{-2}{3}$ and $\frac{2}{3}$ are equivalent forms of the same rational number. Both represent a negative rational number because they involve a quotient of integers with unlike signs. When writing a negative rational number it is customary to place the negative sign on the number in the numerator. For example, $\frac{5}{-8}$ is usually written $\frac{-5}{8}$.

Q6 Find each of the following products:

a. $\frac{-12}{15} \cdot \frac{3}{6}$ **b.** $\frac{-4}{6} \cdot \frac{-3}{18}$ **c.** $\frac{2}{9} \cdot \frac{5}{25}$ **d.** $\frac{7}{12} \cdot \frac{9}{-11}$

e. $\frac{-5}{7} \cdot \frac{6}{-13}$ **f.** $\frac{5}{-28} \cdot \frac{-16}{17}$ **g.** $\frac{7}{3} \cdot \frac{2}{3}$ **h.** $\frac{-4}{7} \cdot \frac{7}{4}$

STOP • STOP • STOP • STOP • STOP • STOP • STOP • STOP • STOP

A6 **a.** $\frac{-2}{5}$ **b.** $\frac{1}{9}$ **c.** $\frac{2}{45}$ **d.** $\frac{-21}{44}$

e. $\frac{30}{91}$ **f.** $\frac{20}{119}$ **g.** $\frac{14}{9}$ **h.** ⁻1

---

5 When finding a product involving mixed numbers, write all mixed numbers in improper-fraction form before using the definition of Frame 1. For example,

$$-2\frac{1}{3}\cdot\frac{5}{14}=\frac{\overset{-1}{-\cancel{7}}}{3}\cdot\frac{5}{\underset{2}{\cancel{14}}}$$

$$=\frac{-5}{6}$$

Q7 Find the product:

a. $5\frac{1}{8}\cdot\frac{-4}{9}$

b. $-1\frac{2}{3}\cdot-4\frac{2}{7}$

STOP • STOP • STOP • STOP • STOP • STOP • STOP • STOP • STOP

A7 **a.** $\frac{-41}{18}$: $5\frac{1}{8}\cdot\frac{-4}{9}=\frac{41}{\underset{2}{\cancel{8}}}\cdot\frac{\overset{-1}{-\cancel{4}}}{9}$ **b.** $\frac{50}{7}$: $-1\frac{2}{3}\cdot-4\frac{2}{7}=\frac{-5}{\underset{1}{\cancel{3}}}\cdot\frac{\overset{-10}{-\cancel{30}}}{7}$

$$=\frac{-41}{18}\qquad\qquad\qquad\qquad=\frac{50}{7}$$

---

6 When finding a product of a rational number and an integer, write the integer with a denominator of 1 before using the definition of Frame 1. For example,

$$5\frac{2}{9}\cdot-3=\frac{47}{\underset{3}{\cancel{9}}}\cdot\frac{\overset{-1}{-\cancel{3}}}{1}$$

$$=\frac{-47}{3}$$

Q8 Find the product:

a. $2\cdot-1\frac{2}{5}$

b. $\frac{1}{5}\cdot5$

STOP • STOP • STOP • STOP • STOP • STOP • STOP • STOP • STOP

A8 **a.** $\frac{-14}{5}$: $2\cdot-1\frac{2}{5}=\frac{2}{1}\cdot\frac{-7}{5}$ **b.** 1: $\frac{1}{5}\cdot5=\frac{1}{\cancel{5}}\cdot\frac{\overset{1}{\cancel{5}}}{1}$

$$=\frac{-14}{5}\qquad\qquad\qquad\qquad=\frac{1}{1}$$

$$=1$$

Q9   Find each of the following products:

a. $\dfrac{-3}{11} \cdot 3$ b. $\dfrac{4}{-5} \cdot \dfrac{-10}{12}$ c. $2\dfrac{1}{3} \cdot -1\dfrac{3}{4}$ d. $-5\dfrac{2}{9} \cdot -12$

e. $\dfrac{-6}{7} \cdot \dfrac{7}{-6}$ f. $-4 \cdot \dfrac{-1}{4}$ g. $-1\dfrac{1}{2} \cdot 3\dfrac{1}{5}$ h. $3\dfrac{1}{5} \cdot -1\dfrac{1}{2}$

STOP • **STOP** • **STOP** • **STOP** • **STOP** • **STOP** • **STOP** • **STOP** • **STOP**

A9   a. $\dfrac{-9}{11}$ b. $\dfrac{2}{3}$ c. $\dfrac{-49}{12}$ d. $\dfrac{188}{3}$

e. 1   f. 1   g. $\dfrac{-24}{5}$ h. $\dfrac{-24}{5}$

---

7   Consider the product of any rational number $\dfrac{a}{b}$ and zero:

$$\frac{a}{b} \cdot 0 = \frac{a}{b} \cdot \frac{0}{1}$$

$$= \frac{a \cdot 0}{b \cdot 1}$$

$$= \frac{0}{b}$$

$$= 0$$

Thus, the product of any rational number and zero is zero. This fact is referred to as the *multiplication property of zero*. The multiplication property of zero states that $\dfrac{a}{b} \cdot 0 = 0 \cdot \dfrac{a}{b} = 0$ for any rational number $\dfrac{a}{b}$.

Q10   Find the product:

a. $4\dfrac{1}{2} \cdot 0$ _____   b. $0 \cdot \dfrac{-5}{8}$ _____

STOP • **STOP** • **STOP** • **STOP** • **STOP** • **STOP** • **STOP** • **STOP** • **STOP**

A10   a. 0   b. 0

---

8   When finding the product of more than two rational numbers, use one of the following procedures:

1. Find the product of two rational numbers at a time working from left to right, or

2. Write all rational numbers in $\dfrac{p}{q}$ form, perform all possible cancellations, and place the product of the numerators over the product of the denominators.

Examples:

$$4\frac{1}{2}\cdot {}^{-}4\cdot \frac{5}{18} \qquad \text{or} \qquad 4\frac{1}{2}\cdot {}^{-}4\cdot \frac{5}{18}$$

$$\left(4\frac{1}{2}\cdot {}^{-}4\right)\cdot \frac{5}{18}$$

$$\left(\frac{9}{\cancel{2}}\cdot \frac{{}^{-}\cancel{4}^{\,-2}}{1}\right)\cdot \frac{5}{18}$$
$$^{1}$$

$$\frac{\overset{1}{\cancel{9}}}{\cancel{2}_1}\cdot \frac{{}^{-}\overset{-1}{\cancel{4}}}{1}\cdot \frac{5}{\underset{1}{\cancel{18}}_2}$$

$$\frac{{}^{-}\cancel{18}}{1}\cdot \frac{5}{\cancel{18}}$$
$$_1$$

$$\frac{{}^{-}\cancel{18}}{1}\cdot \frac{5}{\cancel{18}_1}$$

$$^{-}5$$

$$^{-}5$$

Q11   Find the product:

    a. $\dfrac{4}{7}\cdot \dfrac{{}^{-}3}{8}\cdot \dfrac{14}{15}$ **b.** ${}^{-}3\dfrac{1}{3}\cdot \dfrac{7}{20}\cdot \dfrac{{}^{-}9}{14}$ **c.** $2\dfrac{3}{4}\cdot 0\cdot {}^{-}6$

***

STOP • **STOP** • **STOP** • **STOP** • **STOP** • **STOP** • **STOP** • **STOP** • **STOP**

A11   **a.** $\dfrac{{}^{-}1}{5}$ **b.** $\dfrac{3}{4}$ **c.** 0

---

9   Two rational numbers whose product is 1 are called *reciprocals.** Thus, 5 and $\dfrac{1}{5}$ are reciprocals because

$$5\cdot \frac{1}{5} = \frac{\cancel{5}^{\,1}}{1}\cdot \frac{1}{\cancel{5}}$$
$$_1$$

$$= \frac{1}{1}$$

$$= 1$$

Similarly, $\dfrac{{}^{-}5}{8}$ and $\dfrac{8}{{}^{-}5}$ are reciprocals because

$$\frac{{}^{-}\cancel{5}^{\,1}}{\cancel{8}_1}\cdot \frac{\cancel{8}^{\,1}}{{}^{-}\cancel{5}_1} = \frac{1}{1}$$

$$= 1$$

*Some mathematicians refer to reciprocals as "multiplicative inverses."

Q12   Find the products of the following reciprocals:

    a. $\dfrac{-2}{3} \cdot \dfrac{3}{-2} =$ _____
    b. $2\dfrac{2}{3} \cdot \dfrac{3}{8} =$ _____
    c. $\dfrac{1}{8} \cdot 8 =$ _____

STOP • STOP • STOP • STOP • STOP • STOP • STOP • STOP • STOP

A12   **a.** 1   **b.** 1   **c.** 1

---

10   Notice that to find the reciprocal for a given rational number, it is necessary to interchange or invert the numerator and denominator. That $\dfrac{5}{6}$ and $\dfrac{6}{5}$ are reciprocals is verified by the fact that their product is 1.

$$\dfrac{\overset{1}{\cancel{5}}}{\underset{1}{\cancel{6}}} \cdot \dfrac{\overset{1}{\cancel{6}}}{\underset{1}{\cancel{5}}} = \dfrac{1}{1}$$

$$= 1$$

To find the reciprocal of a mixed number, write the number in improper-fraction form and interchange the numerator and denominator. For example, since $3\dfrac{1}{2} = \dfrac{7}{2}$, its reciprocal is $\dfrac{2}{7}$.

To find the reciprocal of an integer, write the integer with a denominator of 1 and interchange the numerator and denominator. For example, since $9 = \dfrac{9}{1}$, its reciprocal is $\dfrac{1}{9}$.

Q13   Write the reciprocals for each of the following:

    a. $\dfrac{2}{3}$ _____
    b. $1\dfrac{5}{9}$ _____
    c. 15 _____

STOP • STOP • STOP • STOP • STOP • STOP • STOP • STOP • STOP

A13   **a.** $\dfrac{3}{2}$ **b.** $\dfrac{9}{14}$ **c.** $\dfrac{1}{15}$

Q14   What is the reciprocal of $\dfrac{-4}{7}$? _____

STOP • STOP • STOP • STOP • STOP • STOP • STOP • STOP • STOP

A14 $\dfrac{-7}{4}$: $\dfrac{7}{-4}$ is written $\dfrac{-7}{4}$

Q15   Write the reciprocal for each of the following:

    a. $\dfrac{-6}{13}$ _____
    b. $-5\dfrac{1}{2}$ _____
    c. -9 _____

STOP • STOP • STOP • STOP • STOP • STOP • STOP • STOP • STOP

A15   **a.** $\dfrac{-13}{6}$ **b.** $\dfrac{-2}{11}$ **c.** $\dfrac{-1}{9}$

---

11   Consider the replacement that makes the following open sentence a true statement.

$$0 \cdot \underline{\hspace{1cm}} = 1$$

Since $0 \cdot \dfrac{\text{no}}{\text{number}} = 1$, there is no reciprocal for the number zero.

Q16   What replacement makes the open expression $0 \cdot \underline{\hspace{1cm}} = 1$ a true statement?

_____

STOP • **STOP** • **STOP** • **STOP** • **STOP** • **STOP** • **STOP** • **STOP** • **STOP**

A16   no number: Because $0 \cdot \dfrac{\text{no}}{\text{number}} = 1$.

Q17   What is the reciprocal of zero?_____

STOP • **STOP** • **STOP** • **STOP** • **STOP** • **STOP** • **STOP** • **STOP** • **STOP**

A17   There is none.

Q18   **a.** Evaluate both ab and ba for $a = \dfrac{-3}{5}$ and $b = \dfrac{5}{21}$.

b. In part a, are ab and ba equivalent expressions?_____

c. Do you think ab and ba are equivalent expressions for all rational-number replacements of a and b?_____

d. Is $ab = ba$ true for all rational-number replacements of a and b?_____

STOP • **STOP** • **STOP** • **STOP** • **STOP** • **STOP** • **STOP** • **STOP** • **STOP**

A18   **a.** $ab = \dfrac{-3}{5} \cdot \dfrac{5}{21} \qquad ba = \dfrac{5}{21} \cdot \dfrac{-3}{5}$

$\qquad\quad = \dfrac{-1}{7} \qquad\qquad\quad = \dfrac{-1}{7}$

b. yes   **c.** yes (assumption)   **d.** yes (assumption)

---

12   The statement $ab = ba$ is true for all rational-number replacements of a and b. Thus, multiplication is a commutative operation for the set of rational numbers. That is, a change in the *order* of the factors when multiplying two rational numbers does not affect the product.

Q19   Verify that $(ab)c = a(bc)$ is a true statement for $a = \dfrac{-2}{3}$, $b = \dfrac{-6}{35}$, and $c = \dfrac{7}{9}$.

STOP • **STOP** • **STOP** • **STOP** • **STOP** • **STOP** • **STOP** • **STOP** • **STOP**

A19

$$(ab)c = a(bc)$$

$$\left(\frac{^-2}{3} \cdot \frac{^-6}{35}\right)\frac{7}{9} \stackrel{?}{=} \frac{^-2}{3}\left(\frac{^-6}{35} \cdot \frac{7}{9}\right)$$

$$\frac{4}{35} \cdot \frac{7}{9} \stackrel{?}{=} \frac{^-2}{3} \cdot \frac{^-2}{15}$$

$$\frac{4}{45} = \frac{4}{45}$$

---

13   The statement $(ab)c = a(bc)$ is true for all rational-number replacements of a, b, and c. Thus, multiplication is an associative operation for the set of rational numbers. That is, a change in the *grouping* of the factors when multiplying more than two rational numbers docs not affect the product.

This completes the instruction for this section.

3.4   Exercises

1. Write the reciprocal for each of the following rational numbers:

a. $\dfrac{1}{7}$ **b.** $\dfrac{^-3}{4}$ **c.** $2\dfrac{1}{4}$ **d.** 0 **e.** 6 **f.** $^-5\dfrac{1}{4}$

2. Complete the following open sentences to form true statements:

a. $\dfrac{^-2}{3} \cdot \underline{\hspace{1cm}} = 1$ **b.** $4\dfrac{1}{2} \cdot \underline{\hspace{1cm}} = 0$ **c.** $\dfrac{^-4}{7} \cdot \dfrac{^-7}{4} = \underline{\hspace{1cm}}$

d. $0 \cdot \underline{\hspace{1cm}} = 1$ **e.** $1\dfrac{1}{2} \cdot \underline{\hspace{1cm}} = 1$ **f.** $\dfrac{6}{19} \cdot \underline{\hspace{1cm}} = 1$

3. Find the product:

a. $\dfrac{^-3}{4} \cdot 1\dfrac{1}{3}$ **b.** $\dfrac{^-4}{15} \cdot \dfrac{9}{16}$ **c.** $\dfrac{^-4}{5} \cdot \dfrac{^-2}{7}$

d. $\dfrac{^-10}{24} \cdot \dfrac{^-6}{25}$ **e.** $6\dfrac{2}{3} \cdot ^-4\dfrac{1}{2}$ **f.** $^-2\dfrac{1}{7} \cdot ^-25$

g. $0 \cdot 3\dfrac{2}{17}$ **h.** $^-7 \cdot 1\dfrac{4}{7}$ **i.** $5\dfrac{1}{4} \cdot 11\dfrac{1}{3}$

j. $\dfrac{^-6}{17} \cdot \dfrac{^-17}{6}$ **k.** $^-1\dfrac{2}{5} \cdot \dfrac{^-5}{7}$ **l.** $\dfrac{^-3}{8} \cdot \dfrac{3}{8}$

4. Find the product:

a. $\dfrac{1}{4} \cdot \dfrac{^-2}{3} \cdot \dfrac{6}{7}$ **b.** $^-2\dfrac{1}{3} \cdot \dfrac{6}{7} \cdot \dfrac{^-1}{2}$ **c.** $4\dfrac{3}{19} \cdot 0 \cdot ^-1$

d. $\dfrac{^-7}{8} \cdot \dfrac{^-24}{25} \cdot \dfrac{^-5}{21}$ **e.** $4\dfrac{3}{8} \cdot ^-2\dfrac{1}{5} \cdot 1\dfrac{1}{7}$ **f.** $^-4\dfrac{2}{3} \cdot \dfrac{2}{5} \cdot \dfrac{^-3}{14}$

5. True or false:
The set of rational numbers is:
a. commutative for multiplication.
b. associative for multiplication.

3.4   **Exercise Answers**

1. **a.** 7   **b.** $\dfrac{-4}{3}$ **c.** $\dfrac{4}{9}$ **d.** zero does not have a reciprocal

   e. $\dfrac{1}{6}$ **f.** $\dfrac{-4}{21}$

2. **a.** $\dfrac{-3}{2}$ **b.** 0   **c.** 1   **d.** no number

   e. $\dfrac{2}{3}$ **f.** $\dfrac{19}{6}$

3. **a.** $^-1$ **b.** $\dfrac{-3}{20}$ **c.** $\dfrac{8}{35}$ **d.** $\dfrac{1}{10}$ **e.** $^-30$ **f.** $\dfrac{375}{7}$ **g.** 0

   h. $^-11$ **i.** $\dfrac{119}{2}$ **j.** 1   **k.** 1   **l.** $\dfrac{-9}{64}$

4. **a.** $\dfrac{-1}{7}$ **b.** 1   **c.** 0   **d.** $\dfrac{-1}{5}$ **e.** $^-11$ **f.** $\dfrac{2}{5}$

5. **a.** true   **b.** true

3.5   **Division**

---

1   Let $\dfrac{a}{b}$ and $\dfrac{c}{d}$ stand for any two rational numbers with $\dfrac{c}{d} \neq 0$. Consider the simplification of the following quotient:

$$\frac{a}{b} \div \frac{c}{d} = \frac{\dfrac{a}{b}}{\dfrac{c}{d}} = \frac{\dfrac{a}{b} \cdot \dfrac{d}{c}}{\dfrac{c}{d} \cdot \dfrac{d}{c}} = \frac{\dfrac{a}{b} \cdot \dfrac{d}{c}}{1} = \frac{a}{b} \cdot \frac{d}{c}$$

Thus, the *definition of division for two rational numbers* (with a nonzero divisor) is stated:

$$\frac{a}{b} \div \frac{c}{d} = \frac{a}{b} \cdot \frac{d}{c}$$

$$= \frac{ad}{bc} \qquad \frac{c}{d} \neq 0$$

In words, the quotient of two rational numbers with a nonzero divisor is equal to the product of the first (dividend) times the reciprocal of the second (divisor).

Examples:

$$\frac{2}{3} \div \frac{4}{9} = \frac{\overset{1}{\cancel{2}}}{\underset{1}{\cancel{3}}} \cdot \frac{\overset{3}{\cancel{9}}}{\underset{2}{\cancel{4}}} \qquad \left(\text{the reciprocal of } \frac{4}{9} \text{ is } \frac{9}{4}\right)$$

$$= \frac{3}{2}$$

$$\frac{4}{5} \div \frac{^-6}{7} = \frac{\overset{2}{\cancel{4}}}{5} \cdot \frac{^-7}{\underset{3}{\cancel{6}}} \qquad \left(\text{the reciprocal of } \frac{^-6}{7} \text{ is } \frac{^-7}{6}\right)$$

$$= \frac{^-14}{15}$$

Q1   Find the quotient:

a. $\dfrac{^-2}{3} \div \dfrac{3}{4}$ 　　　　　　　　b. $\dfrac{^-7}{18} \div \dfrac{3}{9}$

STOP • **STOP** • **STOP** • **STOP** • **STOP** • **STOP** • **STOP** • **STOP** • **STOP**

A1   a. $\dfrac{^-8}{9}$: $\dfrac{^-2}{3} \div \dfrac{3}{4} = \dfrac{^-2}{3} \cdot \dfrac{4}{3}$ 　　b. $\dfrac{^-7}{6}$: $\dfrac{^-7}{18} \div \dfrac{3}{9} = \dfrac{^-7}{\underset{2}{\cancel{18}}} \cdot \dfrac{\overset{1}{\cancel{9}}}{3}$

$$= \frac{^-8}{9} \qquad\qquad\qquad\qquad\qquad = \frac{^-7}{6}$$

Q2   Find the quotient:

a. $\dfrac{5}{6} \div \dfrac{3}{4}$ 　　b. $\dfrac{^-17}{20} \div \dfrac{3}{14}$ 　　c. $\dfrac{^-3}{7} \div \dfrac{3}{7}$ 　　d. $\dfrac{20}{21} \div \dfrac{15}{42}$

e. $\dfrac{5}{8} \div \dfrac{^-1}{16}$ 　　f. $\dfrac{^-13}{16} \div \dfrac{2}{3}$ 　　g. $\dfrac{^-5}{8} \div \dfrac{^-5}{8}$ 　　h. $0 \div \dfrac{^-5}{9}$

STOP • **STOP** • **STOP** • **STOP** • **STOP** • **STOP** • **STOP** • **STOP** • **STOP**

A2   a. $\dfrac{10}{9}$ 　　b. $\dfrac{^-119}{30}$ 　　c. $^-1$ 　　d. $\dfrac{8}{3}$

e. $^-10$ 　　f. $\dfrac{^-39}{32}$ 　　g. 1 　　h. 0

---

2   When finding a quotient that involves mixed numbers, write all mixed numbers in improper-fraction form *before* using the definition of division (Frame 1). For example,

$$\frac{^-5}{14} \div 3\frac{2}{7}$$

$$\frac{^-5}{14} \div \frac{23}{7}$$

$$\frac{^-5}{\cancel{14}_{2}} \cdot \frac{\cancel{7}^{1}}{23} \qquad \left(\text{the reciprocal of } \frac{23}{7} \text{ is } \frac{7}{23}\right)$$

$$\frac{^-5}{46}$$

Q3   Find the quotient $8\frac{1}{2} \div {}^-1\frac{3}{4}$.

STOP • STOP • STOP • STOP • STOP • STOP • STOP • STOP • STOP

A3 $\dfrac{^-34}{7}$: $8\dfrac{1}{2} \div {}^-1\dfrac{3}{4}$

$$\frac{17}{2} \div \frac{^-7}{4}$$

$$\frac{17}{\cancel{2}_{1}} \cdot \frac{\cancel{^-4}^{^-2}}{7} \qquad \left(\text{the reciprocal of } \frac{^-7}{4} \text{ is } \frac{^-4}{7}\right)$$

$$\frac{^-34}{7}$$

Q4   Find the quotient $\dfrac{^-11}{15} \div {}^-4\dfrac{2}{5}$.

STOP • STOP • STOP • STOP • STOP • STOP • STOP • STOP • STOP

A4 $\dfrac{1}{6}$: $\dfrac{^-11}{15} \div {}^-4\dfrac{2}{5}$

$$\frac{^-11}{15} \div \frac{^-22}{5}$$

$$\frac{\cancel{11}^{^-1}}{\cancel{15}_{3}} \cdot \frac{\cancel{^-5}^{^-1}}{\cancel{22}_{2}} \qquad \left(\text{the reciprocal of } \frac{^-22}{5} \text{ is } \frac{^-5}{22}\right)$$

$$\frac{1}{6}$$

<table>
<tr><td>3</td><td>When an integer is involved in the quotient of two rational numbers, write the integer with a denominator of 1 before using the definition of division (Frame 1). For example,</td></tr>
</table>

$$\frac{-5}{8} \div 12$$

$$\frac{-5}{8} \div \frac{12}{1}$$

$$\frac{-5}{8} \cdot \frac{1}{12} \quad \left(\text{the reciprocal of } 12 \text{ is } \frac{1}{12}\right)$$

$$\frac{-5}{96}$$

Q5   Find the quotient $2\frac{2}{3} \div {}^-8$.

STOP • STOP • STOP • STOP • STOP • STOP • STOP • STOP • STOP

A5 $\frac{-1}{3}$: $2\frac{2}{3} \div {}^-8$

$$\frac{8}{3} \div \frac{-8}{1} \quad \left(\text{write } 2\frac{2}{3} \text{ as } \frac{8}{3} \text{ and } {}^-8 \text{ as } \frac{-8}{1}\right)$$

$$\frac{\overset{1}{\cancel{8}}}{3} \cdot \frac{-1}{\underset{1}{\cancel{8}}} \quad \left(\text{the reciprocal of } \frac{-8}{1} \text{ is } \frac{-1}{8}\right)$$

$$\frac{-1}{3}$$

Q6   Find the quotient ${}^-9 \div {}^-4\frac{2}{7}$.

STOP • STOP • STOP • STOP • STOP • STOP • STOP • STOP • STOP

A6 $\frac{21}{10}$: ${}^-9 \div {}^-4\frac{2}{7}$

$$\frac{\overset{-3}{\cancel{-9}}}{1} \cdot \frac{-7}{\underset{10}{\cancel{30}}}$$

$$\frac{21}{10}$$

Q7   Find the quotient:

a. $\dfrac{-29}{50} \div 3\dfrac{1}{10}$ b. $^{-}4\dfrac{1}{5} \div {}^{-}3\dfrac{1}{3}$ c. $^{-}4\dfrac{1}{5} \div 3$ d. $\dfrac{1}{2} \div {}^{-}2$

e. $4 \div 4\dfrac{5}{8}$ f. $^{-}8 \div \dfrac{-1}{8}$ g. $0 \div 3\dfrac{4}{7}$ h. $^{-}15 \div {}^{-}2\dfrac{5}{8}$

STOP • STOP • STOP • STOP • STOP • STOP • STOP • STOP • STOP

A7   a. $\dfrac{-29}{155}$ b. $\dfrac{63}{50}$ c. $\dfrac{-7}{5}$ d. $\dfrac{-1}{4}$

e. $\dfrac{32}{37}$ f. 64 g. 0 h. $\dfrac{40}{7}$

---

4 | By the definition of division it is possible to find the quotient $\dfrac{a}{b} \div \dfrac{c}{d}$ only if $\dfrac{c}{d} \neq 0$, that is, only if the divisor is not zero. Consider why this is so.

The quotient of two rational numbers is found by multiplying the dividend by the reciprocal of the divisor. Since zero does not have a reciprocal, it is impossible to complete the quotient.

Q8   What is the reciprocal of 0? _____

STOP • STOP • STOP • STOP • STOP • STOP • STOP • STOP • STOP

A8   There is none!

Q9   a. Is it possible to find $\dfrac{2}{3} \div 0$? _____

b. Why? _____

STOP • STOP • STOP • STOP • STOP • STOP • STOP • STOP • STOP

A9   a. no   b. Zero has no reciprocal.

---

5 | Notice that the impossibility of division by zero in the set of rational numbers is consistent with the impossibility of division by zero in the set of integers. That is, $\dfrac{a}{0}$ or $a \div 0$ is undefined for any integer replacement for a, and $a \div 0$ is undefined for any rational-number replacement for a since division by zero is impossible.

Q10   Find the quotient:

a. $\dfrac{-3}{0} =$ _____   b. $\dfrac{5}{8} \div \dfrac{0}{1} =$ _____

STOP • STOP • STOP • STOP • STOP • STOP • STOP • STOP • STOP

A10   **a.** undefined (division by zero is impossible)
            b. undefined

---

6   Although the divisor cannot be zero, the dividend can be. For example, let $\dfrac{a}{b}$ be any nonzero rational number:

$$0 \div \frac{a}{b}$$

$$\frac{0}{1} \cdot \frac{b}{a}$$

$$\frac{0}{a}$$

$$0$$

That is, zero divided by any nonzero number gives a quotient of zero.

Q11   Complete:

         a. $0 \div \dfrac{2}{3} =$ _____   **b.** $\dfrac{2}{3} \div 0 =$ _____

         c. $^{-}2\dfrac{1}{2} \div 0 =$ _____   **d.** $0 \div 4\dfrac{2}{5} =$ _____

STOP • **STOP** • **STOP** • **STOP** • **STOP** • **STOP** • **STOP** • **STOP** • **STOP**

A11   **a.** 0   **b.** undefined
            c. undefined   **d.** 0

Q12   Is multiplication commutative in the set of rational numbers? _____

STOP • **STOP** • **STOP** • **STOP** • **STOP** • **STOP** • **STOP** • **STOP** • **STOP**

A12   yes: the order of the factors does not affect the product.

---

7   In Chapter 2 it was established that $a \div b = b \div a$ is false for most integer replacements of a and b. Since the set of integers is a subset of the set of rational numbers, $a \div b = b \div a$ is also false for most rational-number replacements of a and b. For example, if $a = \dfrac{2}{3}$ and $b = \dfrac{-5}{6}$,

$$a \div b = \frac{2}{3} \div \frac{-5}{6} \qquad \text{while} \qquad b \div a = \frac{-5}{6} \div \frac{2}{3}$$

$$= \frac{2}{3} \cdot \frac{-6}{5} \qquad\qquad\qquad\qquad = \frac{-5}{6} \cdot \frac{3}{2}$$

$$= \frac{-4}{5} \qquad\qquad\qquad\qquad\qquad = \frac{-5}{4}$$

Thus $\dfrac{2}{3} \div \dfrac{-5}{6} = \dfrac{-5}{6} \div \dfrac{2}{3}$ is a false statement.

Q13   **a.** Evaluate $a \div b$ and $b \div a$ for $a = \dfrac{-5}{6}$ and $b = \dfrac{4}{15}$.

      b. Is $a \div b = b \div a$ a true statement for all rational numbers a and b?_____

      c. Are the rational numbers commutative for the operation of division?_____

STOP • **STOP** • **STOP** • **STOP** • **STOP** • **STOP** • **STOP** • **STOP** • **STOP**

A13   **a.** $a \div b = \dfrac{^{-}5}{6} \div \dfrac{4}{15}$ $b \div a = \dfrac{4}{15} \div \dfrac{^{-}5}{6}$

$$= \frac{^{-}5}{6} \cdot \frac{15}{4} \qquad\qquad = \frac{4}{15} \cdot \frac{^{-}6}{5}$$

$$= \frac{^{-}25}{8} \qquad\qquad\quad = \frac{^{-}8}{25}$$

      b. no   **c.** no

---

8   It was also established in Chapter 2 that $(a \div b) \div c = a \div (b \div c)$ is false for most integer replacements of a, b, and c. Since the set of integers is a subset of the set of rational numbers, $(a \div b) \div c = a \div (b \div c)$ is also false for most rational-number replacements of a, b, and c. Thus, the set of rational numbers is not associative for the operation of division.

Q14   Verify that $(a \div b) \div c \neq a \div (b \div c)$ for $a = \dfrac{1}{2}$, $b = \dfrac{3}{8}$, and $c = \dfrac{4}{5}$.

STOP • **STOP** • **STOP** • **STOP** • **STOP** • **STOP** • **STOP** • **STOP** • **STOP**

A14 $(a \div b) \div c = \left(\dfrac{1}{2} \div \dfrac{3}{8}\right) \div \dfrac{4}{5}$ $a \div (b \div c) = \dfrac{1}{2} \div \left(\dfrac{3}{8} \div \dfrac{4}{5}\right)$

$$= \left(\frac{1}{2} \cdot \frac{8}{3}\right) \div \frac{4}{5} \qquad\qquad = \frac{1}{2} \div \left(\frac{3}{8} \cdot \frac{5}{4}\right)$$

$$= \frac{4}{3} \div \frac{4}{5} \qquad\qquad\qquad = \frac{1}{2} \div \frac{15}{32}$$

$$= \frac{4}{3} \cdot \frac{5}{4} \qquad\qquad\qquad = \frac{1}{2} \cdot \frac{32}{15}$$

$$= \frac{5}{3} \qquad\qquad\qquad\quad = \frac{16}{15}$$

Q15   Is the set of rational numbers associative for the operation of division?_____

STOP • **STOP** • **STOP** • **STOP** • **STOP** • **STOP** • **STOP** • **STOP** • **STOP**

A15   no

This completes the instruction for this section.

3.5   Exercises

1. Find the quotient:

a. $\dfrac{^-4}{15} \div \dfrac{3}{2}$ b. $\dfrac{^-7}{8} \div \dfrac{^-3}{4}$ c. $^-2\dfrac{1}{2} \div 5$ d. $^-2 \div \dfrac{^-6}{11}$

e. $^-3\dfrac{2}{3} \div {}^-3\dfrac{1}{3}$ f. $5 \div \dfrac{^-47}{9}$ g. $0 \div \dfrac{^-1}{2}$ h. $\dfrac{2}{5} \div \dfrac{5}{2}$

i. $^-1 \div \dfrac{3}{17}$ j. $\dfrac{3}{17} \div {}^-1$

2. Decide whether the answer is "zero" or "undefined":

a. $\dfrac{3}{0}$ b. $3 \div 0$ c. $0 \div \dfrac{^-4}{7}$ d. $\dfrac{0}{^-6}$

e. $\dfrac{2}{3} \div 0$ f. $\dfrac{^-5}{11} \cdot \dfrac{0}{1}$

3. True or false:
The set of rational numbers is:

a. commutative for the operation of division.
b. associative for the operation of division.
c. associative for the operation of multiplication.
d. commutative for the operation of multiplication.

4. Insert $\in$ or $\notin$ to form a true statement:
N: set of natural numbers
W: set of whole numbers
I: set of integers
Q: set of rational numbers

a. $7\underline{\quad}N$ b. $7\underline{\quad}W$ c. $7\underline{\quad}I$ d. $7\underline{\quad}Q$

e. $0\underline{\quad}N$ f. $0\underline{\quad}W$ g. $0\underline{\quad}I$ h. $0\underline{\quad}Q$

i. $^-2\underline{\quad}N$ j. $^-2\underline{\quad}W$ k. $^-2\underline{\quad}I$ l. $^-2\underline{\quad}Q$

m. $\dfrac{2}{3}\underline{\quad}N$ n. $\dfrac{2}{3}\underline{\quad}W$ o. $\dfrac{2}{3}\underline{\quad}I$ p. $\dfrac{2}{3}\underline{\quad}Q$

q. $\dfrac{^-2}{3}\underline{\quad}N$ r. $\dfrac{^-2}{3}\underline{\quad}W$ s. $\dfrac{^-2}{3}\underline{\quad}I$ t. $\dfrac{^-2}{3}\underline{\quad}Q$

u. $4\dfrac{1}{2}\underline{\quad}N$ v. $4\dfrac{1}{2}\underline{\quad}W$ w. $4\dfrac{1}{2}\underline{\quad}I$ x. $4\dfrac{1}{2}\underline{\quad}Q$

y. $\dfrac{^-7}{3}\underline{\quad}I$ z. $\dfrac{^-7}{3}\underline{\quad}Q$

3.5   Exercise Answers

1. a. $\dfrac{^-8}{45}$ **b.** $\dfrac{7}{6}$ **c.** $\dfrac{^-1}{2}$ **d.** $\dfrac{11}{3}$ **e.** $\dfrac{11}{10}$ **f.** $\dfrac{^-45}{47}$ **g.** 0

h. $\dfrac{4}{25}$ **i.** $\dfrac{^-17}{3}$ **j.** $\dfrac{^-3}{17}$

2. a. undefined   b. undefined   c. 0   d. 0
    e. undefined   f. 0

3. a. false   b. false   c. true   d. true

4. a. $\in$ b. $\in$ c. $\in$ d. $\in$ e. $\notin$ f. $\in$ g. $\in$
   h. $\in$ i. $\notin$ j. $\notin$ k. $\in$ l. $\in$ m. $\notin$ n. $\notin$
   o. $\notin$ p. $\in$ q. $\notin$ r. $\notin$ s. $\notin$ t. $\in$ u. $\notin$
   v. $\notin$ w. $\notin$ x. $\in$ y. $\notin$ z. $\in$

Chapter 3 Sample Test

At the completion of Chapter 3 it is expected that you will be able to work the following problems.

3.1   Introduction

1. Indicate the rational numbers:

$$0, \quad 2\frac{1}{2}, \quad {}^{-}3, \quad \frac{4}{5}, \quad \frac{0}{7}, \quad \frac{5}{0}, \quad \frac{{}^{-}4}{7}, \quad {}^{-}2\frac{2}{9}$$

2. If N denotes the set of natural numbers, W the set of whole numbers, I the set of integers, and Q the set of rational numbers, answer true or false for each of the following:
    a. N is a subset of Q.   b. Q is a subset of I.
    c. I is a subset of Q.   d. W is a subset of Q.

3. Graph the following rational numbers and their opposites on the number line provided:

    a. $-1\frac{1}{2}$ b. 0   c. 4   d. $3\frac{1}{3}$

4. Find the decimal representation for each of the following rational numbers and state whether each is terminating or infinite repeating:

    a. $\frac{1}{5}$ b. $^{-}2\frac{2}{3}$ c. $\frac{22}{7}$ d. $\frac{^{-}3}{8}$

3.2   Addition

5. Find each of the following sums:

    a. $\frac{^{-}3}{5} + \frac{7}{5}$ b. $\frac{8}{15} + \frac{^{-}4}{3}$ c. $\frac{^{-}2}{3} + \frac{^{-}6}{7}$ d. $1\frac{3}{5} + 4\frac{2}{7}$

    e. $^{-}2\frac{4}{9} + 1$ f. $3 + {}^{-}1\frac{2}{5}$ g. $^{-}4\frac{3}{11} + {}^{-}6\frac{2}{3}$ h. $\frac{3}{10} + \frac{^{-}7}{6} + \frac{^{-}5}{12}$

6. True or false

    a. $\dfrac{4}{15} + \dfrac{-3}{7} = \dfrac{-3}{7} + \dfrac{4}{15}$ **b.** $\left(\dfrac{2}{3} + \dfrac{4}{7}\right) + \dfrac{-5}{8} = \dfrac{2}{3} + \left(\dfrac{4}{7} + \dfrac{-5}{8}\right)$

3.3   Subtraction

7. Find each of the following differences:

    a. $\dfrac{3}{5} - \dfrac{6}{7}$ **b.** $\dfrac{4}{5} - \dfrac{-3}{8}$ **c.** $2\dfrac{1}{3} - 1\dfrac{2}{5}$ **d.** $\dfrac{-1}{4} - 2\dfrac{3}{4}$

    e. $3\dfrac{2}{7} - 5\dfrac{4}{7}$ **f.** $3\dfrac{1}{5} - 1\dfrac{1}{3} - \dfrac{7}{30}$

8. True or false:

    a. $\dfrac{2}{7} - \dfrac{3}{5} = \dfrac{3}{5} - \dfrac{2}{7}$

    b. The set of rational numbers is associative for the operation of subtraction.

3.4   Multiplication

9. Find each of the following products:

    a. $\dfrac{-2}{3} \cdot 1\dfrac{4}{11}$ **b.** $\dfrac{-7}{24} \cdot \dfrac{-6}{21}$ **c.** $-3\dfrac{1}{3} \cdot 4\dfrac{2}{5}$ **d.** $0 \cdot \dfrac{-5}{9}$

    e. $\dfrac{-7}{8} \cdot \dfrac{-24}{25} \cdot \dfrac{-5}{21}$ **f.** $-4\dfrac{2}{3} \cdot \dfrac{2}{3} \cdot \dfrac{-3}{14}$

10. True or false:

    a. The set of rational numbers is commutative for the operation of multiplication.

    b. $\left(\dfrac{2}{5} \cdot \dfrac{4}{5}\right) \cdot \dfrac{-6}{7} = \dfrac{2}{5}\left(\dfrac{4}{5} \cdot \dfrac{-6}{7}\right)$

3.5   Division

11. Find each of the following quotients:

    a. $\dfrac{3}{7} \div \dfrac{-14}{9}$ **b.** $\dfrac{-4}{11} \div \dfrac{-8}{22}$ **c.** $0 \div 2\dfrac{1}{3}$ **d.** $\dfrac{1}{7} \div 0$

    e. $4\dfrac{2}{3} \div 7\dfrac{1}{9}$ **f.** $-1\dfrac{1}{5} \div 6\dfrac{3}{12}$

12. True or false:

    a. $\dfrac{1}{5} \div \dfrac{-4}{7} = \dfrac{-4}{7} \div \dfrac{1}{5}$ **b.** $\left(-3 \div \dfrac{1}{5}\right) \div \dfrac{-4}{17} = -3 \div \left(\dfrac{1}{5} \div \dfrac{-4}{17}\right)$

Chapter 3 Sample Test Answers

1. All except $\dfrac{5}{0}$ are rational numbers.

2. a. true   **b.** false   **c.** true   **d.** true

3.

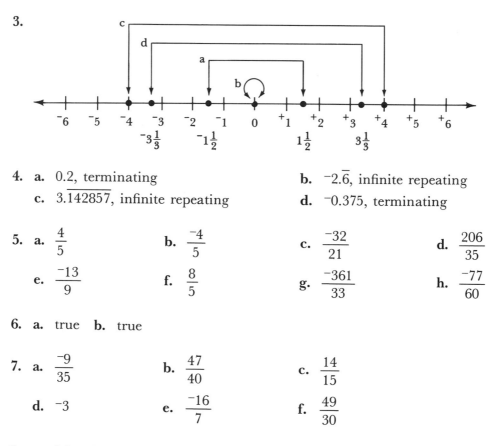

4. a. 0.2, terminating   **b.** $^-2.\overline{6}$, infinite repeating
   c. $3.\overline{142857}$, infinite repeating   **d.** $^-0.375$, terminating

5. a. $\dfrac{4}{5}$ **b.** $\dfrac{^-4}{5}$ **c.** $\dfrac{^-32}{21}$ **d.** $\dfrac{206}{35}$

   e. $\dfrac{^-13}{9}$ **f.** $\dfrac{8}{5}$ **g.** $\dfrac{^-361}{33}$ **h.** $\dfrac{^-77}{60}$

6. a. true   **b.** true

7. a. $\dfrac{^-9}{35}$ **b.** $\dfrac{47}{40}$ **c.** $\dfrac{14}{15}$

   d. $^-3$ **e.** $\dfrac{^-16}{7}$ **f.** $\dfrac{49}{30}$

8. a. false   **b.** false

9. a. $\dfrac{^-10}{11}$ **b.** $\dfrac{1}{12}$ **c.** $\dfrac{^-44}{3}$ **d.** 0

   e. $\dfrac{^-1}{5}$ **f.** $\dfrac{2}{3}$

10. a. true   **b.** true

11. a. $\dfrac{^-27}{98}$ **b.** 1   **c.** 0   **d.** undefined

   e. $\dfrac{21}{32}$ **f.** $\dfrac{^-24}{125}$

12. a. false   **b.** false

Chapter 4

Fundamental Operations with Algebraic Expressions

In Chapter 1 expressions such as $a + 3$, $5y - 2x$, $9(b + 7)$, and $(x - 2)(x + 3)$ were introduced and referred to as "open expressions." Expressions such as these are also correctly called *algebraic expressions*. Since letters (variables) in an algebraic expression can be replaced by rational numbers, all properties known to be valid for the set of rational numbers can be used when simplifying algebraic expressions.* The purpose of this chapter is to develop skill in the use of these properties for the simplification of algebraic expressions.

4.1 Simplifying Algebraic Expressions by Use of the Commutative and Associative Properties of Addition

---

1 The building blocks or components of algebraic expressions are called terms. When an algebraic expression shows only additions, the *terms* are the parts separated by plus signs. Subtraction signs appearing in an algebraic expression can be converted to addition signs using the definition of subtraction.

Examples:	*Terms*
$8y + 3$	$8y$ and 3
$\frac{1}{2}x - 4 + 2y = \frac{1}{2}x + {}^-4 + 2y$	$\frac{1}{2}x$, ${}^-4$, and $2y$
$x - 3y - 6 = x + {}^-3y + {}^-6$	x, ${}^-3y$, and ${}^-6$

Q1 Identify the terms in the algebraic expression $4a + 7ab - 12$. _____

STOP • STOP • STOP • STOP • STOP • STOP • STOP • STOP • STOP

A1 $4a$, $7ab$, and ${}^-12$

---

2 The *like terms* of an algebraic expression are terms that have exactly the same literal co-efficients (letter factors). The like terms of the algebraic expression $7x - 3y - \frac{2}{3}x + 6$ are $\frac{-2}{3}x$ and $7x$, because each has the literal coefficient x.

In the expression $\frac{4}{5}y - 3xy + \frac{2}{3}x - 2y + 8xy$ there are two sets of like terms. They

---

*Unless otherwise stated, the replacement set for all variables will be the set of rational numbers.

121

are $\frac{4}{5}y$ and ^-2y, with the letter factor y, and ^-3xy and $8xy$, with the letter factors x and y.

Numbers without a literal factor are called *constants* and are considered to be like terms. That is, 5, $^-7$, $\frac{7}{8}$, 3, and so on, are constants and like terms.

Q2   Identify the like terms in each of the following algebraic expressions:

a. $4x - 2y + 3 - 6x$ _____

b. $6 - 2y + \frac{5}{9}$ _____

c. $5r - 6s + 2rs - \frac{1}{2}s$ _____

d. $6a - 2b + 3ab$ _____

STOP • **STOP** • **STOP** • **STOP** • **STOP** • **STOP** • **STOP** • **STOP** • **STOP**

A2   a. $4x$, ^-6x b. 6, $\frac{5}{9}$ c. ^-6s, $\frac{^-1}{2}s$

d. There are no like terms in part d.

---

3   Algebraic expressions are simplified by combining (by addition or subtraction) like terms. The justification for combining like terms is the distributive property of multiplication over addition or the distributive property of multiplication over subtraction. For example, $5x - 7x$ is simplified:

Justification

$5x - 7x$
$(5 - 7)x$ right distributive property of multiplication over subtraction
^-2x number fact: $5 - 7 = ^-2$

Therefore, $5x - 7x$ simplifies to ^-2x.

Q3   Use a distributive property to simplify the following expressions:

a. $4a - 7a$ b. $\frac{2}{3}x + \frac{1}{4}x$

STOP • **STOP** • **STOP** • **STOP** • **STOP** • **STOP** • **STOP** • **STOP** • **STOP**

A3   a. ^-3a: $4a - 7a$ b. $\frac{11}{12}x$: $\frac{2}{3}x + \frac{1}{4}x$

$(4 - 7)a$ $\left(\frac{2}{3} + \frac{1}{4}\right)x$

^-3a $\frac{11}{12}x$

---

4   Consider the following simplification:

Justification

$4x - 3x$

$(4 - 3)x$ right distributive property of multiplication over subtraction

$1x$ number fact: $4 - 3 = 1$

x multiplication property of one

Therefore, $4x - 3x = x$.

Q4   Simplify $9y - 8y$.

STOP • STOP • STOP • STOP • STOP • STOP • STOP • STOP • STOP

A4 y: $9y - 8y$

$(9 - 8)y$

$1y$

y

Q5   Simplify $7b - 6b$.

STOP • STOP • STOP • STOP • STOP • STOP • STOP • STOP • STOP

A5 b

---

5   By the multiplication property of one, x also equals $1x$. This idea is used to simplify expressions such as $4x + x$ or $3b - b$.

Example 1: Simplify $4x + x$.

Solution

Justification

$4x + x$

$4x + 1x$ multiplication property of one

$(4 + 1)x$ right distributive property of multiplication over addition

$5x$ number fact: $4 + 1 = 5$

Therefore, $4x + x = 5x$.

Example 2: Simplify $3b - b$.

Solution

$3b - b$

$3b - 1b$ multiplication property of one

$(3 - 1)b$ right distributive property of multiplication over subtraction

$2b$ number fact: $3 - 1 = 2$

Therefore, $3b - b = 2b$.

Q6   Simplify:
    a. $7y + y$ **b.** $x + 2x$

    c. $b - 5b$ **d.** $a - \dfrac{2}{3}a$

STOP • **STOP** • **STOP** • **STOP** • **STOP** • **STOP** • **STOP** • **STOP** • **STOP**

A6   **a.** $8y$: $7y + 7$ **b.** $3x$: $x + 2x$
                         $7y + 1y$ $1x + 2x$
                         $(7 + 1)y$ $(1 + 2)x$
                         $8y$ $3x$

    c. ^-4b: $b - 5b$ **d.** $\dfrac{1}{3}a$: $a - \dfrac{2}{3}a$

                         $1b - 5b$ $1a - \dfrac{2}{3}a$

                         $(1 - 5)b$ $\left(1 - \dfrac{2}{3}\right)a$

                         ^-4b $\dfrac{1}{3}a$

---

6   By the multiplication property of negative one, $^-1x = {}^-x$. Consider its use in the simplification of $x - 2x$.

Justification

$x - 2x$
$1x - 2x$ multiplication property of one
$(1 - 2)x$ right distributive property of multiplication over subtraction
^-1x number fact: $1 - 2 = {}^-1$
^-x multiplication property of negative one

Therefore, $x - 2x = {}^-x$.

Q7   Simplify:
    a. $7x - 8x$ **b.** $12y - 13y$

STOP • **STOP** • **STOP** • **STOP** • **STOP** • **STOP** • **STOP** • **STOP** • **STOP**

A7   **a.** ^-x: $7x - 8x$ **b.** ^-y
                        $(7 - 8)x$
                        ^-1x
                        ^-x

7   Notice that using a distributive property to simplify algebraic expressions is merely the process of combining the numerical coefficients of the like terms. Thus, the simplification process can be shortened as follows:

$$^-5b + 3b = ^-2b \quad \text{because} \quad ^-5 + 3 = ^-2$$

$$\frac{^-1}{2}z - \frac{2}{3}z = \frac{^-7}{6}z \quad \text{because} \quad \frac{^-1}{2} - \frac{2}{3} = \frac{^-7}{6}$$

In algebra, numbers are usually written in $\dfrac{p}{q}$ form rather than as mixed numbers. Thus, in the preceding example, the answer is written as $\dfrac{^-7}{6}z$ rather than as $^-1\dfrac{1}{6}z$.

Q8   Simplify the following algebraic expressions by combining the numerical coefficients of the like terms:

a. $7y - 12y =$ _____   **b.** $4b - b =$ _____

c. $\dfrac{^-2}{3}x + \dfrac{4}{5}x =$ _____   **d.** $y - y =$ _____

STOP • STOP • STOP • STOP • STOP • STOP • STOP • STOP • STOP

A8   **a.** ^-5y: because $7 - 12 = {}^-5$ **b.** $3b$:   because $4 - 1 = 3$

   c. $\dfrac{2}{15}x$:   because $\dfrac{^-2}{3} + \dfrac{4}{5} = \dfrac{2}{15}$ **d.** 0:   because $1 - 1 = 0$

8   To simplify expressions involving more than two terms, a similar procedure is followed. Study the following examples:

$$3y - 12y + 7 \qquad\qquad 8x - 3x + 9 - 12$$
$$^-9y + 7 \qquad\qquad\quad ^-11x - 3$$

Q9   Simplify each of the following expressions:
   a. $^-3m - 7m + 9$ **b.** $11 - 1 + 7x - 3x$

   c. $5y + 7y - 6 + 3$ **d.** $4 - 10 + 8y - \dfrac{3}{2}y$

STOP • STOP • STOP • STOP • STOP • STOP • STOP • STOP • STOP

A9   **a.** $^-10m + 9$ **b.** $10 + 4x$ **c.** $12y - 3$ **d.** $^-6 + \dfrac{13}{2}y$

9   The associative property of addition is used to simplify $(3x - 8) + 4$ as follows:

Justification

$(3x - 8) + 4$

$(3x + {}^-8) + 4$ definition of subtraction

$3x + ({}^-8 + 4)$ associative property of addition

$3x + {}^-4$ number fact

$3x - 4$ definition of subtraction

Recall that by the definition of subtraction,

$a - b = a + {}^-b$

Or, reversing the right and left sides,

$a + {}^-b = a - b$

The form $a - b$ is considered "simpler" than $a + {}^-b$. Therefore, when simplifying algebraic expressions, $5x - 3$ is written rather than $5x + {}^-3$.

Q10   Simplify $(2x - 5) + 3$.

STOP • STOP • STOP • STOP • STOP • STOP • STOP • STOP • STOP

A10 $2x - 2$: $(2x - 5) + 3$

$(2x + {}^-5) + 3$

$2x + ({}^-5 + 3)$

$2x + {}^-2$

$2x - 2$

10   The expression $(3 + 7y) - 16y$ is simplified as follows:

Justification

$(3 + 7y) - 16y$

$(3 + 7y) + {}^-16y$ definition of subtraction

$3 + (7y + {}^-16y)$ associative property of addition

$3 + {}^-9y$ like terms combined

$3 - 9y$ definition of subtraction

Q11   Simplify $(7 - 4x) + 3x$.

STOP • STOP • STOP • STOP • STOP • STOP • STOP • STOP • STOP

A11 $7 - x$: $(7 - 4x) + 3x$
                        $(7 + {}^-4x) + 3x$
                        $7 + ({}^-4x + 3x)$
                        $7 + {}^-x$
                        $7 - x$

Q12   Simplify each of the following algebraic expressions using the associative property of addition where necessary:

             a. $(x + 7) + 4$ **b.** $(3 + y) + 8y$

             c. $(2b - 4b) + 6$ **d.** $4z + (3z - 5)$

             e. $^-5y + (2y + 3)$ **f.** $^-5 + (2 + 4x)$

             g. $\dfrac{4}{5} + \left(\dfrac{2}{3} - x\right)$ **h.** $\left(3x - \dfrac{2}{5}\right) + \dfrac{2}{5}$

             i. $(^-5 - 2x) + 2x$ **j.** $(^-7x + 3) - 3$

STOP • STOP • STOP • STOP • STOP • STOP • STOP • STOP • STOP

A12   **a.** $x + 11$ **b.** $3 + 9y$ **c.** $^-2b + 6$ **d.** $7z - 5$

         e. $^-3y + 3$ **f.** $^-3 + 4x$ **g.** $\dfrac{22}{15} - x$ **h.** $3x$

         i. $^-5$ **j.** ^-7x

---

11   The commutative property of addition is used whenever there is a need to change the order of terms in addition as an aid in the simplification of algebraic expressions. For example, $(3x + 7) - 2x$ is simplified as follows:

Justification

$(3x + 7) - 2x$
$(7 + 3x) - 2x$ commutative property of addition
$(7 + 3x) + {}^{-}2x$ definition of subtraction
$7 + (3x + {}^{-}2x)$ associative property of addition
$7 + x$ like terms combined

Q13   Use the commutative property of addition to change the order of the terms within the parentheses.

$(4y + 9) - 3y = $ _____

STOP • **STOP** • **STOP** • **STOP** • **STOP** • **STOP** • **STOP** • **STOP** • **STOP**

A13 $(9 + 4y) - 3y$

Q14   Use the associative property of addition to complete the simplification of A13.

STOP • **STOP** • **STOP** • **STOP** • **STOP** • **STOP** • **STOP** • **STOP** • **STOP**

A14 $(9 + 4y) - 3y$
$(9 + 4y) + {}^{-}3y$
$9 + (4y + {}^{-}3y)$
$9 + y$

---

12   The expression ${}^{-}5y + (3 - 2y)$ is simplified as follows:

Justification

${}^{-}5y + (3 - 2y)$
${}^{-}5y + (3 + {}^{-}2y)$ definition of subtraction
${}^{-}5y + ({}^{-}2y + 3)$ commutative property of addition
$({}^{-}5y + {}^{-}2y) + 3$ associative property of addition
${}^{-}7y + 3$ like terms combined

Thus, ${}^{-}5y + (3 - 2y)$ simplifies to ${}^{-}7y + 3$.

Q15   Simplify $6t + (7 - 9t)$.

STOP • **STOP** • **STOP** • **STOP** • **STOP** • **STOP** • **STOP** • **STOP** • **STOP**

A15 $^-3t + 7$: $6t + (7 - 9t)$
                          $6t + (7 + {}^-9t)$
                          $6t + ({}^-9t + 7)$
                          $(6t + {}^-9t) + 7$
                          $^-3t + 7$

Q16   Simplify each of the following algebraic expressions:

    a. $4 + (3t - 9)$ **b.** $\left(\frac{2}{3}z - 5\right) - \frac{4}{7}z$

    c. $(3 + 5y) - 5y$ **d.** $4m + (2m - 7m)$

STOP • **STOP** • **STOP** • **STOP** • **STOP** • **STOP** • **STOP** • **STOP** • **STOP**

A16   **a.** $^-5 + 3t$ **b.** $^-5 + \dfrac{2}{21}z$ **c.** 3 **d.** ^-m

---

13	The expression $5x - 7$ may also be written $^-7 + 5x$. This fact is verified below:

Justification

$5x - 7$
$5x + {}^-7$ definition of subtraction
$^-7 + 5x$ commutative property of addition

Q17   Verify that $3x - 4 = {}^-4 + 3x$ by completing the following:

Justification

      $3x - 4$

    a. _____   definition of subtraction

    b. _____   commutative property of addition

STOP • **STOP** • **STOP** • **STOP** • **STOP** • **STOP** • **STOP** • **STOP** • **STOP**

A17   **a.** $3x + {}^-4$ **b.** $^-4 + 3x$

14	The expression $^-8 - 7x$ may also be written $^-7x - 8$. The verification is as follows:

Justification

$^-7x - 8$
$^-7x + {}^-8$ definition of subtraction
$^-8 + {}^-7x$ commutative property of addition
$^-8 - 7x$ definition of subtraction

Q18   Verify that $^-2x - 9 = ^-9 - 2x$ by completing the following:

Justification

$^-2x - 9$

a. _____ definition of subtraction

b. _____ commutative property of addition

c. _____ definition of subtraction

STOP • STOP • STOP • STOP • STOP • STOP • STOP • STOP • STOP

A18   a. $^-2x + ^-9$ b. $^-9 + ^-2x$ c. $^-9 - 2x$

---

15   Frame 13 showed that $5x - 7$ could be written $^-7 + 5x$. Frame 14 showed that $^-8 - 7x = ^-7x - 8$. These frames demonstrate the following useful procedure: *In an algebraic expression, the terms may be rearranged in any order as long as the original sign of each term is left unchanged.* Some examples are:

$$5 - 2x = ^-2x + 5$$

$$3x - \frac{4}{5} = \frac{^-4}{5} + 3x$$

$$\frac{^-3}{4}x - 5 = ^-5 - \frac{3}{4}x$$

$$^-2x + 3 + 5x - 7 = ^-2x + 5x + 3 - 7$$

$$7y - 4 - 3y - 5 + 2y + 6 = 7y - 3y + 2y - 4 - 5 + 6$$

Notice that in the fourth and fifth examples the terms are rearranged so that the like terms are together.

Q19   Write equivalent expressions for each of the following:

a. $3x + 2 =$ _____   b. $^-5x - 4 =$ _____

c. $\frac{^-5}{8}x + 12 =$ _____   d. $^-4x + 3 - 2x =$ _____

e. $^-2x - 7 - 5x + 6 =$ _____

STOP • STOP • STOP • STOP • STOP • STOP • STOP • STOP • STOP

A19   a. $2 + 3x$ b. $^-4 - 5x$ c. $12 - \frac{5}{8}x$

d. $^-4x - 2x + 3$ e. $^-2x - 5x - 7 + 6$

---

16   The expression $5x - 2 - 3x - 9$ may be simplified by rearranging and combining the like terms as follows:

$$5x - 2 - 3x - 9$$
$$5x - 3x - 2 - 9$$
$$2x - 11$$

Q20   Simplify $7 - 3t + 9 - 6t$ by rearranging and combining the like terms.

STOP • STOP • STOP • STOP • STOP • STOP • STOP • STOP • STOP

A20 $^-9t + 16$ or $16 - 9t$: $7 - 3t + 9 - 6t$
$^-3t - 6t + 7 + 9$
$^-9t + 16$

Q21   Simplify $4x - 3 - 5x + 6 - x$.

STOP • **STOP** • **STOP** • **STOP** • **STOP** • **STOP** • **STOP** • **STOP** • **STOP**

A21 $^-2x + 3$ or $3 - 2x$: $4x - 3 - 5x + 6 - x$
$4x - 5x - x - 3 + 6$
$^-2x + 3$

Q22   Simplify the following algebraic expressions:
a. $^-3 + 5x + 7$ b. $5y - 6 + 2y$

c. $^-3x - 4 - x$ d. $x + 3x + 7 - 9$

e. $^-2z + 3 - 5z - 1$ f. $6t + 3 - 7t - 6 + t$

STOP • **STOP** • **STOP** • **STOP** • **STOP** • **STOP** • **STOP** • **STOP** • **STOP**

A22   a. $5x + 4$ or $4 + 5x$ b. $7y - 6$ or $^-6 + 7y$
c. $^-4x - 4$ or $^-4 - 4x$ d. $4x - 2$ or $^-2 + 4x$
e. $^-7z + 2$ or $2 - 7z$ f. $^-3$

This completes the instruction for this section.

4.1   Exercises

1. Simplify the following algebraic expressions by combining like terms:
a. $^-3x + 7x$ b. $2x - 5 - x$
c. $5y - 7 - 6y$ d. $3 - 6 + 5z$
e. $4b - 3b + 2$ f. $^-6a + 3 + 6a$
g. $\dfrac{2}{3}x + \dfrac{7}{8}x$ h. $5y - \dfrac{3}{5}y + 7$

2. Simplify using the commutative and/or associative properties of addition:
   a. $(4y + 7) + 3y$
   b. $6 + (2x - 9)$
   c. $b + \left(\dfrac{1}{3}b + 2\right)$
   d. $\left(x - \dfrac{5}{8}x\right) + 3x$
   e. $(2x - 3) + 7x$
   f. $(5 - 4x) + 9$

3. Simplify:
   a. $3x + {}^-7$
   b. $5 - {}^-4y$
   c. $3 + {}^-7x + 5$
   d. ${}^-2y + 6 + \dfrac{-2}{3}y$

4. Simplify:
   a. $3x - 4y + 7x - 6$
   b. $\dfrac{2}{5}z - \dfrac{3}{4}z + 8$
   c. $\dfrac{5}{2}b - 7a + \dfrac{3}{5}$
   d. $6 - 4r + 6s - 2r$
   e. ${}^-x + 6 - 3x + {}^-5$
   f. $7 - 2y + y$
   g. ${}^-4m + 6 + 4m$
   h. $\dfrac{-2}{3}x + x - \dfrac{1}{3}x + 2$
   i. $4xy - 6y + 7 + 2y$
   j. $3m + 4n - 6m + n - 6$

4.1 Exercise Answers

1. a. $4x$
   b. $x - 5$
   c. ${}^-y - 7$
   d. $5z - 3$
   e. $b + 2$
   f. 3
   g. $\dfrac{37}{24}x$
   h. $\dfrac{22}{5}y + 7$

2. a. $7y + 7$
   b. $2x - 3$
   c. $\dfrac{4}{3}b + 2$
   d. $\dfrac{27}{8}x$
   e. $9x - 3$
   f. ${}^-4x + 14$ or $14 - 4x$

3. a. $3x - 7$
   b. $4y + 5$
   c. ${}^-7x + 8$ or $8 - 7x$
   d. $\dfrac{-8}{3}y + 6$ or $6 - \dfrac{8}{3}y$

4. a. $10x - 4y - 6$
   b. $\dfrac{-7}{20}z + 8$ or $8 - \dfrac{7}{20}z$
   c. $\dfrac{5}{2}b - 7a + \dfrac{3}{5}$ cannot be simplified because there are no like terms.
   d. $6s - 6r + 6$
   e. ${}^-4x + 1$ or $1 - 4x$
   f. $7 - y$
   g. 6
   h. 2
   i. $4xy - 4y + 7$
   j. $5n - 3m - 6$

4.2 Simplifying Algebraic Expressions by Use of the Commutative and Associative Properties of Multiplication

---

1 The associative and commutative properties of multiplication are used whenever there is a need to change the grouping or order of multiplication as an aid in the simplification of algebraic expressions. The associative property of multiplication is used to simplify $3(4x)$ as follows:

Justification

$3(4x)$

$(3 \cdot 4)x$ associative property of multiplication

$12x$ number fact

Thus, $3(4x) = 12x$.

Q1 Use the associative property of multiplication to simplify $^-5(7y)$.

STOP • STOP • STOP • STOP • STOP • STOP • STOP • STOP • STOP

A1 ^-35y: $^-5(7y)$

$(^-5 \cdot 7)y$

^-35y

---

2 The expression $7\left(\dfrac{^-3}{7}t\right)$ is simplified:

Justification

$7\left(\dfrac{^-3}{7}t\right)$

$\left(7 \cdot \dfrac{^-3}{7}\right)t$ associative property of multiplication

^-3t number fact: $\dfrac{7}{1} \cdot \dfrac{^-3}{7} = {}^-3$

Thus, $7\left(\dfrac{^-3}{7}t\right) = {}^-3t$.

Q2 Simplify $^-9\left(\dfrac{^-1}{9}x\right)$.

STOP • STOP • STOP • STOP • STOP • STOP • STOP • STOP • STOP

A2 x: $-9\left(\dfrac{-1}{9}x\right)$

$\left(-9 \cdot \dfrac{-1}{9}\right)x$

$1x$

x

---

3   The expression $\dfrac{3}{5}\left(\dfrac{-4}{9}y\right)$ is simplified:

Justification

$\dfrac{3}{5}\left(\dfrac{-4}{9}y\right)$

$\left(\dfrac{3}{5} \cdot \dfrac{-4}{9}\right)y$ associative property of multiplication

$\dfrac{-4}{15}y$ number fact

Thus, $\dfrac{3}{5}\left(\dfrac{-4}{9}y\right) = \dfrac{-4}{15}y$.

Q3   Simplify $\dfrac{-3}{4}\left(\dfrac{-2}{7}z\right)$.

STOP • **STOP** • **STOP** • **STOP** • **STOP** • **STOP** • **STOP** • **STOP** • **STOP**

A3 $\dfrac{3}{14}z$: $\dfrac{-3}{4}\left(\dfrac{-2}{7}z\right)$

$\left(\dfrac{-3}{4} \cdot \dfrac{-2}{7}\right)z$

$\dfrac{3}{14}z$

Q4   Simplify each of the following algebraic expressions:

a. $4\left(\dfrac{3}{4}x\right)$ b. $\dfrac{2}{3}\left(\dfrac{3}{2}y\right)$

c. $-5\left(\dfrac{-2}{15}z\right)$ d. $\dfrac{-6}{7}\left(\dfrac{5}{12}t\right)$

e. $\dfrac{-5}{9}\left(\dfrac{-9}{5}m\right)$ **f.** $\dfrac{1}{7}(7x)$

STOP • STOP • STOP • STOP • STOP • STOP • STOP • STOP • STOP

A4   **a.** $3x$ **b.** y **c.** $\dfrac{2}{3}z$ **d.** $\dfrac{-5}{14}t$ **e.** m **f.** x

---

4   Both the commutative and associative properties of multiplication are used to simplify $\left(\dfrac{3}{4}x\right)\dfrac{2}{3}$ as follows:

Justification

$\left(\dfrac{3}{4}x\right)\dfrac{2}{3}$

$\dfrac{2}{3}\left(\dfrac{3}{4}x\right)$ commutative property of multiplication

$\left(\dfrac{2}{3}\cdot\dfrac{3}{4}\right)x$ associative property of multiplication

$\dfrac{1}{2}x$ number fact

Q5   Use the commutative and associative properties of multiplication to simplify $\left(\dfrac{2}{5}x\right)5$.

STOP • STOP • STOP • STOP • STOP • STOP • STOP • STOP • STOP

A5 $2x$: $\left(\dfrac{2}{5}x\right)5$

          $5\left(\dfrac{2}{5}x\right)$

          $\left(5\cdot\dfrac{2}{5}\right)x$

          $2x$

---

5   The expression $\left(\dfrac{5}{8}y\right)\cdot\dfrac{-2}{3}$ is simplified as follows:

Justification

$\left(\dfrac{5}{8}y\right)\cdot\dfrac{-2}{3}$

$\dfrac{-2}{3}\left(\dfrac{5}{8}y\right)$ commutative property of multiplication

$$\left(\frac{-2}{3} \cdot \frac{5}{8}\right)y \qquad \text{associative property of multiplication}$$

$$\frac{-5}{12}y \qquad \text{number fact}$$

Thus, $\left(\frac{5}{8}y\right) \cdot \frac{-2}{3} = \frac{-5}{12}y.$

Q6   Simplify $\left(\frac{-3}{5}x\right) \cdot \frac{-5}{3}$ by use of the commutative and associative properties.

STOP • STOP • STOP • STOP • STOP • STOP • STOP • STOP • STOP

A6 x: $\left(\frac{-3}{5}x\right) \cdot \frac{-5}{3}$

$$\frac{-5}{3}\left(\frac{-3}{5}x\right)$$

$$\left(\frac{-5}{3} \cdot \frac{-3}{5}\right)x$$

$$1x$$

$$x$$

6   The expression $\left(\frac{8}{13}z\right)13$ is simplified:

Justification

$$\left(\frac{8}{13}z\right)13$$

$$13\left(\frac{8}{13}z\right) \qquad \text{commutative property of multiplication}$$

$$\left(13 \cdot \frac{8}{13}\right)z \qquad \text{associative property of multiplication}$$

$$8z \qquad \text{number fact}$$

Thus, $\left(\frac{8}{13}z\right)13 = 8z.$

Q7   Simplify the expression $\left(\frac{-4}{5}t\right)5.$

STOP • STOP • STOP • STOP • STOP • STOP • STOP • STOP • STOP

A7 $^-4t:$ $\left(\dfrac{^-4}{5}t\right)5$

$5\left(\dfrac{^-4}{5}t\right)$

$\left(5\cdot\dfrac{^-4}{5}\right)t$

^-4t

Q8   Simplify each of the following algebraic expressions:

a. $\left(\dfrac{^-3}{4}x\right)\cdot{}^-4$ b. $\left(\dfrac{5}{7}y\right)\dfrac{7}{5}$ c. $(5z)\dfrac{1}{5}$

d. $\left(\dfrac{^-2}{3}t\right)\cdot\dfrac{^-9}{12}$ e. $\left(\dfrac{^-1}{7}x\right)\cdot{}^-7$ f. $\left(\dfrac{3}{5}y\right)\dfrac{4}{7}$

STOP • STOP • STOP • STOP • STOP • STOP • STOP • STOP • STOP

A8   a. $3x$ b. y c. z d. $\dfrac{1}{2}t$ e. x f. $\dfrac{12}{35}y$

This completes the instruction for this section.

4.2   Exercises

1. Simplify each of the following algebraic expressions:

a. $2\left(\dfrac{3}{4}x\right)$ b. $\dfrac{^-1}{2}(8y)$ c. $\dfrac{^-4}{5}\left(\dfrac{^-5}{4}z\right)$ d. $\dfrac{^-1}{5}(^-5c)$

e. $15\left(\dfrac{4}{5}x\right)$ f. $\dfrac{3}{4}\left(\dfrac{^-2}{5}y\right)$

2. Simplify:

a. $(2x)\dfrac{5}{8}$ b. $\left(\dfrac{^-2}{3}x\right)\cdot\dfrac{^-3}{2}$ c. $(^-7y)\cdot\dfrac{^-1}{7}$ d. $\left(\dfrac{^-2}{5}z\right)10$

e. $\left(\dfrac{^-7}{4}z\right)12$ f. $\left(\dfrac{^-3}{25}x\right)25$ g. $\left(\dfrac{^-2}{3}y\right)\dfrac{5}{7}$ h. $\left(\dfrac{^-1}{9}z\right)9$

4.2   Exercise Answers

1. a. $\dfrac{3}{2}x$ b. ^-4y c. z d. c e. $12x$ f. $\dfrac{^-3}{10}y$

2. a. $\dfrac{5}{4}x$ b. x c. y d. ^-4z e. ^-21z f. ^-3x

g. $\dfrac{^-10}{21}y$ h. ^-z

4.3   Using the Distributive Properties to Simplify Algebraic Expressions

---

1   The left distributive property of multiplication over addition and the left distributive property of multiplication over subtraction state that

$$a(b + c) = ab + ac$$

and

$$a(b - c) = ab - ac$$

for all rational-number replacements of a, b, and c. Two examples are:

$$3(7 + 5) = 3 \cdot 7 + 3 \cdot 5$$
$$2(1 - 9) = 2 \cdot 1 - 2 \cdot 9$$

Q1   Use the left distributive property of multiplication over addition to fill in the blanks.

$$5(2 + 9) = \underline{\hspace{2cm}} + \underline{\hspace{2cm}}$$

STOP • STOP • STOP • STOP • STOP • STOP • STOP • STOP • STOP

A1 $5(2 + 9) = 5 \cdot 2 + 5 \cdot 9$

Q2   Use the left distributive property of multiplication over subtraction to fill in the blanks.

$$4(8 - 5) = \underline{\hspace{2cm}} - \underline{\hspace{2cm}}$$

STOP • STOP • STOP • STOP • STOP • STOP • STOP • STOP • STOP

A2 $4(8 - 5) = 4 \cdot 8 - 4 \cdot 5$

---

2   When parentheses are removed from an algebraic expression, the same procedure is used. For example, to remove the parentheses from $3(x + 2)$, proceed as follows:

$$3(x + 2) = 3 \cdot x + 3 \cdot 2$$
$$= 3x + 6$$

Q3   Use the left distributive property of multiplication over addition to remove the parentheses from $7(y + 3)$.

STOP • STOP • STOP • STOP • STOP • STOP • STOP • STOP • STOP

A3 $7y + 21$: $7(y + 3) = 7 \cdot y + 7 \cdot 3$
$$= 7y + 21$$

Q4   Remove the parentheses:
a. $5(x + 1)$ b. $8(2 + c)$

STOP • STOP • STOP • STOP • STOP • STOP • STOP • STOP • STOP

A4   a. $5x + 5$: $5(x + 1) = 5 \cdot x + 5 \cdot 1$ b. $16 + 8c$: $8(2 + c) = 8 \cdot 2 + 8 \cdot c$
$$= 5x + 5 \qquad\qquad\qquad\qquad\qquad = 16 + 8c$$

---

3   The parentheses may be removed from $4(t - 3)$ by use of the left distributive property of multiplication over subtraction as follows:

$$4(t - 3) = 4 \cdot t - 4 \cdot 3$$
$$= 4t - 12$$

Q5   Use the left distributive property of multiplication over subtraction to remove the parentheses from $7(a - 5)$.

STOP • **STOP** • **STOP** • **STOP** • **STOP** • **STOP** • **STOP** • **STOP** • **STOP**

A5 $7a - 35$: $7(a - 5)$
          $7 \cdot a - 7 \cdot 5$
          $7a - 35$

Q6   Remove the parentheses:
a. $2(x - 9)$ **b.** $6(2 - y)$

STOP • **STOP** • **STOP** • **STOP** • **STOP** • **STOP** • **STOP** • **STOP** • **STOP**

A6   **a.** $2x - 18$ **b.** $12 - 6y$

---

4   To remove the parentheses from $(a + 3)$, recall that by the multiplication property of one:

$$(a + 3) = 1 \cdot (a + 3)$$

Thus,

$$(a + 3) = 1 \cdot (a + 3)$$
$$= 1 \cdot a + 1 \cdot 3$$
$$= a + 3$$

Notice that the result is exactly the expression within the parentheses. For example,

$$(b + 2) = b + 2$$
$$(7 - y) = 7 - y$$
$$(2x - 3) = 2x - 3$$

Q7   Remove the parentheses:

a. $(x - 9) = $ _____   **b.** $(3 - x) = $ _____
c. $(5x + 6) = $ _____

STOP • **STOP** • **STOP** • **STOP** • **STOP** • **STOP** • **STOP** • **STOP** • **STOP**

A7   **a.** $x - 9$ **b.** $3 - x$ **c.** $5x + 6$

Q8   Remove the parentheses from each of the following algebraic expressions:
a. $2(x + 7)$ **b.** $5(b - 1)$ **c.** $3(4 + y)$

d. $(x - 1)$ **e.** $8(7 - z)$ **f.** $4(m - 2)$

g. $9(3 + t)$ **h.** $(4 - a)$

STOP • **STOP** • **STOP** • **STOP** • **STOP** • **STOP** • **STOP** • **STOP** • **STOP**

A8   **a.** $2x + 14$ **b.** $5b - 5$ **c.** $12 + 3y$ **d.** $x - 1$
      e. $56 - 8z$ **f.** $4m - 8$ **g.** $27 + 9t$ **h.** $4 - a$

5	The right distributive property of multiplication over addition and the right distributive property of multiplication over subtraction state that

$$(a + b)c = ac + bc$$

and

$$(a - b)c = ac - bc$$

for all rational-number replacements of a, b, and c. Two examples are:

$$(3 + 5)2 = 3 \cdot 2 + 5 \cdot 2$$

$$(7 - 4)\frac{3}{8} = 7 \cdot \frac{3}{8} - 4 \cdot \frac{3}{8}$$

Q9   Use the right distributive property of multiplication over addition to fill in the blanks.

$$(4 + 9)7 = \underline{\hspace{1.5cm}} + \underline{\hspace{1.5cm}}$$

STOP • **STOP** • **STOP** • **STOP** • **STOP** • **STOP** • **STOP** • **STOP** • **STOP**

A9 $(4 + 9)7 = 4 \cdot 7 + 9 \cdot 7$

Q10   Use the right distributive property of multiplication over subtraction to fill in the blanks.

$$\left(\frac{5}{6} - 1\right)6 = \underline{\hspace{1.5cm}} - \underline{\hspace{1.5cm}}$$

STOP • **STOP** • **STOP** • **STOP** • **STOP** • **STOP** • **STOP** • **STOP** • **STOP**

A10 $\left(\dfrac{5}{6} - 1\right)6 = \dfrac{5}{6} \cdot 6 - 1 \cdot 6$

6	The right distributive properties are also used to simplify algebraic expressions. Two examples are:

$$(x - 2)3 = x \cdot 3 - 2 \cdot 3$$
$$ = 3x - 6 \quad \text{(it is customary to rewrite } x \cdot 3 \text{ as } 3x)$$

$$(5 + r)3 = 5 \cdot 3 + r \cdot 3$$
$$ = 15 + 3r$$

Q11   Use the right distributive property of multiplication over addition to remove the parentheses from $(2 + x)7$.

STOP • **STOP** • **STOP** • **STOP** • **STOP** • **STOP** • **STOP** • **STOP** • **STOP**

A11 $14 + 7x$: $(2 + x)7 = 2 \cdot 7 + x \cdot 7$
$$= 14 + 7x$$

Q12   Use the right distributive property of multiplication over subtraction to remove the parentheses from $(y - 4)5$.

STOP • STOP • STOP • STOP • STOP • STOP • STOP • STOP • STOP

A12 $5y - 20$: $(y - 4)5 = y \cdot 5 - 4 \cdot 5$
$$= 5y - 20$$

Q13   Remove the parentheses from $(b - 7)9$.

STOP • STOP • STOP • STOP • STOP • STOP • STOP • STOP • STOP

A13 $9b - 63$

---

7   To remove the parentheses in the algebraic expression $3(4x - 7)$, the following steps are used:

$$3(4x - 7) = 3 \cdot 4x - 3 \cdot 7$$
$$= 12x - 21$$

Q14   Remove the parentheses from $7(6x - 4)$.

STOP • STOP • STOP • STOP • STOP • STOP • STOP • STOP • STOP

A14 $42x - 28$: $7(6x - 4) = 7 \cdot 6x - 7 \cdot 4$
$$= 42x - 28$$

Q15   Remove the parentheses from $4\left(12 - \dfrac{3}{4}x\right)$.

STOP • STOP • STOP • STOP • STOP • STOP • STOP • STOP • STOP

A15 $48 - 3x$

---

8   To remove the parentheses from $(3x - 2)5$, the procedure is as follows:

$$(3x - 2)5 = 3x \cdot 5 - 2 \cdot 5$$
$$= 5 \cdot 3x - 2 \cdot 5$$
$$= 15x - 10$$

Q16   Remove the parentheses from $(7x - 4)9$.

STOP • STOP • STOP • STOP • STOP • STOP • STOP • STOP • STOP

A16 $63x - 36$: $(7x - 4)9 = 7x \cdot 9 - 4 \cdot 9$
$$= 9 \cdot 7x - 4 \cdot 9$$
$$= 63x - 36$$

Q17   Remove the parentheses from $(5a + 7)2$.

STOP • STOP • STOP   STOP • STOP • STOP • STOP • STOP • STOP

A17 $10a + 14$

9	To remove the parentheses from $4(^-x + 7)$, recall that ^-x means ^-1x (multiplication property of negative one). Thus,

$$4(^-x + 7) = 4(^-1x + 7)$$
$$= 4 \cdot {}^-1x + 4 \cdot 7$$
$$= {}^-4x + 28 \quad \text{or} \quad 28 - 4x$$

Q18   Remove the parentheses from $3(^-x - 5)$.

STOP • STOP • STOP • STOP • STOP • STOP • STOP • STOP • STOP

A18 $^-3x - 15$: $3(^-x - 5) = 3(^-1x - 5)$
$$= 3 \cdot {}^-1x - 3 \cdot 5$$
$$= {}^-3x - 15$$

10	When removing parentheses by use of the distributive properties, it is convenient to be able to find the result mentally (without showing work). For example, to remove parentheses from $5(3x - 6)$, write only $5(3x - 6) = 15x - 30$.

Q19   Mentally remove the parentheses from $2(8 - 5t)$.

STOP • STOP • STOP • STOP • STOP • STOP • STOP • STOP • STOP

A19 $16 - 10t$

Q20   Mentally remove the parentheses from $3\left(7 - \dfrac{4}{3}y\right)$.

STOP • STOP • STOP • STOP • STOP • STOP • STOP • STOP • STOP

A20 $21 - 4y$

Q21   Remove the parentheses from $(6z + 1)9$.

STOP • STOP • STOP • STOP • STOP • STOP • STOP • STOP • STOP

A21 $54z + 9$

---

11   To remove the parentheses from $^-5(3x - 4)$, the following steps can be used:

$$^-5(3x - 4) = {}^-5 \cdot 3x - {}^-5 \cdot 4$$
$$= {}^-15x - {}^-20$$
$$= {}^-15x + 20$$

You should notice that $^-15x - {}^-20$ is equivalent to $^-15x + 20$ because of the definition of subtraction. The expression $^-15x + 20$ is considered to be in simplest form. The expression $20 - 15x$ may also be written.

Q22   Remove the parentheses from $^-4(7x - 9)$ and write in simplest form.

STOP • STOP • STOP • STOP • STOP • STOP • STOP • STOP • STOP

A22 $^-28x + 36$ or $36 - 28x$: $^-4(7x - 9) = {}^-4 \cdot 7x - {}^-4 \cdot 9$
$$= {}^-28x - {}^-36$$
$$= {}^-28x + 36$$

Q23   Remove the parentheses from $^-1(7 + 3y)$ and write in simplest form.

STOP • STOP • STOP • STOP • STOP • STOP • STOP • STOP • STOP

A23 $^-7 - 3y$: $^-1(7 + 3y) = {}^-1 \cdot 7 + {}^-1 \cdot 3y$
$$= {}^-7 + {}^-3y$$
$$= {}^-7 - 3y$$

Q24   Remove the parentheses from $^-5(^-3x + 4)$ and write in simplest form.

STOP • STOP • STOP • STOP • STOP • STOP • STOP • STOP • STOP

A24 $15x - 20$: $^-5(^-3x + 4) = {}^-5 \cdot {}^-3x + {}^-5 \cdot 4$
$$= 15x + {}^-20$$
$$= 15x - 20$$

---

12   The expression $(^-4x - 7) \cdot {}^-8$ is simplified:

$$(^-4x - 7) \cdot {}^-8 = {}^-4x \cdot {}^-8 - 7 \cdot {}^-8$$
$$= 32x - {}^-56$$
$$= 32x + 56$$

Q25   Simplify $(3y + 9) \cdot {}^-6$.

STOP • STOP • STOP • STOP • STOP • STOP • STOP • STOP • STOP

A25 $^-18y - 54$: $(3y + 9) \cdot {}^-6 = 3y \cdot {}^-6 + 9 \cdot {}^-6$
$$= {}^-18y + {}^-54$$
$$= {}^-18y - 54$$

Q26   Simplify $({}^-a - 5) \cdot {}^-2$.

STOP • **STOP** • **STOP** • **STOP** • **STOP** • **STOP** • **STOP** • **STOP** • **STOP**

A26 $2a + 10$: $({}^-a - 5) \cdot {}^-2 = {}^-a \cdot {}^-2 - 5 \cdot {}^-2$
$$= 2a - {}^-10$$
$$= 2a + 10$$

---

13   By the multiplication property of negative one, the expression $^-(3x + 7)$ is equivalent to $^-1(3x + 7)$. Thus $^-(3x + 7)$ is simplified:

$^-(3x + 7) = {}^-1(3x + 7)$
$$= {}^-1 \cdot 3x + {}^-1 \cdot 7$$
$$= {}^-3x + {}^-7$$
$$= {}^-3x - 7$$

An alternative method to the above is to notice that $^-(3x + 7)$ means the *opposite of* $(3x + 7)$, which can be found by taking the opposite of each term within the parentheses. Thus,

$^-(3x + 7) = {}^-3x + {}^-7$
$$= {}^-3x - 7$$

Q27   Write an equivalent expression for $^-(5 + 2y)$ by forming the opposite of each term within the parentheses.

STOP • **STOP** • **STOP** • **STOP** • **STOP** • **STOP** • **STOP** • **STOP** • **STOP**

A27 $^-5 - 2y$: $^-(5 + 2y) = {}^-5 + {}^-2y$

Q28   Simplify $^-1(5 + 2y)$ by removing the parentheses.

STOP • **STOP** • **STOP** • **STOP** • **STOP** • **STOP** • **STOP** • **STOP** • **STOP**

A28 $^-5 - 2y$: $^-1(5 + 2y) = {}^-1 \cdot 5 + {}^-1 \cdot 2y$
$$= {}^-5 + {}^-2y$$
$$= {}^-5 - 2y$$

Q29   Simplify $^-(4b - 5)$.

STOP • **STOP** • **STOP** • **STOP** • **STOP** • **STOP** • **STOP** • **STOP** • **STOP**

A29
$$\begin{aligned}
^-4b + 5 \text{ or } 5 - 4b: \quad ^-(4b - 5) &= {}^-1(4b - 5)\\
&= {}^-1 \cdot 4b - {}^-1 \cdot 5\\
&= {}^-4b - {}^-5\\
&= {}^-4b + 5 \text{ or } 5 - 4b
\end{aligned}$$

or
$$\begin{aligned}
^-(4b - 5) &= {}^-4b - {}^-5\\
&= {}^-4b + 5 \text{ or } 5 - 4b
\end{aligned}$$

Q30   Simplify $^-(x - 7)$.

STOP • **STOP** • **STOP** • **STOP** • **STOP** • **STOP** • **STOP** • **STOP** • **STOP**

A30 $^-x + 7$ or $7 - x$

This completes the instruction for this section.

4.3   Exercises

Simplify each of the following algebraic expressions:

1. $2(x + 3)$

2. $15\left(\dfrac{2}{3}y - \dfrac{4}{5}\right)$

3. $(4u - 3)$

4. $^-1(2x + 7)$

5. $(3t + 1)2$

6. $(7 + 3b) \cdot {}^-2$

7. $(4 - 9x) \cdot {}^-3$

8. $^-(9 + {}^-4c)$

9. $^-2(7t - 4)$

10. $4\left(^-3y + \dfrac{7}{4}\right)$

11. $^-5(8x - 3)$

12. $^-3(^-5 - 7z)$

13. $^-(t - 1)$

14. $^-(3 - 5r)$

15. $^-3(^-x - 7)$

4.3   Exercise Answers

1. $2x + 6$
2. $10y - 12$
3. $4a - 3$
4. $^-2x - 7$
5. $6t + 2$
6. $^-14 - 6b$
7. $^-12 + 27x$ or $27x - 12$
8. $^-9 + 4c$ or $4c - 9$
9. $^-14t + 8$ or $8 - 14t$
10. $^-12y + 7$ or $7 - 12y$
11. $^-40x + 15$ or $15 - 40x$
12. $15 + 21z$
13. $^-t + 1$ or $1 - t$
14. $^-3 + 5r$ or $5r - 3$
15. $3x + 21$

4.4 Simplifying Algebraic Expressions by Use of the Associative, Commutative, and Distributive Properties

---

1

An algebraic expression is said to be in "simplest" form when:

1. All parentheses have been removed.
2. All like terms have been combined.
3. The definition of subtraction has been applied to remove all the "raised" negative signs that can be removed.

Examples: *Simplest form?*

Example	Simplest form?
$3x + 5 - 2x$	no, like terms not combined
$7x - 3 + 2y$	yes
$4(x - 2) + 3x$	no, parentheses not removed
$3 + {}^-5y$	no, raised negative sign can be removed by writing $3 - 5y$
$9t - {}^-4$	no, raised negative sign can be removed by writing $9t + 4$
${}^-5x + 2$	yes

Q1 Indicate whether each of the following expressions is in simplest form. If the expression is not in simplest form, briefly state why.

   a. $4x - 2y + 9$ _____   _____

   b. $3(2 - 4y) - 1$ _____   _____

   c. $5t - 6 + 9t$ _____   _____

   d. $8x - 5 + {}^-y$ _____   _____

   e. $5x + 7y - 3xy$ _____   _____

STOP • STOP • STOP • STOP • STOP • STOP • STOP • STOP • STOP

A1
   a. yes
   b. no, parentheses not removed
   c. no, like terms not combined
   d. no, raised negative sign can be removed by writing $8x - 5 - y$
   e. yes

2

In Section 4.1 algebraic expressions such as ${}^-2x + 3 + 5x - 7$ were simplified by rearranging and combining like terms as follows:

$${}^-2x + 3 + 5x - 7$$
$${}^-2x + 5x + 3 - 7$$
$$3x - 4$$

Q2 Simplify $5 - 6t + 3 - 9t$ by rearranging and combining like terms.

STOP • STOP • STOP • STOP • STOP • STOP • STOP • STOP • STOP

A2 ${}^-15t + 8$ or $8 - 15t$: $5 - 6t + 3 - 9t$
                                 ${}^-6t - 9t + 5 + 3$
                                   ${}^-15t + 8$ or $8 - 15t$

Q3   Simplify $^-3x + 7x - 5 + x$.

STOP   •   STOP   •   STOP   •   STOP   •   STOP   •   STOP   •   STOP   •   STOP   •   STOP

A3 $5x - 5$: $^-3x + 7x - 5 + x$
             $^-3x + 7x + x - 5$
             $5x - 5$

Q4   Simplify each of the following:

    a. $4y - 7 + 6y - 2$ **b.** $^-1 - \dfrac{2}{3}t + 7 - t$

    c. $10b + 6 - 3b$ **d.** $\dfrac{4}{5}z - \dfrac{1}{2}z + 9$

STOP   •   STOP   •   STOP   •   STOP   •   STOP   •   STOP   •   STOP   •   STOP   •   STOP

A4   **a.** $10y - 9$ **b.** $\dfrac{^-5}{3}t + 6$

    c. $7b + 6$ **d.** $\dfrac{3}{10}z + 9$

---

3   To simplify expressions that involve parentheses:

    1. Use the distributive properties to remove parentheses.
    2. Rearrange and combine the like terms.

    Examples:

    1. $2(3y - 7) + 4$ **2.** $^-2x + \dfrac{4}{9}(x - 4)$
        $6y - 14 + 4$
        $6y - 10$ $^-2x + \dfrac{4}{9}x - \dfrac{16}{9}$

                                   $\dfrac{^-14}{9}x - \dfrac{16}{9}$

Q5   Simplify $5(2x - 1) + 7$.

STOP   •   STOP   •   STOP   •   STOP   •   STOP   •   STOP   •   STOP   •   STOP   •   STOP

A5 $10x + 2$: $5(2x - 1) + 7$
                              $10x - 5 + 7$
                              $10x + 2$

Q6   Simplify $3x + 7(^-x + 4)$.

STOP • **STOP** • **STOP** • **STOP** • **STOP** • **STOP** • **STOP** • **STOP** • **STOP**

A6 $^-4x + 28$ or $28 - 4x$: $3x + 7(^-x + 4)$
                                        $3x + {}^-7x + 7 \cdot 4$
                                        $^-4x + 28$ or $28 - 4x$

Q7   Simplify $\left(^-3y + \dfrac{1}{8}\right) - \dfrac{2}{5} + 2y$.

STOP • **STOP** • **STOP** • **STOP** • **STOP** • **STOP** • **STOP** • **STOP** • **STOP**

A7 $^-y - \dfrac{11}{40}$: $\left(^-3y + \dfrac{1}{8}\right) - \dfrac{2}{5} + 2y$

                            $^-3y + \dfrac{1}{8} - \dfrac{2}{5} + 2y$

                            $^-y - \dfrac{11}{40}$

Q8   Simplify each of the following:

         a. $(7x + 8) - 3$ **b.** $3(2t - 4) + 7$

         c. $8b + 2(b - 3)$ **d.** $(^-5 + 2x) - 3x$

         e. $(^-3x + 4) - 9$ **f.** $^-7(4 - x) + 3x$

         g. $\dfrac{5}{12}(z - 4) + 8$ **h.** $^-1(3 + y) + \dfrac{1}{2}y$

i. $4(^-3x + 2) + 12x$ **j.** $^-4y + 2(y + 8)$

k. $4a + ^-3(a + 2)$ **l.** $\dfrac{^-2}{7}(y + 7) - 8$

m. $x + ^-1(3x - 5)$ **n.** $(5y + 4) - 4$

o. $^-2(4t \quad 8) - 5t$ **p.** $\dfrac{^-1}{3}(3x + 7) - \dfrac{5}{6}x$

STOP • STOP • STOP • STOP • STOP • STOP • STOP • STOP • STOP

A8   **a.** $7x + 5$ **b.** $6t - 5$ **c.** $10b - 6$
        d. $^-x - 5$ **e.** $^-3x - 5$ **f.** $10x - 28$
        g. $\dfrac{5}{12}z + \dfrac{19}{3}$ **h.** $\dfrac{^-1}{2}y - 3$ **i.** 8
        j. $^-2y + 16$ or $16 - 2y$ **k.** $a - 6$ **l.** $\dfrac{^-2}{7}y - 10$
        m. $^-2x + 5$ or $5 - 2x$ **n.** $5y$
        o. $^-13t + 16$ or $16 - 13t$ **p.** $\dfrac{^-11}{6}x - \dfrac{7}{3}$

4   Recall that the expression $^-(x + 3)$ could be simplified in either of two ways:

$$^-(x + 3) = ^-1(x + 3) \qquad \text{or} \qquad ^-(x + 3) = ^-x + ^-3$$
$$= ^-1 \cdot x + ^-1 \cdot 3 \qquad\qquad\qquad = ^-x - 3$$
$$= ^-x + ^-3$$
$$= ^-x - 3$$

$[^-(x + 3)$ means the opposite of $(x + 3)$, which is the same as the opposite of each term within the parentheses]

Q9   Simplify:

a. $^-(y + 9) =$ _____   **b.** $^-(2x + 7) =$ _____

c. $^-\left(4z - \dfrac{4}{3}\right) =$ _____   **d.** $^-(b - 12) =$ _____

e. $^-(5y + 7) =$ _____

STOP • STOP • STOP • STOP • STOP • STOP • STOP • STOP • STOP

A9   **a.** $^-y - 9$ **b.** $^-2x - 7$

        c. $^-4z + \dfrac{4}{3}$ or $\dfrac{4}{3} - 4z$ **d.** $^-b + 12$ or $12 - b$

        e. $^-5y - 7$

---

5   To simplify $4 - (x + 7)$ the following steps are used:

$$4 - (x + 7) = 4 + {}^-(x + 7)$$
$$= 4 + {}^-x + {}^-7$$
$$= {}^-x + {}^-3$$
$$= {}^-x - 3$$

Q10   Simplify $15 - (2x + 7)$.

STOP • STOP • STOP • STOP • STOP • STOP • STOP • STOP • STOP

A10 $^-2x + 8$ or $8 - 2x$: $15 - (2x + 7) = 15 + {}^-(2x + 7)$
$$= 15 + {}^-2x + {}^-7$$
$$= {}^-2x + 8 \text{ or } 8 - 2x$$

Q11   Simplify $^-4 - (2b - 3)$.

STOP • STOP • STOP • STOP • STOP • STOP • STOP • STOP • STOP

A11 $^-2b - 1$: $^-4 - (2b - 3) = {}^-4 + {}^-(2b - 3)$
$$= {}^-4 + {}^-2b - {}^-3$$
$$= {}^-4 - 2b + 3$$
$$= {}^-2b - 1$$

Q12   Simplify each of the following:

        a. $4 - (y + 3)$ **b.** $2z - (z - 5)$

        c. $^-5 - \left(\dfrac{2}{3} + 4x \right)$ **d.** $\dfrac{3}{2} + \left(2a - \dfrac{4}{5} \right)$

        e. $9 - (2a - 4)$ **f.** $6y - (2y + 7)$

STOP • STOP • STOP • STOP • STOP • STOP • STOP • STOP • STOP

A12   **a.** $^-y + 1$ or $1 - y$ **b.** $z + 5$ **c.** $^-4x - \dfrac{17}{3}$

d. $2a + \dfrac{7}{10}$ **e.** $^-2a + 13$ or $13 - 2a$ **f.** $4y - 7$

---

6   To simplify $(2x + 7) - (3x - 4)$, remove the parentheses and combine like terms as follows:

$$(2x + 7) - (3x - 4) = (2x + 7) + ^-(3x - 4)$$
$$= (2x + 7) + ^-3x - ^-4$$
$$= 2x + 7 + ^-3x + 4$$
$$= ^-x + 11$$

Notice that parentheses preceded by no sign (or a "$+$" sign) are removed by just dropping them, whereas parentheses preceded by a "$-$" sign are removed by rewriting the subtraction problem as an addition problem and taking the opposite of the terms within the parentheses.

Q13   Remove the parentheses only (do not simplify): $(x - 5) - (2x + 7)$

STOP   •   STOP   •   STOP   •   STOP   •   STOP   •   STOP   •   STOP   •   STOP   •   STOP

A13 $x - 5 - 2x - 7$: $(x - 5) - (2x + 7) = (x - 5) + ^-(2x + 7)$
$$= (x - 5) + ^-2x + ^-7$$
$$= x - 5 - 2x - 7$$

Q14   Simplify the result in A13.

STOP   •   STOP   •   STOP   •   STOP   •   STOP   •   STOP   •   STOP   •   STOP   •   STOP

A14 $^-x - 12$

Q15   Remove the parentheses only (do not simplify): $(1 - 3b) - (4b - 7)$

STOP   •   STOP   •   STOP   •   STOP   •   STOP   •   STOP   •   STOP   •   STOP   •   STOP

A15 $1 - 3b - 4b + 7$

Q16   Simplify the result in A15.

STOP   •   STOP   •   STOP   •   STOP   •   STOP   •   STOP   •   STOP   •   STOP   •   STOP

A16 $^-7b + 8$ or $8 - 7b$

Q17   Simplify $(x + 3) - (x - 3)$.

STOP   •   STOP   •   STOP   •   STOP   •   STOP   •   STOP   •   STOP   •   STOP   •   STOP

A17   6

Q18   Simplify $(^-z - 4) - (2z - 4)$.

STOP • STOP • STOP • STOP • STOP • STOP • STOP • STOP • STOP

A18 ^-3z

Q19   Simplify each of the following expressions:

           a. $(5 - 4x) - (x + 3)$ **b.** $(2x - 5) - (3 - 5x)$

           c. $(^-5 + b) - (b - 5)$ **d.** $^-(4y + 3) - (7y + 3)$

           e. $(z - 9) - (^-z - 5)$ **f.** $^-(x + 5) - (x - 5)$

STOP • STOP • STOP • STOP • STOP • STOP • STOP • STOP • STOP

A19   **a.** $^-5x + 2$ or $2 - 5x$ **b.** $7x - 8$ **c.** 0
                d. $^-11y - 6$ **e.** $2z - 4$ **f.** ^-2x

---

7   To simplify expressions such as $3(x - 2) - 4(x + 3)$, the following steps are used:

$$
\begin{aligned}
3(x - 2) - 4(x + 3) &= 3(x - 2) + {}^-4(x + 3) \\
&= 3(x - 2) + {}^-4x + {}^-12 \\
&= 3x - 6 - 4x - 12 \\
&= {}^-x - 18
\end{aligned}
$$

Note: The work is usually shortened to:

$$
\begin{aligned}
3(x - 2) - 4(x + 3) &= 3x - 6 - 4x - 12 \\
&= {}^-x - 18
\end{aligned}
$$

Q20   Remove the parentheses only (do not simplify): $5(y + 7) - 2(y - 4)$

STOP • STOP • STOP • STOP • STOP • STOP • STOP • STOP • STOP

A20 $5y + 35 - 2y + 8$

Q21   Complete the simplification of A20.

STOP • STOP • STOP • STOP • STOP • STOP • STOP • STOP • STOP

A21 $3y + 43$

Q22 Remove the parentheses only (do not simplify): $2(2x - 3) - 4(x + 5)$

STOP • **STOP** • **STOP** • **STOP** • **STOP** • **STOP** • **STOP** • **STOP** • **STOP**

A22 $4x - 6 - 4x - 20$

Q23 Complete the simplification of A22.

STOP • **STOP** • **STOP** • **STOP** • **STOP** • **STOP** • **STOP** • **STOP** • **STOP**

A23 $^-26$

Q24 Remove the parentheses only (do not simplify): $^-3(^-2z + 7) - 4(z + 5)$

STOP • **STOP** • **STOP** • **STOP** • **STOP** • **STOP** • **STOP** • **STOP** • **STOP**

A24 $6z - 21 - 4z - 20$

Q25 Complete the simplification of A24.

STOP • **STOP** • **STOP** • **STOP** • **STOP** • **STOP** • **STOP** • **STOP** • **STOP**

A25 $2z - 41$

Q26 Simplify each of the following algebraic expressions:
a. $(3x - 4) - 2(2x + 5)$ **b.** $^-2(b + 7) - (3b - 5)$

c. $2(y + 3) + 3(^-2y - 7)$ **d.** $(y + 7) - 2(y + 5)$

e. $^-(5x - 3) - \dfrac{4}{5}(3x - 1)$ **f.** $^-4(t - 3) - 4(2t - 1)$

STOP • **STOP** • **STOP** • **STOP** • **STOP** • **STOP** • **STOP** • **STOP** • **STOP**

A26
     a. $^-x - 14$ **b.** $^-5b - 9$
     c. $^-4y - 15$ **d.** $^-y - 3$
     e. $\dfrac{-37}{5}x + \dfrac{19}{5}$ or $\dfrac{19}{5} - \dfrac{37}{5}x$ **f.** $^-12t + 16$ or $16 - 12t$

This completes the instruction for this section.

4.4   Exercises

1. An algebraic expression is said to be in simplest form when what three conditions have been met?
2. Simplify each of the following by rearranging and combining like terms:
     a. $^-3x - 5 - x$ **b.** $4 - 2x + 7 + 3x$
     c. $5y - 2y - 6 + 11$ **d.** $^-7 - 5z + 6 - 3z$
     e. $1 - t - 1 - t$ **f.** $3t - 4t + 6$
     g. $7 - 6b + 3 + \dfrac{15}{4}b$ **h.** $^-5 - 4 + 7z - 3z$
     i. $\dfrac{3}{8}y - \dfrac{2}{3} - \dfrac{4}{7}y$ **j.** $4z - 5 - 4z + 5$

3. Remove the parentheses only (do not simplify):
     a. $(x + 8)$ **b.** $^-(x + 8)$
     c. $^-(y - 4)$ **d.** $(^-3x - 5)$
     e. $3(^-4y + 7)$ **f.** $^-(z + 3) + (2z - 4)$
     g. $(5b - 3) - (4 - 6b)$ **h.** $^-\left(\dfrac{1}{2}x + 3\right) - (4 + x)$
     i. $2(a + 7) - 3(4 - 3a)$ **j.** $^-4\left(\dfrac{1}{4}x - 5\right) + 7(3x - 4)$

4. Simplify each of the following algebraic expressions:
     a. $^-3(x + 7)$ **b.** $(4 + 3y) + (^-5y - 2)$
     c. $^-(a - 5) + (3a - 5)$ **d.** $2(3x - 5) - (^-7x + 5)$
     e. $\dfrac{-3}{4}(y - 4) - 3(^-y - 7)$ **f.** $9(^-3 + z) - 3(^-3z - 9)$
     g. $^-3x - (4 + 7x) - 6$ **h.** $5y - 2(2 - 7y) - 3y$
     i. $4 - 7b - \dfrac{3}{4}(4 + b)$ **j.** $^-3(x + 5) + 5(^-4x - 7)$

4.4   Exercise Answers

1.   1. All parentheses are removed.
     2. All like terms are combined.
     3. The definition of subtraction has been applied to remove "raised" negative signs.

2.   **a.** $^-4x - 5$ **b.** $x + 11$ **c.** $3y + 5$
     d. $^-8z - 1$ **e.** ^-2t **f.** $^-t + 6$ or $6 - t$
     g. $\dfrac{-9}{4}b + 10$ or $10 - \dfrac{9}{4}b$ **h.** $4z - 9$ **i.** $\dfrac{-11}{56}y - \dfrac{2}{3}$
     j. 0

3. **a.** $x + 8$ **b.** $^-x - 8$ **c.** $^-y + 4$
   d. $^-3x - 5$ **e.** $^-12y + 21$ **f.** $^-z - 3 + 2z - 4$

   g. $5b - 3 - 4 + 6b$ **h.** $\dfrac{^-1}{2}x - 3 - 4 - x$ **i.** $2a + 14 - 12 + 9a$

   j. $^-x + 20 + 21x - 28$

4. **a.** $^-3x - 21$ **b.** $^-2y + 2$ **c.** $2a$

   d. $13x - 15$ **e.** $\dfrac{9}{4}y + 24$ **f.** $18z$

   g. $^-10x - 10$ **h.** $16y - 4$ **i.** $\dfrac{^-31}{4}b + 1$

   j. $^-23x - 50$

Chapter 4 Sample Test

At the completion of Chapter 4 it is expected that you will be able to work the following problems.

4.1   Simplifying Algebraic Expressions by Use of the Commutative and Associative Properties of Addition

1. Identify the like terms in each of the following algebraic expressions:
   a. $3x - 5y - 6x + 4$ **b.** $3r - 2 + 7s - 6$

2. Simplify by combining like terms:
   a. $4x - 2x$ **b.** $m - 3 - 7m$

   c. $7 - 4y - 6 - 3y$ **d.** $\dfrac{3}{5}x - \dfrac{4}{7}x - 2$

3. Simplify by use of the commutative and/or associative properties of addition:
   a. $(3x - 2) + 5x$ **b.** $^-15 + (7 - 8a)$

   c. $^-x + \left(7x - \dfrac{2}{5}x\right)$ **d.** $\left(\dfrac{3}{5}b - \dfrac{2}{3}b\right) + b$

4. Simplify:
   a. $4y + {}^-9$ **b.** $5 - 2r + 2s - 6r$

   c. $\dfrac{1}{3}x + y - 6xy - 4y$ **d.** $4m + 7 - 3m + {}^-7 - m$

4.2   Simplifying Algebraic Expressions by Use of the Commutative and Associative Properties of Multiplication

5. Simplify:
   a. $2\left(\dfrac{3}{4}x\right)$ **b.** $\dfrac{^-1}{5}\left(\dfrac{^-4}{9}y\right)$

   c. $\dfrac{^-1}{7}(7m)$ **d.** $\dfrac{5}{7}\left(\dfrac{7}{5}x\right)$

e. $\left(\dfrac{-1}{9}z\right) \cdot {}^-9$

f. $\left(\dfrac{-3}{5}x\right)\dfrac{2}{3}$

g. $(4y)\dfrac{5}{4}$

h. $\left(\dfrac{-3}{5}x\right)20$

4.3   Using the Distributive Properties to Simplify Algebraic Expressions

6. Simplify:
   a. $2(x - 3)$
   b. $(y + 7) \cdot {}^-3$
   c. ${}^-4(2x - 3)$
   d. $(4 - 2x)5$
   e. $(7x - 9)$
   f. ${}^-(7x - 9)$
   g. ${}^-({}^-t + 2)$
   h. ${}^-5({}^-7 - 3x)$
   i. $(6x - 5) \cdot {}^-8$
   j. ${}^-7(3y + {}^-9)$

4.4   Simplifying Algebraic Expressions by Use of the Associative, Commutative, and Distributive Properties

7. Simplify:

   a. ${}^-4y + 9 + 2y + 5$

   b. $\dfrac{4}{5}x - \dfrac{2}{3}x + 2 - 9$

   c. $(7x + 8) - 8$

   d. ${}^-2(x + 7) - 3$

   e. $\dfrac{-4}{5}(y + 10) + 7$

   f. $4 - (z + 7)$

   g. $3(x + 2) + 7(x - 5)$

   h. $(y - 6) - (4y + 7)$

   i. ${}^-7(x + 2) - 2(3x - 7)$

   j. ${}^-\left(\dfrac{1}{3}x + 5\right) - (5 - x)$

Chapter 4 Sample Test Answers

1. a. $3x$ and ${}^-6x$
   b. ${}^-2$ and ${}^-6$

2. a. $2x$
   b. ${}^-6m - 3$
   c. ${}^-7y + 1$ or $1 - 7y$
   d. $\dfrac{1}{35}x - 2$

3. a. $8x - 2$
   b. ${}^-8a - 8$
   c. $\dfrac{28}{5}x$
   d. $\dfrac{14}{15}b$

4. a. $4y - 9$
   b. ${}^-8r + 2s + 5$
   c. $\dfrac{1}{3}x - 3y - 6xy$
   d. 0

5. a. $\dfrac{3}{2}x$ b. $\dfrac{4}{45}y$ c. ${}^-m$ d. x e. z f. $\dfrac{-2}{5}x$ g. $5y$
   h. ${}^-12x$

6. **a.** $2x - 6$ **b.** $^-3y - 21$
   c. $^-8x + 12$ or $12 - 8x$ **d.** $20 - 10x$
   e. $7x - 9$ **f.** $^-7x + 9$ or $9 - 7x$
   g. $t - 2$ **h.** $15x + 35$
   i. $^-48x + 40$ or $40 - 48x$ **j.** $^-21y + 63$ or $63 - 21y$

7. **a.** $^-2y + 14$ or $14 - 2y$ **b.** $\dfrac{2}{15}x - 7$ **c.** $7x$

   d. $^-2x - 17$ **e.** $\dfrac{^-4}{5}y - 1$ **f.** $^-z - 3$

   g. $10x - 29$ **h.** $^-3y - 13$ **i.** ^-13x

   j. $\dfrac{2}{3}x - 10$

Chapter 5

Solving Equations

Chapter 4 dealt with the procedures involved in simplifying open algebraic expressions. These included the use of the fundamental operations (addition, subtraction, multiplication, and division) in combining like terms and the application of the commutative, associative, and distributive properties. The skills developed in Chapter 4 will now be utilized in the study of equations.

5.1   Equations, Open Sentences, Replacement Set, and Solution Set

---

1   An equation is a statement that the expressions on opposite sides of an equal sign represent the same number. Some examples of equations are:

$3 + 4 = 9 - 2$
$15 = y + 4$
$x - 2 = 6$
$2x - 3 = 5x + 9$

The expressions on opposite sides of the equal sign are referred to as the left and right *sides* of the equation. For example,

$$\underbrace{4x - 7}_{\text{left side}} = \underbrace{2 - 6x}_{\text{right side}}$$

Q1   Identify the left and right sides of the following equations:

a. $14 - 9 = 5$ left side _____ ; right side _____

b. $3 = y + 1$ left side _____ ; right side _____

c. $3x - 7 = 8x + 13$ left side _____ ; right side _____

STOP • STOP • STOP • STOP • STOP • STOP • STOP • STOP • STOP

A1   **a.** left side, $14 - 9$; right side, 5
b. left side, 3; right side, $y + 1$
c. left side, $3x - 7$; right side, $8x + 13$

---

2   Equations that do not contain enough information to be judged as either true or false are often referred to as *open sentences*. Thus,

$18 - y = 14$ and $x + 6 = 9$

are examples of open sentences, because they cannot be judged true or false until numbers are replaced for the unknown quantities y and x. If y is replaced by 10 in the open sentence $18 - y = 14$, the resulting statement $18 - 10 = 14$ is false. If x is replaced by 3 in the open sentence $x + 6 = 9$, the resulting statement $3 + 6 = 9$ is true.

Q2   **a.** If x is replaced by 7 in the open sentence $x - 2 = 5$, is the resulting statement true or false?   _____

b. If y is replaced by 4 in the open sentence $12 + y = 15$, is the resulting statement true or false?   _____

STOP • STOP • STOP • STOP • STOP • STOP • STOP • STOP • STOP

A2   **a.** true: $7 - 2 = 5$
b. false: $12 + 4 \neq 15$ ($\neq$ means "is not equal to")

3   The set of all numbers that may be used to replace the variable in an open sentence is called the *replacement set* of the variable. If the replacement set is not stated, it will be understood to be the set of rational numbers.

Consider the open sentence $x - 5 = 15$. If the replacement set for the variable x is the set $\{17, 18, 19, 20, 21\}$, the result of all possible replacements for x is four *false* statements,

$17 - 5 = 15$
$18 - 5 = 15$
$19 - 5 = 15$
$21 - 5 = 15$

and one *true* statement,

$20 - 5 = 15$

Q3   If the replacement set for the variable x is $\{0, 6, {}^{-}3, 7\}$, find the result of all possible replacements for x in the open sentence $x + 11 = 17$. Label each statement true or false.

STOP • STOP • STOP • STOP • STOP • STOP • STOP • STOP • STOP

A3 $0 + 11 = 17$ false ${}^{-}3 + 11 = 17$ false
$6 + 11 = 17$ true $7 + 11 = 17$ false

4   To solve an equation is to find all values of the variable from the replacement set which convert the open sentence into a true statement. The values are called the *solutions* of the equation. Thus, the solution of the equation $x + 11 = 17$ is $x = 6$, because 6 is the only rational number that will convert $x + 11 = 17$ into a true statement, namely, $6 + 11 = 17$.

The *solution set* or *truth set* of an equation is the set of all values from the replacement set which convert the open sentence into a true statement. For example, the solution set for the equation $x + 11 = 17$ is $\{6\}$.

Q4   Use the given set as the replacement set to determine the solution (truth) set for each equation:

a. $\{4, 9, 13, 15\}$: $y + 4 = 17$ _____

b. $\{0, {}^{-}1, 2, {}^{-}3, 4, {}^{-}6, 7\}$: ${}^{-}3x = {}^{-}21$ _____

    c. $\{0, 5, {}^-4, {}^-7\}$: $9 + x = 9$ _____

    d. $\{6, {}^-3, 5, 1, 0\}$: $4x + 3 = 23$ _____

STOP • STOP • STOP • STOP • STOP • STOP • STOP • STOP • STOP

A4   **a.** $\{13\}$: $13 + 4 = 17$ **b.** $\{7\}$: ${}^-3(7) = {}^-21$
             c. $\{0\}$: $9 + 0 = 9$ **d.** $\{5\}$: $4(5) + 3 = 23$

This completes the instruction for this section.

5.1   Exercises

Use $\{0, 1, 2, 3, 4, 5, 6, 7, 8, 9, 10\}$ as the replacement set and find the truth set for each of the following equations:

1. $y + 7 = 15$ 2. $2 + x = 7$ 3. $6 = 2 + y$ 4. $x - 5 = 1$
5. $24 - y = 15$ 6. $6 = y - 2$ 7. $4y = 16$ 8. $2x = 0$
9. $56 = 8y$ 10. $2x + 5 = 7$ 11. $3 + 2x = 11$ 12. $3y - 1 = 5$
13. $8 + 7y = 8$ 14. $3x + 3 = x + 7$

5.1   Exercise Answers

1. $\{8\}$: $8 + 7 = 15$ 2. $\{5\}$: $2 + 5 = 7$
3. $\{4\}$: $6 = 2 + 4$ 4. $\{6\}$: $6 - 5 = 1$
5. $\{9\}$: $24 - 9 = 15$ 6. $\{8\}$: $6 = 8 - 2$
7. $\{4\}$: $4(4) = 16$ 8. $\{0\}$: $2(0) = 0$
9. $\{7\}$: $56 = 8(7)$ 10. $\{1\}$: $2(1) + 5 = 7$
11. $\{4\}$: $3 + 2(4) = 11$ 12. $\{2\}$: $3(2) - 1 = 5$
13. $\{0\}$: $8 + 7(0) = 8$ 14. $\{2\}$: $3(2) + 3 = 2 + 7$

5.2   Addition and Subtraction Principles of Equality

1   Since it is not always possible to guess the solution to an equation, it is necessary to study some basic procedures that can be used to solve equations. Recall that an equation is a statement that the expressions on opposite sides of the equal sign represent the same number. Since it is necessary to maintain this equality between sides, a basic rule in solving equations is that _whatever operation is performed on one side of an equation must also be performed on the other side of the equation._

In general, solving an equation is like untying a knot, in that you always do the opposite of what has been done to form the equation. The equation has been solved when it has been changed to the form "a variable = a number" or "a number = a variable." If the variable is x, the form is "$x =$ a number" or "a number $= x$."

In the equation $x - 2 = 10$, for example, 2 has been subtracted from x to equal 10. Since _the opposite of subtracting 2 is adding 2,_ the equation can be solved by adding 2 to both sides as follows:

$$x - 2 = 10 \quad \text{(add 2 to both sides)}$$
$$x - 2 + 2 = 10 + 2$$
$$x = 12$$

The solution can be checked by seeing if it converts the original open sentence into a true statement.

Check: $x - 2 = 10$
$$12 - 2 \stackrel{?}{=} 10$$
$$10 = 10$$

So 12 is the correct solution, because 12 converts $x - 2 = 10$ into the true statement $10 = 10$.

The solution of the preceding equation demonstrates the *addition principle of equality: If the same number is added to both sides of an equation, the result is another equation with the same truth set.* In general, if $a = b$, then $a + c = b + c$ for any numbers a, b, and c.

Q1   Solve the following equations using the addition principle of equality and check the solutions:

a. $x - 3 = 5$ **b.** $y - 12 = {}^-7$ **c.** $3 = x - 6$

STOP • STOP • STOP • STOP • STOP • STOP • STOP • STOP • STOP

A1   **a.** $x - 3 = 5$ (3 was subtracted from x, so add 3 to both sides)
$$x - 3 + 3 = 5 + 3$$
$$x = 8*$$
Check:
$$x - 3 = 5$$
$$8 - 3 \stackrel{?}{=} 5$$
$$5 = 5$$

b. $y - 12 = {}^-7$ (12 was subtracted from x, so add 12 to both sides)
$$y - 12 + 12 = {}^-7 + 12$$
$$y = 5$$
Check: $y - 12 = {}^-7$
$$5 - 12 \stackrel{?}{=} {}^-7$$
$${}^-7 = {}^-7$$

c. $3 = x - 6$ (6 was subtracted from x, so add 6 to both sides)
$$3 + 6 = x - 6 + 6$$
$$9 = x$$
Check: $3 = x - 6$
$$3 \stackrel{?}{=} 9 - 6$$
$$3 = 3$$

*The truth or solution set would be $\{8\}$. However, in this and the remaining sections of this chapter, solutions to equations will be left as $x = 5$, $y = 7$, and so on.

2

Observe that in the equation $x + 4 = 13$, 4 has been added to x to equal 13. The opposite of adding 4 is subtracting 4, so the equation can be solved by subtracting 4 from both sides as follows:

$$x + 4 = 13$$
$$x + 4 - 4 = 13 - 4$$
$$x = 9$$

Check: $x + 4 = 13$
$$9 + 4 \overset{?}{=} 13$$
$$13 = 13$$

The solution of the preceding equation demonstrates a second principle useful in solving equations, the *subtraction principle of equality: If the same number is subtracted from both sides of an equation, the result is another equation with the same truth set.* In general, if $a = b$, then $a - c = b - c$ for any numbers a, b, and c.

Q2

Solve the following equations using the subtraction principle of equality and check each of the solutions:

a. $y + 1 = 16$ 　　　　　　　b. $x + 14 = {}^-29$ 　　　　　　　c. $34 = x + 11$

STOP • **STOP** • **STOP** • **STOP** • **STOP** • **STOP** • **STOP** • **STOP** • **STOP**

A2

a. 　　$y + 1 = 16$ 　　(1 was added to y, so subtract 1 from both sides)
$$y + 1 - 1 = 16 - 1$$
$$y = 15$$
Check: 　$y + 1 = 16$
$$15 + 1 \overset{?}{=} 16$$
$$16 = 16$$

b. 　　$x + 14 = {}^-29$ 　　(14 was added to x, so subtract 14 from both sides)
$$x + 14 - 14 = {}^-29 - 14$$
$$x = {}^-43$$
Check: 　$x + 14 = {}^-29$
$${}^-43 + 14 \overset{?}{=} {}^-29$$
$${}^-29 = {}^-29$$

c. 　　$34 = x + 11$ 　　(11 was added to x, so subtract 11 from both sides)
$$34 - 11 = x + 11 - 11$$
$$23 = x$$
Check: 　$34 = x + 11$
$$34 \overset{?}{=} 23 + 11$$
$$34 = 34$$

---

3   It is often necessary to simplify one or both sides of an equation by combining like terms before proceeding with the solution. For example,

$$x - 5 = 15 - 2 \quad \text{(simplify by combining like terms)}$$
$$x - 5 = 13$$
$$x - 5 + 5 = 13 + 5$$
$$x = 18$$

Q3   Solve the equation by first combining like terms: $y - 7 + 2 = 13$

STOP • STOP • STOP • STOP • STOP • STOP • STOP • STOP • STOP

A3 $y - 7 + 2 = 13$
$$y - 5 = 13$$
$$y - 5 + 5 = 13 + 5$$
$$y = 18$$

This completes the instruction for this section.

5.2   Exercises

Use the addition and subtraction principles of equality to solve the following equations and check each of the solutions:

1. $x - 2 = 5$ 2. $y + 7 = 11$
3. $1 = x - 7$ 4. $x + 3 = 3$
5. $12 = x + 8$ 6. $x - 7 = -5$
7. $11 + y = -7$ 8. $x + 6 = 5$
9. $^-3 = y + 9$ 10. $7 + x = 2$
11. $^-1 = x + 1$ 12. $9 - 7 = x + 5$
13. $y + 7 = 2 - 11$ 14. $^-4 + x = 7 - 14$

5.2   Exercise Answers

1. $x = 7$ 2. $y = 4$
3. $8 = x \,(x = 8)$ 4. $x = 0$
5. $4 = x \,(x = 4)$ 6. $x = 2$
7. $y = {}^-18$ 8. $x = {}^-1$
9. $^-12 = y \,(y = {}^-12)$ 10. $x = {}^-5$
11. $^-2 = x \,(x = {}^-2)$ 12. $^-3 = x \,(x = {}^-3)$
13. $y = {}^-16$ 14. $x = {}^-3$

5.3   Multiplication and Division Principles of Equality

---

1   In Section 5.2, equations such as $x - 2 = 5$ and $^-3 = x + 9$ were solved. In each of these equations the understood coefficient of the variable is 1. That is, x is understood to be the same as $1x$. The purpose of this section is to develop skill in solving equations in which the coefficient of the variable is a number other than 1. Some examples of this type of equation are $^-5x = 20$ and $\frac{3}{7}y = 27$.

Consider, first, the equation

$$4x = 24$$

Recall that the term $4x$ means 4 times x. Since the variable x has been multiplied by 4 and the opposite of multiplying by 4 is dividing by 4, the equation can be solved by dividing both sides of the equation by 4, as follows:

$$4x = 24 \qquad \text{Check:} \qquad 4x = 24$$
$$\frac{4x}{4} = \frac{24}{4} \qquad\qquad\qquad 4(6) \overset{?}{=} 24$$
$$1x = 6 \qquad\qquad\qquad\qquad 24 = 24$$
$$x = 6$$

The procedure used in the preceding equation demonstrates the *division principle of equality: If both sides of an equation are divided by the same nonzero number, the result is another equation with the same truth set.* (Zero is excluded since division by zero is impossible.) In general,

if $a = b$, then $\frac{a}{c} = \frac{b}{c}$ for any numbers a, b, and c, $c \neq 0$.

Two examples of the division principle of equality are as follows:

$$^-3x = 75 \qquad (x \text{ was multiplied by } ^-3, \text{ so divide both sides by } ^-3)$$
$$\frac{^-3x}{^-3} = \frac{75}{^-3}$$
$$1x = ^-25$$
$$x = ^-25$$

Check:
$$^-3x = 75$$
$$^-3(^-25) \overset{?}{=} 75$$
$$75 = 75$$

$$^-17y = ^-29 \qquad (y \text{ was multiplied by } ^-17, \text{ so divide both sides by } ^-17)$$
$$\frac{^-17y}{^-17} = \frac{^-29}{^-17}$$
$$1y = \frac{29}{17}$$
$$y = \frac{29}{17}$$

Check:
$$^-17y = ^-29$$
$$^-17\left(\frac{29}{17}\right) \overset{?}{=} ^-29$$
$$^-29 = ^-29$$

Notice that the number used to divide both sides is exactly the same as the coefficient of the variable.

Q1   Solve the following equations using the division principle of equality and check each of the solutions:

a. $2x = 10$ **b.** $^-4y = 12$

c. $4y = {}^-8$ **d.** $^-3x = {}^-7$

STOP • **STOP** • **STOP** • **STOP** • **STOP** • **STOP** • **STOP** • **STOP** • **STOP**

A1

a. $2x = 10$ Check: $2x = 10$

$$\frac{2x}{2} = \frac{10}{2}$$

$$2(5) \overset{?}{=} 10$$
$$10 = 10$$

$$1x = 5$$
$$x = 5$$

b. $^-4y = 12$ Check: $^-4y = 12$

$$\frac{^-4y}{^-4} = \frac{12}{^-4}$$

$$^-4(^-3) \overset{?}{=} 12$$
$$12 = 12$$

$$1y = {}^-3$$
$$y = {}^-3$$

c. $4y = {}^-8$ Check: $4y = {}^-8$

$$\frac{4y}{4} = \frac{^-8}{4}$$

$$4(^-2) \overset{?}{=} {}^-8$$
$$^-8 = {}^-8$$

$$1y = {}^-2$$
$$y = {}^-2$$

d. $^-3x = {}^-7$ Check: $^-3x = {}^-7$

$$\frac{^-3x}{^-3} = \frac{^-7}{^-3}$$

$$\frac{^-3}{1}\left(\frac{7}{3}\right) \overset{?}{=} {}^-7$$
$$^-7 = {}^-7$$

$$1x = \frac{7}{3}$$

$$x = \frac{7}{3}$$

---

2   When the coefficient of the variable in an equation is a fraction, a similar procedure can be followed. For example, in the equation $\frac{3}{4}x = 12$, because the variable x has been multiplied by $\frac{3}{4}$, the equation can be solved by dividing both sides by $\frac{3}{4}$.

$$\frac{3}{4}x = 12 \quad \left(\text{divide by } \frac{3}{4}\right)$$

$$\frac{\frac{3}{4}x}{\frac{3}{4}} = \frac{12}{\frac{3}{4}}$$

$$1x = 12 \div \frac{3}{4}$$

$$x = 12 \cdot \frac{4}{3}$$

$$x = 16$$

The above solution can be simplified if it is recalled that dividing by $\frac{3}{4}$ is the same as multiplying by its reciprocal, $\frac{4}{3}$. Thus, the value of $1x$ (or x) can be found by multiplying both sides of the equation by $\frac{4}{3}$ as follows:

$$\frac{3}{4}x = 12 \qquad\qquad \text{Check:} \qquad \frac{3}{4}x = 12$$

$$\frac{4}{3}\left(\frac{3}{4}x\right) = \frac{4}{3}(12) \qquad\qquad\qquad \frac{3}{4}(16) \stackrel{?}{=} 12$$

$$1x = \frac{4}{3}\left(\frac{12}{1}\right) \qquad\qquad\qquad\qquad 12 = 12$$

$$x = 16$$

The procedure used to solve the preceding equation demonstrates the fourth principle useful in solving equations, the *multiplication principle of equality: If both sides of an equation are multiplied by the same nonzero number, the result is another equation with the same truth set.* In general, if $a = b$, then $ac = bc$ for any numbers a, b, and c, $c \neq 0$.

Study the following examples of the multiplication principle of equality before proceeding to the problems of Q2.

$$\frac{5}{7}y = 10 \qquad\qquad\qquad\qquad \frac{^-2}{3}x = \frac{4}{5}$$

$$\frac{7}{5}\left(\frac{5}{7}y\right) = \frac{7}{5}(10) \qquad\qquad \frac{^-3}{2}\left(\frac{^-2}{3}x\right) = \frac{^-3}{2}\left(\frac{4}{5}\right)$$

$$1y = \frac{7}{5} \cdot \frac{10}{1} \qquad\qquad\qquad 1x = \frac{^-6}{5}$$

$$y = 14 \qquad\qquad\qquad\qquad x = \frac{^-6}{5} \text{ or } ^-1\frac{1}{5}$$

$$\text{Check:} \qquad \frac{5}{7}y = 10 \qquad\qquad \text{Check:} \qquad \frac{^-2}{3}x = \frac{4}{5}$$

$$\frac{5}{7}(14) \stackrel{?}{=} 10 \qquad\qquad\qquad \frac{^-2}{3}\left(\frac{^-6}{5}\right) \stackrel{?}{=} \frac{4}{5}$$

$$10 = 10 \qquad\qquad\qquad\qquad \frac{4}{5} = \frac{4}{5}$$

Q2   Use the multiplication principle of equality to solve the following equations and check each of the solutions:

a. $\frac{1}{2}x = 12$ b. $\frac{4}{5}y = {}^-40$

c. $\dfrac{^{-}3}{7}x = \dfrac{5}{12}$ 　　　　　　　　　　　**d.** $\dfrac{^{-}6}{7}y = {}^{-}3$

STOP • **STOP** • **STOP** • **STOP** • **STOP** • **STOP** • **STOP** • **STOP** • **STOP**

A2 　　　**a.** 　$\dfrac{1}{2}x = 12$ 　　　　　　　**b.** 　$\dfrac{4}{5}y = {}^{-}40$

$$\dfrac{2}{1}\left(\dfrac{1}{2}x\right) = \dfrac{2}{1}(12)$$ 　　　　　$$\dfrac{5}{4}\left(\dfrac{4}{5}y\right) = \dfrac{5}{4}(-40)$$

$$1x = \dfrac{2}{1}\left(\dfrac{12}{1}\right)$$ 　　　　　　$$1y = \dfrac{5}{4}\left(\dfrac{^{-}40}{1}\right)$$

$$x = 24$$ 　　　　　　　　　　$$y = {}^{-}50$$

Check: 　$\dfrac{1}{2}x = 12$ 　　　　　Check: 　$\dfrac{4}{5}y = {}^{-}40$

$$\dfrac{1}{2}(24) \overset{?}{=} 12$$ 　　　　　　　$$\dfrac{4}{5}({}^{-}50) \overset{?}{=} {}^{-}40$$

$$12 = 12$$ 　　　　　　　　　$${}^{-}40 = {}^{-}40$$

c. 　$\dfrac{^{-}3}{7}x = \dfrac{5}{12}$ 　　　　　**d.** 　$\dfrac{^{-}6}{7}y = {}^{-}3$

$$\dfrac{^{-}7}{3}\left(\dfrac{^{-}3}{7}x\right) = \dfrac{^{-}7}{3}\left(\dfrac{5}{12}\right)$$ 　　　$$\dfrac{^{-}7}{6}\left(\dfrac{^{-}6}{7}y\right) = \dfrac{^{-}7}{6}({}^{-}3)$$

$$1x = \dfrac{^{-}35}{36}$$ 　　　　　　　　$$1y = \dfrac{7}{2}$$

$$x = \dfrac{^{-}35}{36}$$ 　　　　　　　　$$y = \dfrac{7}{2}$$

Check: 　$\dfrac{^{-}3}{7}x = \dfrac{5}{12}$ 　　　　Check: 　$\dfrac{^{-}6}{7}y = {}^{-}3$

$$\dfrac{^{-}3}{7}\left(\dfrac{^{-}35}{36}\right) \overset{?}{=} \dfrac{5}{12}$$ 　　　　　$$\dfrac{^{-}6}{7}\left(\dfrac{7}{2}\right) \overset{?}{=} {}^{-}3$$

$$\dfrac{5}{12} = \dfrac{5}{12}$$ 　　　　　　　　$${}^{-}3 = {}^{-}3$$

3 　　The equation $\dfrac{x}{7} = 3$ can be solved in a manner similar to that used with the preceding equations. Using the understood coefficient 1 for x, the solution proceeds as follows:

$$\frac{x}{7} = 3$$

$$\frac{1x}{7} = 3$$

$$\frac{1}{7}x = 3$$

$$\frac{7}{1}\left(\frac{1}{7}x\right) = \frac{7}{1}(3)$$

$$1x = 21 \qquad \text{(this step is usually omitted)}$$

$$x = 21$$

Q3   Solve the following equations using the understood coefficient 1 for the variable:

a. $\dfrac{x}{5} = 2$ **b.** $\dfrac{^{-}y}{8} = 2$

STOP • STOP • STOP • STOP • STOP • STOP • STOP • STOP • STOP

A3 **a.** $\dfrac{x}{5} = 2$ **b.** $\dfrac{^{-}y}{8} = 2$

$$\frac{1x}{5} = 2 \qquad\qquad \frac{^{-}1y}{8} = 2$$

$$\frac{1}{5}x = 2 \qquad\qquad \frac{^{-}1}{8}y = 2$$

$$\frac{5}{1}\cdot\frac{1}{5}x = \frac{5}{1}(2) \qquad\qquad \frac{^{-}8}{1}\cdot\frac{^{-}1}{8}y = \frac{^{-}8}{1}(2)$$

$$x = 10 \qquad\qquad y = {}^{-}16$$

4   It is often necessary to solve equations of the form $^{-}x = a$ for some number a, that is, equations with a coefficient of $^{-}1$ on the variable. These can be solved using the multiplication principle of equality. For example,

$$^{-}x = 9$$
$$^{-}1x = 9$$
$$(^{-}1)(^{-}1x) = (^{-}1)(9)$$
$$x = {}^{-}9$$

Q4   Use the procedure of Frame 4 to solve each of the following equations:
a. $^{-}x = 5$ **b.** $^{-}y = {}^{-}7$

STOP • STOP • STOP • STOP • STOP • STOP • STOP • STOP • STOP

A4 **a.** $\begin{aligned} {}^-x &= 5 \\ {}^-1x &= 5 \\ (^-1)(^-1x) &= (^-1)(5) \\ x &= {}^-5 \end{aligned}$ **b.** $\begin{aligned} {}^-y &= {}^-7 \\ {}^-1y &= {}^-7 \\ (^-1)(^-1y) &= {}^-1(^-7) \\ y &= 7 \end{aligned}$

Q5 Solve the following equations (do step 2 mentally):

a. ${}^-x = \dfrac{-3}{5}$ **b.** ${}^-y = 0$

STOP • STOP • STOP • STOP • STOP • STOP • STOP • STOP • STOP

A5 **a.** ${}^-x = \dfrac{-3}{5}$

$(^-1)(^-x) = (^-1)\left(\dfrac{-3}{5}\right)$

$x = \dfrac{3}{5}$

b. ${}^-y = 0$

$(^-1)(^-y) = (^-1)(0)$

$y = 0$

This completes the instruction for this section.

5.3   Exercises

1. Use the multiplication and division principles of equality to solve the following equations, and check each of the solutions:

a. $2x = 10$ **b.** $\dfrac{2}{5}x = 20$ **c.** ${}^-4y = 12$ **d.** $\dfrac{3}{8}x = 24$

e. $\dfrac{x}{5} = {}^-3$ **f.** $5y = {}^-15$ **g.** ${}^-4x = {}^-8$ **h.** ${}^-y = 12$

i. $\dfrac{x}{4} = {}^-4$ **j.** $\dfrac{-5}{7}y = 10$ **k.** $5y = {}^-6$ **l.** $\dfrac{-3}{11}x = \dfrac{2}{3}$

m. ${}^-x = {}^-1$ **n.** $13x = {}^-26$ **o.** $\dfrac{x}{4} = \dfrac{3}{4}$ **p.** ${}^-7y = {}^-5$

q. $\dfrac{9}{16} = \dfrac{-3}{4}x$ **r.** $12 = \dfrac{x}{2}$ **s.** $32y = {}^-4$ **t.** $\dfrac{7}{8}y = 0$

u. $\dfrac{x}{5} = \dfrac{1}{5}$ **v.** $\dfrac{-5}{6}x = \dfrac{-2}{3}$ **w.** $8 = {}^-y$ **x.** $3y = {}^-4$

y. ${}^-7x = \dfrac{3}{5}$ **z.** $\dfrac{-4}{9} = {}^-3y$

2. Use the addition, subtraction, multiplication, and division principles of equality to solve the following equations, and check each of the solutions:

a. $x - 3 = 7$

b. $4y = {}^-12$

c. $\dfrac{-2}{3}x = \dfrac{-4}{5}$

d. $y + 9 = 2$

e. $3 + x = 5$

f. $^-5y = 25$

g. $\dfrac{1}{2}x = 7$

h. $\dfrac{3}{4}y = {}^-12$

i. $15 = x - 7$

j. $42 = {}^-6y$

k. $10 = \dfrac{-2}{5}y$

l. $0 = \dfrac{4}{7}x$

m. $x - 3 = {}^-5$

n. $^-12 + x = 5$

o. $x + 7 = 11 - 5$

p. $\dfrac{-4}{3}x = 2$

q. $5y = \dfrac{3}{7}$

r. $13 = 4 + x$

s. $5 + x = {}^-3$

t. $^-7y = \dfrac{14}{15}$

5.3   Exercise Answers

1. a. $x = 5$

b. $x = 50$

c. $y = {}^-3$

d. $x = 64$

e. $x = {}^-15$

f. $y = {}^-3$

g. $x = 2$

h. $y = {}^-12$

i. $x = {}^-16$

j. $y = {}^-14$

k. $y = \dfrac{-6}{5}$

l. $x = \dfrac{-22}{9}$

m. $x = 1$

n. $y = {}^-2$

o. $x = 3$

p. $y = \dfrac{5}{7}$

q. $x = \dfrac{-3}{4}$

r. $x = 24$

s. $y = \dfrac{-1}{8}$

t. $y = 0$

u. $x = 1$

v. $x = \dfrac{4}{5}$

w. $y = {}^-8$

x. $y = \dfrac{-4}{3}$

y. $x = \dfrac{-3}{35}$

z. $y = \dfrac{4}{27}$

2. a. $x = 10$

b. $y = {}^-3$

c. $x = \dfrac{6}{5}$

d. $y = {}^-7$

e. $x = 2$

f. $x = {}^-5$

g. $x = 14$

h. $y = {}^-16$

i. $x = 22$

j. $y = {}^-7$

k. $y = {}^-25$

l. $x = 0$

m. $x = {}^-2$

n. $x = 17$

o. $x = {}^-1$

p. $x = \dfrac{-3}{2}$

q. $y = \dfrac{3}{35}$

r. $x = 9$

s. $x = {}^-8$

t. $y = \dfrac{-2}{15}$

5.4   Solving Equations by the Use of Two or More Steps

1   Frequently it is necessary to use more than one step in solving an equation. Recall that if the variable is x, an equation is solved when it is changed to the form "$x = $ a number" or "a number $= x$." Thus the aim is to *isolate all terms involving variables on one side of the equation and all numbers on the opposite side.* By "isolate" it is meant that the *only* terms involving variables are alone on one side of the equation and *only* number terms are alone on the other side of the equation.

In the equation $3x - 4 = 11$ the objective is to isolate the x term on the left side. Since 4 has been subtracted from $3x$, add 4 (the opposite of subtracting 4) to both sides of the equation.

$$3x - 4 = 11$$
$$3x - 4 + 4 = 11 + 4$$
$$3x = 15$$

With the number and variable terms isolated on opposite sides, reduce the $3x$ to $1x$ by use of the division principle of equality.

$$3x = 15 \qquad \text{Check:} \quad 3x - 4 = 11$$
$$\frac{3x}{3} = \frac{15}{3} \qquad\qquad 3(5) - 4 \overset{?}{=} 11$$
$$x = 5 \qquad\qquad\qquad 11 = 11$$

Q1   **a.** Solve the equation $5x - 3 = 7$ by first adding 3 to both sides.

b. Check the solution to the equation.

STOP • STOP • STOP • STOP • STOP • STOP • STOP • STOP • STOP

A1   **a.** $\qquad 5x - 3 = 7 \qquad$ (3 was subtracted, so add 3 to both sides)
$$5x - 3 + 3 = 7 + 3$$
$$5x = 10 \qquad (x \text{ was multiplied by 5, so divide both sides by 5})$$
$$\frac{5x}{5} = \frac{10}{5}$$
$$x = 2$$

b. Check: $\quad 5x - 3 = 7$
$$5(2) - 3 \overset{?}{=} 7$$
$$7 = 7$$

2   In the equation $^-9 = 5x + 6$, because 6 has been added to $5x$, do the opposite and subtract 6 from both sides.

$$^-9 = 5x + 6$$
$$^-9 - 6 = 5x + 6 - 6$$
$$^-15 = 5x$$

With the number and variable terms isolated on opposite sides, reduce the $5x$ to $1x$ by use of the division principle of equality.

$$^-15 = 5x$$

$$\frac{^-15}{5} = \frac{5x}{5}$$

$$^-3 = x$$

Check: $^-9 = 5x + 6$

$$^-9 \overset{?}{=} 5(^-3) + 6$$

$$^-9 \overset{?}{=} {}^-15 + 6$$

$$^-9 = {}^-9$$

Q2 **a.** Solve $12 = {}^-4x + 8$.

b. Check the solution to the equation.

STOP • STOP • STOP • STOP • STOP • STOP • STOP • STOP • STOP

A2 **a.**
$$12 = {}^-4x + 8$$
$$12 - 8 = {}^-4x + 8 - 8$$
$$4 = {}^-4x$$
$$\frac{4}{^-4} = \frac{^-4x}{^-4}$$
$$^-1 = x$$

b. Check:
$$12 = {}^-4x + 8$$
$$12 \overset{?}{=} {}^-4(^-1) + 8$$
$$12 \overset{?}{=} 4 + 8$$
$$12 = 12$$

3 In the equation $3 + 2x = 7$, the variable term will be isolated on the left side if 3 is subtracted from both sides:

$$3 + 2x = 7$$
$$3 + 2x - 3 = 7 - 3$$
$$2x = 4$$

The solution can now be completed by use of the division principle of equality:

$$\frac{2x}{2} = \frac{4}{2}$$
$$x = 2$$

Q3 Solve $7 - 5x = 12$ by first isolating the variable term.

STOP • STOP • STOP • STOP • STOP • STOP • STOP • STOP • STOP

A3
$$7 - 5x = 12$$
$$7 - 5x - 7 = 12 - 7$$
$$^-5x = 5$$
$$\frac{^-5x}{^-5} = \frac{5}{^-5}$$
$$x = ^-1$$

---

4   In the equation $^-17 = ^-8 - 3x$, the variable term will be isolated on the right side if 8 is added to both sides:

$$^-17 = ^-8 - 3x$$
$$^-17 + 8 = ^-8 - 3x + 8$$
$$^-9 = ^-3x$$

The solution can now be completed by use of the division principle of equality:

$$\frac{^-9}{^-3} = \frac{^-3x}{^-3}$$
$$3 = x$$

Q4   Solve $0 = ^-5 + 7x$ by first isolating the variable term.

STOP • **STOP** • **STOP** • **STOP** • **STOP** • **STOP** • **STOP** • **STOP** • **STOP**

A4
$$0 = ^-5 + 7x$$
$$0 + 5 = ^-5 + 7x + 5$$
$$5 = 7x$$
$$\frac{5}{7} = \frac{7x}{7}$$
$$\frac{5}{7} = x$$

---

5   In the preceding equations, it is important to notice that the *addition or subtraction principles of equality* are used *first* with the *multiplication or division principles of equality* used in the *final step* of the problem. Study the following two examples before proceeding to Q5.

$$\frac{^-2}{3}x + 7 = 1 \qquad \text{Check:} \qquad \frac{^-2}{3}x + 7 = 1$$

$$\frac{^-2}{3}x + 7 - 7 = 1 - 7 \qquad\qquad \frac{^-2}{3}(9) + 7 \overset{?}{=} 1$$

$$\frac{^-2}{3}x = ^-6 \qquad\qquad ^-6 + 7 = 1$$

$$\frac{^-3}{2} \cdot \frac{^-2}{3}x = \frac{^-3}{2}(^-6) \qquad\qquad 1 = 1$$

$$x = 9$$

$$^-3 = {}^-3 + 4x$$
$$^-3 + 3 = {}^-3 + 4x + 3$$
$$0 = 4x$$
$$\frac{0}{4} = \frac{4x}{4}$$
$$0 = x$$

Check: $^-3 = {}^-3 + 4x$
$$^-3 \overset{?}{=} {}^-3 + 4(0)$$
$$^-3 \overset{?}{=} {}^-3 + 0$$
$$^-3 = {}^-3$$

Q5   Solve the following equations by first applying the addition or subtraction principle of equality and then the multiplication or division principle of equality:

a. $3x - 7 = {}^-19$

b. $5 = \dfrac{4}{5}y - 3$

c. $0 = {}^-6 + 12x$

d. $25 = 7 - 2x$

STOP • **STOP** • **STOP** • **STOP** • **STOP** • **STOP** • **STOP** • **STOP** • **STOP**

A5   **a.**
$$3x - 7 = {}^-19$$
$$3x - 7 + 7 = {}^-19 + 7$$
$$3x = {}^-12$$
$$\frac{3x}{3} = \frac{^-12}{3}$$
$$x = {}^-4$$

b.
$$5 = \frac{4}{5}y - 3$$
$$5 + 3 = \frac{4}{5}y - 3 + 3$$
$$8 = \frac{4}{5}y$$
$$\frac{5}{4}(8) = \frac{5}{4} \cdot \frac{4}{5}y$$
$$10 = y$$

c.
$$0 = {}^-6 + 12x$$
$$0 + 6 = {}^-6 + 12x + 6$$
$$6 = 12x$$
$$\frac{6}{12} = \frac{12x}{12}$$
$$\frac{1}{2} = x$$

d.
$$25 = 7 - 2x$$
$$25 - 7 = 7 - 2x - 7$$
$$18 = {}^-2x$$
$$\frac{18}{^-2} = \frac{^-2x}{^-2}$$
$$^-9 = x$$

6   In the preceding equations a variable term occurred on only one side of the equation. If variable terms occur on both sides of the equation, the general procedure used in solving the equation is the same. That is, *isolate all terms involving variables on one side of the equation and all numbers on the opposite side.*

For example, consider the equation

$$3x - 2 = x + 6$$

To isolate the variable terms on the left side, subtract x from both sides:

$$3x - 2 - x = x + 6 - x$$
$$2x - 2 = 6$$

To isolate the numbers on the right, add 2 to both sides:

$$2x - 2 + 2 = 6 + 2$$
$$2x = 8$$

The final step involves the division principle of equality:

$$\frac{2x}{2} = \frac{8}{2}$$
$$x = 4$$

Q6   **a.** Solve $5x - 4 = 2x + 5$ by isolating the terms that involve the variables on the left side and the numbers on the right side.

      b. Check the solution.

STOP • STOP • STOP • STOP • STOP • STOP • STOP • STOP • STOP

A6   **a.**

$$5x - 4 = 2x + 5$$
$$5x - 4 - 2x = 2x + 5 - 2x$$
$$3x - 4 = 5$$
$$3x - 4 + 4 = 5 + 4$$
$$3x = 9$$
$$\frac{3x}{3} = \frac{9}{3}$$
$$x = 3$$

b.

$$5x - 4 = 2x + 5$$
$$5(3) - 4 \overset{?}{=} 2(3) + 5$$
$$15 - 4 \overset{?}{=} 6 + 5$$
$$11 = 11$$

7   Consider the equation $x + 3 = 15 - 5x$. If you decide to isolate the terms involving variables on the left side and the numbers on the right side, you must add $5x$ and subtract 3 from both sides.

$$x + 3 = 15 - 5x$$
$$x + 3 + 5x = 15 - 5x + 5x$$
$$6x + 3 = 15$$
$$6x + 3 - 3 = 15 - 3$$
$$6x = 12$$
$$\frac{6x}{6} = \frac{12}{6}$$
$$x = 2$$

Check:
$$x + 3 = 15 - 5x$$
$$2 + 3 \overset{?}{=} 15 - 5(2)$$
$$5 = 5$$

Q7   **a.** Solve $4x + 7 = 2x + 8$ by isolating the terms involving variables on the left and the numbers on the right.

b. Check the solution.

STOP • STOP • STOP • STOP • STOP • STOP • STOP • STOP • STOP

A7   **a.**

$$4x + 7 = 2x + 8$$
$$4x + 7 - 2x = 2x + 8 - 2x$$
$$2x + 7 = 8$$
$$2x + 7 - 7 = 8 - 7$$
$$2x = 1$$
$$\frac{2x}{2} = \frac{1}{2}$$
$$x = \frac{1}{2}$$

b.

$$4x + 7 = 2x + 8$$
$$4 \cdot \frac{1}{2} + 7 \stackrel{?}{=} 2 \cdot \frac{1}{2} + 8$$
$$2 + 7 \stackrel{?}{=} 1 + 8$$
$$9 = 9$$

8   The preceding problems have been solved by isolating the terms that involve variables on the left side. However, equations can be solved by isolating the variable on either side. You may wish to isolate the variable on the side that makes the coefficient of the variable positive.

Example 1: Solve $3x - 1 = x + 5$ by isolating the variable on the left side.

Solution

Subtract x from and add 1 to both sides:

$$3x - 1 = x + 5$$
$$3x - 1 - x = x + 5 - x$$
$$2x - 1 = 5$$
$$2x - 1 + 1 = 5 + 1$$
$$2x = 6$$
$$x = 3$$

Example 2: Solve $3x - 1 = x + 5$ by isolating the variable on the right side.

Solution

Subtract $3x$ and 5 from both sides:

$$3x - 1 = x + 5$$
$$3x - 1 - 3x = x + 5 - 3x$$
$$^-1 = {}^-2x + 5$$
$$^-1 - 5 = {}^-2x + 5 - 5$$
$$^-6 = {}^-2x$$
$$3 = x$$

The same solution was obtained by isolating the variable on either side.

Q8   Solve $5x - 3 = 6x + 9$ by isolating the variable on the left side.

STOP   •   STOP   •   STOP   •   STOP   •   STOP   •   STOP   •   STOP   •   STOP   •   STOP

A8
$$5x - 3 = 6x + 9$$
$$5x - 3 - 6x = 6x + 9 - 6x$$
$$^-x - 3 = 9$$
$$^-x - 3 + 3 = 9 + 3$$
$$^-x = 12$$
$$x = {}^-12$$

Q9   Solve $5x - 3 = 6x + 9$ by isolating the variable on the right side.

STOP   •   STOP   •   STOP   •   STOP   •   STOP   •   STOP   •   STOP   •   STOP   •   STOP

A9
$$5x - 3 = 6x + 9$$
$$5x - 3 - 5x = 6x + 9 - 5x$$
$$^-3 = x + 9$$
$$^-3 - 9 = x + 9 - 9$$

Q10   Solve by isolating the variable on the side that makes the coefficient of the variable positive:

a. $2x + 1 = {}^-x - 2$ b. $^-4x + 2 = 7x + 3$

STOP   •   STOP   •   STOP   •   STOP   •   STOP   •   STOP   •   STOP   •   STOP   •   STOP

A10   **a.** $2x + 1 = {}^-x - 2$

$2x + 1 + x = {}^-x - 2 + x$

$3x + 1 = {}^-2$

$3x + 1 - 1 = {}^-2 - 1$

$3x = {}^-3$

$x = {}^-1$

b. ${}^-4x + 2 = 7x + 3$

${}^-4x + 2 + 4x = 7x + 3 + 4x$

$2 = 11x + 3$

$2 - 3 = 11x + 3 - 3$

${}^-1 = 11x$

$\dfrac{{}^-1}{11} = x$

---

9   To solve equations that involve parentheses, first simplify both sides of the equation wherever possible, and then proceed as before.

Examples:

1. $2x - (x + 2) = 7$

$2x - x - 2 = 7$ (remove parentheses)

$x - 2 = 7$ (combine like terms)

$x - 2 + 2 = 7 + 2$

$x = 9$

2. $x + 5 = x + (2x - 3)$

$x + 5 = x + 2x - 3$ (remove parentheses)

$x + 5 = 3x - 3$ (combine like terms)

$x + 5 - 3x = 3x - 3 - 3x$

${}^-2x + 5 = {}^-3$

${}^-2x + 5 - 5 = {}^-3 - 5$

${}^-2x = {}^-8$

$\dfrac{{}^-2x}{{}^-2} = \dfrac{{}^-8}{{}^-2}$

$x = 4$

Q11   Solve by first simplifying both sides of the equation:

a. $5x - (2x + 7) = 8$ **b.** $6x = 8 + (2x - 4)$

STOP • **STOP** • **STOP** • **STOP** • **STOP** • **STOP** • **STOP** • **STOP** • **STOP**

A11   **a.** $5x - (2x + 7) = 8$

$5x - 2x - 7 = 8$

$3x - 7 = 8$

$3x - 7 + 7 = 8 + 7$

$3x = 15$

$\dfrac{3x}{3} = \dfrac{15}{3}$

$x = 5$

b. $6x = 8 + (2x - 4)$

$6x = 8 + 2x - 4$

$6x = 4 + 2x$

$6x - 2x = 4 + 2x - 2x$

$4x = 4$

$\dfrac{4x}{4} = \dfrac{4}{4}$

$x = 1$

---

10   To solve $2(5x + 3) - 3(x - 5) = 7$, the following steps are used:

$$2(5x + 3) - 3(x - 5) = 7$$
$$10x + 6 - 3x + 15 = 7$$
$$7x + 21 = 7$$
$$7x + 21 - 21 = 7 - 21$$
$$7x = {}^-14$$
$$\frac{7x}{7} = \frac{{}^-14}{7}$$
$$x = {}^-2$$

Q12   Solve:

a. $5(x + 6) = 45$

b. $3(x - 2) = x - 2(3x - 1)$

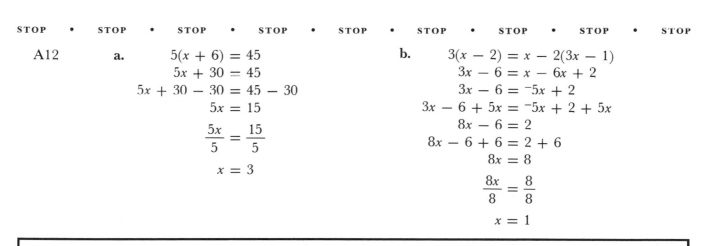

STOP • STOP • STOP • STOP • STOP • STOP • STOP • STOP • STOP

A12   **a.**
$$5(x + 6) = 45$$
$$5x + 30 = 45$$
$$5x + 30 - 30 = 45 - 30$$
$$5x = 15$$
$$\frac{5x}{5} = \frac{15}{5}$$
$$x = 3$$

b.
$$3(x - 2) = x - 2(3x - 1)$$
$$3x - 6 = x - 6x + 2$$
$$3x - 6 = {}^-5x + 2$$
$$3x - 6 + 5x = {}^-5x + 2 + 5x$$
$$8x - 6 = 2$$
$$8x - 6 + 6 = 2 + 6$$
$$8x = 8$$
$$\frac{8x}{8} = \frac{8}{8}$$
$$x = 1$$

---

11   When solving equations such as $5 - 4(x + 3) = 1$, it is again important to first simplify both sides of the equation by removing parentheses and combining like terms. For example,

$$5 - 4(x + 3) = 1$$
$$5 - 4x - 12 = 1$$
$${}^-4x - 7 = 1$$
$${}^-4x - 7 + 7 = 1 + 7$$
$${}^-4x = 8$$
$$\frac{{}^-4x}{{}^-4} = \frac{8}{{}^-4}$$
$$x = {}^-2$$

Q13   Solve:

   a. $5 - 3(x - 2) = 14$ **b.** $x + 8 = {}^-10 - 6(2x - 3)$

STOP • **STOP** • **STOP** • **STOP** • **STOP** • **STOP** • **STOP** • **STOP** • **STOP**

A13   **a.** $5 - 3(x - 2) = 14$ **b.** $x + 8 = {}^-10 - 6(2x - 3)$

$5 - 3x + 6 = 14$ $x + 8 = {}^-10 - 12x + 18$

${}^-3x + 11 = 14$ $x + 8 = 8 - 12x$

${}^-3x + 11 - 11 = 14 - 11$ $x + 8 + 12x = 8 - 12x + 12x$

$\dfrac{{}^-3x}{{}^-3} = \dfrac{3}{{}^-3}$ $13x + 8 = 8$

$x = {}^-1$ $13x + 8 - 8 = 8 - 8$

$13x = 0$

$\dfrac{13x}{13} = \dfrac{0}{13}$

$x = 0$

This completes the instruction for this section.

5.4   Exercises

1. Solve and check each of the following:

   a. $2x - 3 = 9$ **b.** $15 = 4y + 7$

   c. $6x + 11 = {}^-7$ **d.** ${}^-5y + 1 = {}^-4$

   e. $\dfrac{2}{5}x - 3 = 11$ **f.** $6 = 6 - 7y$

   g. $5 - y = 12$ **h.** $6 = 2x + 5$

   i. $7 - \dfrac{1}{2}x = 3$ **j.** ${}^-3y = {}^-5$

   k. $4 + 3y = 0$ **l.** $2 = {}^-3 + \dfrac{5}{7}y$

   m. $6 = 9 - x$ **n.** $\dfrac{{}^-3}{4}x - 3 = 5$

2. Solve the following equations by isolating the terms that involve variables on one side and the numbers on the opposite side:

   a. $2x + 3 = x + 4$ **b.** $7 + 3x = 4x - 2$

   c. $12 + x = 2x + 3$ **d.** $5x + 3 = x - 1$

   e. $4y = y + 9$ **f.** $2 - 3y = 2y + 2$

   g. $x + 11 = {}^-x + 5$ **h.** ${}^-2y + 3 = {}^-5y + 5$

    i. $^-7x + 3 - 5x = 0$ **j.** $9 = 3x - 15$

    k. $4x - 11 = 2 - x$ **l.** $3 + 4y = {}^-2y + 1$

3. Solve:

    a. $2(x - 4) = 10$ **b.** $2x + 3 = 4 + (x - 6)$

    c. $6x - (x - 7) = 22$ **d.** $3(x - 5) = {}^-7 + 2(x - 4)$

    e. $2(x - 3) - 3(2x + 5) = {}^-25$ **f.** $4 - 2(x + 1) = 2x + 4$

    g. $3x + 10 = 7 - 5(2x + 15)$ **h.** $2(3x - 6) + 4 = 22 - 2(x - 1)$

5.4   Exercise Answers

1. **a.** $x = 6$ **b.** $y = 2$ **c.** $x = {}^-3$ **d.** $y = 1$

  e. $x = 35$ **f.** $y = 0$ **g.** $y = {}^-7$ **h.** $x = \dfrac{1}{2}$

  i. $x = 8$ **j.** $y = \dfrac{5}{3}$ **k.** $y = \dfrac{-4}{3}$ **l.** $y = 7$

  m. $x = 3$ **n.** $x = \dfrac{-32}{3}$

2. **a.** $x = 1$ **b.** $x = 9$ **c.** $x = 9$ **d.** $x = {}^-1$

  e. $y = 3$ **f.** $y = 0$ **g.** $x = {}^-3$ **h.** $y = \dfrac{2}{3}$

  i. $x = \dfrac{1}{4}$ **j.** $x = 8$ **k.** $x = \dfrac{13}{5}$ **l.** $y = \dfrac{-1}{3}$

3. **a.** $x = 9$ **b.** $x = {}^-5$ **c.** $x = 3$ **d.** $x = 0$

  e. $x = 1$ **f.** $x = \dfrac{-1}{2}$ **g.** $x = {}^-6$ **h.** $x = 4$

5.5   **Equations with Fractions**

1

Many times it is necessary to solve equations that involve several rational numbers (fractions). This is usually done by eliminating the fractions from the equation by use of the multiplication property of equality, and solving the resulting equation using the procedures of previous sections.

The fractions can be eliminated from the equation $\dfrac{x}{2} - 4 = \dfrac{x}{3}$ by multiplying both sides of the equation by each of the denominators, 2 and 3. For example,

$$\frac{x}{2} - 4 = \frac{x}{3}$$

$$2\left(\frac{x}{2} - 4\right) = 2\left(\frac{x}{3}\right)$$

$$2\left(\frac{x}{2}\right) - 2(4) = 2\left(\frac{x}{3}\right)$$

$$x - 8 = \frac{2x}{3}$$

The first fraction has now been eliminated.

$$3(x - 8) = 3\left(\frac{2x}{3}\right)$$

$$3(x) - 3(8) = 3\left(\frac{2x}{3}\right)$$

$$3x - 24 = 2x$$

All fractions have now been eliminated and the solution can be completed using the procedures already studied.

$$\begin{aligned} 3x - 24 &= 2x \\ 3x - 24 - 2x &= 2x - 2x \\ x - 24 &= 0 \\ x - 24 + 24 &= 0 + 24 \\ x &= 24 \end{aligned}$$

Check: $\dfrac{x}{2} - 4 = \dfrac{x}{3}$

$$\frac{24}{2} - 4 \stackrel{?}{=} \frac{24}{3}$$

$$12 - 4 \stackrel{?}{=} 8$$

$$8 = 8$$

Q1   **a.** Eliminate the fractions from the equation $\dfrac{x}{2} = 5 + \dfrac{x}{3}$ by first multiplying both sides of the equation by 2 and then by 3.

b. Solve the resulting equation.   **c.** Check the solution.

STOP • STOP • STOP • STOP • STOP • STOP • STOP • STOP • STOP

A1   **a.** $\dfrac{x}{2} = 5 + \dfrac{x}{3}$

$$2\left(\frac{x}{2}\right) = 2\left(5 + \frac{x}{3}\right)$$

$$2\left(\frac{x}{2}\right) = 2(5) + 2\left(\frac{x}{3}\right)$$

$$x = 10 + \frac{2x}{3}$$

$$3(x) = 3\left(10 + \frac{2x}{3}\right)$$

$$3(x) = 3(10) + 3 \cdot \frac{2x}{3}$$

$$3x = 30 + 2x$$

b. $\begin{aligned} 3x - 2x &= 30 + 2x - 2x \\ x &= 30 \end{aligned}$

c. $\dfrac{x}{2} = 5 + \dfrac{x}{3}$

$$\frac{30}{2} \stackrel{?}{=} 5 + \frac{30}{3}$$

$$15 \stackrel{?}{=} 5 + 10$$

$$15 = 15$$

2   A much shorter procedure for eliminating several fractions from an equation is to multiply both sides by just *one* number. For example, rather than to multiply the equation $\frac{x}{4} - 3 = \frac{x}{5}$ by the two numbers 4 and 5, the fractions can be eliminated by multiplying both sides by *one* number which has both factors 4 and 5. The smallest number with both factors 4 and 5 is 20. Therefore, multiply both sides of the equation by 20.

$$20\left(\frac{x}{4} - 3\right) = 20\left(\frac{x}{5}\right)$$

$$20\left(\frac{x}{4}\right) - 20(3) = 20\left(\frac{x}{5}\right)$$

$$5x - 60 = 4x$$

With the fractions eliminated, the solution of the equation can now be completed:

$$5x - 60 = 4x$$
$$5x - 60 - 4x = 4x - 4x$$
$$x - 60 = 0$$
$$x - 60 + 60 = 0 + 60$$
$$x = 60$$

Check: $\frac{x}{4} - 3 = \frac{x}{5}$

$$\frac{60}{4} - 3 \stackrel{?}{=} \frac{60}{5}$$

$$15 - 3 \stackrel{?}{=} 12$$

$$12 = 12$$

Q2   **a.** Eliminate the fractions from the equation $\frac{x}{3} - 4 = \frac{x}{5}$ by multiplying both sides by the smallest number with the factors 3 and 5.

b. Solve the resulting equation.   **c.** Check the solution.

STOP • STOP • STOP • STOP • STOP • STOP • STOP • STOP • STOP

A2   **a.** The smallest number is 15.

$$15\left(\frac{x}{3} - 4\right) = 15\left(\frac{x}{5}\right)$$

$$15\left(\frac{x}{3}\right) - 15(4) = 15\left(\frac{x}{5}\right)$$

$$5x - 60 = 3x$$

b. $5x - 60 - 3x = 3x - 3x$
$$2x - 60 = 0$$
$$2x - 60 + 60 = 0 + 60$$
$$2x = 60$$
$$\frac{2x}{2} = \frac{60}{2}$$
$$x = 30$$

c. Check: $\dfrac{x}{3} - 4 = \dfrac{x}{5}$

$$\dfrac{30}{3} - 4 \overset{?}{=} \dfrac{30}{5}$$

$$10 - 4 \overset{?}{=} 6$$

$$6 = 6$$

---

3 | The procedure of eliminating the fractions from an equation is called "clearing an equation of fractions." An equation can be cleared of fractions by multiplying both sides by the smallest number that all the denominators in the equation will divide into evenly. This number is called the least common denominator (LCD) for the fractions in the equation.

Q3   Solve the following equations by first clearing them of fractions using the LCD.

a. $\dfrac{2x}{3} - 5 = \dfrac{x}{4}$ b. $\dfrac{2}{3}x - \dfrac{2}{5} = \dfrac{2}{5}x$ c. $\dfrac{x}{2} + \dfrac{5}{2} = \dfrac{2x}{3}$

STOP • **STOP** • **STOP** • **STOP** • **STOP** • **STOP** • **STOP** • **STOP** • **STOP**

A3   **a.** The LCD is 12.

$$12\left(\dfrac{2x}{3} - 5\right) = 12\left(\dfrac{x}{4}\right)$$

$$12\left(\dfrac{2x}{3}\right) - 12(5) = 12\left(\dfrac{x}{4}\right)$$

$$8x - 60 = 3x$$

$$8x - 60 - 3x = 3x - 3x$$

$$5x - 60 = 0$$

$$5x - 60 + 60 = 0 + 60$$

$$5x = 60$$

$$\dfrac{5x}{5} = \dfrac{60}{5}$$

$$x = 12$$

b. The LCD is 15.

$$15\left(\dfrac{2}{3}x - \dfrac{2}{5}\right) = 15\left(\dfrac{2}{5}x\right)$$

$$15\left(\dfrac{2}{3}x\right) - 15\left(\dfrac{2}{5}\right) = 15\left(\dfrac{2}{5}x\right)$$

$$10x - 6 = 6x$$

$$10x - 6 - 6x = 6x - 6x$$

$$4x - 6 = 0$$

$$4x - 6 + 6 = 0 + 6$$

$$4x = 6$$

$$\dfrac{4x}{4} = \dfrac{6}{4}$$

$$x = \dfrac{3}{2}$$

c. The LCD is 6. The solution is $x = 15$.

---

4   The steps used when solving an equation with fractions are:

Step 1: Determine the LCD.

Step 2: Multiply both sides of the equation by the LCD. This step "clears" the equation of all fractions.

Step 3: Solve the resulting equation.

Step 4: Check the solution.

Q4   Solve each of the following equations using the preceding four-step procedure:

a. $\dfrac{x}{2} = 7 - \dfrac{2x}{3}$

b. $\dfrac{7x}{8} + \dfrac{5}{6} = \dfrac{1}{12}$

c. $\dfrac{9}{10} = \dfrac{-3}{4}x + \dfrac{2}{5}$

d. $\dfrac{3y}{5} + \dfrac{5}{2} = \dfrac{-y}{5} - \dfrac{3}{2}$

STOP • STOP • STOP • STOP • STOP • STOP • STOP • STOP • STOP

A4   **a.**   The LCD is 6.

$$6\left(\dfrac{x}{2}\right) = 6\left(7 - \dfrac{2x}{3}\right)$$

$$6\left(\dfrac{x}{2}\right) = 6(7) - 6\left(\dfrac{2x}{3}\right)$$

$$3x = 42 - 4x$$

$$3x + 4x = 42 - 4x + 4x$$

$$7x = 42$$

$$\dfrac{7x}{7} = \dfrac{42}{7}$$

$$x = 6$$

Check: $\dfrac{x}{2} = 7 - \dfrac{2x}{3}$

$$\dfrac{6}{2} \overset{?}{=} 7 - \dfrac{2(6)}{3}$$

$$3 \overset{?}{=} 7 - \dfrac{12}{3}$$

$$3 = 3$$

b. The LCD is 24. The solution is $x = \dfrac{^-6}{7}$.

c. The LCD is 20. The solution is $x = \dfrac{^-2}{3}$.

d. The LCD is 10.

$$10\left(\frac{3y}{5} + \frac{5}{2}\right) = 10\left(\frac{^-y}{5} - \frac{3}{2}\right)$$

$$10\left(\frac{3y}{5}\right) + 10\left(\frac{5}{2}\right) = 10\left(\frac{^-y}{5}\right) - 10\left(\frac{3}{2}\right)$$

$$6y + 25 = ^-2y - 15$$

$$6y + 25 + 2y = ^-2y - 15 + 2y$$

$$8y + 25 = ^-15$$

$$8y + 25 - 25 = ^-15 - 25$$

$$8y = ^-40$$

$$\frac{8y}{8} = \frac{^-40}{8}$$

$$y = ^-5$$

Check:

$$\frac{3y}{5} + \frac{5}{2} = \frac{^-y}{5} - \frac{3}{2}$$

$$\frac{3(^-5)}{5} + \frac{5}{2} \stackrel{?}{=} \frac{^-(^-5)}{5} - \frac{3}{2}$$

$$^-3 + \frac{5}{2} \stackrel{?}{=} 1 - \frac{3}{2}$$

$$\frac{^-1}{2} = \frac{^-1}{2}$$

---

5 To solve equations such as $\dfrac{x + 1}{2} = \dfrac{x - 2}{5}$ the LCD is again used to clear the equation of fractions. For example,

$$10\left(\frac{x + 1}{2}\right) = 10\left(\frac{x - 2}{5}\right)$$

$$5(x + 1) = 2(x - 2)$$

The solution can now be completed by removing parentheses and proceeding as in previous sections.

$$5(x + 1) = 2(x - 2)$$

$$5x + 5 = 2x - 4$$

$$5x + 5 - 2x = 2x - 4 - 2x$$

$$3x + 5 = ^-4$$

$$3x + 5 - 5 = ^-4 - 5$$

$$3x = ^-9$$

$$\frac{3x}{3} = \frac{^-9}{3}$$

$$x = ^-3$$

Q5 Solve the equation $\dfrac{x - 1}{3} = \dfrac{x + 2}{4}$ by first multiplying both sides by the LCD.

STOP • STOP • STOP • STOP • STOP • STOP • STOP • STOP • STOP

A5
$$12\left(\frac{x-1}{3}\right) = 12\left(\frac{x+2}{4}\right)$$
$$4(x-1) = 3(x+2)$$
$$4x - 4 = 3x + 6$$
$$4x - 4 - 3x = 3x + 6 - 3x$$
$$x - 4 = 6$$
$$x - 4 + 4 = 6 + 4$$
$$x = 10$$

Q6   Solve $\dfrac{y-6}{2} = \dfrac{2y-3}{3}$.

STOP • STOP • STOP • STOP • STOP • STOP • STOP • STOP • STOP

A6
$$6\left(\frac{y-6}{2}\right) = 6\left(\frac{2y-3}{3}\right)$$
$$3(y-6) = 2(2y-3)$$
$$3y - 18 = 4y - 6$$
$$3y - 18 - 3y = 4y - 6 - 3y$$
$$^{-}18 = y - 6$$
$$^{-}18 + 6 = y - 6 + 6$$
$$^{-}12 = y$$

6   Study the solution of the following equation before proceeding to Q7.

$$\frac{3x+2}{4} - 1 = \frac{2x-3}{6}$$

$$12\left(\frac{3x+2}{4} - 1\right) = 12\left(\frac{2x-3}{6}\right)$$

$$12\left(\frac{3x+2}{4}\right) - 12(1) = 12\left(\frac{2x-3}{6}\right)$$

$$3(3x+2) - 12 = 2(2x-3)$$

$$9x + 6 - 12 = 4x - 6$$

$$9x - 6 = 4x - 6$$

$$9x - 6 - 4x = 4x - 6 - 4x$$

$$5x - 6 = {}^{-}6$$

$$5x - 6 + 6 = {}^{-}6 + 6$$

$$5x = 0$$

$$\frac{5x}{5} = \frac{0}{5}$$

$$x = 0$$

Check:

$$\frac{3x+2}{4} - 1 = \frac{2x-3}{6}$$

$$\frac{3(0)+2}{4} - 1 \stackrel{?}{=} \frac{2(0)-3}{6}$$

$$\frac{2}{4} - 1 \stackrel{?}{=} \frac{^{-}3}{6}$$

$$\frac{^{-}1}{2} = \frac{^{-}1}{2}$$

Q7   Solve $\dfrac{3x - 2}{4} - 2 = \dfrac{2x - 2}{5}$.

STOP   •   STOP   •   STOP   •   STOP   •   STOP   •   STOP   •   STOP   •   STOP   •   STOP

A7 $$20\left(\dfrac{3x - 2}{4} - 2\right) = 20\left(\dfrac{2x - 2}{5}\right)$$

$$20\left(\dfrac{3x - 2}{4}\right) - 20(2) = 20\left(\dfrac{2x - 2}{5}\right)$$

$$5(3x - 2) - 40 = 4(2x - 2)$$
$$15x - 10 - 40 = 8x - 8$$
$$15x - 50 = 8x - 8$$
$$15x - 50 - 8x = 8x - 8 - 8x$$
$$7x - 50 = {}^-8$$
$$7x - 50 + 50 = {}^-8 + 50$$
$$7x = 42$$
$$\dfrac{7x}{7} = \dfrac{42}{7}$$
$$x = 6$$

Q8   Solve each of the following:

a. $\dfrac{x - 2}{3} = \dfrac{x + 1}{4}$

b. $\dfrac{x}{4} - \dfrac{x + 4}{2} = \dfrac{1}{2}$

c. $\dfrac{x}{5} - \dfrac{x + 2}{3} = \dfrac{6}{5}$

STOP   •   STOP   •   STOP   •   STOP   •   STOP   •   STOP   •   STOP   •   STOP   •   STOP

A8   **a.** $x = 11$ **b.** $x = {}^-10$ **c.** $x = {}^-14$

This completes the instruction for this section.

5.5   Exercises

Solve each of the following equations:

1. $\frac{4}{3}x - 2 = 10$

2. $\frac{x}{2} - 6 = \frac{x}{4}$

3. $\frac{3}{4}y - \frac{2}{3} = \frac{5}{12}$

4. $\frac{3}{2} - \frac{x}{3} = 5 + \frac{x}{6}$

5. $\frac{x}{2} - 7 = \frac{2x}{3}$

6. $\frac{3y}{5} + \frac{5}{2} = \frac{{}^-y}{5} - \frac{3}{2}$

7. $\frac{x + 2}{3} = \frac{2x - 7}{5}$

8. $\frac{2x - 3}{4} - 1 = \frac{x - 5}{3}$

9. $\frac{3x}{4} - \frac{1}{2} = \frac{x}{4} + \frac{11}{2}$

10. $\frac{y - 7}{2} - \frac{1}{3} = 1 + \frac{y + 9}{7}$

11. $\frac{3 - x}{2} = \frac{{}^-6 - 5x}{7}$

12. $\frac{1 - x}{2} = 1 - \frac{x + 4}{8}$

13. $\frac{x}{4} - \frac{x + 4}{2} = \frac{1}{2}$

14. $\frac{2x - 3}{6} - \frac{2}{3} - \frac{x + 5}{9} = \frac{1}{2}$

5.5   Exercise Answers

1. $x = 9$

2. $x = 24$

3. $y = \frac{13}{9}$

4. $x = {}^-7$

5. $x = {}^-42$

6. $y = {}^-5$

7. $x = 31$

8. $x = \frac{1}{2}$

9. $x = 12$

10. $y = \frac{257}{15}$

11. $x = {}^-11$

12. $x = 0$

13. $x = {}^-10$

14. $x = 10$

Chapter 5 Sample Test

At the completion of Chapter 5 it is expected that you will be able to work the following problems.

5.1   **Equations, Open Sentences, Replacement Set, and Solution Set**

1. What word best describes each phrase?
    a. a statement that two expressions are equal
    b. an equation that is neither true nor false

    c. the set of all numbers that may replace the variable in an open sentence

    d. the values from the replacement set which convert the open sentence into a true statement

2. Identify the right and left sides of each of the following equations:

    a. $3x = 7 - 2$ **b.** $y - 4 + 3y = 0$

3. If the replacement set for the variable x is $\{^-5, 0, 1, ^-1, 4\}$, find the result of all possible replacements for x in the open sentence $9 - x = 8$. Label each statement as true or false.

5.2   Addition and Subtraction Principles of Equality

4. Label each of the following as demonstrating the addition principle of equality or the subtraction principle of equality:

    a. $x + 7 = 12$

         $x + 7 - 7 = 12 - 7$

             $x = 5$

    b. $^-3 = ^-5 + y$

         $^-3 + 5 = ^-5 + y + 5$

             $2 = y$

    c. $2y = y + 4$

         $2y - y = y + 4 - y$

             $y = 4$

    d. $^-4 = x - 4$

         $^-4 + 4 = x - 4 + 4$

             $0 = x$

5. Solve and check each of the following equations:

    a. $^-3 + x = 5$ **b.** $^-7 = y + 9$

    c. $x + 8 = 8$ **d.** $7 + y = 6 - 13$

    e. $15 = ^-25 + x$ **f.** $17 = y - 5$

5.3   Multiplication and Division Principles of Equality

6. Label each of the following as demonstrating the multiplication principle of equality or the division principle of equality:

    a. $^-5x = 30$

         $\dfrac{^-5x}{^-5} = \dfrac{30}{^-5}$

           $x = ^-6$

    b. $^-12 = \dfrac{3}{4}y$

         $\dfrac{4}{3}(^-12) = \dfrac{4}{3}\left(\dfrac{3}{4}y\right)$

            $^-16 = y$

    c. $\dfrac{y}{7} = \dfrac{5}{14}$

        $7\left(\dfrac{y}{7}\right) = 7\left(\dfrac{5}{14}\right)$

           $y = \dfrac{5}{2}$

    d. $^-18 = ^-5x$

         $\dfrac{^-18}{^-5} = \dfrac{^-5x}{^-5}$

          $3\dfrac{3}{5} = x$

7. Solve and check each of the following equations:

    a. $^-5x = 25$ **b.** $\dfrac{5}{7}x = 35$

    c. $\dfrac{^-2}{3}y = \dfrac{5}{9}$ **d.** $^-24 = \dfrac{3}{4}x$

    e. $\dfrac{y}{8} = 2$ **f.** $^-x = \dfrac{7}{9}$

    g. $\dfrac{^-4}{9} = ^-3y$ **h.** $^-7y = \dfrac{14}{15}$

5.4   **Solving Equations by the Use of Two or More Steps**

8. Solve and check each of the following equations:

    a. $2x - 5 = 55$ **b.** $^-5y + 1 = {}^-4$

    c. $\dfrac{2}{5}y - 3 = 11$ **d.** $7x - 5 = 9x - 5$

    e. $2 - 3y = 2y + 2$ **f.** $x + 11 = {}^-x + 5$

    g. $6 - (2x + 3) = 9 + (x + 15)$ **h.** $5x - 2(x - 7) = 8(x - 2) + 10$

5.5   **Equations with Fractions**

9. Solve and check each of the following equations:

    a. $\dfrac{2}{3}x - 5 = \dfrac{1}{4}x$ **b.** $\dfrac{3y}{5} + \dfrac{5}{2} = \dfrac{{}^-y}{5} - \dfrac{3}{2}$

    c. $\dfrac{x - 1}{3} = \dfrac{x + 2}{4}$ **d.** $\dfrac{1 - y}{2} = 1 - \dfrac{y + 4}{8}$

Chapter 5 Sample Test Answers

1. **a.** equation   **b.** open sentence   **c.** replacement set   **d.** solutions
2. **a.** right side, $7 - 2$; left side, $3x$ **b.** right side, 0; left side, $y - 4 + 3y$
3. $9 - {}^-5 = 8$, false
    $9 - 0 = 8$, false
    $9 - 1 = 8$, true
    $9 - {}^-1 = 8$, false
    $9 - 4 = 8$, false
4. **a.** subtraction principle of equality   **b.** addition principle of equality
    c. subtraction principle of equality   **d.** addition principle of equality
5. **a.** $x = 8$ **b.** $y = {}^-16$ **c.** $x = 0$ **d.** $y = {}^-14$
    e. $x = 40$ **f.** $y = 22$
6. **a.** division principle of equality   **b.** multiplication principle of equality
    c. multiplication principle of equality   **d.** division principle of equality
7. **a.** $x = {}^-5$ **b.** $x = 49$ **c.** $y = \dfrac{^-5}{6}$ **d.** $x = {}^-32$

    e. $y = 16$ **f.** $x = \dfrac{^-7}{9}$ **g.** $y = \dfrac{4}{27}$ **h.** $y = \dfrac{^-2}{15}$
8. **a.** $x = 30$ **b.** $y = 1$ **c.** $y = 35$ **d.** $x = 0$
    e. $y = 0$ **f.** $x = {}^-3$ **g.** $x = {}^-7$ **h.** $x = 4$
9. **a.** $x = 12$ **b.** $y = {}^-5$ **c.** $x = 10$ **d.** $y = 0$

Chapter 6

Ratio and Proportion

6.1   **Introduction to Ratio and Proportion**

---

1   A fraction can be used to compare two quantities or numbers. Suppose that Sally has $13 and Linda has $39. Sally then has $\frac{13}{39}$ or $\frac{1}{3}$ as much money as Linda. When a fraction is used in this way to compare two numbers, the fraction is called a ratio. A *ratio* is the quotient of two quantities or numbers. The ratio $\frac{1}{3}$ is read "one to three" and can be written using a colon as $1:3$.

Q1   **a.** 4 of 7 square regions are shaded. Write the ratio of the shaded region to the total region. _____

b. Of the 12 rectangular regions, 5 are shaded. Write the ratio of shaded regions to the total region. _____

c. 6 of the 11 hexagonal regions are shaded. Write the indicated ratio. _____

d. 5 of the 11 hexagonal regions are *not* shaded. What ratio represents this statement? _____

e. _____ of 6 circular regions are shaded.

f. Write the ratio. _____

g. 4 of _____ squares are circled.

h. Write the ratio. _____

i. _____ of _____ triangular regions are shaded.

j. Write the indicated ratio._____

STOP • STOP • STOP • STOP • STOP • STOP • STOP • STOP • STOP

A1 **a.** $\frac{4}{7}$ or $4:7$ **b.** $\frac{5}{12}$ or $5:12$ **c.** $\frac{6}{11}$ or $6:11$ **d.** $\frac{5}{11}$ or $5:11$

e. 5 **f.** $\frac{5}{6}$ or $5:6$ **g.** 9 **h.** $\frac{4}{9}$ or $4:9$

i. 9, 18 **j.** $\frac{9}{18}$ or $9:18\left(\frac{1}{2}\text{ or }1:2\right)$

Q2 The comparison of one quantity to another by division is called a _____.

STOP • STOP • STOP • STOP • STOP • STOP • STOP • STOP • STOP

A2 ratio

Q3 **a.** What is the ratio 24 years to 9 years? _____

b. What is the ratio of 9 years to 24 years? _____

STOP • STOP • STOP • STOP • STOP • STOP • STOP • STOP • STOP

A3 **a.** $\frac{8}{3}$: $\frac{24}{9} = \frac{8}{3}$ **b.** $\frac{3}{8}$

Q4 Find the following ratios as fractions:

a. 25 to 125 **b.** 25 to 10 **c.** 25 to 5 **d.** 25 to $\frac{1}{5}$

e. $\frac{1}{4}$ to $\frac{1}{2}$ **f.** $\frac{1}{4}$ to 2

STOP • STOP • STOP • STOP • STOP • STOP • STOP • STOP • STOP

A4 **a.** $\frac{1}{5}$ **b.** $\frac{5}{2}$ **c.** $\frac{5}{1}$ **d.** $\frac{125}{1}$

e. $\frac{1}{2}$ **f.** $\frac{1}{8}$

Q5 The ratio $\frac{1}{2}$ is read "_____."

STOP • STOP • STOP • STOP • STOP • STOP • STOP • STOP • STOP

A5 one to two

---

2 Writing the ratio of 2 feet to 15 inches as $\frac{2}{15}$ would be incorrect, because 2 feet is not $\frac{2}{15}$ of 15 inches. Two feet is actually 24 inches, so the ratio is really $\frac{24}{15}$ or $\frac{8}{5}$.

> When a comparison between two quantities is desired, the quantities should be expressed in the same unit of measurement. However, the ratio is an abstract number (written without a unit of measurement).

Q6   Determine the ratio of $5 to 40 cents.

STOP • STOP • STOP • STOP • STOP • STOP • STOP • STOP • STOP

A6 $\dfrac{25}{2}$: $5 = 500$ cents, hence, $\dfrac{500}{40}$

Q7   Express as a ratio:
    a. 15 inches to 2 feet   **b.** 10 ounces to 1 pound

    c. 45 minutes to 1 hour   **d.** a nickel to a dime

STOP • STOP • STOP • STOP • STOP • STOP • STOP • STOP • STOP

A7   **a.** $\dfrac{5}{8}$:   2 feet = 24 inches   **b.** $\dfrac{5}{8}$:   1 pound = 16 ounces

          c. $\dfrac{3}{4}$:   1 hour = 60 minutes   **d.** $\dfrac{1}{2}$

> **3**   A statement of equality between two ratios (or fractions) is called a *proportion*. A proportion is actually a special type of equation. The statement $\dfrac{1}{3} = \dfrac{13}{39}$ is a proportion and could be written
>
> $1:3 = 13:39$
>
>   A proportion consists of four terms labeled as follows:
>
> 1st term : 2nd term = 3rd term : 4th term
>
> In the proportion $\dfrac{1}{3} = \dfrac{13}{39}$, the first term is 1, the second term is 3, the third term is 13, and the fourth term is 39.

Q8   **a.** A statement of equality between two ratios is called a _____.

          b. The four parts of a proportion are called _____.

          c. In the proportion $\dfrac{2}{3} = \dfrac{10}{15}$, what is the third term? _____

          d. In the proportion $1:8 = 2:16$, what is the fourth term? _____

STOP • STOP • STOP • STOP • STOP • STOP • STOP • STOP • STOP

A8   **a.** proportion   **b.** terms   **c.** 10   **d.** 16

Q9   Use colons to write the proportion $\dfrac{7}{2} = \dfrac{21}{6}$.

STOP • **STOP** • **STOP** • **STOP** • **STOP** • **STOP** • **STOP** • **STOP** • **STOP**

A9 $7:2 = 21:6$

Q10   Write the proportion $25:5 = 5:1$ in fraction form.

STOP • **STOP** • **STOP** • **STOP** • **STOP** • **STOP** • **STOP** • **STOP** • **STOP**

A10 $\dfrac{25}{5} = \dfrac{5}{1}$

---

4   In a proportion, the first and fourth terms are called the *extremes* and the second and third terms are called the *means*. In the proportion

$$\frac{2}{5} = \frac{14}{35}$$

2 and 35 are the extremes and 5 and 14 are the means.

$$2:5 = 14:35$$

means

extremes

An important property of proportions is that *the product of the means is equal to the product of the extremes.* In the proportion $\dfrac{2}{5} = \dfrac{14}{35}$ the product of the means is $5 \cdot 14 = 70$ and the product of the extremes is $2 \cdot 35 = 70$.

Q11   **a.** In a proportion the first and fourth terms are called the _____.

b. In a proportion the product of the _____ is equal to the product of the

_____.

c. Name the means of $\dfrac{7}{21} = \dfrac{1}{3}$. _____

d. What is the product of the extremes in $\dfrac{2}{5} = \dfrac{8}{20}$? _____

STOP • **STOP** • **STOP** • **STOP** • **STOP** • **STOP** • **STOP** • **STOP** • **STOP**

A11   **a.** extremes   **b.** means, extremes   **c.** 21 and 1   **d.** 40: $2 \cdot 20$

---

5   The truth of a proportion can be determined by using the property that the product of the means is equal to the product of the extremes. The statement of equality between the ratios will be true only if the above property is satisfied. The proportion $\dfrac{7}{2} = \dfrac{14}{4}$ is true because

$$\underbrace{2 \cdot 14}_{} = \underbrace{7 \cdot 4}_{}$$

product of means   product of extremes

However, the proportion $\dfrac{6}{7} = \dfrac{5}{8}$ is false because

$7 \cdot 5 \neq 6 \cdot 8$

Q12   **a.** In the proportion $\dfrac{7}{8} = \dfrac{8}{9}$ the product of the means is _____ and

b. the product of the extremes is _____.

c. Is the proportion true or false? _____.

STOP   •   STOP   •   STOP   •   STOP   •   STOP   •   STOP   •   STOP   •   STOP   •   STOP

A12   **a.** $8 \cdot 8 = 64$ **b.** $7 \cdot 9 = 63$ **c.** false

Q13   Express $\dfrac{7}{3} = \dfrac{21}{9}$ as a product of its means and extremes.

STOP   •   STOP   •   STOP   •   STOP   •   STOP   •   STOP   •   STOP   •   STOP   •   STOP

A13 $3 \cdot 21 = 7 \cdot 9$

6   The property that the product of the means is equal to the product of the extremes, in a proportion, can be developed and generalized using the principles of equality. That is, for the proportion

$$\frac{a}{b} = \frac{c}{d}$$

the means are b and c and the extremes are a and d. The proportion can be rewritten as follows:

$$\frac{a}{b} = \frac{c}{d}$$

$$bd\left(\frac{a}{b}\right) = bd\left(\frac{c}{d}\right) \quad \text{(multiplying both sides by the LCD, } bd\text{)}$$

Hence,

$ad = bc$

which states that the product of the extremes is equal to the product of the means. The last equation could also be written

$bc = ad$

Q14   Is $\dfrac{7}{3} = \dfrac{21}{9}$ true or false? _____

STOP   •   STOP   •   STOP   •   STOP   •   STOP   •   STOP   •   STOP   •   STOP   •   STOP

A14   true: $3 \cdot 21 = 7 \cdot 9$
                      $63 = 63$

Q15   Is each of the following a true proportion?

    a. $\dfrac{7}{14} = \dfrac{21}{42}$ _____

                                                                 b. $\dfrac{4}{6} = \dfrac{20}{24}$ _____

    c. $3 : 8 = 15 : 40$ _____   **d.** $6 : 9 = 48 : 72$ _____

    e. $\dfrac{15}{20} = \dfrac{5}{4}$ _____   **f.** $\dfrac{9}{15} = \dfrac{72}{120}$ _____

STOP • **STOP** • **STOP** • **STOP** • **STOP** • **STOP** • **STOP** • **STOP** • **STOP**

A15   **a.** yes: $14 \cdot 21 = 7 \cdot 42$ **b.** no: $6 \cdot 20 \neq 4 \cdot 24$
           c. yes: $8 \cdot 15 = 3 \cdot 40$ **d.** yes: $9 \cdot 48 = 6 \cdot 72$
           e. no: $20 \cdot 5 \neq 15 \cdot 4$ **f.** yes: $15 \cdot 72 = 9 \cdot 120$

Q16   Write the proportion $\dfrac{m}{n} = \dfrac{x}{y}$ so that the product of the means is equal to the product
              of the extremes.

STOP • **STOP** • **STOP** • **STOP** • **STOP** • **STOP** • **STOP** • **STOP** • **STOP**

A16 $nx = my$

This completes the instruction for this section.

6.1   Exercises

    1. The comparison of one quantity to another by division is called a _____.
    2. Find the following ratios in fraction form:

        a. 16 to 8   **b.** $\dfrac{9}{27}$ **c.** 25 to 10   **d.** 6 to 3

        e. $\dfrac{1}{4}$ to $\dfrac{1}{12}$ **f.** 0.2 to 2   **g.** 16 to 64   **h.** 9 to $\dfrac{1}{3}$

        i. 25 to 5   **j.** 6 to 2   **k.** $\dfrac{1}{4}$ to 2   **l.** 0.2 to $\dfrac{1}{10}$

    3. Find the following ratios in fraction form:
        a. a quart to a pint   **b.** a quart to a gallon
        c. a yard to two inches   **d.** an inch to a foot
        e. five minutes to an hour   **f.** a day to an hour

4. The statement of equality between two ratios is called a _____.

5. Name the means in the proportion $\dfrac{6}{13} = \dfrac{12}{26}$.

6. Name the extremes in the proportion $1:5 = 7:35$.

7. In a proportion the product of the _____ is equal to the product of the _____.

8. Which of the following represent true proportions?

    a. $\dfrac{7}{2} = \dfrac{35}{10}$ **b.** $\dfrac{2}{7} = \dfrac{3}{11}$

    c. $6:2 = 23:8$ **d.** $9:11 = 63:77$

    e. $\dfrac{7}{15} = \dfrac{8}{14}$ **f.** $\dfrac{3}{18} = \dfrac{1}{6}$

6.1 Exercise Answers

1. ratio

2. **a.** $\dfrac{2}{1}$ **b.** $\dfrac{1}{3}$ **c.** $\dfrac{5}{2}$ **d.** $\dfrac{2}{1}$ **e.** $\dfrac{3}{1}$ **f.** $\dfrac{1}{10}$ **g.** $\dfrac{1}{4}$

     h. $\dfrac{27}{1}$ **i.** $\dfrac{5}{1}$ **j.** $\dfrac{3}{1}$ **k.** $\dfrac{1}{8}$ **l.** $\dfrac{2}{1}$

3. **a.** $\dfrac{2}{1}$ **b.** $\dfrac{1}{4}$ **c.** $\dfrac{18}{1}$ **d.** $\dfrac{1}{12}$ **e.** $\dfrac{1}{12}$ **f.** $\dfrac{24}{1}$

4. proportion

5. 13 and 12

6. 1 and 35

7. means, extremes

8. a, d, and f

6.2 Solution of a Proportion

1 Since a proportion is a special type of equation, it can be solved using the principles established for solving equations. For example, the proportion

$$\frac{n}{3} = \frac{4}{21}$$

may be solved as follows:

$$\frac{n}{3} = \frac{4}{21}$$

$$3\left(\frac{n}{3}\right) = 3\left(\frac{4}{21}\right) \qquad \text{(multiplying both sides by 3)}$$

$$n = \frac{4}{7}$$

Q1   Solve the proportion $\dfrac{n}{14} = \dfrac{51}{7}$.

STOP • STOP • STOP • STOP • STOP • STOP • STOP • STOP • STOP

A1   102: $14\left(\dfrac{n}{14}\right) = 14\left(\dfrac{51}{7}\right)$

$n = 2 \cdot 51$

Q2   Solve the following proportions:

a. $\dfrac{4}{7} = \dfrac{n}{28}$ b. $\dfrac{5}{14} = \dfrac{n}{42}$

STOP • STOP • STOP • STOP • STOP • STOP • STOP • STOP • STOP

A2   a. 16: $28\left(\dfrac{4}{7}\right) = 28\left(\dfrac{n}{28}\right)$ b. 15

Q3   Solve the following proportions:

a. $\dfrac{4}{17} = \dfrac{n}{34}$ b. $\dfrac{n}{30} = \dfrac{70}{175}$

STOP • STOP • STOP • STOP • STOP • STOP • STOP • STOP • STOP

A3   a. 8   b. 12

---

2   The missing term in a proportion can be any of the four terms of the proportion. For example, in the proportion

$$\frac{15}{540} = \frac{2}{n}$$

the missing term is the fourth term. This proportion can be solved using principles of equality; however, the solution is simplified by using the property that the product of the means is equal to the product of the extremes. That is,

$$540 \cdot 2 = 15 \cdot n$$

$$\frac{540 \cdot 2}{15} = n$$

$$72 = n$$

Q4 Solve the proportion $\dfrac{12}{n} = \dfrac{72}{30}$.

STOP • STOP • STOP • STOP • STOP • STOP • STOP • STOP • STOP

A4 5: $n \cdot 72 = 12 \cdot 30$

$$n = \frac{12 \cdot 30}{72}$$

Q5 Solve the proportion $\dfrac{2}{n} = \dfrac{6}{15}$.

STOP • STOP • STOP • STOP • STOP • STOP • STOP • STOP • STOP

A5 5: $n \cdot 6 = 2 \cdot 15$

$$n = \frac{2 \cdot 15}{6}$$

Q6 Solve the following proportions:

    a. $\dfrac{18}{63} = \dfrac{440}{n}$ **b.** $\dfrac{108}{27} = \dfrac{52}{n}$

STOP • STOP • STOP • STOP • STOP • STOP • STOP • STOP • STOP

A6 **a.** 1,540: $63 \cdot 440 = 18 \cdot n$ **b.** 13

$$\frac{63 \cdot 440}{18} = n$$

Q7 Solve the following proportions:

    a. $\dfrac{n}{12} = \dfrac{2.5}{5}$ **b.** $\dfrac{0.75}{n} = \dfrac{19}{114}$

STOP • STOP • STOP • STOP • STOP • STOP • STOP • STOP • STOP

A7 **a.** 6   **b.** 4.5

3   Various types of problems can be solved by writing the conditions of the problem as a proportion and solving for the missing term. Two general statements will aid in properly placing the conditions of a problem in the proportion.

1. Each ratio should be a comparison of similar things.
2. In every proportion both ratios must be written in the same order of value.

Example: If 6 pencils cost 25 cents, how much will 12 pencils cost?

Solution

Let c represent the cost of 12 pencils and substitute the conditions of the problem in the following proportion:

$$\frac{\text{small number of pencils}}{\text{large number of pencils}} = \frac{\text{small cost}}{\text{large cost}}$$

Each ratio is a comparison of similar things. Each ratio is written in the same order of value. Hence,

$$\frac{6}{12} = \frac{25}{c}$$

$$12 \cdot 25 = 6 \cdot c$$

$$50 = c$$

Therefore, 12 pencils will cost 50 cents. The quantities in the above proportion are *directly* related because an increase in the number of pencils causes a corresponding increase in the cost.

Q8   If 3 cans of cola cost 25 cents, how much will 15 cans cost?

$$\frac{\text{large number}}{\text{small number}} = \frac{\text{large cost}}{\text{small cost}}$$

(Note that the order of the quantities in the ratios has been reversed.)

STOP • **STOP** • **STOP** • **STOP** • **STOP** • **STOP** • **STOP** • **STOP** • **STOP**

A8 $1.25: $\dfrac{15}{3} = \dfrac{c}{0.25}$ ⟵ large cost because 15 cans will cost more

Q9   In a proportion, if an increase in one quantity causes a corresponding increase in another quantity, the quantities are _____ related.

STOP • **STOP** • **STOP** • **STOP** • **STOP** • **STOP** • **STOP** • **STOP** • **STOP**

A9   directly

Q10   An increase in the height of an object will cause a corresponding _____ in
the length of its shadow.

increase/decrease

STOP • **STOP** • **STOP** • **STOP** • **STOP** • **STOP** • **STOP** • **STOP** • **STOP**

A10	increase

Q11	If an 18-foot pole casts a 15-foot shadow, what will be the length of the shadow cast by a 50-foot pole?

STOP • **STOP** • **STOP** • **STOP** • **STOP** • **STOP** • **STOP** • **STOP** • **STOP**

A11	$41\frac{2}{3}$ feet: $\frac{18}{50} = \frac{15}{s}$ (s is the unknown length and will be larger than 15 feet)

Q12	A car can travel 47 miles on 3 gallons of gas. How far can it go on 18 gallons?

STOP • **STOP** • **STOP** • **STOP** • **STOP** • **STOP** • **STOP** • **STOP** • **STOP**

A12	282 miles: $\frac{18}{3} = \frac{N}{47}$ or $\frac{3}{18} = \frac{47}{N}$

Q13	If $1,000 worth of insurance costs $20.80, what will $15,000 of insurance cost?

STOP • **STOP** • **STOP** • **STOP** • **STOP** • **STOP** • **STOP** • **STOP** • **STOP**

A13	$312: $\frac{1}{15} = \frac{20.80}{N}$

Q14	If valve stems cost 3 for 10 cents, what is the cost of 11 valve stems?

STOP • **STOP** • **STOP** • **STOP** • **STOP** • **STOP** • **STOP** • **STOP** • **STOP**

A14	37 cents, rounded off to the nearest cent: $\frac{3}{11} = \frac{10}{c}$

Q15	If 3.5 tons of top soil cost $20, what will 12.5 tons of top soil cost?

STOP • **STOP** • **STOP** • **STOP** • **STOP** • **STOP** • **STOP** • **STOP** • **STOP**

A15 $71.43, rounded off to the nearest cent

---

4   If an increase in one quantity causes a corresponding decrease in another quantity or a decrease in one quantity causes a corresponding increase in the other, the quantities are *indirectly* or inversely related. For example, if the number of workers on a job is increased, the time required to do the work is decreased. Problems involving indirect relationships can be solved using the principles established earlier.

Example: If 2 workers can build a garage in 5 days, how long will it take 6 workers, assuming that they all work at the same rate?

Solution

Since the number of days required to build the garage becomes smaller as the number of workers becomes larger, the quantities vary indirectly. Thus,

$$\frac{2 \text{ (smaller no. workers)}}{6 \text{ (larger no. workers)}} = \frac{N \text{ (smaller no. days)}}{5 \text{ (larger no. days)}}$$

$$6N = 10$$

$$N = 1\frac{2}{3} \text{ days}$$

Q16   Three workers can build a house in 10 days. Will it take 8 workers working at the same rate more or less time to build the house? _____

STOP • STOP • STOP • STOP • STOP • STOP • STOP • STOP • STOP

A16   less

Q17   A plane flies from Detroit to Philadelphia in 1 hour and 40 minutes at 440 miles per hour. Will it take more or less time to make the trip flying at 380 miles per hour? _____

STOP • STOP • STOP • STOP • STOP • STOP • STOP • STOP • STOP

A17   more

Q18   Eight students sell 3,800 tickets to a football game. Will it take more or fewer students to sell 5,200 tickets selling at the same rate? _____

STOP • STOP • STOP • STOP • STOP • STOP • STOP • STOP • STOP

A18   more

Q19   If 4 men can clear the snow near a school in 6 hours, how long will it take 12 men working at the same rate to do the job?

$$\frac{4}{12} = \frac{(\ \ )}{(\ \ )}$$

(Let N equal the unknown time and ask yourself, "Does the time get smaller or larger?" The smaller time must be placed in the numerator and the larger time in the denominator.)

STOP • STOP • STOP • STOP • STOP • STOP • STOP • STOP • STOP

A19   2:   smaller number of men $\longrightarrow$ $\dfrac{4}{12} = \dfrac{N}{6}$ $\longleftarrow$ smaller time

Q20   A plane takes 3 hours at a speed of 320 miles per hour to go from Chicago to New York. How fast must the plane fly to make the trip in 2.5 hours?

STOP • STOP • STOP • STOP • STOP • STOP • STOP • STOP • STOP

A20   384: $\text{greatest time} \longrightarrow \dfrac{3}{2.5} = \dfrac{N}{320} \longleftarrow \text{greatest speed}$

(The problem could have been solved by placing the smaller numbers on top; that is,

$\dfrac{2.5}{3} = \dfrac{320}{N}$.)

Q21   When two pulleys are belted together, the revolutions per minute (rpm) vary inversely as the size of the pulleys. A 20-inch pulley running at 180 rpm drives an 8-inch pulley. Find the rpm of the 8-inch pulley.

STOP • STOP • STOP • STOP • STOP • STOP • STOP • STOP • STOP

A21   450 rpm: $\dfrac{20}{8} = \dfrac{N}{180}$

This completes the instruction for this section.

6.2   Exercises

1. If an increase in one quantity causes a decrease in another quantity, the quantities are _____ related.
2. If a decrease in one quantity causes a decrease in another quantity, the quantities are _____ related.
3. Solve the following proportions:

     a. $\dfrac{5}{7} = \dfrac{N}{49}$ **b.** $\dfrac{49}{98} = \dfrac{7}{N}$ **c.** $\dfrac{6}{18} = \dfrac{N}{21}$ **d.** $\dfrac{5}{2} = \dfrac{15}{N}$

     e. $\dfrac{4}{3} = \dfrac{N}{21}$ **f.** $\dfrac{25}{N} = \dfrac{15}{24}$ **g.** $\dfrac{2}{N} = \dfrac{4}{8}$ **h.** $\dfrac{3}{N} = \dfrac{18}{27}$

     i. $\dfrac{9}{2} = \dfrac{5}{x}$ **j.** $\dfrac{\frac{2}{3}}{9} = \dfrac{6}{x}$ **k.** $\dfrac{\frac{1}{2}}{\frac{1}{4}} = \dfrac{x}{\frac{3}{4}}$ **l.** $\dfrac{0.3}{0.8} = \dfrac{x}{\frac{1}{8}}$

4. If 30 men do a job in 12 days, how long will it take 20 men to do the same work?
5. If the wing area of a model airplane is 3 square feet and it can lift 10 pounds, how much can a wing whose area is 149 square feet lift?
6. A swimming pool large enough for 6 people must hold 5,000 gallons of water. How much water must a pool large enough for 15 people hold?
7. If a car goes 52 miles on 3 gallons of gas, how far will it go on 11 gallons?
8. If lobster tails are 3 boxes for $3.60, how much will 14 boxes cost?
9. A train takes 24 hours at a rate of 17 miles per hour to travel a certain distance. How fast would a plane have to fly in order to cover the same distance in 3 hours?
10. If electricity costs 10 cents for 4 kilowatt-hours, how many kilowatt-hours could you use for $7.20?

6.2   Exercise Answers

1. indirectly
2. directly
3. **a.** 35   **b.** 14   **c.** 7   **d.** 6

   e. 28   **f.** 40   **g.** 4   **h.** $\dfrac{9}{2}$ or $4\dfrac{1}{2}$

   i. $\dfrac{10}{9}$ or $1\dfrac{1}{9}$ **j.** 81   **k.** $\dfrac{3}{2}$ or $1\dfrac{1}{2}$ **l.** 0.046875

4. 18 days
5. $496\dfrac{2}{3}$ pounds
6. 12,500 gallons
7. $190\dfrac{2}{3}$ miles
8. $16.80
9. 136 miles per hour
10. 288 kilowatt-hours

Chapter 6 Sample Test

At the completion of Chapter 6 it is expected that you will be able to work the following problems.

6.1   Introduction to Ratio and Proportion

1. Determine the ratio of (write in fraction form):
   a. 15 years to 10 years   **b.** $5.00 to 60 cents

   c. 1 foot to 4 yards   **d.** $2\dfrac{1}{2}$ to $3\dfrac{1}{4}$

2. **a.** Name the means in the proportion $\dfrac{3}{8} = \dfrac{12}{32}$.

   b. What is the product of the extremes in the proportion $7 : 12 = 14 : 24$?

   c. Is $\dfrac{2}{3} = \dfrac{4}{5}$ a true proportion?

   d. Why?

6.2 **Solution of a Proportion**

3. Solve the following proportions:

   a. $\dfrac{16}{52} = \dfrac{M}{36}$ **b.** $\dfrac{5}{12} = \dfrac{M}{36}$ **c.** $\dfrac{M}{12} = \dfrac{2.5}{5}$ **d.** $\dfrac{4}{5} = \dfrac{7}{M}$

4. Solve the following problems:
   a. Four bricklayers can brick a building in 15 hours. How long would it take with 6 men working at the same rate?
   b. If rain is falling at the rate of 0.5 inch per hour, how many inches will fall in 15 minutes?
   c. A well produces 4,800 gallons of water in 120 minutes. How long will it take to produce 5,700 gallons of water?
   d. An airplane travels from one city to another in 160 minutes at the rate of 270 miles per hour. How long will it take to make the return trip, traveling at 240 miles per hour?

Chapter 6 Sample Test Answers

1. **a.** $\dfrac{3}{2}$ **b.** $\dfrac{25}{3}$ **c.** $\dfrac{1}{12}$ **d.** $\dfrac{10}{13}$

2. **a.** 8 and 12   **b.** 168   **c.** no   **d.** $3 \times 4 \neq 2 \times 5$

3. **a.** $\dfrac{144}{13}$ or $11\dfrac{1}{13}$ **b.** 15   **c.** 6   **d.** $\dfrac{35}{4}$ or $8\dfrac{3}{4}$

4. **a.** 10 hours   **b.** $\dfrac{1}{8}$ inch   **c.** 142 minutes 30 seconds, or $142\dfrac{1}{2}$ minutes

   d. 180 minutes

Chapter 7

Uses of Algebra

The purpose of much of the mathematical training you receive is to permit the solving of numerical problems encountered in real life. These situations might be relatively simple, such as adjusting a recipe for a different number of people, or much more complicated, such as determining the orbit of a satellite. The ability to translate a situation into an appropriate mathematical sentence (equation) enables one to cope with many problems in everyday life, as well as the problems encountered in science and industry, in an orderly, logical manner.

This chapter will discuss two fundamental approaches to solving numerical problems. One involves writing and solving an equation which represents the arithmetic of the situation. The other involves the use of formulas which govern the relationships present in the situation. In either case, correct interpretation and translation of English phrases into algebraic expressions is essential.

7.1   Expressing English Phrases As Algebraic Expressions

---

1   Before you can solve mathematical situations, you must be able to translate English phrases into algebraic expressions. For example, letting a letter represent the number, the algebraic expressions on the right represent the English phrases on the left.

English phrase	Algebraic expression
five more than a number	$x + 5$
a number more than five	$5 + y$
three times a number	$3c$
a number minus two	$h - 2$
two minus a number	$2 - n$

Q1   Let n represent the number.*   Write an algebraic expression for:

   a. seven more than the number   _____

   b. the number more than seven   _____

   c. four times the number   _____

   d. the number minus six   _____

   e. six minus the number   _____

STOP • STOP • STOP • STOP • STOP • STOP • STOP • STOP • STOP

A1   **a.** $n + 7$ **b.** $7 + n$ **c.** $4n$ **d.** $n - 6$
     e. $6 - n$

*n will be used throughout this section to represent "the number."

---

2   Key words in the phrase indicate the operation involved. Some common words or phrases that indicate addition are plus, added to, more than, sum of, and increased by. These are the clues to help decide what operation to use.

Q2   Write an algebraic expression for:

    a. the number increased by one   _____

    b. the number added to nine   _____

    c. eight plus the number   _____

    d. the sum of the number and four   _____

    e. ten more than the number   _____

STOP • **STOP** • **STOP** • **STOP** • **STOP** • **STOP** • **STOP** • **STOP** • **STOP**

A2   **a.** $n + 1$ **b.** $9 + n$ **c.** $8 + n$ **d.** $n + 4$

    e. $n + 10$

---

3   Since addition is a commutative operation, $n + 1 = 1 + n$. However, $n + 1$ and $1 + n$ come from different English phrases. That is, "the number plus one" translates $n + 1$, while "one plus the number" translates $1 + n$. It is important to translate precisely so that the meaning of the phrase does not become distorted. The sum of two and the number is translated $2 + n$. The sum of the number and two is translated $n + 2$.

Q3   Translate each phrase into an algebraic expression:

    a. the sum of the number and ten   _____

    b. the sum of ten and the number   _____

STOP • **STOP** • **STOP** • **STOP** • **STOP** • **STOP** • **STOP** • **STOP** • **STOP**

A3   **a.** $n + 10$ **b.** $10 + n$

---

4   Some common words or phrases that indicate subtraction are difference of, difference between, take away, taken from, reduced by, less, less than, diminished by, subtracted from, decreased by, smaller than, minus, depreciate, and borrowed from.

Q4   Translate each phrase into an algebraic expression:

    a. five taken from the number   _____

    b. the number taken from five   _____

    c. the number less seven   _____

    d. seven less the number   _____

    e. the number diminished by nine   _____

    f. nine diminished by the number   _____

STOP • **STOP** • **STOP** • **STOP** • **STOP** • **STOP** • **STOP** • **STOP** • **STOP**

A4   **a.** $n - 5$ **b.** $5 - n$ **c.** $n - 7$

    d. $7 - n$ **e.** $n - 9$ **f.** $9 - n$

---

5   Since subtraction is not commutative, $n - 5 \neq 5 - n$. If you write a translation of subtraction in the wrong order, it will not only be a distortion, it will be wrong. This serves as a good example to emphasize the importance of translating precisely. "Less" and "less than" require special attention. "Five less two" translates $5 - 2$. "The number less two" translates $n - 2$. "Two less than five" translates $5 - 2$. "Two less than the number" translates $n - 2$.

Q5   Translate each phrase into an algebraic expression:

    a. the number less one   _____

    b. one less than the number   _____

    c. one less the number   _____

    d. the number less than one   _____

STOP • **STOP** • **STOP** • **STOP** • **STOP** • **STOP** • **STOP** • **STOP** • **STOP**

A5   **a.** $n - 1$ **b.** $n - 1$ **c.** $1 - n$ **d.** $1 - n$

6   In the phrase "the difference between" the values are listed in the order in which they are used. For example, "the difference between the number and two" is $n - 2$. "The difference between two and the number" is $2 - n$.

Q6   Translate each phrase into an algebraic expression:

    a. the difference between the number and three   _____

    b. the difference between three and the number   _____

STOP • **STOP** • **STOP** • **STOP** • **STOP** • **STOP** • **STOP** • **STOP** • **STOP**

A6   **a.** $n - 3$ **b.** $3 - n$

7   Some common words or phrases that indicate multiplication are product of, times, double, twice, triple, and of.

Q7   Translate each phrase into an algebraic expression:

    a. twice the number   _____

    b. five times the number   _____

    c. the product of three and the number   _____

    d. two-thirds of the number   _____

    e. triple the number   _____

STOP • **STOP** • **STOP** • **STOP** • **STOP** • **STOP** • **STOP** • **STOP** • **STOP**

A7   **a.** $2n$ **b.** $5n$ **c.** $3n$ **d.** $\frac{2}{3}n$ **e.** $3n$

8   It is sometimes necessary to express a percent of a number. For example, 60 percent of a number. Since 60 percent is equivalent to $\frac{3}{5}$ or 0.6, 60 percent of a number can be written $\frac{3}{5}n$ or $0.6n$.

Q8   Translate each phrase into an algebraic expression:

    a. 20% of a number   _____

b. 15% of a number   _____

c. 50% of a number   _____

d. 75% of a number   _____

e. 100% of a number   _____

f. 125% of a number   _____

STOP • STOP • STOP • STOP • STOP • STOP • STOP • STOP • STOP

A8   **a.** $\frac{1}{5}n$ or $0.2n$ **b.** $\frac{3}{20}n$ or $0.15n$ **c.** $\frac{1}{2}n$ or $0.5n$

    d. $\frac{3}{4}n$ or $0.75n$ **e.** $1n$ or n **f.** $1\frac{1}{4}n$ or $\frac{5}{4}n$ or $1.25n$

---

9   Some common words or phrases that indicate division are divided into, quotient of, divided by, and divides.

Q9   Translate each phrase into an algebraic expression:

    a. the number divided by six   _____

    b. the number divides six   _____

    c. nine divided into the number   _____

    d. the quotient of the number by two   _____

    e. nine divided by the number   _____

STOP • STOP • STOP • STOP • STOP • STOP • STOP • STOP • STOP

A9   **a.** $n \div 6$ or $\frac{n}{6}$ **b.** $6 \div n$ or $\frac{6}{n}$ **c.** $n \div 9$ or $\frac{n}{9}$ **d.** $n \div 2$ or $\frac{n}{2}$

    e. $9 \div n$ or $\frac{9}{n}$

---

10   Skill in writing algebraic expressions involving more than one operation is also necessary.

English phrase	Algebraic expression
three more than five times the number	$5n + 3$
twice the sum of the number and four	$2(n + 4)$

Q10   Translate each phrase into an algebraic expression:

    a. the sum of the number, and five times the number   _____

    b. three times as much as two more than the number   _____

    c. twice the number, minus five   _____

    d. the difference between three times the number and the number   _____

    e. two less than the number doubled   _____

    f. the number divided by three, plus eight times the number   _____

    g. the number depreciated by five percent of the number   _____

    h. the difference between fifteen and one-half of the number   _____

STOP • STOP • STOP • STOP • STOP • STOP • STOP • STOP • STOP

A10

    a. $n + 5n$ **b.** $3(n + 2)$ **c.** $2n - 5$

    d. $3n - n$ **e.** $2n - 2$ **f.** $\dfrac{n}{3} + 8n$

    g. $n - \dfrac{1}{20}n$ or $n - 0.05n$ **h.** $15 - \dfrac{1}{2}n$

---

11   The word "and" plays an important role.

Example 1: Write the sum of 2, 3, *and* 4.

Solution

$2 + 3 + 4$.

Example 2: Write the difference of 4 *and* 3.

Solution

$4 - 3$.

Example 3: Write the product of 2, 3, *and* 4.

Solution

$2(3)(4)$.

Example 4: Write the quotient of 15 *and* 3.

Solution

$15 \div 3$ or $\dfrac{15}{3}$

    In each example, "and" simply separates the last number listed from the previous numbers. The word "and" does not indicate an operation. The phrases sum, difference of, product, and quotient indicate the operation to be performed.

Example 5: The product of one more than a number *and* one less than the same number.

Solution

$(n + 1)(n - 1)$

Example 6: The quotient of a number increased by six and the same number decreased by four.

Solution

$\dfrac{n + 6}{n - 4}$

Q11   Write an algebraic expression for:

    a. the sum of a number, twice the number, and three times the number   _____

    b. the product of six, a number less two, and two less than the same number   _____

STOP • **STOP** • **STOP** • **STOP** • **STOP** • **STOP** • **STOP** • **STOP** • **STOP**

| A11 | **a.** $n + 2n + 3n$ | **b.** $6(n - 2)(n - 2)$ |

This completes the instruction for this section.

7.1 Exercises

Write each phrase as an algebraic expression. Use n for the number.

1. the number added to seven
2. the sum of the number and twelve
3. three increased by the number
4. the number doubled decreased by seven
5. the number less five
6. six less than the number
7. the difference between twice the number and nine
8. three times the number, diminished by eight
9. five times the number
10. the product of three and the number, depreciated by ten
11. sixteen divided into the number
12. two-fifths of the number
13. twelve percent of the number
14. five percent of the sum of the number and fifty
15. the sum of the number and one, divided by four
16. the product of five, the number plus one, and the number minus one

7.1 Exercise Answers

1. $7 + n$	2. $n + 12$	3. $3 + n$	4. $2n - 7$
5. $n - 5$	6. $n - 6$	7. $2n - 9$	8. $3n - 8$
9. $5n$	10. $3n - 10$	11. $n \div 16$ or $\dfrac{n}{16}$	12. $\dfrac{2}{5}n$

13. $\dfrac{3}{25}n$ or $0.12n$ 14. $\dfrac{1}{20}(n + 50)$ or $0.05(n + 50)$

15. $(n + 1) \div 4$ or $\dfrac{n + 1}{4}$ 16. $5(n + 1)(n - 1)$

7.2 Solving Simple Word Problems

| 1 | A simple word statement such as "twice the number, minus three, is equal to seven" may be translated into an equation. Letting n be the number: Twice the number minus three is equal to seven becomes $2n - 3 = 7$. "is equal to" translates as "$=$." |

| Q1 | Translate each word statement into an equation. Let x be the number. |

    a. Seventeen subtracted from the number is equal to nine. _____

    b. The sum of four times the number and six is equal to thirty-eight. _____

c. Three times the number, plus two more than the number, is equal to eighteen. _____

d. Three more than twice a certain number is seven less than that number. (*Note:* "is" also translates as "=.") _____

e. A number increased by six is two less than three times as large as the same number. _____

STOP • **STOP** • **STOP** • **STOP** • **STOP** • **STOP** • **STOP** • **STOP** • **STOP**

A1 a. $x - 17 = 9$ b. $4x + 6 = 38$ c. $3x + x + 2 = 18$
    d. $2x + 3 = x - 7$ e. $x + 6 = 3x - 2$

---

2 Any of the English statements above could be expressed as a problem. For example, "If the sum of four times a number and six is equal to thirty-eight, what is the number?" To solve this problem and others of this type, use the procedure stated below.

Step 1: Let some letter (variable), such as n, represent the number.

Step 2: Translate the statement of the problem into an equation.

Step 3: Solve the equation for the value of the letter.

Step 4: Check the solution against the statement of the problem.

The solution to the problem stated above would be:

Step 1: Let n = the number.

Step 2: $4n + 6 = 38$

Step 3: $4n = 32$
            $n = 8$

Step 4: Since the sum of four times eight (32) and six is equal to thirty-eight, eight is the correct solution.

Q2 Complete the four steps outlined in Frame 2 to solve this problem. If seventeen subtracted from a number is nine, what is the number?

a. Step 1: Let $x = $ _____ .

b. Step 2: _____ .

c. Step 3:

d. Step 4: Since seventeen subtracted from _____ is nine, _____ is the correct solution.

STOP • **STOP** • **STOP** • **STOP** • **STOP** • **STOP** • **STOP** • **STOP** • **STOP**

A2 a. the number b. $x - 17 = 9$ c. $x - 17 + 17 = 9 + 17$ (optional)
                                        $x = 26$

d. twenty-six (26), twenty-six (26)

Q3 Use the four-step procedure to solve each of the following word problems. (Do not forget the check.) Use n for the number.

a. Three times the number, plus two more than the number, is equal to eighteen. What is the number?

b. If three more than twice a certain number is seven less than that number, what is the number?

STOP • STOP • STOP • STOP • STOP • STOP • STOP • STOP • STOP

A3   **a.** 4: let n = the number

$$3n + (n + 2) = 18$$
$$4n + 2 = 18$$
$$4n = 16$$
$$n = 4$$

Since three times four (12) plus two more than four (6) is equal to eighteen, four (4) is the correct solution.

b. ⁻10: let n = the number

$$2n + 3 = n - 7$$
$$2n + 3 - n = n - 7 - n$$
$$n + 3 = {}^-7$$
$$n = {}^-10$$

Since three more than twice negative ten ($-20 + 3$ or ⁻17) is seven less than negative ten (⁻10 − 7 or ⁻17), negative ten (⁻10) is the correct solution.

3	The purpose of the word statements in the check is to emphasize that the solution should be checked against the statement of the problem. If an incorrect equation is written, checking the equation will only tell you that you solved the equation correctly. In this case, you still have the incorrect solution for the problem.

Q4   Solve this problem. A number increased by six is two less than three times as large as the same number. What is the number? Use x for the number.

STOP • STOP • STOP • STOP • STOP • STOP • STOP • STOP • STOP

A4   4: let x = the number

$$x + 6 = 3x - 2$$
$$x + 6 - x = 3x - 2 - x$$
$$6 = 2x - 2$$
$$8 = 2x$$
$$4 = x$$

Since four increased by six (10) is two less than three times four (10), the solution 4 is correct.

Q5   Eighty percent of a number diminished by two is ten. What is the number? Use n for the number.

STOP • **STOP** • **STOP** • **STOP** • **STOP** • **STOP** • **STOP** • **STOP** • **STOP**

A5   15: $0.8n - 2 = 10$ or $\dfrac{4}{5}n - 2 = 10$

$$0.8n = 12$$
$$\frac{0.8n}{0.8} = \frac{12}{0.8}$$
$$n = \frac{120}{8} = 15$$

$4n - 10 = 50$ (multiplying both sides by 5)

$$4n = 60$$
$$n = 15$$

This completes the instruction for this section.

7.2   Exercises

Solve each of these word problems by use of the four-step procedure outlined in this section. Use x for the number.

1. Twice a number, minus two more than the number, is equal to fifteen. What is the number?
2. If five less than a number is divided by four, the result is seven. Find the number.
3. Five times the sum of a number and six is forty-five. What is the number?
4. If the product of five and two less than a certain number is six more than that number, what is the number?
5. If a number is added to five less than twice the number, the result is six times the original number. What is the number?
6. If one-half is subtracted from three-fifths of a number, the result is three-fourths. What is the number?
7. One-half of a number increased by five is eight less than two-thirds of the same number. What is the number?
8. A number plus twenty percent of the same number is 36. Find the number.
*9. One-fourth of the sum of the number and five, increased by three, is the same as the number less nine, divided by five. Find the number.
*10. The ratio of six less than a number to two is equal to the ratio of four more than three times the number to seven. Find the number.

7.2   Exercise Answers

1. 17: $2x - (x + 2) = 15$

2. 33: $\dfrac{x - 5}{4} = 7$

3. 3: $5(x + 6) = 45$

4. 4: $5(x - 2) = x + 6$

5. $\dfrac{-5}{3}$: $2x - 5 + x = 6x$

6. $\dfrac{25}{12}$ or $2\dfrac{1}{12}$: $\dfrac{3}{5}x - \dfrac{1}{2} = \dfrac{3}{4}$

7. 78: $\dfrac{1}{2}x + 5 = \dfrac{2}{3}x - 8$

8. 30: $x + 0.2x = 36$ or $x + \dfrac{1}{5}x = 36$

*9. $^-121$: $\dfrac{1}{4}(x + 5) + 3 = \dfrac{x - 9}{5}$

*10. 50: $\dfrac{x - 6}{2} = \dfrac{3x + 4}{7}$

7.3   **Expressing More Than One Unknown in Terms of the Same Variable**

1	Two unknowns are frequently present in a word problem. If so, it must be possible to express one unknown in terms of the other. This section will analyze techniques for expressing more than one unknown in terms of the same variable. Consider the statement: "The larger number is five times the smaller number." Letting $n =$ the smaller number, $5n =$ the larger number. In this example the larger number, $5n$, is expressed in terms of the smaller number, n.

Q1   Complete each of the following:

  a. The larger number is five more than the smaller number.

      Let $x =$ the smaller number and _____ = the larger number.

  b. The smaller number is one half of the larger number.

      Let $n =$ the larger number and _____ = the smaller number.

  c. The smaller number is six less than the larger number.

      Let $a =$ the larger number and _____ = the smaller number.

  d. The larger number is five less than three times the smaller number.

      Let $y =$ the smaller number and _____ = the larger number.

STOP • STOP • STOP • STOP • STOP • STOP • STOP • STOP • STOP

A1   **a.** $x + 5$ **b.** $\dfrac{1}{2}n$ **c.** $a - 6$ **d.** $3y - 5$

2	When expressing unknowns, it is necessary to analyze which unknown is being expressed in terms of the other. For example, in the statement "The larger of two numbers is five more than the smaller," the larger number is being expressed in terms of the smaller. Hence, the variable is used to represent the smaller number first. That is, let $n =$ the smaller number and $n + 5 =$ the larger number

Q2   A second number is one-half of the first number.

  a. In the above statement, the _____ number is expressed in terms of the
                                              first/second

      _____ number.
      first/second

b. Let $n = $ _____ number, and $\frac{1}{2}n = $ _____ number.

      first/second   first/second

STOP • STOP • STOP • STOP • STOP • STOP • STOP • STOP • STOP

A2   **a.** second, first   **b.** first, second

Q3   The smaller of two numbers is six less than the larger number.

      a. In the statement above, the _____ number is expressed in terms of the

                              smaller/larger

      _____ number.

      smaller/larger

      b. Let $y = $ _____ number, and $y - 6 = $ _____ number.

                 smaller/larger   smaller/larger

STOP • STOP • STOP • STOP • STOP • STOP • STOP • STOP • STOP

A3   **a.** smaller, larger   **b.** larger, smaller

Q4   Complete each of the following:

      a. A second number is two more than the first number. Let $x = $ _____ number,

                                          first/second

      and $x + 2 = $ _____ number.

                 first/second

      b. The larger number is twice the smaller number. Let $y = $ _____ number,

                                          smaller/larger

      and $2y = $ _____ number.

                 smaller/larger

      c. Of two numbers, the first is five more than three times the second. Let $n = $

      _____ number, and $3n + 5 = $ _____ number.

      first/second   first/second

      d. The smaller of two numbers is five less than the larger number. Let $x = $ _____

                                          smaller/larger

      number, and $x - 5 = $ _____ number.

                    smaller/larger

STOP • STOP • STOP • STOP • STOP • STOP • STOP • STOP • STOP

A4   **a.** first, second   **b.** smaller, larger   **c.** second, first   **d.** larger, smaller

---

3   When three unknowns are present in a problem, it is often possible to express all three unknowns in terms of the same variable.

      Of three numbers, the first is three times the second and the third is ten more than the second. From this statement you must analyze:

      1. The first number is expressed in terms of the second number.
      2. The third number is expressed in terms of the second.

      Hence, it is possible to write: Let

$$n = \text{second number}$$
$$3n = \text{first number}$$
$$n + 10 = \text{third number}$$

Of three numbers, the second is two less than the third and the first is five more than the second. Observe:

1. The second number is expressed in terms of the third, and
2. The first number is expressed in terms of the second number.

Hence, it is possible to write: Let

$$n = \text{third number}$$
$$n - 2 = \text{second number}$$
$$(n - 2) + 5 = n + 3 = \text{first number}$$

Q5   Complete each of the following:
   a. Of three numbers, the second is twice the first, and the third is three less than the first.

   Let $n =$ _____ number,

   _____ = _____ number, and

   _____ = _____ number.

   b. Of three numbers, the first is five times the second, and the third is twelve more than the second.

   Let $n =$ _____ number,

   _____ = _____ number, and

   _____ = _____ number.

   c. Of three numbers, the first is two more than the third, and the second is three less than the first.

   Let $n =$ _____ number,

   _____ = _____ number, and

   _____ = _____ number.

   d. Of three numbers, the first has been increased by two more than the second, and the third is triple the first.

   Let $n =$ _____ number,

   _____ = _____ number, and

   _____ = _____ number.

STOP   •   STOP   •   STOP   •   STOP   •   STOP   •   STOP   •   STOP   •   STOP   •   STOP

A5   a. $n = \text{first number}$ b. $n = \text{second number}$
            $2n = \text{second number}$ $5n = \text{first number}$
            $n - 3 = \text{third number}$ $n + 12 = \text{third number}$

      c. $n = \text{third number}$ d. $n = \text{second number}$
            $n + 2 = \text{first number}$ $n + 2 = \text{first number}$
            $n - 1 = \text{second number}$ $3(n + 2) = \text{third number}$

4   The statement "The sum of two numbers is ten" involves this open sentence: $\triangle +$ $\square = 10$, where $\triangle$ is one of the numbers and $\square$ is the other number. Consider these examples.

$$\triangle + \square = 10$$

When $\triangle$ is 2:	$2 + \square = 10$
	$\square = 10 - 2$
When $\triangle$ is 7:	$7 + \square = 10$
	$\square = 10 - 7$
When $\triangle$ is n:	$n + \square = 10$
	$\square = 10 - n$
When $\square$ is 2:	$\triangle + 2 = 10$
	$\triangle = 10 - 2$
When $\square$ is 7:	$\triangle + 7 = 10$
	$\triangle = 10 - 7$
When $\square$ is n:	$\triangle + n = 10$
	$\triangle = 10 - n$

These examples show that when the sum of two numbers is ten, if one of the numbers is n, the other number is $10 - n$.

Q6   Complete each of the following:
   a. The sum of two numbers is twenty-five. Let $n =$ one of the numbers and

   _____ = the other number.
   b. The sum of two numbers is one hundred. Let $x =$ the first number and

   _____ = the other number.

   c. Separate seventy-two into two parts. Let $a =$ one part and _____ = the other part.

STOP • **STOP** • **STOP** • **STOP** • **STOP** • **STOP** • **STOP** • **STOP** • **STOP**

A6   **a.** $25 - n$ **b.** $100 - x$ **c.** $72 - a$

5   To eliminate confusion, it is often important to note the relative size of two unknowns. For example, "the first number is five more than a second number." Let $n =$ the second number (smaller) and $n + 5 =$ the first number (larger).

Q7   Indicate larger or smaller:
   a. The second number is three less than the first number. Let $x =$ the first number

   (_____) and $x - 3 =$ the second number (_____).
   b. The second number is two more than the first number. Let $n =$ the first number

   (_____) and $n + 2 =$ the second number (_____).

STOP • **STOP** • **STOP** • **STOP** • **STOP** • **STOP** • **STOP** • **STOP** • **STOP**

A7   **a.** larger, smaller   **b.** smaller, larger

6   In some cases it is impossible to determine which number is larger or smaller. Consider "the second number is seven less than twice the first." Let $n =$ the first number and $2n - 7 =$ the second number. If n is any number greater than 7, the first number is smaller than the second number. However, if n is 7, the two numbers are equal. If n is any number less than 7, the first number is larger than the second number.

Q8   The sum of two numbers is ten. Let $n =$ one of the numbers and $10 - n =$ the other

   number. Which number, n or $10 - n$, is larger?_____

STOP • **STOP** • **STOP** • **STOP** • **STOP** • **STOP** • **STOP** • **STOP** • **STOP**

A8   impossible to determine

This completes the instruction for this section.

7.3   Exercises

1. Let n represent one of the unknowns. Express all unknowns in terms of n.
   a. One number is two more than another number.
   b. The sum of two numbers is five.
   c. One number is ten times as great as another number.
   d. The smaller number is six less than the larger number.
   e. The second number is three greater than the first number.
   f. One number is thirteen percent of another.
   g. Of three numbers, the second number is four less than the third and the first is twice the second.
   h. Of three numbers, the second is twice the first and the third is two more than the sum of the first and second.
   i. Of three numbers, the first is twice the second and the third is three less than the second.
   *j. Of four numbers, the third is three less the second, the first is four times the second, and the fourth is equal to the difference between the first and third.
2. If the second number is two less than the first number, indicate the larger and smaller number.
3. If thirty is separated into two parts, n and $30 - n$, which part is larger?

7.3   Exercise Answers

1. a. Let $n =$ one number and $n + 2 =$ the other number.
   b. Let $n =$ one number and $5 - n =$ the other number.
   c. Let $n =$ one number and $10n =$ the other number.
   d. Let $n =$ the larger number and $n - 6 =$ the smaller number.
   e. Let $n =$ the first number and $n + 3 =$ the second number.
   f. Let $n =$ one number and $0.13n =$ the other number.
   g. Let $n =$ the third number, $n - 4 =$ the second number, and $2(n - 4) =$ the first number.
   h. Let $n =$ the first number, $2n =$ the second number, and $n + 2n + 2 = 3n + 2 =$ the third number.
   i. Let $n =$ the second number, $2n =$ the first number, and $n - 3 =$ the third number.
   *j. Let $n =$ the second number, $3 - n =$ the third number, $4n =$ the first number, and $4n - (3 - n) = 5n - 3 =$ the fourth number.
2. The first number is larger and the second number is smaller.
3. impossible to determine

7.4   Solving More-Complicated Word Problems

---

1   There is a basic procedure for analyzing word problems.

Step 1: Read the problem *carefully,* several times if necessary.

Step 2: Determine what is the unknown in the problem. Chose a variable to represent the unknown and state clearly in writing what the variable represents. If more than one unknown is present, express all unknowns in terms of one variable, stating clearly what each expression represents.

Step 3: Analyze the facts presented and write an equation using the expression(s) from step 2.

Step 4: Solve the equation.

Step 5: Determine the numerical value(s) of the unknown(s).

Step 6: Check the solution(s) against the statement of the original problem.

Consider the following problem.

A television repairman charges $15 for a house call plus $10 per hour for the time worked. Following a house call, a bill for $60 was received. How long did the call last?

Step 1: Have you read the problem carefully?

Step 2: The unknown is the length of time (in hours) the call lasted. Let n = the number of hours the call lasted.

Step 3: Since the charge is $10 per hour plus $15, the charge may be represented as $10n + 15$. The bill came to $60. Therefore, the equation $10n + 15 = 60$ represents the facts presented.

Step 4: $10n + 15 = 60$
$$10n = 45$$
$$n = 4\frac{1}{2}$$

Step 5: The call lasted $4\frac{1}{2}$ hours.

Step 6: $10 per hour for $4\frac{1}{2}$ hours is $45. $45 plus $15 for house call is $60. Therefore, $4\frac{1}{2}$ hours is correct.

Q1   To promote sales a magazine publisher offers a one-year subscription for $8 with each subsequent gift subscription only $6. A customer was billed for $80. How many gift subscriptions were purchased? Step 1 assumed!

a. (Step 2) Let n = _____.

b. (Step 3) Equation:_____.

c. (Step 4) Solve equation from part b.

d. (Step 5) The customer purchased _____ gift subscriptions.

e. (Step 6) Show the arithmetic that verifies that the solution is correct.

A1   **a.** the number of gift subscriptions purchased

              b. $6n + 8 = 80$ or $8 + 6n = 80$

              c. $6n + 8 = 80$ **d.** 12   **e.** 12

$$6n = 72$$
$$n = 12$$

$$\begin{array}{r} 12 \\ 6 \\ \hline 72 \\ +8 \\ \hline 80 \end{array}$$

---

2   In Q1 there was only one unknown. Therefore, only one fact about that unknown was needed in order to write an equation. If two unknowns are present, two facts are necessary: one to relate the unknowns and one to write an equation.

Example: One number is five times another. Their sum is 48. What are the numbers? Either fact may be used to express the unknown. The other then will determine the equation.

Using "one number is five times another," let $n =$ one number and $5n =$ the other. Using "their sum is 48," the equation would be: $n + 5n = 48$. Solving the equation:

$$n + 5n = 48$$
$$6n = 48$$
$$n = 8$$

Therefore, $n = 8$ and $5n = 40$. The check requires that you verify that the numbers 8 and 40 satisfy both facts in the original problem. 40 is 5 times 8 and the sum of 40 and 8 is 48.

Q2   **a.** Use "Their sum is 48" to express the unknowns.

           _____ = one number (smaller)*

           _____ = the other number (larger)

      b. Use "One number is five times another" to write the equation. _____

STOP • **STOP** • **STOP** • **STOP** • **STOP** • **STOP** • **STOP** • **STOP** • **STOP**

A2   **a.** Let $n =$ one number (smaller) and $48 - n =$ the other number (larger).

      b. $5n = 48 - n$

---

3   Notice the importance of labeling the smaller and larger number in Q2. If this had not been done, it would not have been clear which expression should be multiplied by five. Completing the solution:

$$5n = 48 - n$$
$$6n = 48$$
$$n = 8$$

Hence, $n = 8$ and $48 - n = 40$, the same solutions as before.

Q3   <u>The second of two numbers is six more than the first.</u> The larger number is twice the smaller number.

      a. Use the underlined sentence above to express the unknowns. Label as larger or smaller. Use n. Let _____ = the first number (_____) and _____ = the second number (_____).

      b. Use the other fact to write an equation._____

*The assignment of (smaller) and (larger) is arbitrary. Frame 3 will explain why this is helpful.

c. Solve the equation.

d. Determine both unknowns._____

e. Check the solution against the statement of the original problem.

STOP • **STOP** • **STOP** • **STOP** • **STOP** • **STOP** • **STOP** • **STOP** • **STOP**

A3
    a. n (smaller), $n + 6$ (larger)   **b.** $2n = n + 6$ **c.** $n = 6$
    d. The first number (n) = 6 and the second number ($n + 6$) = 12.
    e. The second number (12) is six more than the first number (6), and the larger number (12) is twice the smaller number (6).

Q4
    The second of two numbers is six more than the first. The larger number is twice the smaller number. Use n.
    a. Use the underlined sentence above to express the unknowns.

        Let _____ = the smaller number (first) and _____ = the larger number (second).

    b. Use the other fact to write an equation._____

    c. Solve the equation.

    d. Determine both unknowns._____

    e. Check the solution against the statement of the original problem.

STOP • **STOP** • **STOP** • **STOP** • **STOP** • **STOP** • **STOP** • **STOP** • **STOP**

A4
    a. $n, 2n$ **b.** $2n = n + 6$ **c.** $2n - n = n + 6 - n$ (optional)
                                                          $n = 6$
    d. The smaller number (n) = 6 and the larger number ($2n$) = 12.
    e. The second number (12) is six more than the first number (6) and the larger number (12) is twice the smaller number (6).

4
    A problem with three unknowns will now be illustrated to summarize the procedure presented thus far.

    Example: Of three numbers, the first is five times the second and the third is twelve less than the second. The sum of the first and third is sixty. What are the three numbers?

    Solution

    Let n = the second number, $5n$ = the first number, and $n - 12$ = the third number.

$$5n + (n - 12) = 60$$
$$6n - 12 = 60$$
$$6n = 72$$
$$n = 12$$

second number $(n) = 12$
first number $(5n) = 60$
third number $(n - 12) = 0$

Check: The first number (60) is five times the second (12). The third number (0) is twelve less than the second (12). The sum of the first (60) and third (0) is sixty.

Q5   Use the procedure summarized above to solve the following problem. Of three numbers, the second is three less than twice the first and the third is the sum of the first two. The sum of all three numbers is forty-eight. Find the numbers. Use n.

STOP • STOP • STOP • STOP • STOP • STOP • STOP • STOP • STOP

A5   Let $n =$ the first number, $2n - 3 =$ the second number, and $3n - 3 =$ the third number.

$$n + (2n - 3) + (3n - 3) = 48$$
$$6n - 6 = 48$$
$$n = 9$$

first number $(n) = 9$
second number $(2n - 3) = 15$
third number $(3n - 3) = 24$

Check: The second number (15) is three less than twice the first ($2 \cdot 9 - 3$), and the third (24) is the sum of the first two ($9 + 15$). The sum of all three ($9 + 15 + 24$) is forty-eight.

5   The check to a problem is often simply the arithmetic necessary to prove to oneself that the correct solution has been obtained. In Q5 the numbers 9, 15, and 24 were found using the relationships presented in the statement of the problem. Actually, the only remaining fact to be checked is whether the sum of all numbers is forty-eight. Therefore, the check might be shown

first number (n) $= 9$
second number $(2n - 3) = 15$
third number $(3n - 3)$ $= \underline{24}$
                             48

In the exercise with this section, simplify the check to only the essential arithmetic required to verify that the numbers found satisfy the facts presented in the original problem.

This completes the instruction for this section.

7.4   Exercises

1. After working thirty hours, a crew of men finished three fourths of a job. If the crew works at the same rate, how many hours will be required to finish the job?
2. Seventy percent of a certain number is eighty-four. Find the number.
3. Taking one fiftieth of a certain number gives the same result as subtracting the number from twenty-seven. What is the number?
4. The sum of two angles is a right angle (90°). If one angle is four times as large as the other, how many degrees are there in each angle?
5. The sum of two numbers is forty-five. If the first number is decreased by two-thirds of the second number, the result is ten. What are the two numbers?
6. The combined cost of a bat and ball is $2.55. The bat costs five cents more than the ball. How much did each item cost? (*Hint:* $2.55 = 255 cents.)
7. The profits of a business amount to $1,800. If they are to be divided among three partners so that the first two get equal shares and the third gets twice as much as each of the other two, determine how much each is to receive.
8. One number is 34 more than another. Their sum is 108. What are the numbers?
9. The sum of three numbers is 180. The first number is twice as large as the third and the second is twenty more than the third number. What are the numbers?
10. One number is four times another. If five is added to the larger number, the result is thirty-three. What are the numbers?
*11. One number is five times another. If twelve is added to both numbers, the larger number would only be twice as large as the smaller number. What are the original numbers?

7.4   Exercise Answers

1. 10 hours: $\dfrac{3}{4}n = 30$

2. 120: $0.7n = 84$

3. $26\dfrac{8}{17}$: $\dfrac{1}{50}n = 27 - n$

4. 18 and 72: $n + 4n = 90$

5. First number is 24 and second number is 21; let n = first number and $45 - n$ = second number.

$$n - \frac{2}{3}(45 - n) = 10$$

6. Ball costs $1.25 and the bat costs $1.30

7. $450, $450, and $900

8. 37 and 71

9. 80, 60, and 40

10. 7 and 28

*11. 4 and 20: let n = the smaller number and $5n$ = the larger number
$$2(n + 12) = 5n + 12$$

7.5   Solving Word Problems That Involve Formulas

---

1 | Formulas are useful in many situations in which certain information is needed.

Example: Determine the perimeter of a rectangle given the length and width to be 17.0 centimeters and 2.1 centimeters, respectively.

Solution

The formula $p = 2l + 2w$ for the perimeter of a rectangle can be used where $p = $ perimeter, $l = $ length, and $w = $ width.

$p = 2l + 2w$
$p = 2(17.0) + 2(2.1)$
$p = 34.0 + 4.2$
$p = 38.2$

Therefore, the perimeter of the desired rectangle is 38.2 centimeters.

Q1   Determine the area of a trapezoid if the parallel sides (bases) are 12 inches and $7\frac{1}{2}$ inches and the altitude (height) is $3\frac{3}{4}$ inches. (Do not forget to record the final answer in square inches.) Use the formula $A = \frac{1}{2}h(b_1 + b_2)$ for the area of a trapezoid where $A = $ area, $h = $ altitude (height), and b_1 and $b_2 = $ parallel sides (bases).*

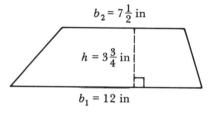

$$b_2 = 7\frac{1}{2} \text{ in}$$
$$h = 3\frac{3}{4} \text{ in}$$
$$b_1 = 12 \text{ in}$$

STOP • STOP • STOP • STOP • STOP • STOP • STOP • STOP • STOP

A1 $36\frac{9}{16}$ square inches or 36.5625 square inches:

$$A = \frac{1}{2}h(b_1 + b_2)$$

$$A = \frac{1}{2}\left(3\frac{3}{4}\right)\left(12 + 7\frac{1}{2}\right) \quad \text{or} \quad A = 0.5(3.75)(12 + 7.5)$$

---

2   In the formula, $p = 2l + 2w$, p is called the *subject* of the formula. In this formula, the perimeter, p, is expressed in terms of the length, l, and the width, w. The subject of any formula is expressed in terms of the other variables present.

Q2   What is the subject of each of the following formulas?

   a. $A = \frac{1}{2}bh$ _____   **b.** $c = 2\pi r$ _____

   c. $V = \frac{1}{3}\pi r^2 h$ _____   **d.** $a + b + c = p$ _____

STOP • STOP • STOP • STOP • STOP • STOP • STOP • STOP • STOP

*In b_1 (read "b sub-one") and b_2 (read "b sub-two"), the 1 and 2 are called *subscripts* and are used to distinguish between the two bases.

A2   **a.** A **b.** c **c.** V **d.** p

3

Often it is necessary to find the value of a variable that is not the subject of the formula. This can be done as long as values for all other variables in the formula are known.

Example: Find the length of a rectangle if the perimeter of the rectangle is 52 meters and the width of the rectangle is 12 meters.

Solution

Step 1: Determine the appropriate formula. The formula for the perimeter of a rectangle is $p = 2l + 2w$, where p = perimeter, l = length, and w = width.

Step 2: Identify the given and unknown information from the statement of the problem.

$p = 52$ meters
$l =$ unknown
$w = 12$ meters

Step 3: Substitute the values of the known information into the formula.

$p = 2l + 2w$
$52 = 2l + 2(12)$

Step 4: Solve the resulting equation.

$p = 2l + 2w$
$52 = 2l + 2(12)$
$52 = 2l + 24$
$28 = 2l$
$14 = l$

Step 5: Check the solution.

$2(14) + 2(12) \stackrel{?}{=} 52$
$28 + 24 \stackrel{?}{=} 52$
$52 = 52$

Therefore, the required length of the rectangle is 14 meters.

Q3

Find the height of a rectangular prism (box) if its volume is 315 cubic inches, the length of its base is 25 inches, and the width of its base is 3 inches. The formula for the volume of a rectangular prism is $V = lwh$, where V = volume, l = length of the base, w = width of the base, and h = height of the prism.

a. Identify the given and unknown information.

$V =$ _____ $l =$ _____

$w =$ _____ $h =$ _____

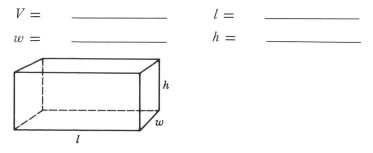

b. Substitute the values of the known information into the formula $V = lwh$.

    c. Solve the resulting equation.

    d. Check the solution.

    e. Therefore, the required height is _____.

STOP • **STOP** • **STOP** • **STOP** • **STOP** • **STOP** • **STOP** • **STOP** • **STOP**

A3   **a.** $V = 315$ cubic inches   **b.** $V = lwh$ **c.** $315 = 75h$
             $l = 25$ inches $315 = 25(3)h$ $4.2 = h$
             $w = 3$ inches
             $h =$ unknown
         d. $315 \overset{?}{=} 25(3)(4.2)$ **e.** 4.2 inches
             $315 = 315$

Q4   Determine the required height for a cylindrical tank if the volume is to be 660 cubic feet and the diameter of the tank is to be 7 feet. $\left(Hint:\ V = \pi r^2 h,\ \text{where}\ r = \dfrac{d}{2}.\ \text{Use}\ \pi \doteq \dfrac{22}{7}.*\right)$

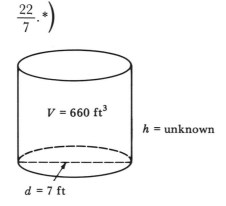

$V = 660\ \text{ft}^3$ $h =$ unknown

$d = 7\ \text{ft}$

STOP • **STOP** • **STOP** • **STOP** • **STOP** • **STOP** • **STOP** • **STOP** • **STOP**

*π is an irrational number approximately equal to $\dfrac{22}{7}$. $\dfrac{22}{7} = 3.\overline{142857} = 3.14$ rounded to the nearest hundredth, whereas $\pi = 3.14159\cdots$ ($\doteq$ means "is approximately equal to").

A4 $17\frac{1}{7}$ feet: $V = \pi r^2 h$

$$660 = \frac{22}{7}\left(\frac{7}{2}\right)\left(\frac{7}{2}\right)h$$

$$660 = \frac{77}{2}h$$

$$\frac{2}{77}(660) = \frac{2}{77}\left(\frac{77}{2}h\right)$$

$$17\frac{1}{7} = h$$

Therefore, the required height is $17\frac{1}{7}$ feet $\left(17 \text{ ft } 1\frac{5}{7} \text{ in}\right)$.

Q5   Find the other parallel side of a trapezoid if the area is 24 square meters, the altitude is 3 meters, and the smaller parallel side is 2 meters. Use $A = \frac{1}{2}h(b_1 + b_2)$.

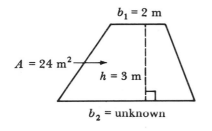

$b_1 = 2$ m

$A = 24$ m^2

$h = 3$ m

$b_2 = $ unknown

STOP • STOP • STOP • STOP • STOP • STOP • STOP • STOP • STOP

A5   14 meters: $A = \frac{1}{2}h(b_1 + b_2)$

$$24 = \frac{1}{2}(3)(2 + b_2)$$

$$24 = \frac{3}{2}(2 + b_2)$$

$$\frac{2}{3}(24) = \frac{2}{3}\cdot\frac{3}{2}(2 + b_2)$$

$$16 = 2 + b_2$$

$$14 = b_2$$

4   Whenever a formula can be used to determine missing information, the technique of substituting the known values and solving the resulting equation is often convenient. The

examples used thus far have all concerned geometric objects. However, the technique being discussed can be applied to any situation in which a formula is available.

Example: Find the time required to accumulate $300 interest if $20,000 is invested at 6 percent simple interest.

Solution

The formula $i = prt$, where $i =$ interest, $p =$ principal (amount invested or borrowed), $r =$ annual rate of interest, and $t =$ time in years, could be used.

As before, it is important to identify the given and unknown information. That is, $i = \$300$, $p = \$20,000$, $r = 6$ percent, and $t =$ unknown. Hence,

$$i = prt$$
$$300 = 20,000(6\%)t$$
$$300 = 20,000\left(\frac{6}{100}\right)t*$$
$$300 = 1,200t$$
$$\frac{1}{4} = t$$

Therefore, the time required is $\frac{1}{4}$ of a year, or 3 months.

*Six percent could be expressed as 0.06, or $\frac{3}{50}$.

Q6   Given $d = rt$, where $d =$ distance, $r =$ rate, and $t =$ time, determine the rate at which a car must travel to cover a distance of 135 kilometers in 3 hours. Answer should be expressed in kilometers per hour (km/hr).

STOP • **STOP** • **STOP** • **STOP** • **STOP** • **STOP** • **STOP** • **STOP** • **STOP**

A6   45 km/hr: $d = rt$
$$135 = r \cdot 3$$
$$135 = 3r$$
$$45 = r$$

This completes the instruction for this section.

7.5   Exercises

1. Given $p = a + b + c$, find a when $p = 32$, $b = 9$, and $c = 5$.
2. Given $p = 2l + 2w$, find w when $p = 29$ and $l = 4.5$.
3. Given $A = \frac{1}{2}bh$, find b when $A = 54$ and $h = 9$.
4. Given $c = 2\pi r$, find r when $c = 37.68$. Use $\pi \doteq 3.14$.
5. Given $V = \frac{1}{3}Bh$, find B when $V = 2,500$ and $h = 75$.

6. Given $A = p(1 + rt)$, find t when $A = 1,000$, $p = 750$, and $r = 0.06$.
7. Given $V_1 = V_0 + at$, find a when $V_1 = 50$, $V_0 = 42$, and $t = 2$.
8. Given $\dfrac{P_1 V_1}{T_1} = \dfrac{P_2 V_2}{T_2}$, find P_1 when $V_1 = 30$, $T_1 = 75$, $P_2 = 5$, $V_2 = 20$, and $T_2 = 60$.
9. Given $V = lwh$, find h when $V = 25$, $l = 10$, and $w = 10$.
10. Given $i = prt$, find p when $i = \$1,500$, $r = 5$ percent, and $t = 3$ years.
11. Given $p = 2l + 2w$, determine the length of a rectangle if the perimeter is 35 centimeters and the width is 5.5 centimeters.
12. Given $A = bh$, where A = the area of a parallelogram, b = the base of the parallelogram, and h = the altitude of the parallelogram, determine the base of a parallelogram whose area is 51.8 square yards and whose height (altitude) is 14 yards.
13. Given $V = \pi r^2 h$, where V = the volume of a cylinder, r = the radius of the cylinder, and h = the height of the cylinder, find the height of a cylinder whose volume is 1,884 cubic centimeters when the radius of the cylinder is 10 centimeters. Use $\pi \doteq 3.14$.
14. Given $A = \dfrac{1}{2} h(b_1 + b_2)$, find the height of a trapezoid if the area of the trapezoid is 168 square meters when the two parallel bases are 15 and 17 meters.
15. Given $i = prt$, determine the time (in years) required to accumulate $300 interest if $1,000 is invested at $7\dfrac{1}{2}$ percent simple interest.

7.5   Exercise Answers

1. 18	2. 10	3. 12	4. 6
5. 100	6. $\dfrac{50}{9}$ or $5\dfrac{5}{9}$	7. 4	8. $\dfrac{25}{6}$ or $4\dfrac{1}{6}$
9. $\dfrac{1}{4}$ or 0.25	10. $10,000	11. 12 centimeters	12. 3.7 yards
13. 6 centimeters	14. 10.5 meters	15. 4 years	

Chapter 7 Sample Test

At the completion of Chapter 7 it is expected that you will be able to work the following problems.

7.1   Expressing English Phrases As Algebraic Expressions

1. Write each phrase as an algebraic expression. Use n for the number.
   a. the sum of fifteen and twice a number
   b. the difference between three times the number and the number
   c. five less than the number
   d. the product of nine and the number, less two
   e. the quotient of the number divided by the quantity one minus the number
   f. the product of five times the number, the number increased by three, and the number decreased by eight

> **g.** twelve percent of the number added to the number
> **h.** two-thirds of the number

7.2   Solving Simple Word Problems

2. Solve each of these word problems using the four-step procedure outlined in this section.
   a. The sum of five times the number and three times the number, minus four, equals twenty-eight. What is the number?
   b. Two-thirds of the number taken from the number is five less than the number. Find the number.
   c. What number when increased by three, and the result multiplied by negative two, produces ten?
   d. A number less six, subtracted from five, is four. What is the number?
   e. The sum of one-half of a number and nine is equal to four-fifths of the difference between the number and three. Find the number.

7.3   Expressing More Than One Unknown in Terms of the Same Variable

3. Let n represent one of the unknowns. Express all unknowns in terms of n.
   a. The smaller number is ten less than twice the larger.
   b. The sum of two numbers is twelve.
   c. A second number is one half of a first number.
   d. Of three numbers, the third is twice the second and the first is nine less than the second.
   e. Of three numbers, the third is two less than one third the first and the second is two less than the third.

7.4   Solving More-Complicated Word Problems

4. a. Fifteen percent of an amount of money is $1.05. What is the number?
   b. A number increased by 6 percent of that number produces 33.92. What is the number?
   c. The sum of two numbers is twenty-one. If the larger number is subtracted from the smaller, the difference is negative one. What are the two numbers?
   d. One number is five less than three times the other. Their sum is eighteen. Find the two numbers.
   e. Of three numbers, the first is twice the second. The third is two less than the sum of the first two. The sum of all three numbers is twenty-two. Find all three numbers.

7.5   Solving Word Problems That Involve Formulas

5. When solving these problems, substitute the given information into the appropriate formula, and solve the resulting equation for the desired value:
   a. Given $V_1 = V_0 + at$, find t when $V_1 = 18.5$, $V_0 = 3.5$, and $a = 5$.

b. Given $\dfrac{P_1 V_1}{T_1} = \dfrac{P_2 V_2}{T_2}$, find V_2 when $P_1 = 70$, $V_1 = 30$, $T_1 = 90$, $P_2 = 85$, and $T_2 = 102$.

c. Given the simple interest formula $i = prt$, where i = interest, p = principal, r = annual rate of interest, and t = time in years, determine the annual rate of interest required to accumulate \$165 interest if \$1,500 is invested for 2 years at simple interest.

d. Given $p = 2l + 2w$, where p = perimeter of a rectangle, l = length of the rectangle, w = width of the rectangle, find the width of a rectangle if the perimeter is 41 centimeters and the length is 13 centimeters.

e. Given $T = 2\pi r^2 + 2\pi r h$, where T = total surface of a cylinder, r = radius of the circular base, and h = height (altitude) of the cylinder, determine the height of a cylinder if the total surface of the cylinder is 748 square millimeters and the radius of the base is 7 millimeters. Use $\pi \doteq \dfrac{22}{7}$.

Chapter 7 Sample Test Answers

1. a. $15 + 2n$ **b.** $3n - n$ **c.** $n - 5$

    d. $9n - 2$ **e.** $\dfrac{n}{1 - n}$ **f.** $5n(n + 3)(n - 8)$

    g. $n + 0.12n$ **h.** $\dfrac{2}{3} n$

2. a. 4 **b.** $\dfrac{15}{2}$ or $7\dfrac{1}{2}$ or 7.5 **c.** $^-8$

    d. 7 **e.** 38

3. a.   Let n = larger number
       $2n - 10$ = smaller number

    b.   Let n = one of the numbers
       $12 - n$ = the other number

    c.   Let n = first number

       $\dfrac{1}{2} n$ = second number

    d.   Let n = second number
       $2n$ = third number
       $n - 9$ = first number

    e.   Let n = first number

       $\dfrac{1}{3} n - 2$ = third number

       $\dfrac{1}{3} n - 4$ = second number

4. a. \$7.00   **b.** 32   **c.** 10 and 11

    d. $\dfrac{23}{4}$ and $\dfrac{49}{4}$ $\left(5\dfrac{3}{4} \text{ and } 12\dfrac{1}{4}\right)$ **e.** 8, 4, and 10

5. a. 3   **b.** 28   **c.** 5.5 percent   **d.** 7.5 centimeters
    e. 10 millimeters

Chapter 8

Inequalities

8.1   Order and the Symbols of Inequality

---

1   For any two rational numbers, one of the numbers is

1. less than,
2. equal to, or
3. greater than

the other number. For example,

0 is less than 25
5 is equal to 5
9 is greater than 2

The study of the order of a given set of numbers is a study of the three relationships above that exist between the numbers of the set.

Q1   Underline one of the words in parentheses to form a true statement:
    a. 4 is (greater, less) than 6.
    b. 3 is (greater, less) than 1.
    c. 19 is (greater, less) than 0.

STOP • STOP • STOP • STOP • STOP • STOP • STOP • STOP • STOP

A1   **a.** less   **b.** greater   **c.** greater

Q2   True or false:

    a. 5 is less than 3.   _____   **b.** 6 is greater than 2.   _____

    c. 7 is less than 10.   _____   **d.** 13 is less than 10.   _____

STOP • STOP • STOP • STOP • STOP • STOP • STOP • STOP • STOP

A2   **a.** false   **b.** true   **c.** true   **d.** false

---

2   The symbol for "greater than" is "$>$":

Examples:

$8 > 3$ means 8 is greater than 3
$4 > 0$ means 4 is greater than 0

The statement $12 > 5$ is true because 12 is greater than 5. The statement $7 > 9$ is a false statement because 7 is *not* greater than 9.

Q3   The statement $17 > 5$ means_____.

STOP • STOP • STOP • STOP • STOP • STOP • STOP • STOP • STOP

A3   17 is greater than 5.

Q4   Write an abbreviated statement for 13 is greater than 8._____

STOP • STOP • STOP • STOP • STOP • STOP • STOP • STOP • STOP

A4 $13 > 8$

Q5   True or false:

    a. $3 > 5$ _____   **b.** $17 > 12$ _____

    c. $0 > 1$ _____   **d.** $9 > 6$ _____

STOP • STOP • STOP • STOP • STOP • STOP • STOP • STOP • STOP

A5   **a.** false   **b.** true   **c.** false   **d.** true

---

3   For two numbers x and y, if x is greater than y, then y is less than x. Thus, if $7 > 2$, then 2 is less than 7.

The symbol for "less than" is "$<$":

Examples:

$2 < 3$ means 2 is less than 3
$15 < 105$ means 15 is less than 105

The statement $4 < 7$ is true because 4 is less than 7. The statement $5 < 0$ is false because 5 is *not* less than 0.

Q6   The statement $8 < 2$ means_____

STOP • STOP • STOP • STOP • STOP • STOP • STOP • STOP • STOP

A6   8 is less than 2.

Q7   Write an abbreviated statement for 1 is less than 9._____

STOP • STOP • STOP • STOP • STOP • STOP • STOP • STOP • STOP

A7 $1 < 9$

Q8   True or false:

    a. $4 < 12$ _____   **b.** $6 < 3$ _____

    c. $0 < 1$ _____   **d.** $14 < 41$ _____

STOP • STOP • STOP • STOP • STOP • STOP • STOP • STOP • STOP

A8   **a.** true   **b.** false   **c.** true   **d.** true

---

4   When placing $>$ or $<$ between two numbers it is sometimes helpful to notice that the symbol always "opens" toward the larger number.

Examples:

$5 < 7$ means 5 is less than 7 and the symbol opens toward the 7
$12 > 4$ means 12 is greater than 4 and the symbol opens toward the 12

Q9   Place $>$ or $<$ between each of the following pairs of numbers to form a true statement:

a. 3_____2   **b.** 0_____10   **c.** 4_____18   **d.** 5_____2

STOP • STOP • STOP • STOP • STOP • STOP • STOP • STOP • STOP

A9   **a.** $>$ **b.** $<$ **c.** $<$ **d.** $>$

Q10   Write abbreviated statements for each of the following:

a. 0 is less than 6.   _____

b. 6 is greater than 0.   _____

c. 25 is greater than 13.   _____

d. 13 is less than 25.   _____

STOP • STOP • STOP • STOP • STOP • STOP • STOP • STOP • STOP

A10   **a.** $0 < 6$ **b.** $6 > 0$ **c.** $25 > 13$ **d.** $13 < 25$

---

5   Consider the order of the numbers on the following number line:

Numbers *increase* in size as one moves to the *right* on the number line and *decrease* in size as one moves to the *left* on the number line. Thus the larger of any two numbers is always the one to the right on the number line.

Examples:

$7 > 1$ because 7 is to the right of 1 on the number line
$0 > {}^-2$ because 0 is to the right of $^-2$ on the number line
$^-5 < {}^-4$ because $^-4$ is to the right of $^-5$ on the number line

In general, for any two real numbers a and b, positioned as follows on the number line, $b > a$ or $a < b$.

That is, the number to the right is greater. Also, the number to the left is smaller.

Q11   **a.** Which number is greater, $^-3$ or 4?_____

b. Why?_____

STOP • STOP • STOP • STOP • STOP • STOP • STOP • STOP • STOP

A11   **a.** 4   **b.** because 4 is to the right of $^-3$ on the number line

---

Q12   Which statement is true, $^-3 > 4$ or $4 > {}^-3$?_____

STOP • STOP • STOP • STOP • STOP • STOP • STOP • STOP • STOP

A12	$4 > {}^-3$

Q13	Which number is to the right on the number line, ${}^-8$ or ${}^-10?$ _____

STOP • **STOP** • **STOP** • **STOP** • **STOP** • **STOP** • **STOP** • **STOP** • **STOP**

A13	${}^-8$

Q14	True or false: ${}^-8 > {}^-10$ _____

STOP • **STOP** • **STOP** • **STOP** • **STOP** • **STOP** • **STOP** • **STOP** • **STOP**

A14	true

Q15	Place $>$ or $<$ between each of the following pairs of numbers to form true statements:
	a. ${}^-3$ ___ 0 **b.** ${}^-2$ ___ ${}^-3$ **c.** 0 ___ 5 **d.** 5 ___ 10

STOP • **STOP** • **STOP** • **STOP** • **STOP** • **STOP** • **STOP** • **STOP** • **STOP**

A15	**a.** $<$ **b.** $>$ **c.** $<$ **d.** $<$

Q16	Which number, $\dfrac{3}{8}$ or $\dfrac{7}{8}$, is to the right on the number line? _____

STOP • **STOP** • **STOP** • **STOP** • **STOP** • **STOP** • **STOP** • **STOP** • **STOP**

A16	$\dfrac{7}{8}$

Q17	True or false: $\dfrac{3}{8} < \dfrac{7}{8}$ _____

STOP • **STOP** • **STOP** • **STOP** • **STOP** • **STOP** • **STOP** • **STOP** • **STOP**

A17	true

Q18	True or false:
	a. $\dfrac{-1}{2} > \dfrac{1}{2}$ _____ **b.** $2\dfrac{5}{9} < 2\dfrac{7}{9}$ _____
	c. ${}^-3.5 > {}^-3.9$ _____ **d.** $2.7 < 2.3$ _____

STOP • **STOP** • **STOP** • **STOP** • **STOP** • **STOP** • **STOP** • **STOP** • **STOP**

A18	**a.** false **b.** true **c.** true **d.** false

6	The symbol "$\leqslant$" means "less than *or* equal to." The statement $x \leqslant y$ is true if *either* $x < y$ or $x = y$ is true. Thus, $5 \leqslant 7$ is true because $5 < 7$. $3 \leqslant 3$ is true because $3 = 3$.

Q19	The statement $2 \leqslant 6$ means _____

STOP • **STOP** • **STOP** • **STOP** • **STOP** • **STOP** • **STOP** • **STOP** • **STOP**

A19	2 is less than or equal to 6.

Q20	True or false:
	a. $4 \leqslant 5$ _____ **b.** $2 \leqslant 2$ _____

    c. $^-3 \leqslant ^-9$ _____   **d.** $^-4 \leqslant ^-4$ _____

STOP • STOP • STOP • STOP • STOP • STOP • STOP • STOP • STOP

A20   **a.** true   **b.** true   **c.** false   **d.** true

---

7   The symbol "$\geqslant$" means "greater than or equal to." The statement $x \geqslant y$ is true if *either* $x > y$ or $x = y$ is true. Thus, $5 \geqslant 2$ is true because $5 > 2$. $^-9 \geqslant ^-9$ is true because $^-9 = ^-9$.

Q21   The statement $^-12 \geqslant ^-15$ means _____

STOP • STOP • STOP • STOP • STOP • STOP • STOP • STOP • STOP

A21 $^-12$ is greater than or equal to $^-15$

Q22   True or false:

    a. $^-5 \geqslant ^-9$ _____   **b.** $19 \geqslant 19$ _____

    c. $12.5 \geqslant 12.7$ _____   **d.** $^-3\frac{1}{2} \geqslant 3\frac{1}{2}$ _____

    e. $^-5 \leqslant 3$ _____   **f.** $^-4 \leqslant 0$ _____

STOP • STOP • STOP • STOP • STOP • STOP • STOP • STOP • STOP

A22   **a.** true   **b.** true   **c.** false   **d.** false   **e.** true   **f.** true

---

8   The symbols $<$, $\leqslant$, $>$, and $\geqslant$ are called *symbols of inequality*. An inequality symbol is negated by drawing a slanted line, /, through the symbol. Thus,

    $\not<$ means "is *not* less than"
    $\not\leqslant$ means "is *not* less than or equal to"
    $\not>$ means "is *not* greater than"
    $\not\geqslant$ means "is *not* greater than or equal to"

The statement $5 \not< 2$ is true because 5 is not less than 2. The statement $2 \not\geqslant 0$ is false because 2 is greater than or equal to 0.

Q23   The statement $7 \not\geqslant 4$ means _____.

STOP • STOP • STOP • STOP • STOP • STOP • STOP • STOP • STOP

A23   7 is not greater than or equal to 4.

Q24   Is $7 \not\geqslant 4$ true or false? _____

STOP • STOP • STOP • STOP • STOP • STOP • STOP • STOP • STOP

A24   false: because 7 is greater than 4.

Q25 $^-3 \not< ^-5$ means _____.

STOP • STOP • STOP • STOP • STOP • STOP • STOP • STOP • STOP

A25 $^-3$ is not less than $^-5$.

Q26   Is $^-3 \not< ^-5$ true or false? _____

STOP • STOP • STOP • STOP • STOP • STOP • STOP • STOP • STOP

A26   true: because $^-3$ is greater than $^-5$.

Q27   True or false:

    a. $5 \not< 2$ _____   **b.** $0 \not> {}^-3$ _____

    c. ${}^-4 \not< {}^-4$ _____   **d.** ${}^-3 \not< {}^-1$ _____

STOP • *STOP* • *STOP* • *STOP* • *STOP* • *STOP* • *STOP* • *STOP* • *STOP*

A27   **a.** true   **b.** false   **c.** false   **d.** false

---

9    $x > 4$ is an open sentence that is neither true nor false until the variable x is replaced with a number. If the replacement set for x is $\{2, {}^-3, 7, 0, 5\}$, the following statements and their truth values result from all possible replacements:

    $2 > 4$ false
    ${}^-3 > 4$ false
    $7 > 4$ true
    $0 > 4$ false
    $5 > 4$ true

Since the values 7 and 5 produce a true statement, they are said to be elements of the truth set for $x > 4$.

Q28   Which numbers in the replacement set $\{0, 4, {}^-3, 2\}$ are in the truth set of the open sentence $x < 2$? _____

STOP • *STOP* • *STOP* • *STOP* • *STOP* • *STOP* • *STOP* • *STOP* • *STOP*

A28   0 and ${}^-3$:   because $0 < 2$ is true and ${}^-3 < 2$ is true ($4 \not< 2$ and $2 \not< 2$).

Q29   Which numbers in the replacement set $\{{}^-1, 7, {}^-5, 3, 0\}$ are in the truth set of $x \geqslant 0$? _____

STOP • *STOP* • *STOP* • *STOP* • *STOP* • *STOP* • *STOP* • *STOP* • *STOP*

A29   7, 3, and 0:   because $7 \geqslant 0$, $3 \geqslant 0$, and $0 \geqslant 0$ are all true statements (${}^-1 \not\geqslant 0$, ${}^-5 \not\geqslant 0$).

---

10   When *set builder notation* is used, the truth set of $x \leqslant 3$ is written $\{x \mid x \leqslant 3\}$. $\{x \mid x \leqslant 3\}$ is read "the set of all numbers x such that x is less than or equal to 3." (The vertical bar, $\mid$, means "such that.") 2 is an element of $\{x \mid x \leqslant 3\}$ because $2 \leqslant 3$ is true. Five is not an element of $\{x \mid x \leqslant 3\}$ because $5 \not\leqslant 3$.

Q30   Is $1 \in \{x \mid x \leqslant 3\}$? _____

STOP • *STOP* • *STOP* • *STOP* • *STOP* • *STOP* • *STOP* • *STOP* • *STOP*

A30   yes: because $1 \leqslant 3$.

Q31   Is $4 \in \{x \mid x \leqslant 3\}$? _____

STOP • *STOP* • *STOP* • *STOP* • *STOP* • *STOP* • *STOP* • *STOP* • *STOP*

A31   no: because $4 \not\leqslant 3$.

Q32   True or false:

    a. $5 \in \{x \mid x \geqslant 5\}$ _____   **b.** ${}^-3 \in \{x \mid x \geqslant 5\}$ _____

    c. ${}^-5 \notin \{x \mid x \geqslant 5\}$ _____   **d.** $0 \notin \{x \mid x \geqslant 5\}$ _____

STOP • *STOP* • *STOP* • *STOP* • *STOP* • *STOP* • *STOP* • *STOP* • *STOP*

A32   **a.** true: because $5 \geqslant 5$ **b.** false: because $^-3 \not\geqslant 5$
      c. true: because $^-5 \not\geqslant 5$ **d.** true: because $0 \not\geqslant 5$

11	$x > 5$ or $x < 1$ is a compound open sentence using the connecting word "or." A number is in the truth set of an "or" sentence if it is in the truth set of *either* part of the sentence.

Examples:

7 is in the truth set of $x > 5$ or $x < 1$ because $7 > 5$
$^-2$ is in the truth set of $x > 5$ or $x < 1$ because $^-2 < 1$
3 is *not* in the truth set of $x > 5$ or $x < 1$ because $3 \not> 5$ and $3 \not< 1$

Q33   Is 4 in the truth set of $x > 0$ or $x < 7$?_____

STOP • STOP • STOP • STOP • STOP • STOP • STOP • STOP • STOP

A33   yes: $4 < 7$.

Q34   Is $^-4$ in the truth set of $x \geqslant 2$ or $x < ^-2$?_____

STOP • STOP • STOP • STOP • STOP • STOP • STOP • STOP • STOP

A34   yes: $^-4 < ^-2$.

Q35   Which of the numbers $^-2, 0, 3, 9,$ and $^-10$ are in the truth set of $x < ^-3$ or $x \geqslant 9$?_____

STOP • STOP • STOP • STOP • STOP • STOP • STOP • STOP • STOP

A35   9 and $^-10$: because $9 \geqslant 9$ and $^-10 < ^-3$.

12	The truth set in set builder notation of $x > 3$ or $x < ^-5$ is written $\{x \mid x > 3$ or $x < ^-5\}$. $^-1 \notin \{x \mid x > 3$ or $x < ^-5\}$ because $^-1$ makes neither part of the "or" sentence true. That is, $^-1 \not> 3$ and $^-1 \not< ^-5$. $^-7 \in \{x \mid x > 3$ or $x < ^-5\}$ because $^-7 < ^-5$.

Q36   Is $^-8 \in \{x \mid x < 2$ or $x > 5\}$?_____

STOP • STOP • STOP • STOP • STOP • STOP • STOP • STOP • STOP

A36   yes: because $^-8 < 2$.

Q37   Is $0 \in \{x \mid x \leqslant ^-1$ or $x \geqslant 1\}$?_____

STOP • STOP • STOP • STOP • STOP • STOP • STOP • STOP • STOP

A37   no: because $0 \not\leqslant ^-1$ and $0 \not\geqslant 1$.

Q38   True or false:

a. $2 \in \{x \mid x < ^-5$ or $x \geqslant 2\}$ _____

b. $^-4 \in \{x \mid x \geqslant ^-3$ or $x < ^-6)$ _____

c. $^-3 \notin \{^-x \mid x \geqslant ^-3$ or $x < ^-6\}$ _____

d. $8 \in \{x \mid x \geqslant 5$ or $x < 3\}$ _____

STOP • STOP • STOP • STOP • STOP • STOP • STOP • STOP • STOP

A38   **a.** true: because $2 \geqslant 2$.   **b.** false: because $^-4 \not\geqslant ^-3$ and $^-4 \not< ^-6$.
      c. false: because $^-3 \geqslant ^-3$.   **d.** true: because $8 \geqslant 5$.

13 $x \geqslant 3$ and $x \leqslant 7$ is a compound open sentence joined by the connecting word "and." A number is in the truth set of an "and" sentence only if it is in the truth set of *both* parts of the sentence.

Examples:

5 is in the truth set of $x \geqslant 3$ and $x \leqslant 7$ because $5 \geqslant 3$ and $5 \leqslant 7$

$^-3$ is *not* in the truth set $x \geqslant 3$ and $x \leqslant 7$ because $^-3 \not\geqslant 3$

Q39   Is 2 in the truth set of $x \geqslant 5$ and $x \leqslant 2$? _____

STOP • STOP • STOP • STOP • STOP • STOP • STOP • STOP STOP

A39   no: because $2 \not\geqslant 5$. Notice that 2 is in the truth set of $x \leqslant 2$. However, 2 is in the truth set of $x \geqslant 5$ and $x \leqslant 2$ *only* if it is in the truth set of *both* parts of the "and" sentence.

Q40   Is $^-1$ in the truth set of $x \leqslant {}^-1$ and $x \geqslant {}^-6$? _____

STOP • STOP • STOP • STOP • STOP • STOP • STOP • STOP • STOP

A40   yes: because $^-1 \leqslant {}^-1$ and $^-1 \geqslant {}^-6$.

14   The truth set in set builder notation of $x \geqslant 5$ and $x < 11$ is written $\{x \mid x \geqslant 5$ and $x < 11\}$. $6\frac{1}{2} \in \{x \mid x \geqslant 5$ and $x < 11\}$ because $6\frac{1}{2} \geqslant 5$ and $6\frac{1}{2} < 11$. $2 \notin \{x \mid x \geqslant 5$ and $x < 11\}$ because although $2 < 11$, $2 \not\geqslant 5$.

Q41   Is $3.5 \in \{x \mid x \leqslant 7$ and $x \geqslant 3\}$? _____

STOP • STOP • STOP • STOP • STOP • STOP • STOP • STOP • STOP

A41   yes: because $3.5 \leqslant 7$ and $3.5 \geqslant 3$.

Q42   Is $4 \in \{x \mid x \leqslant 5$ and $x \geqslant 7\}$? _____

STOP • STOP • STOP • STOP • STOP • STOP • STOP • STOP • STOP

A42   no: because $4 \not\geqslant 7$.

Q43   Which of the numbers 0, $^-3$, 5, $^-1$, 7 are elements of $\{x \mid x \geqslant {}^-3$ and $x \leqslant 3\}$? _____

STOP • STOP • STOP • STOP • STOP • STOP • STOP • STOP • STOP

A43   0, $^-3$, and $^-1$

15   Notice that there is a definite difference in the "or" and "and" sentences.

Examples:

$2 \notin \{x \mid x \leqslant 2$ *and* $x \geqslant 5\}$ because, although $2 \leqslant 2$ is true, $2 \geqslant 5$ is false

$2 \in \{x \mid x \leqslant 2$ *or* $x \geqslant 5\}$ because $2 \leqslant 2$ is true and it thus makes no difference that $2 \geqslant 5$ is false

A number is in the truth set of
1. an "and" sentence if it is in the truth set of *both* parts of the sentence.
2. an "or" sentence if it is in the truth set of *either* one of the parts of the sentence.

Q44   True or false:

  a. $0 \in \{x \mid x \leqslant 2 \text{ and } x \geqslant 5\}$ _____

  b. $0 \in \{x \mid x \leqslant 2 \text{ or } x \geqslant 5\}$ _____

  c. $^-2 \in \{x \mid x < {}^-1 \text{ and } x > 3\}$ _____

  d. $^-2 \in \{x \mid x < {}^-1 \text{ or } x > 3\}$ _____

STOP • **STOP** • **STOP** • **STOP** • **STOP** • **STOP** • **STOP** • **STOP** • **STOP**

A44   **a.** false: because $0 \ngeqslant 5$.   **b.** true: because $0 \leqslant 2$.
           c. false: because $^-2 \ngtr 3$.   **d.** true: because $^-2 < {}^-1$.

This completes the instruction for this section.

8.1   Exercises

1. Place $>$ or $<$ between each of the following pairs of numbers to form a true statement:

    a. 3 _____ 0   **b.** $^-3$ _____ 3   **c.** $^-5$ _____ $^-7$

    d. 12 _____ 6   **e.** $^-18$ _____ $^-17$ **f.** 0 _____ $^-1$

2. True or false:
    a. $^-3 > {}^-5$ **b.** $0 < {}^-8$ **c.** $^-6 \nleqslant {}^-6$
    d. $7 \leqslant 7$ **e.** $12 \ngeqslant 15$ **f.** $2 \nleqslant 10$

3. Indicate which numbers in the replacement set are in the truth set for each of the following open sentences:
    a. $x > {}^-5$, $\{^-2, 0, {}^-7, 3\}$
    b. $x \leqslant 2$, $\{^-3, 0, 5, 2, 7\}$
    c. $x \leqslant {}^-1$, $\{0, 1, {}^-1, 2, {}^-2\}$
    d. $x > 3$, $\left\{3.2, {}^-4\frac{1}{2}, 3\frac{1}{5}, 3, {}^-3\right\}$

4. True or false:
    a. $3 \in \{x \mid x \leqslant 3\}$ **b.** $^-5 \in \{x \mid x \geqslant {}^-6\}$
    c. $0 \in \{x \mid x > 0\}$ **d.** $^-3 \in \{x \mid x > 5 \text{ or } x \leqslant {}^-2\}$
    e. $5 \in \{x \mid x \leqslant 5 \text{ and } x \geqslant 9\}$ **f.** $^-6 \notin \{x \mid x < {}^-1\}$
    g. $5 \notin \{x \mid x > 5\}$ **h.** $^-7 \notin \{x \mid x > {}^-10 \text{ or } x < 10\}$
    i. $2 \in \{x \mid x \leqslant 3 \text{ and } x \geqslant {}^-3\}$ **j.** $^-1 \notin \{x \mid x \geqslant {}^-5 \text{ and } x \leqslant 5\}$
    k. $0 \notin \{x \mid x < 0 \text{ and } x \geqslant 0\}$

8.1   Exercise Answers

1. **a.** $>$ **b.** $<$ **c.** $>$ **d.** $>$ **e.** $<$ **f.** $>$
2. **a.** true   **b.** false   **c.** false   **d.** true   **e.** true   **f.** false

3. **a.** $^-2$, 0, and 3   **b.** $^-3$, 0, and 2   **c.** $^-1$ and $^-2$ **d.** 3.2 and $3\frac{1}{5}$

4. **a.** true   **b.** true   **c.** false   **d.** true   **e.** false   **f.** false   **g.** true
    h. false   **i.** true   **j.** false   **k.** true

8.2   Graphing Linear Inequalities in One Variable

---

1 The open sentences $x > 3, y \leqslant {}^-4$, and $z < 0$ are called linear inequalities in one variable. The term *linear* is used because each of their graphs is represented by a portion of a *straight line*.

The graph of an inequality is a picture representation of its *truth set.** The graph of the truth set for $x > 3$ is shown on the following number line:

Since the truth set of $x > 3$ contains all numbers greater than 3, the portion of the number line to the right of 3 is shaded. The circle around the endpoint 3 is not shaded, indicating that the endpoint is not a point of the graph. A graph whose endpoint is not included is said to be *open*.

*The truth set of an open sentence is a set of all values from the replacement set which convert the open sentence into a true statement.

Q1 The graph of an inequality is a picture representation of its _____.

STOP • STOP • STOP • STOP • STOP • STOP • STOP • STOP • STOP

A1 truth set

Q2 Graph the truth set of $x > {}^-2$.

STOP • STOP • STOP • STOP • STOP • STOP • STOP • STOP • STOP

A2

Q3 Graph the truth set of $x > 0$.

STOP • STOP • STOP • STOP • STOP • STOP • STOP • STOP • STOP

A3

---

2 The graph of the truth set of $y < {}^-2$ is shown on the following number line:

Since the truth set of $y < {}^-2$ contains all numbers less than ${}^-2$, the portion of the number line to the *left* of ${}^-2$ is shaded. (The circle at ${}^-2$ is unshaded, indicating that ${}^-2$ *is not included* in the truth set.)

Q4 Graph the truth set of $x < 5$.

STOP • STOP • STOP • STOP • STOP • STOP • STOP • STOP • STOP

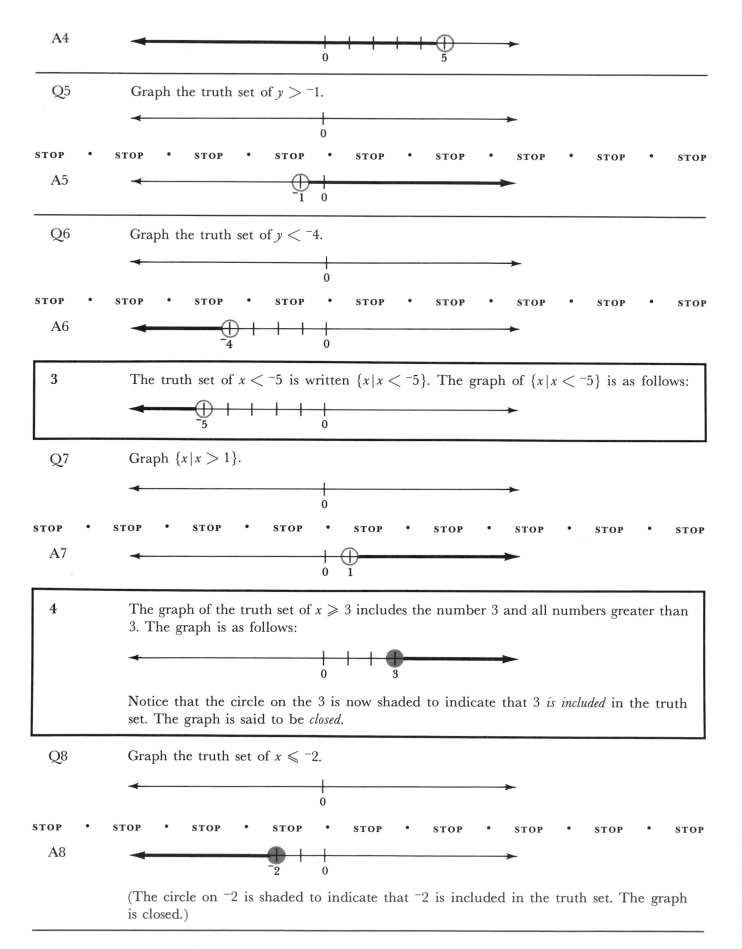

A4

Q5   Graph the truth set of $y > {}^-1$.

STOP • STOP • STOP • STOP • STOP • STOP • STOP • STOP • STOP

A5

Q6   Graph the truth set of $y < {}^-4$.

STOP • STOP • STOP • STOP • STOP • STOP • STOP • STOP • STOP

A6

3   The truth set of $x < {}^-5$ is written $\{x \mid x < {}^-5\}$. The graph of $\{x \mid x < {}^-5\}$ is as follows:

Q7   Graph $\{x \mid x > 1\}$.

STOP • STOP • STOP • STOP • STOP • STOP • STOP • STOP • STOP

A7

4   The graph of the truth set of $x \geqslant 3$ includes the number 3 and all numbers greater than 3. The graph is as follows:

Notice that the circle on the 3 is now shaded to indicate that 3 *is included* in the truth set. The graph is said to be *closed*.

Q8   Graph the truth set of $x \leqslant {}^-2$.

STOP • STOP • STOP • STOP • STOP • STOP • STOP • STOP • STOP

A8

(The circle on ${}^-2$ is shaded to indicate that ${}^-2$ is included in the truth set. The graph is closed.)

Q9   Graph $\{x\,|\,x \geqslant {}^-6\}$.

STOP   •   STOP   •   STOP   •   STOP   •   STOP   •   STOP   •   STOP   •   STOP   •   STOP

A9

---

5   An important part of graphing inequalities is deciding when to shade and when not to shade the circle. Consider the graphs of the truth sets $x \geqslant 1$ and $x > 1$:

$x \geqslant 1$:

$x > 1$:

Notice that the circle *is* shaded for the graph of $x \geqslant 1$ and *is not* shaded for the graph of $x > 1$. In general, the endpoint is shaded for $\leqslant$ and $\geqslant$ and is not shaded for $<$ and $>$.

Q10   True or false:
          a. The graph for the truth set of $x < {}^-4$ will include an unshaded circle around $^-4$.

          _____

          b. The graph for the truth set of $x \leqslant {}^-4$ will include a shaded circle around $^-4$.

          _____

STOP   •   STOP   •   STOP   •   STOP   •   STOP   •   STOP   •   STOP   •   STOP   •   STOP

A10   **a.** true   **b.** true

Q11   Graph: **a.** $\{x\,|\,x < {}^-4\}$

          b. $\{x\,|\,x \leqslant {}^-4\}$

STOP   •   STOP   •   STOP   •   STOP   •   STOP   •   STOP   •   STOP   •   STOP   •   STOP

A11   **a.**
b.

Q12   Sketch a number line and graph each of the following:
          a. $\{x\,|\,x > 2\}$ **b.** $\{x\,|\,x \geqslant 2\}$

          c. $\{x\,|\,x < {}^-3\}$ **d.** $\{x\,|\,x \geqslant {}^-1\}$

e. $\{x\,|\,x \geqslant 0\}$ f. $\{x\,|\,x < 5\}$

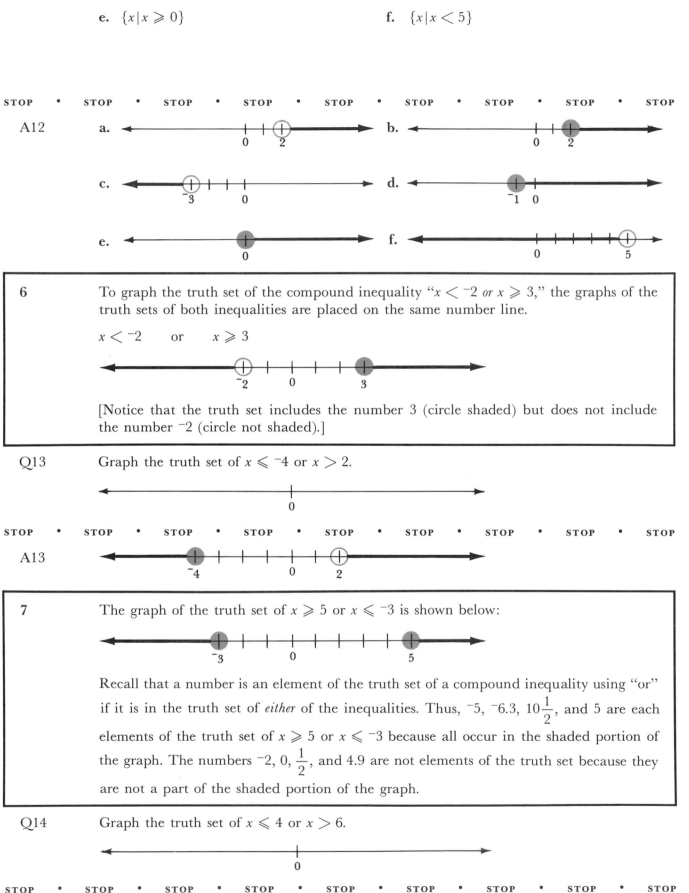

STOP • STOP • STOP • STOP • STOP • STOP • STOP • STOP • STOP

A12   a.   b.

c.   d.

e.   f.

6   To graph the truth set of the compound inequality "$x < ^-2$ *or* $x \geqslant 3$," the graphs of the truth sets of both inequalities are placed on the same number line.

$x < ^-2$ or $x \geqslant 3$

[Notice that the truth set includes the number 3 (circle shaded) but does not include the number $^-2$ (circle not shaded).]

Q13   Graph the truth set of $x \leqslant ^-4$ or $x > 2$.

STOP • STOP • STOP • STOP • STOP • STOP • STOP • STOP • STOP

A13

7   The graph of the truth set of $x \geqslant 5$ or $x \leqslant ^-3$ is shown below:

Recall that a number is an element of the truth set of a compound inequality using "or" if it is in the truth set of *either* of the inequalities. Thus, $^-5$, $^-6.3$, $10\frac{1}{2}$, and 5 are each elements of the truth set of $x \geqslant 5$ or $x \leqslant ^-3$ because all occur in the shaded portion of the graph. The numbers $^-2$, 0, $\frac{1}{2}$, and 4.9 are not elements of the truth set because they are not a part of the shaded portion of the graph.

Q14   Graph the truth set of $x \leqslant 4$ or $x > 6$.

STOP • STOP • STOP • STOP • STOP • STOP • STOP • STOP • STOP

A14

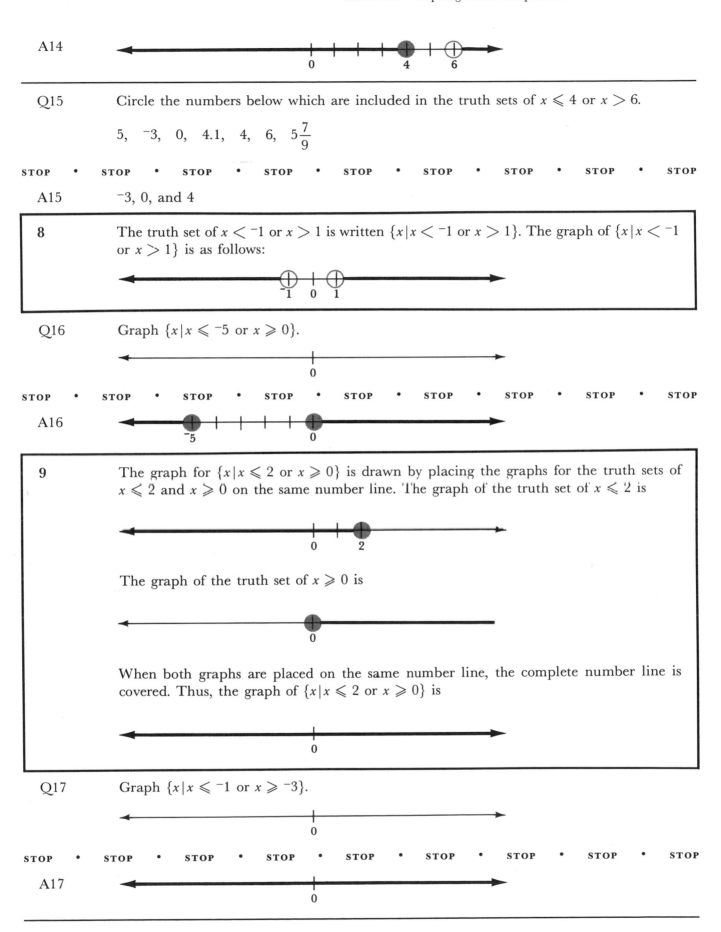

Q15   Circle the numbers below which are included in the truth sets of $x \leqslant 4$ or $x > 6$.

5, $^-3$,   0,   4.1,   4,   6, $5\frac{7}{9}$

STOP   •   STOP   •   STOP   •   STOP   •   STOP   •   STOP   •   STOP   •   STOP   •   STOP

A15 $^-3$, 0, and 4

8   The truth set of $x < {}^-1$ or $x > 1$ is written $\{x \mid x < {}^-1$ or $x > 1\}$. The graph of $\{x \mid x < {}^-1$ or $x > 1\}$ is as follows:

Q16   Graph $\{x \mid x \leqslant {}^-5$ or $x \geqslant 0\}$.

STOP   •   STOP   •   STOP   •   STOP   •   STOP   •   STOP   •   STOP   •   STOP   •   STOP

A16

9   The graph for $\{x \mid x \leqslant 2$ or $x \geqslant 0\}$ is drawn by placing the graphs for the truth sets of $x \leqslant 2$ and $x \geqslant 0$ on the same number line. The graph of the truth set of $x \leqslant 2$ is

The graph of the truth set of $x \geqslant 0$ is

When both graphs are placed on the same number line, the complete number line is covered. Thus, the graph of $\{x \mid x \leqslant 2$ or $x \geqslant 0\}$ is

Q17   Graph $\{x \mid x \leqslant {}^-1$ or $x \geqslant {}^-3\}$.

STOP   •   STOP   •   STOP   •   STOP   •   STOP   •   STOP   •   STOP   •   STOP   •   STOP

A17

Q18   Graph $\{x \mid x < 5 \text{ or } x \geqslant 1\}$.

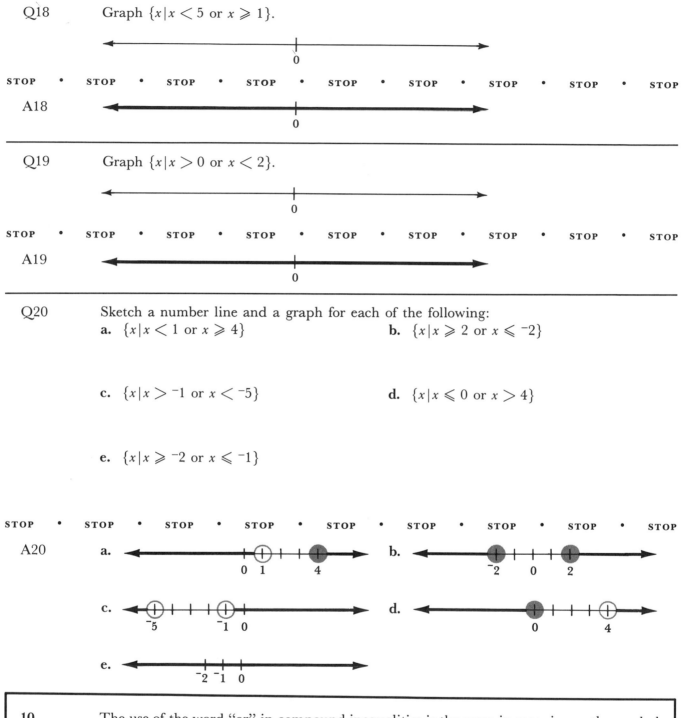

STOP • **STOP** • **STOP** • **STOP** • **STOP** • **STOP** • **STOP** • **STOP** • **STOP**

A18

Q19   Graph $\{x \mid x > 0 \text{ or } x < 2\}$.

STOP • **STOP** • **STOP** • **STOP** • **STOP** • **STOP** • **STOP** • **STOP** • **STOP**

A19

Q20   Sketch a number line and a graph for each of the following:
             a. $\{x \mid x < 1 \text{ or } x \geqslant 4\}$ **b.** $\{x \mid x \geqslant 2 \text{ or } x \leqslant {}^-2\}$

             c. $\{x \mid x > {}^-1 \text{ or } x < {}^-5\}$ **d.** $\{x \mid x \leqslant 0 \text{ or } x > 4\}$

             e. $\{x \mid x \geqslant {}^-2 \text{ or } x \leqslant {}^-1\}$

STOP • **STOP** • **STOP** • **STOP** • **STOP** • **STOP** • **STOP** • **STOP** • **STOP**

A20

10   The use of the word "or" in compound inequalities is the same in meaning as the symbol
             $\cup$ when finding the union of two sets. Thus, just as the union of two sets consists of the
             elements of one set joined with the elements of the second set, the truth set of a compound
             inequality using "or" is formed by joining the truth set of one inequality with the truth
             set of the second inequality. For example:
                 The   graph   of $\{x \mid x \geqslant {}^-3 \text{ or } x \leqslant {}^-5\}$ is   the   same   as   the   graph   of
             $\{x \mid x \geqslant {}^-3\} \cup \{x \mid x \leqslant {}^-5\}$. It is as follows:

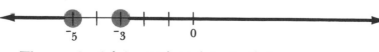

             The graph of $\{x \mid x < 0\} \cup \{x \mid x \geqslant 4\}$ is

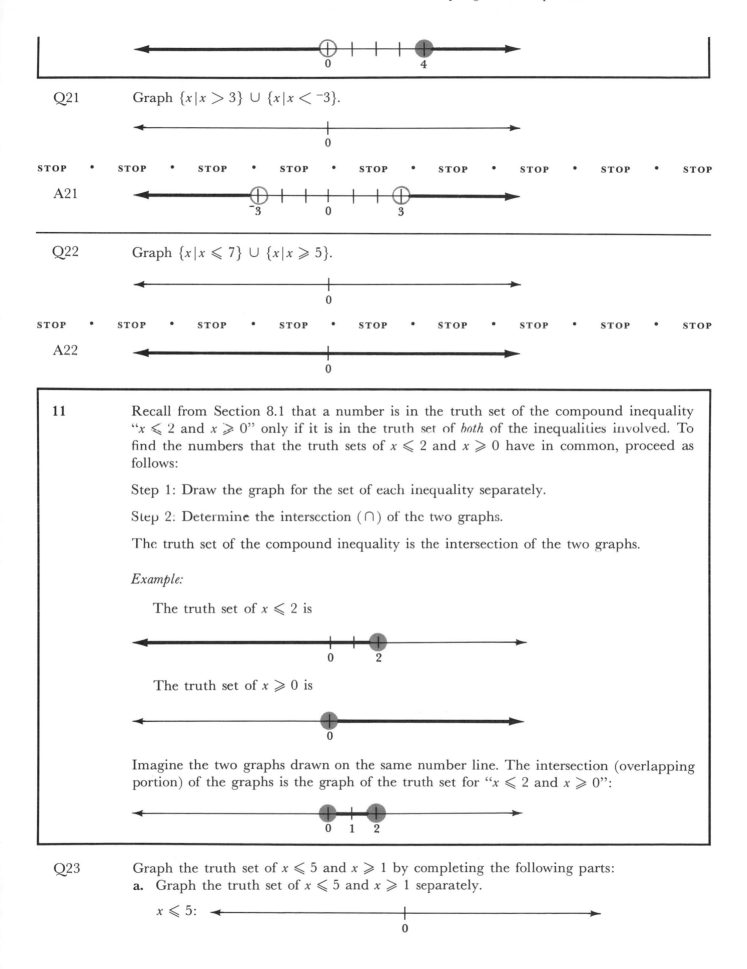

Q21   Graph $\{x \mid x > 3\} \cup \{x \mid x < {}^-3\}$.

STOP • STOP • STOP • STOP • STOP • STOP • STOP • STOP • STOP

A21

Q22   Graph $\{x \mid x \leqslant 7\} \cup \{x \mid x \geqslant 5\}$.

STOP • STOP • STOP • STOP • STOP • STOP • STOP • STOP • STOP

A22

11   Recall from Section 8.1 that a number is in the truth set of the compound inequality "$x \leqslant 2$ and $x \geqslant 0$" only if it is in the truth set of *both* of the inequalities involved. To find the numbers that the truth sets of $x \leqslant 2$ and $x \geqslant 0$ have in common, proceed as follows:

Step 1: Draw the graph for the set of each inequality separately.

Step 2: Determine the intersection ($\cap$) of the two graphs.

The truth set of the compound inequality is the intersection of the two graphs.

Example:

The truth set of $x \leqslant 2$ is

The truth set of $x \geqslant 0$ is

Imagine the two graphs drawn on the same number line. The intersection (overlapping portion) of the graphs is the graph of the truth set for "$x \leqslant 2$ and $x \geqslant 0$":

Q23   Graph the truth set of $x \leqslant 5$ and $x \geqslant 1$ by completing the following parts:
          a. Graph the truth set of $x \leqslant 5$ and $x \geqslant 1$ separately.

          $x \leqslant 5$:

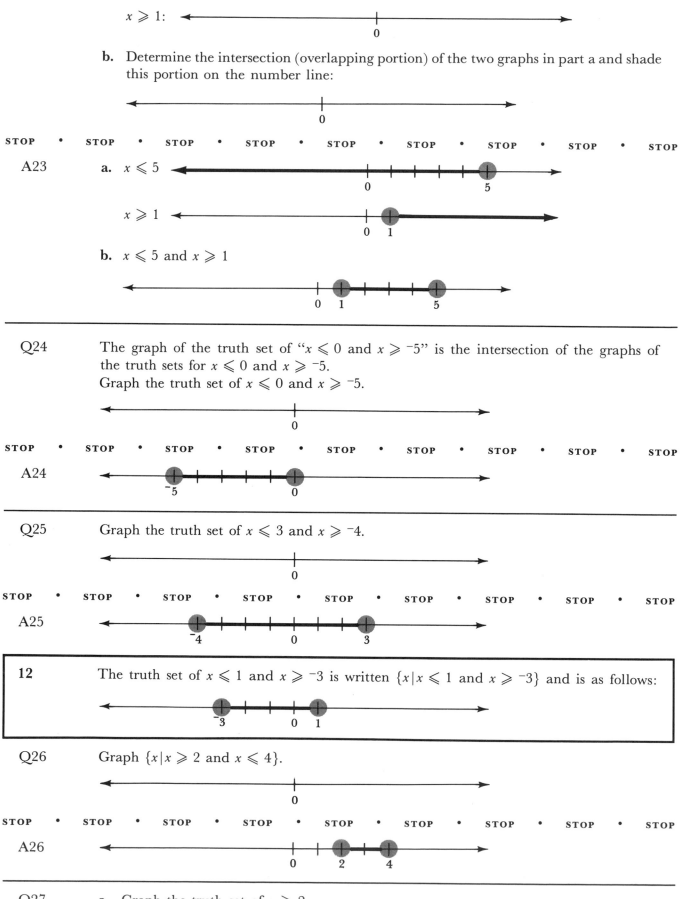

$x \geqslant 1$:

b. Determine the intersection (overlapping portion) of the two graphs in part a and shade this portion on the number line:

STOP • STOP • STOP • STOP • STOP • STOP • STOP • STOP • STOP

A23   **a.** $x \leqslant 5$

$x \geqslant 1$

b. $x \leqslant 5$ and $x \geqslant 1$

Q24   The graph of the truth set of "$x \leqslant 0$ and $x \geqslant {}^-5$" is the intersection of the graphs of the truth sets for $x \leqslant 0$ and $x \geqslant {}^-5$.
Graph the truth set of $x \leqslant 0$ and $x \geqslant {}^-5$.

STOP • STOP • STOP • STOP • STOP • STOP • STOP • STOP • STOP

A24

Q25   Graph the truth set of $x \leqslant 3$ and $x \geqslant {}^-4$.

STOP • STOP • STOP • STOP • STOP • STOP • STOP • STOP • STOP

A25

12   The truth set of $x \leqslant 1$ and $x \geqslant {}^-3$ is written $\{x \mid x \leqslant 1$ and $x \geqslant {}^-3\}$ and is as follows:

Q26   Graph $\{x \mid x \geqslant 2$ and $x \leqslant 4\}$.

STOP • STOP • STOP • STOP • STOP • STOP • STOP • STOP • STOP

A26

Q27   **a.** Graph the truth set of $x \geqslant 2$.

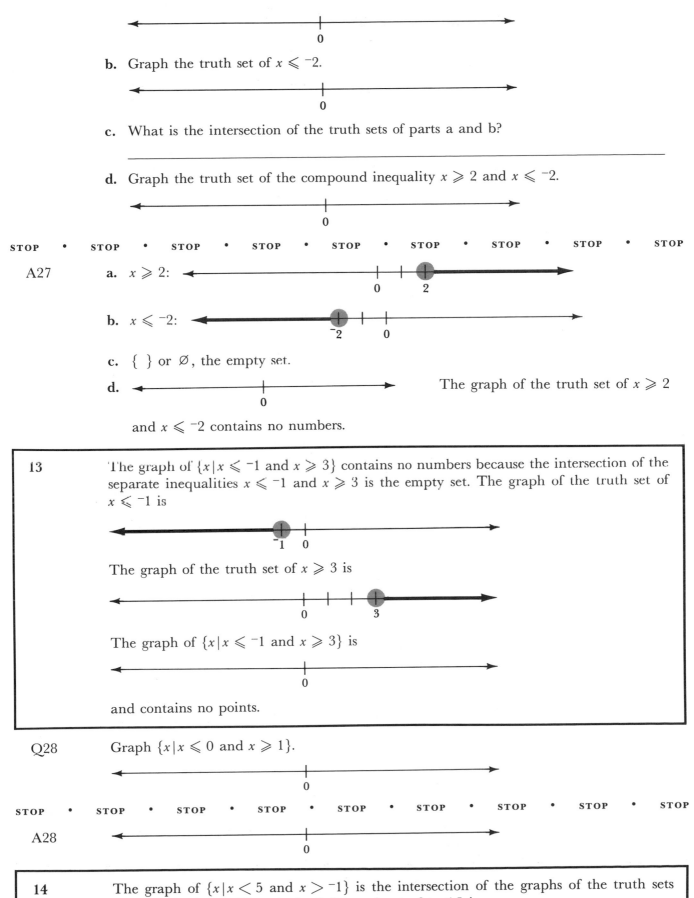

b. Graph the truth set of $x \leqslant {}^-2$.

c. What is the intersection of the truth sets of parts a and b?

_____

d. Graph the truth set of the compound inequality $x \geqslant 2$ and $x \leqslant {}^-2$.

STOP • **STOP** • **STOP** • **STOP** • **STOP** • **STOP** • **STOP** • **STOP** • **STOP**

A27   a. $x \geqslant 2$:

b. $x \leqslant {}^-2$:

c. { } or ∅, the empty set.

d.   The graph of the truth set of $x \geqslant 2$

and $x \leqslant {}^-2$ contains no numbers.

13   The graph of $\{x \mid x \leqslant {}^-1 \text{ and } x \geqslant 3\}$ contains no numbers because the intersection of the separate inequalities $x \leqslant {}^-1$ and $x \geqslant 3$ is the empty set. The graph of the truth set of $x \leqslant {}^-1$ is

The graph of the truth set of $x \geqslant 3$ is

The graph of $\{x \mid x \leqslant {}^-1 \text{ and } x \geqslant 3\}$ is

and contains no points.

Q28   Graph $\{x \mid x \leqslant 0 \text{ and } x \geqslant 1\}$.

STOP • **STOP** • **STOP** • **STOP** • **STOP** • **STOP** • **STOP** • **STOP** • **STOP**

A28

14   The graph of $\{x \mid x < 5 \text{ and } x > {}^-1\}$ is the intersection of the graphs of the truth sets $x < 5$ and $x > {}^-1$. The graph of the truth set of $x < 5$ is

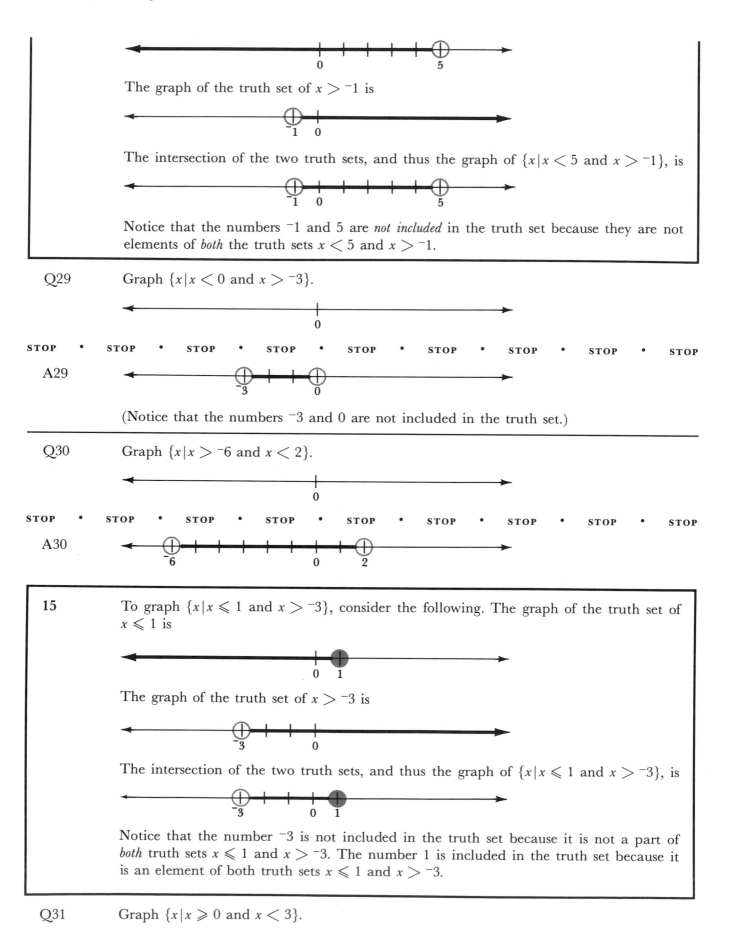

The graph of the truth set of $x > ^-1$ is

The intersection of the two truth sets, and thus the graph of $\{x \mid x < 5 \text{ and } x > ^-1\}$, is

Notice that the numbers $^-1$ and 5 are *not included* in the truth set because they are not elements of *both* the truth sets $x < 5$ and $x > ^-1$.

Q29   Graph $\{x \mid x < 0 \text{ and } x > ^-3\}$.

STOP • STOP • STOP • STOP • STOP • STOP • STOP • STOP • STOP

A29

(Notice that the numbers $^-3$ and 0 are not included in the truth set.)

Q30   Graph $\{x \mid x > ^-6 \text{ and } x < 2\}$.

STOP • STOP • STOP • STOP • STOP • STOP • STOP • STOP • STOP

A30

15   To graph $\{x \mid x \leqslant 1 \text{ and } x > ^-3\}$, consider the following. The graph of the truth set of $x \leqslant 1$ is

The graph of the truth set of $x > ^-3$ is

The intersection of the two truth sets, and thus the graph of $\{x \mid x \leqslant 1 \text{ and } x > ^-3\}$, is

Notice that the number $^-3$ is not included in the truth set because it is not a part of *both* truth sets $x \leqslant 1$ and $x > ^-3$. The number 1 is included in the truth set because it is an element of both truth sets $x \leqslant 1$ and $x > ^-3$.

Q31   Graph $\{x \mid x \geqslant 0 \text{ and } x < 3\}$.

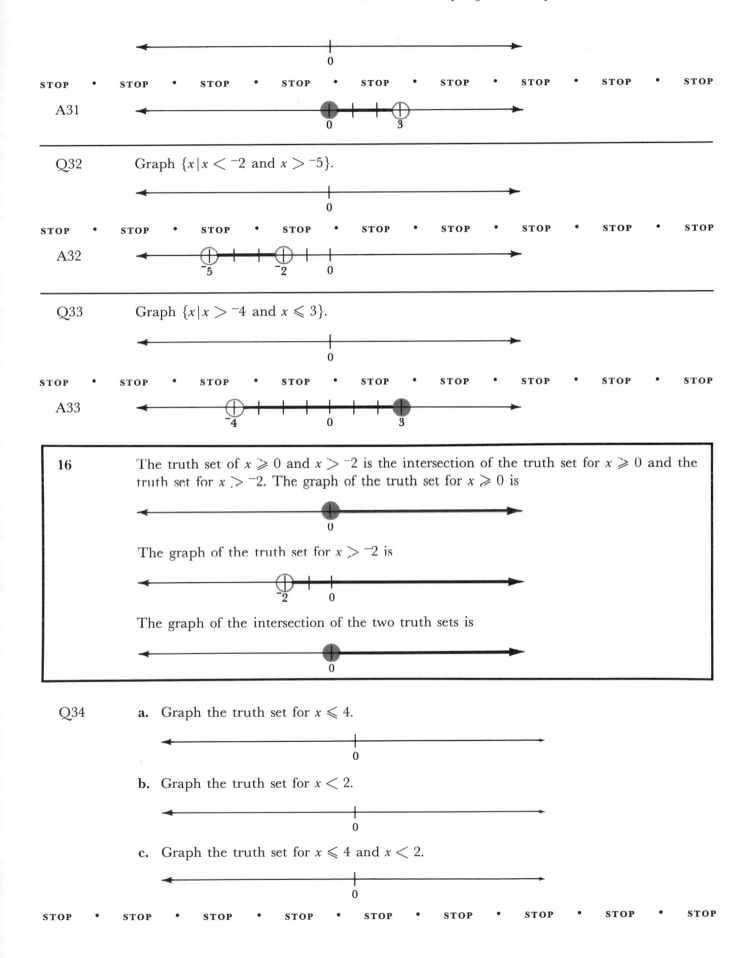

STOP • STOP • STOP • STOP • STOP • STOP • STOP • STOP • STOP

A31

Q32   Graph $\{x \mid x < {}^-2 \text{ and } x > {}^-5\}$.

STOP • STOP • STOP • STOP • STOP • STOP • STOP • STOP • STOP

A32

Q33   Graph $\{x \mid x > {}^-4 \text{ and } x \leqslant 3\}$.

STOP • STOP • STOP • STOP • STOP • STOP • STOP • STOP • STOP

A33

16   The truth set of $x \geqslant 0$ and $x > {}^-2$ is the intersection of the truth set for $x \geqslant 0$ and the truth set for $x > {}^-2$. The graph of the truth set for $x \geqslant 0$ is

The graph of the truth set for $x > {}^-2$ is

The graph of the intersection of the two truth sets is

Q34   **a.** Graph the truth set for $x \leqslant 4$.

b. Graph the truth set for $x < 2$.

c. Graph the truth set for $x \leqslant 4$ and $x < 2$.

STOP • STOP • STOP • STOP • STOP • STOP • STOP • STOP • STOP

A34

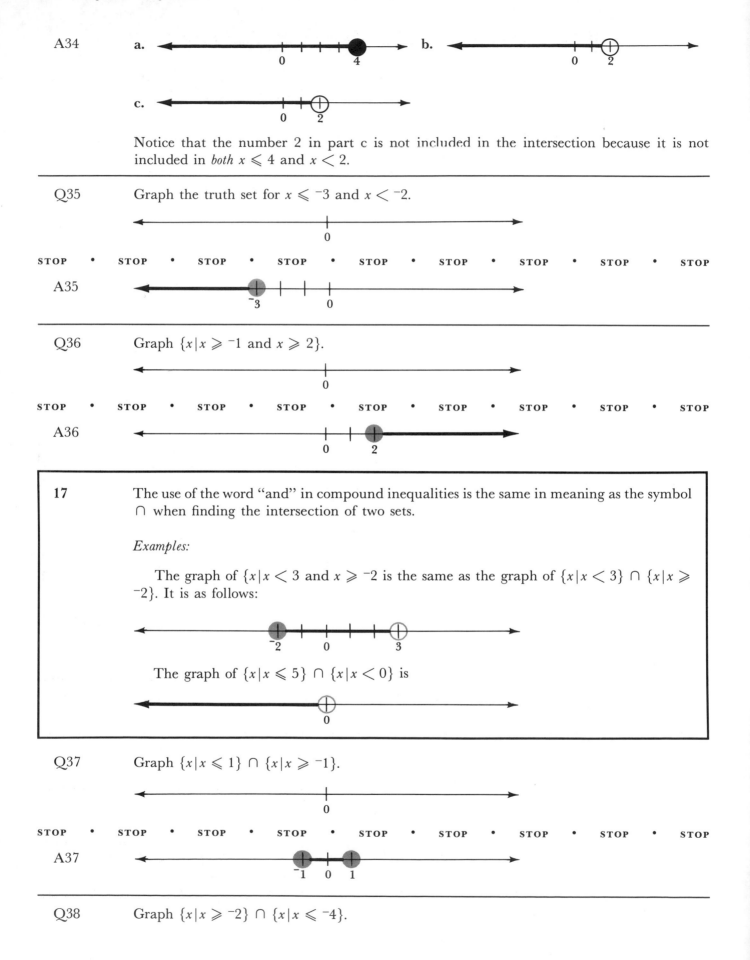

a.

b.

c.

Notice that the number 2 in part c is not included in the intersection because it is not included in *both* $x \leqslant 4$ and $x < 2$.

Q35   Graph the truth set for $x \leqslant {}^-3$ and $x < {}^-2$.

STOP • STOP • STOP • STOP • STOP • STOP • STOP • STOP • STOP

A35

Q36   Graph $\{x \mid x \geqslant {}^-1$ and $x \geqslant 2\}$.

STOP • STOP • STOP • STOP • STOP • STOP • STOP • STOP • STOP

A36

17   The use of the word "and" in compound inequalities is the same in meaning as the symbol ∩ when finding the intersection of two sets.

Examples:

The graph of $\{x \mid x < 3$ and $x \geqslant {}^-2\}$ is the same as the graph of $\{x \mid x < 3\} \cap \{x \mid x \geqslant {}^-2\}$. It is as follows:

The graph of $\{x \mid x \leqslant 5\} \cap \{x \mid x < 0\}$ is

Q37   Graph $\{x \mid x \leqslant 1\} \cap \{x \mid x \geqslant {}^-1\}$.

STOP • STOP • STOP • STOP • STOP • STOP • STOP • STOP • STOP

A37

Q38   Graph $\{x \mid x \geqslant {}^-2\} \cap \{x \mid x \leqslant {}^-4\}$.

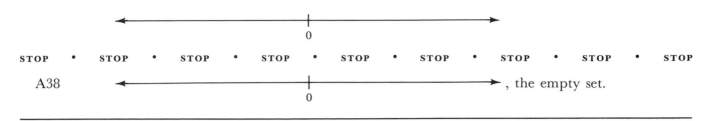

STOP   •   STOP   •   STOP   •   STOP   •   STOP   •   STOP   •   STOP   •   STOP   •   STOP

A38   , the empty set.

This completes the instruction for this section.

8.2   Exercises

1. Graph the truth set for each of the following:
    a. $x > {}^-3$ **b.** $y \leqslant 5$
    c. $x < 0$ **d.** $z \geqslant {}^-1$
    e. $x \leqslant {}^-2$ or $x \geqslant 2$ **f.** $x > 3$ and $x < 5$
    g. $x > {}^-1$ or $x \leqslant {}^-4$ **h.** $x \leqslant 0$ and $x > 3$

2. Graph each of the following:
    a. $\{x \mid x \geqslant {}^-2$ or $x < 0\}$ **b.** $\{x \mid x > 1$ and $x < 3\}$
    c. $\{x \mid x \geqslant 4$ and $x > 1\}$ **d.** $\{x \mid x < 3$ or $x \leqslant 1\}$
    e. $\{x \mid x < {}^-5$ and $x > 1\}$ **f.** $\{x \mid x > {}^-3$ or $x \geqslant 4\}$
    g. $\{x \mid x \geqslant 2$ and $x \geqslant 5\}$ **h.** $\{x \mid x < 4$ and $x > {}^-1\}$
    i. $\{x \mid x < {}^-2$ and $x \leqslant {}^-3\}$ **j.** $\{x \mid x > {}^-4$ or $x > 2\}$
    k. $\{x \mid x \leqslant 1$ or $x > {}^-1\}$ **l.** $\{x \mid x \geqslant 3$ and $x \leqslant {}^-3\}$

3. Graph each of the following:
    a. $\{x \mid x < 1\} \cup \{x \mid x > 2\}$ **b.** $\{x \mid x \geqslant 2\} \cap \{x \mid x \leqslant 4\}$
    c. $\{x \mid x \geqslant 5\} \cap \{x \mid x > 2\}$ **d.** $\{x \mid x < {}^-4\} \cup \{x \mid x \geqslant 0\}$
    e. $\{x \mid x < 0\} \cup \{x \mid x > {}^-1\}$ **f.** $\{x \mid x \leqslant {}^-5\} \cap \{x \mid x \geqslant 5\}$

8.2   Exercise Answers

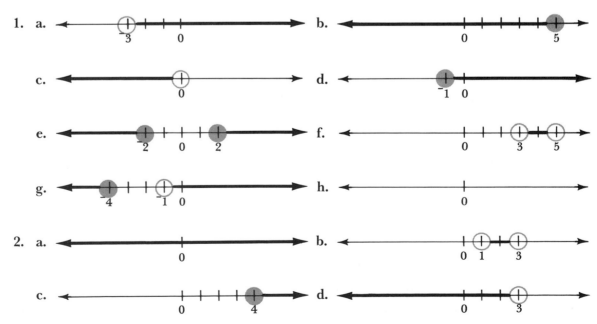

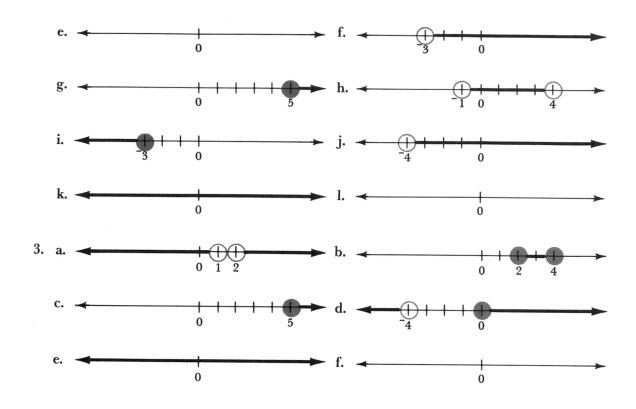

8.3 Solving Linear Inequalities in One Variable

---

1 The procedures used in solving inequalities are very similar to those used in solving equations. For example, consider the true inequality $2 < 7$. If 4 is added to both sides, the result remains a true inequality:

$$2 < 7$$
$$2 + 4 < 7 + 4$$
$$6 < 11$$

Similarly, if 5 is subtracted from both sides of the true inequality $2 < 7$, the result is a true inequality:

$$2 < 7$$
$$2 - 5 < 7 - 5$$
$$^-3 < 2$$

Q1 **a.** Add 3 to both sides of the true inequality $^-4 \leqslant 6$.

b. Is the result a true inequality?_____

STOP • STOP • STOP • STOP • STOP • STOP • STOP • STOP • STOP

A1 **a.** $^-1 \leqslant 9$ **b.** yes

Q2   **a.** Subtract 7 from both sides of the true inequality $5 > 2$.

       b. Is the result a true inequality? _____

STOP • **STOP** • **STOP** • **STOP** • **STOP** • **STOP** • **STOP** • **STOP** • **STOP**

A2   **a.** $^-2 > {}^-5$ **b.** yes

---

2   The addition and subtraction properties of equality are used to solve equations. The previous two questions demonstrate similar properties for solving inequalities.

       The *addition property of inequality* states: For any rational numbers a, b, and c,

$$\text{if} \quad a < b \quad \text{then} \quad a + c < b + c$$
$$\text{if} \quad a \leqslant b \quad \text{then} \quad a + c \leqslant b + c$$
$$\text{if} \quad a > b \quad \text{then} \quad a + c > b + c$$
$$\text{if} \quad a \geqslant b \quad \text{then} \quad a + c \geqslant b + c$$

       The *subtraction property of inequality* states: For any rational numbers a, b, and c,

$$\text{if} \quad a < b \quad \text{then} \quad a - c < b - c$$
$$\text{if} \quad a \leqslant b \quad \text{then} \quad a - c \leqslant b - c$$
$$\text{if} \quad a > b \quad \text{then} \quad a - c > b - c$$
$$\text{if} \quad a \geqslant b \quad \text{then} \quad a - c \geqslant b - c$$

     The addition and subtraction properties of inequality state that if any number is added or subtracted from both sides of a true inequality, the result is another true inequality.

Q3   **a.** Add 5 to both sides of the true inequality $^-5 < {}^-3$.

       b. Is the result a true inequality? _____

       c. What property is demonstrated in parts a and b? _____

       _____

STOP • **STOP** • **STOP** • **STOP** • **STOP** • **STOP** • **STOP** • **STOP** • **STOP**

A3   **a.** $0 < 2$ **b.** yes   **c.** addition property of inequality

---

Q4   **a.** Subtract 9 from both sides of the true inequality $7 \geqslant 7$.

       b. Is the result a true inequality? _____

       c. What property is demonstrated in parts a and b? _____

       _____

STOP • **STOP** • **STOP** • **STOP** • **STOP** • **STOP** • **STOP** • **STOP** • **STOP**

A4   **a.** $^-2 \geqslant {}^-2$ **b.** yes   **c.** subtraction property of inequality

3	The addition and subtraction properties of inequality are used to solve inequalities in the same way that the addition and subtraction properties of equality are used to solve equations. For example,

$$x - 5 < 7 \qquad\qquad\qquad y + 3 \geqslant 2$$
$$x - 5 + 5 < 7 + 5 \quad \text{(add 5)} \qquad y + 3 - 3 \geqslant 2 - 3 \quad \text{(subtract 3)}$$
$$x < 12 \qquad\qquad\qquad\qquad y \geqslant {}^-1$$

Q5 Solve each of the following inequalities:
  a. $x + 7 > 2$ **b.** $y - 4 \leqslant 4$

STOP • **STOP** • **STOP** • **STOP** • **STOP** • **STOP** • **STOP** • **STOP** • **STOP**

A5 **a.** $x > {}^-5$: $x + 7 > 2$ **b.** $y \leqslant 8$: $y - 4 \leqslant 4$
$$x + 7 - 7 > 2 - 7 \qquad\qquad y - 4 + 4 \leqslant 4 + 4$$
$$x > {}^-5 \qquad\qquad\qquad\qquad y \leqslant 8$$

4	Since the truth set of an inequality usually contains more than one number, it is customary to display the truth set with a graph after an inequality has been solved. For example,

$$2x - 2 \leqslant x + 7 \quad \text{(add 2)} \qquad\qquad 5x + 6 > 4x + 2 \quad \text{(subtract 6)}$$
$$2x - 2 + 2 \leqslant x + 7 + 2 \qquad\qquad 5x + 6 - 6 > 4x + 2 - 6$$
$$2x \leqslant x + 9 \quad \text{(subtract } x) \qquad\qquad 5x > 4x - 4 \quad \text{(subtract } 4x)$$
$$2x - x \leqslant x + 9 - x \qquad\qquad 5x - 4x > 4x - 4 - 4x$$
$$x \leqslant 9 \qquad\qquad\qquad\qquad x > {}^-4$$

Q6 Solve and graph $2y - 3 < y + 1$.

STOP • **STOP** • **STOP** • **STOP** • **STOP** • **STOP** • **STOP** • **STOP** • **STOP**

A6 $y < 4$,
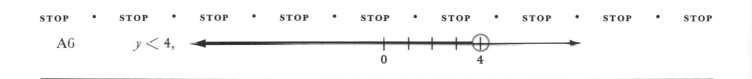

Q7   Solve and graph $3x + 1 \geqslant 2x + 1$.

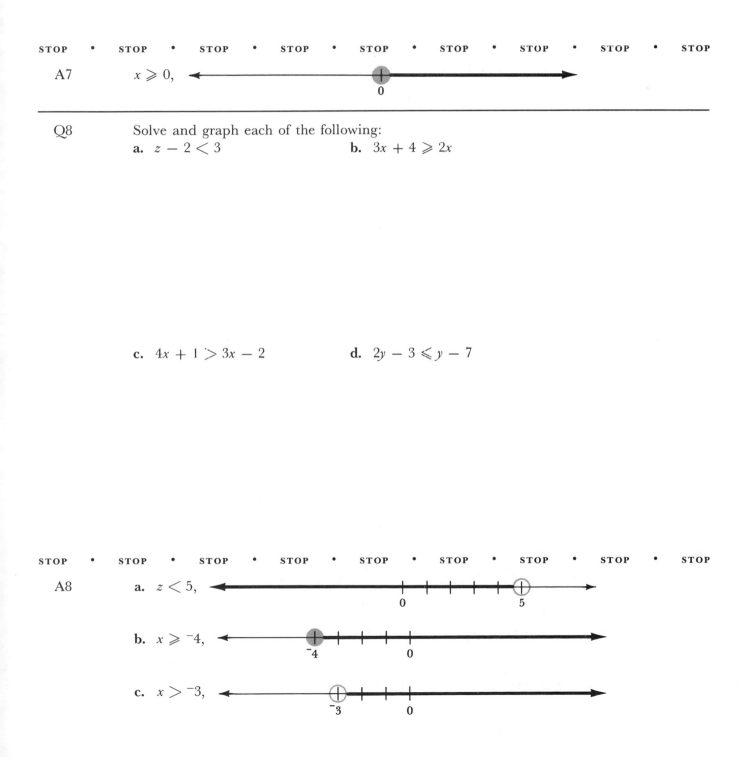

STOP   •   STOP   •   STOP   •   STOP   •   STOP   •   STOP   •   STOP   •   STOP   •   STOP

A7 $x \geqslant 0$,

Q8   Solve and graph each of the following:

a. $z - 2 < 3$ **b.** $3x + 4 \geqslant 2x$

c. $4x + 1 > 3x - 2$ **d.** $2y - 3 \leqslant y - 7$

STOP   •   STOP   •   STOP   •   STOP   •   STOP   •   STOP   •   STOP   •   STOP   •   STOP

A8   **a.** $z < 5$,

          b. $x \geqslant {}^-4$,

          c. $x > {}^-3$,

d. $y \leqslant {}^{-}4,$

---

5   The inequalities $x < 5$ and $5 > x$ mean exactly the same thing. (Notice that the symbol opens toward the 5 in each case.) Confusion often results, however, because of the way $5 > x$ is read. If read left to right $5 > x$ means "5 is greater than x." But if 5 is greater than some number x, the number x must be less than 5. Thus, $5 > x$ means the same as $x < 5$. To avoid confusion, mathematicians usually read solutions to inequalities beginning with the variable. Thus,

$3 < y$ is read y greater than 3
$x \leqslant 7$ is read x less than or equal to 7
${}^{-}1 \leqslant z$ is read z greater than or equal to ${}^{-}1$
$5 > t$ is read t less than 5

Q9 ${}^{-}2 \leqslant x$ is read _____.

STOP • STOP • STOP • STOP • STOP • STOP • STOP • STOP • STOP

A9 x greater than or equal to ${}^{-}2$

Q10   **a.** $0 > y$ is read _____.
b. Graph the truth set of $0 > y$.

STOP • STOP • STOP • STOP • STOP • STOP • STOP • STOP • STOP

A10   **a.** y less than 0

b.

Q11   **a.** ${}^{-}1 \leqslant z$ is read _____.
b. Graph the truth set of ${}^{-}1 \leqslant z$.

STOP • STOP • STOP • STOP • STOP • STOP • STOP • STOP • STOP

A11   **a.** z greater than or equal to ${}^{-}1$.

b.

---

6   If a true inequality is multiplied or divided by a *positive* number, the result is another true inequality.

Examples:

1. ${}^{-}2 > {}^{-}5$ true
     $3({}^{-}2) > 3({}^{-}5)$ (both sides multiplied by 3)
     ${}^{-}6 > {}^{-}15$ true

2. $-4 \leqslant 26$ true

$\dfrac{-4}{2} \leqslant \dfrac{26}{2}$ (both sides divided by 2)

$-2 \leqslant 13$ true

Q12   **a.** Multiply both sides of the true inequality $-1 \geqslant -2$ by 10.

b. Is the result a true inequality?_____

STOP • **STOP** • **STOP** • **STOP** • **STOP** • **STOP** • **STOP** • **STOP** • **STOP**

A12   **a.** $-10 \geqslant -20$ **b.** yes: if both sides of a true inequality are multiplied by a *positive* number, the result is another true inequality.

Q13   **a.** Divide both sides of the true inequality $15 \geqslant 15$ by 3.

b. Is the result a true inequality?_____

STOP • **STOP** • **STOP** • **STOP** • **STOP** • **STOP** • **STOP** • **STOP** • **STOP**

A13   **a.** $5 \geqslant 5$
b. yes: if both sides of a true inequality are divided by a *positive* number, the result is another true inequality.

Q14   **a.** Multiply both sides of the true inequality $6 > 2$ by -2.

b. Is the result a true inequality?_____
c. Place $<$ or $>$ between -12_____-4 to form a true statement.

STOP • **STOP** • **STOP** • **STOP** • **STOP** • **STOP** • **STOP** • **STOP** • **STOP**

A14   **a.** $-12 > -4$ **b.** no   **c.** $<$

Q15   **a.** Divide both sides of the true inequality $-8 \leqslant 4$ by -4.

b. Is the result a true inequality?_____
c. Place $\geqslant$ or $\leqslant$ between 2_____-1 to form a true statement.

STOP • **STOP** • **STOP** • **STOP** • **STOP** **STOP** • **STOP** • **STOP** • **STOP**

A15   **a.** $2 \leqslant -1$ **b.** no   **c.** $\geqslant$

---

7   Questions 14 and 15 demonstrate that if a true inequality is multiplied or divided by a *negative* number, the result is a true inequality *only if the inequality symbol is reversed*. For example, if both sides of the true inequality $^-2 < 3$ are multiplied by $^-5$, the resulting inequality $10 < {}^-15$ is false. However, if the inequality symbol is reversed from $10 < {}^-15$ to $10 > {}^-15$, the result is again a true statement.

Q16   **a.** Multiply both sides of $2 \geqslant {}^-1$ by $^-3$.

b. Is the result a true inequality? _____

c. Reverse the inequality symbol to make the statement true. _____

STOP • **STOP** • **STOP** • **STOP** • **STOP** • **STOP** • **STOP** • **STOP** • **STOP**

A16   **a.** $^-6 \geqslant 3$ **b.** no   **c.** $^-6 \leqslant 3$

Q17   **a.** Divide both sides of $4 > {}^-3$ by $^-4$.

b. Is the result a true inequality? _____

c. Reverse the inequality symbol to make the statement true. _____

STOP • **STOP** • **STOP** • **STOP** • **STOP** • **STOP** • **STOP** • **STOP** • **STOP**

A17   **a.** $^-1 > \dfrac{3}{4}$ **b.** no   **c.** $^-1 < \dfrac{3}{4}$

---

8   The preceding questions demonstrate the following properties of equality. The *multiplication property of inequality* states: For all rational numbers a, b, and any *positive* rational number c,

if $a < b$ then $ac < bc$
if $a \leqslant b$ then $ac \leqslant bc$
if $a > b$ then $ac > bc$
if $a \geqslant b$ then $ac \geqslant bc$

For any *negative* rational number c,

if $a < b$ then $ac > bc$
if $a \leqslant b$ then $ac \geqslant bc$
if $a > b$ then $ac < bc$
if $a \geqslant b$ then $ac \leqslant bc$

The *division property of inequality* states: For any rational numbers a, b, and any *positive* rational number c,

if $a < b$ then $\dfrac{a}{c} < \dfrac{b}{c}$

if $a \leqslant b$ then $\dfrac{a}{c} \leqslant \dfrac{b}{c}$

if $a > b$ then $\dfrac{a}{c} > \dfrac{b}{c}$

if $a \geqslant b$ then $\dfrac{a}{c} \geqslant \dfrac{b}{c}$

For any *negative* rational number c,

if $a < b$ then $\dfrac{a}{c} > \dfrac{b}{c}$

if $a \leqslant b$ then $\dfrac{a}{c} \geqslant \dfrac{b}{c}$

if $a > b$ then $\dfrac{a}{c} < \dfrac{b}{c}$

if $a \geqslant b$ then $\dfrac{a}{c} \leqslant \dfrac{b}{c}$

Examples:

1. $8 \geqslant 6$ true
    $(^-3)8 \leqslant (^-3)6$ (both sides multiplied by $^-3$)
    $^-24 \leqslant ^-18$ true

2. $^-6 < ^-2$ true

    $\dfrac{^-6}{^-1} > \dfrac{^-2}{^-1}$ (divided both sides by $^-1$)

    $6 > 2$ true

3. $^-5 \leqslant 2$ true
    $4(^-5) \leqslant 4(2)$ (multiplied both sides by 4)
    $^-20 \leqslant 8$ true

The multiplication and division properties state that if both sides of a true inequality are multiplied or divided by a *positive* rational number, the result is another true inequality. If both sides of a true inequality are multiplied or divided by a *negative* rational number, the result is a true inequality *only if the inequality symbol is reversed.*

Q18   Use the multiplication or division property of inequality to produce a true statement:

a. Multiply both sides of $^-1 \geqslant ^-3$ by 4.   _____

b. Multiply both sides of $^-1 \geqslant ^-3$ by $^-4$.   _____

c. Divide both sides of $10 < 20$ by 2.   _____

d. Divide both sides of $10 < 20$ by $^-2$.   _____

STOP • **STOP** • **STOP** • **STOP** • **STOP** • **STOP** • **STOP** • **STOP** • **STOP**

A18   **a.** $^-4 \geqslant ^-12$ **b.** $4 \leqslant 12$ **c.** $5 < 10$ **d.** $^-5 > ^-10$

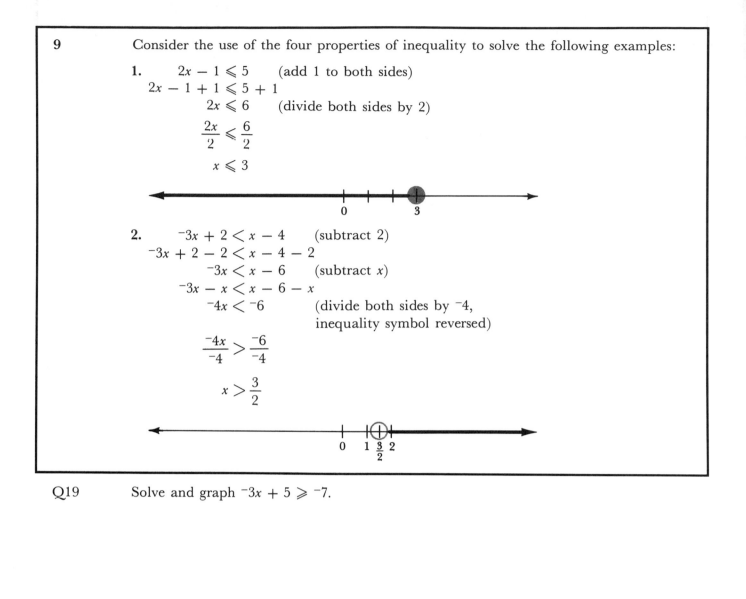

9   Consider the use of the four properties of inequality to solve the following examples:

1. $2x - 1 \leqslant 5$ (add 1 to both sides)

$$2x - 1 + 1 \leqslant 5 + 1$$
$$2x \leqslant 6 \qquad \text{(divide both sides by 2)}$$
$$\frac{2x}{2} \leqslant \frac{6}{2}$$
$$x \leqslant 3$$

2. $^-3x + 2 < x - 4$ (subtract 2)

$$^-3x + 2 - 2 < x - 4 - 2$$
$$^-3x < x - 6 \qquad \text{(subtract } x\text{)}$$
$$^-3x - x < x - 6 - x$$
$$^-4x < ^-6 \qquad \text{(divide both sides by } ^-4,$$
$$\text{inequality symbol reversed)}$$
$$\frac{^-4x}{^-4} > \frac{^-6}{^-4}$$
$$x > \frac{3}{2}$$

Q19   Solve and graph $^-3x + 5 \geqslant ^-7$.

STOP • STOP • STOP • STOP • STOP • STOP • STOP • STOP • STOP

A19 $x \leqslant 4$:

$$^-3x + 5 \geqslant ^-7 \qquad \text{(subtract 5 from both sides)}$$
$$^-3x + 5 - 5 \geqslant ^-7 - 5$$
$$^-3x \geqslant ^-12 \qquad \text{(divide both sides by } ^-3, \text{ inequality}$$
$$\text{symbol reversed)}$$
$$\frac{^-3x}{^-3} \leqslant \frac{^-12}{^-3}$$
$$x \leqslant 4$$

Q20   Solve and graph $2x + 2 > 8$.

STOP   •   STOP   •   STOP   •   STOP   •   STOP   •   STOP   •   STOP   •   STOP   •   STOP

A20 $x > 3$:

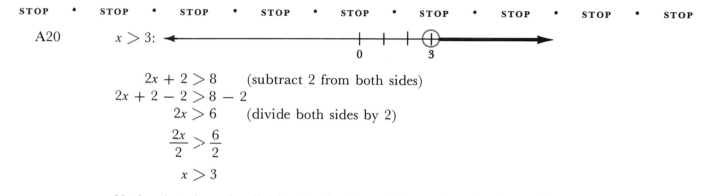

$$2x + 2 > 8 \quad \text{(subtract 2 from both sides)}$$
$$2x + 2 - 2 > 8 - 2$$
$$2x > 6 \quad \text{(divide both sides by 2)}$$
$$\frac{2x}{2} > \frac{6}{2}$$
$$x > 3$$

Notice that since the division involved a *positive* number, the inequality symbol remained the same.

Q21   Solve and graph each of the following:
           a. $^{-}5x < 10$ **b.** $5x < ^{-}10$

           c. $^{-}3x \leqslant x + 12$ **d.** $4x - 1 \leqslant x + 8$

e. $5x - 1 \leqslant 3x + 7$ **f.** $^{-}4x < x + 5$

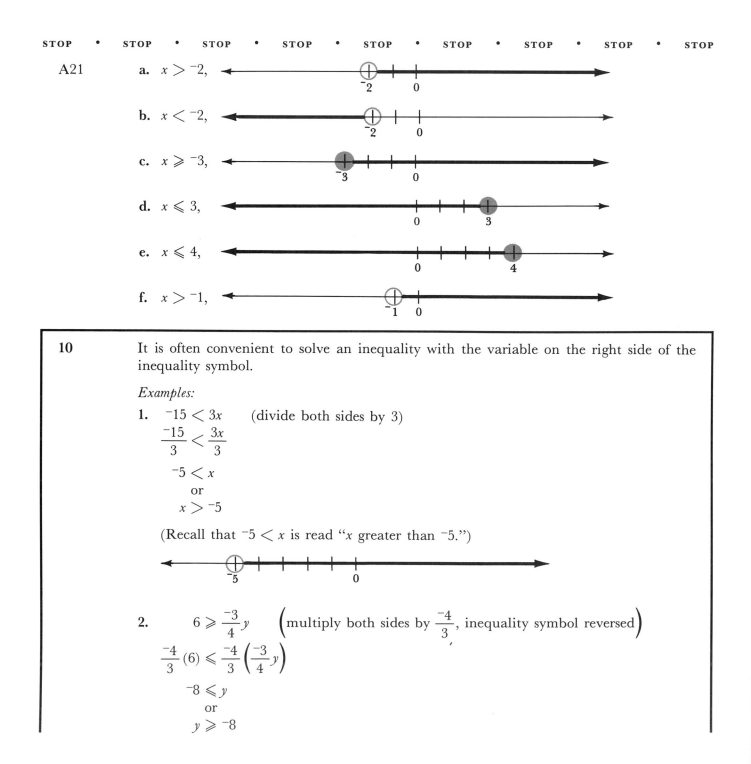

STOP   •   STOP   •   STOP   •   STOP   •   STOP   •   STOP   •   STOP   •   STOP   •   STOP

A21   **a.** $x > {}^{-}2$,

b. $x < {}^{-}2$,

c. $x \geqslant {}^{-}3$,

d. $x \leqslant 3$,

e. $x \leqslant 4$,

f. $x > {}^{-}1$,

---

10   It is often convenient to solve an inequality with the variable on the right side of the inequality symbol.

Examples:

1. $^{-}15 < 3x$ (divide both sides by 3)

$\dfrac{^{-}15}{3} < \dfrac{3x}{3}$

$^{-}5 < x$
      or
  $x > {}^{-}5$

(Recall that $^{-}5 < x$ is read "x greater than $^{-}5$.")

2. $6 \geqslant \dfrac{^{-}3}{4}y$ $\left(\text{multiply both sides by } \dfrac{^{-}4}{3}, \text{ inequality symbol reversed}\right)$

$\dfrac{^{-}4}{3}(6) \leqslant \dfrac{^{-}4}{3}\left(\dfrac{^{-}3}{4}y\right)$

$^{-}8 \leqslant y$
    or
  $y \geqslant {}^{-}8$

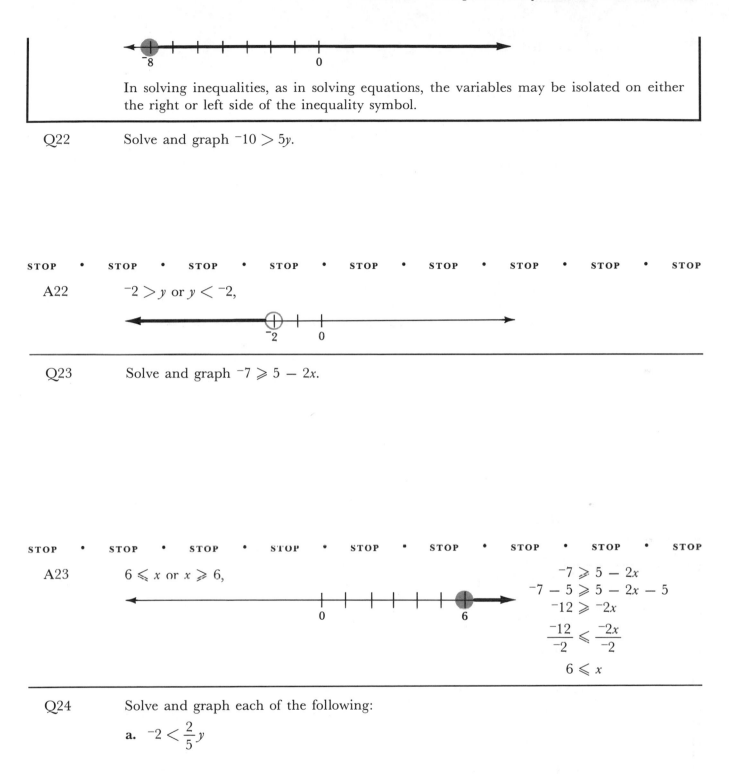

In solving inequalities, as in solving equations, the variables may be isolated on either the right or left side of the inequality symbol.

Q22   Solve and graph $^-10 > 5y$.

STOP • **STOP** • **STOP** • **STOP** • **STOP** • **STOP** • **STOP** • **STOP** • **STOP**

A22 $^-2 > y$ or $y < ^-2$,

Q23   Solve and graph $^-7 \geqslant 5 - 2x$.

STOP • **STOP** • **STOP** • **STOP** • **STOP** • **STOP** • **STOP** • **STOP** • **STOP**

A23 $6 \leqslant x$ or $x \geqslant 6$,

$$^-7 \geqslant 5 - 2x$$
$$^-7 - 5 \geqslant 5 - 2x - 5$$
$$^-12 \geqslant ^-2x$$
$$\frac{^-12}{^-2} \leqslant \frac{^-2x}{^-2}$$
$$6 \leqslant x$$

Q24   Solve and graph each of the following:

a. $^-2 < \frac{2}{5}y$

b. $^-4 \geqslant x + 7$

c. $3 + 2x \leqslant x - 5$

d. $^-21 < ^-5x - 6$

STOP • STOP • STOP • STOP • STOP • STOP • STOP • STOP • STOP

A24

a. $^-5 < y$
  or
  $y > ^-5$

b. $^-11 \geqslant x$
  or
  $x \leqslant ^-11$

c. $x \leqslant ^-8$

d. $3 > x$
  or
  $x < 3$

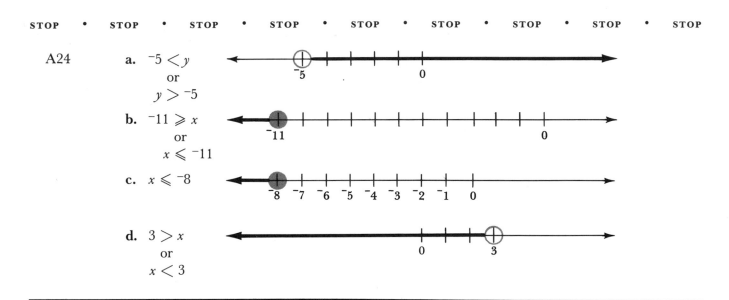

---

11   To solve an inequality that involves fractions:

1. Clear the inequality of fractions by multiplying each term of both sides by the LCD of the denominators present in the inequality.

2. Solve the inequality using the procedures previously developed.

Examples:

1. $\dfrac{x}{3} - 4 \leqslant \dfrac{x}{2}$ (multiply each term of both sides by the LCD 6)

$$6\left(\dfrac{x}{3}\right) - 6(4) \leqslant 6\left(\dfrac{x}{2}\right)$$

$$2x - 24 \leqslant 3x \quad \text{(subtract } 2x \text{ from both sides)}$$
$$2x - 24 - 2x \leqslant 3x - 2x$$
$$^-24 \leqslant x$$
$$\text{or}$$
$$x \geqslant ^-24$$

2. $\qquad \dfrac{-5x}{7} + 1 > \dfrac{-3x}{14} - 2$ (multiply each term of both sides by the LCD, 14)

$$14\left(\dfrac{-5x}{7}\right) + 14(1) > 14\left(\dfrac{-3x}{14}\right) - 14(2)$$

$$-10x + 14 > -3x - 28 \qquad \text{(add } 10x \text{ to both sides)}$$
$$-10x + 14 + 10x > -3x - 28 + 10x$$
$$14 > 7x - 28 \qquad \text{(add 28 to both sides)}$$
$$14 + 28 > 7x - 28 + 28$$
$$42 > 7x \qquad \text{(divide both sides by 7)}$$
$$\dfrac{42}{7} > \dfrac{7x}{7}$$
$$6 > x \quad \text{or} \quad x < 6$$

In the preceding example you should note that the variable term was isolated on the right side so that its coefficient would be positive. If this is done, it will not be necessary to divide later by a negative number, which would necessitate a reversal of the inequality symbol.

Q25   Solve and graph $\dfrac{x}{3} - 2 \leqslant 1$.

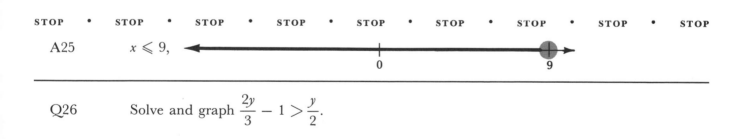

STOP   •   STOP   •   STOP   •   STOP   •   STOP   •   STOP   •   STOP   •   STOP   •   STOP

A25 $x \leqslant 9$,

Q26   Solve and graph $\dfrac{2y}{3} - 1 > \dfrac{y}{2}$.

STOP   •   STOP   •   STOP   •   STOP   •   STOP   •   STOP   •   STOP   •   STOP   •   STOP

A26 $y > 6,$

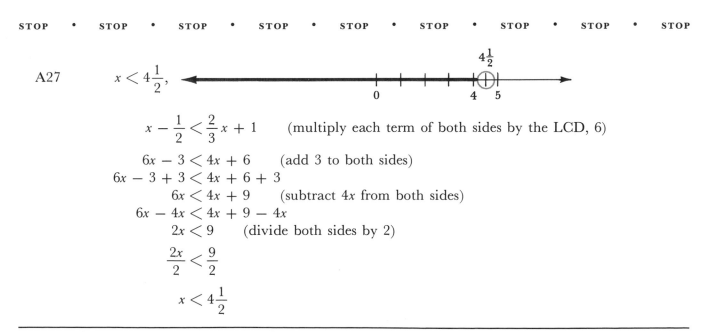

$$\frac{2y}{3} - 1 > \frac{y}{2} \qquad \text{(multiply each term of both sides by the LCD, 6)}$$

$$6\left(\frac{2y}{3}\right) - 6(1) > 6\left(\frac{y}{2}\right)$$

$$4y - 6 > 3y \qquad \text{(add 6 to both sides)}$$
$$4y - 6 + 6 > 3y + 6$$
$$\qquad 4y > 3y + 6 \qquad \text{(subtract } 3y \text{ from both sides)}$$
$$4y - 3y > 3y + 6 - 3y$$
$$\qquad y > 6$$

Q27 Solve and graph $x - \dfrac{1}{2} < \dfrac{2}{3}x + 1$.

STOP • **STOP** • **STOP** • **STOP** • **STOP** • **STOP** • **STOP** • **STOP** • **STOP**

A27 $x < 4\dfrac{1}{2},$

$$x - \frac{1}{2} < \frac{2}{3}x + 1 \qquad \text{(multiply each term of both sides by the LCD, 6)}$$

$$6x - 3 < 4x + 6 \qquad \text{(add 3 to both sides)}$$
$$6x - 3 + 3 < 4x + 6 + 3$$
$$\qquad 6x < 4x + 9 \qquad \text{(subtract } 4x \text{ from both sides)}$$
$$6x - 4x < 4x + 9 - 4x$$
$$\qquad 2x < 9 \qquad \text{(divide both sides by 2)}$$

$$\frac{2x}{2} < \frac{9}{2}$$

$$x < 4\frac{1}{2}$$

This completes the instruction for this section.

8.3 Exercises

1. Solve and graph each of the following:
   a. $x + 7 < 3$ **b.** $2x \geqslant 6$ **c.** $^-3y > 6$
   d. $4 + t < 4$ **e.** $^-10 < ^-2x$ **f.** $^-3 > y + 1$

g. $\dfrac{-1}{5}x \leqslant 1$ **h.** $\dfrac{-3}{2} \geqslant \dfrac{3}{4}x$

2. Solve and graph each of the following:

     a. $2x - 1 \leqslant x - 5$ **b.** $^{-}3x + 4 < 2x - 1$

     c. $5x + 4 > x + 12$ **d.** $5 + x \geqslant 3x$

     e. $2y - 7 \leqslant y - 7$ **f.** $\dfrac{-2}{3}x \leqslant 0$

3. Solve and graph each of the following:

     a. $\dfrac{y}{5} - 1 \leqslant \dfrac{2}{5}$ **b.** $\dfrac{2}{3} - \dfrac{x}{5} < \dfrac{4}{15}$

     c. $\dfrac{x}{2} \geqslant 5 + \dfrac{x}{3}$ **d.** $\dfrac{2x}{3} - 5 > \dfrac{x}{4}$

     e. $\dfrac{2x}{3} - \dfrac{2}{5} \leqslant \dfrac{2x}{5}$ **f.** $\dfrac{x}{3} - \dfrac{3}{4} < \dfrac{-2}{3} + \dfrac{x}{6}$

     g. $\dfrac{3x}{2} \leqslant 3x + \dfrac{3}{2}$ **h.** $\dfrac{x}{5} - \dfrac{9}{10} \geqslant \dfrac{x}{2}$

8.3   Exercise Answers

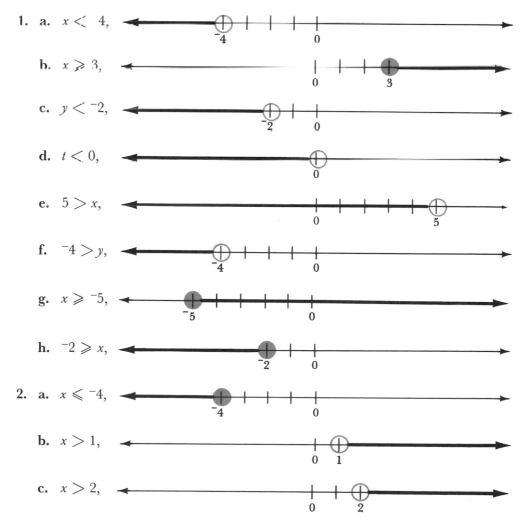

1. **a.** $x < 4,$

     b. $x \geqslant 3,$

     c. $y < {}^{-}2,$

     d. $t < 0,$

     e. $5 > x,$

     f. $^{-}4 > y,$

     g. $x \geqslant {}^{-}5,$

     h. $^{-}2 \geqslant x,$

2. **a.** $x \leqslant {}^{-}4,$

     b. $x > 1,$

     c. $x > 2,$

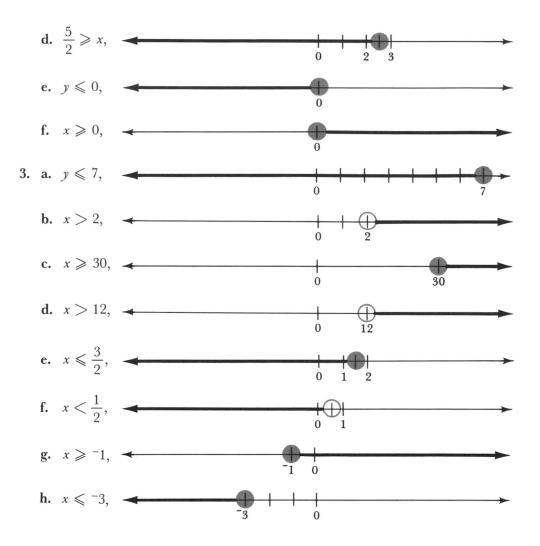

d. $\dfrac{5}{2} \geqslant x,$

e. $y \leqslant 0,$

f. $x \geqslant 0,$

3. a. $y \leqslant 7,$

b. $x > 2,$

c. $x \geqslant 30,$

d. $x > 12,$

e. $x \leqslant \dfrac{3}{2},$

f. $x < \dfrac{1}{2},$

g. $x \geqslant {}^{-}1,$

h. $x \leqslant {}^{-}3,$

Chapter 8 Sample Test

At the completion of Chapter 8 it is expected that you will be able to work the following problems.

8.1   Order and the Symbols of Inequality

1. Write the correct symbol for:
   a. is less than
   b. is greater than
   c. is less than or equal to
   d. is not greater than or equal to
2. True or false:
   a. $4 < 7$
   b. ${}^{-}3 > {}^{-}5$
   c. $3 \geqslant 3$
   d. ${}^{-}5 \leqslant 0$
   e. ${}^{-}2 \not\leqslant 5$
   f. $6 \not> 12$
3. Place $>$ or $<$ between each of the following pairs of numbers to form a true statement:
   a. ${}^{-}3 \underline{\hspace{1cm}} 3$
   b. ${}^{-}10 \underline{\hspace{1cm}} {}^{-}9$
   c. $0 \underline{\hspace{1cm}} {}^{-}2$
   d. $8 \underline{\hspace{1cm}} 6$

4. Write abbreviated statements for each of the following:
  a. $^-2$ is less than $^-1$
  b. 7 is greater than 5
  c. 8 is less than or equal to 8
  d. 5 is not greater than or equal to 9

5. Which numbers in the replacement set $\left\{^-4,\ 3,\ 0,\ ^-2,\ ^-2\frac{1}{2},\ 5,\ ^-1\right\}$ are in the truth set of the open sentence $x \geqslant\ ^-2$?

6. True or false:
  a. $^-2 \in \{x\,|\,x < 0\}$
  b. $5 \in \{x\,|\,x \geqslant 7\}$
  c. $7 \in \{x\,|\,x \leqslant 7\}$
  d. $^-1 \in \{x\,|\,x < 0 \text{ or } x > 5\}$
  e. $10 \in \{x\,|\,x \geqslant 4 \text{ and } x \leqslant 8\}$
  f. $4 \in \{x\,|\,x \geqslant 2 \text{ and } x > 5\}$

8.2   Graphing Linear Inequalities in One Variable

7. Graph the truth set for each of the following:
  a. $x >\ ^-2$
  b. $x \leqslant 3$
  c. $x \leqslant 2 \text{ or } x > 4$
  d. $x >\ ^-5 \text{ and } x < 5$

8. Graph each of the following:
  a. $\{x\,|\,x > 4 \text{ or } x \geqslant\ ^-1\}$
  b. $\{x\,|\,x >\ ^-2 \text{ and } x < 2\}$
  c. $\{x\,|\,x >\ ^-3 \text{ and } x <\ ^-5\}$
  d. $\{x\,|\,x \leqslant 0 \text{ or } x \geqslant\ ^-1\}$
  e. $\{x\,|\,x \geqslant\ ^-3\} \cap \{x\,|\,x < 3\}$
  f. $\{x\,|\,x < 1\} \cup \{x\,|\,x \geqslant 3\}$

8.3   Solving Linear Inequalities in One Variable

9. Solve and graph each of the following:
  a. $x - 3 \leqslant 5$
  b. $2y + 3 > y - 2$
  c. $2y > 6$
  d. $\dfrac{3x}{5} \geqslant\ ^-15$
  e. $\dfrac{x}{2} + 3 \geqslant \dfrac{5}{4}$
  f. $\dfrac{x}{5} + \dfrac{9}{20} \leqslant \dfrac{3x}{4} - \dfrac{x}{10}$

Chapter 8 Sample Test Answers

1. **a.** $<$ **b.** $>$ **c.** $\leqslant$ **d.** $\not\geqslant$
2. **a.** true   **b.** true   **c.** true   **d.** true   **e.** false   **f.** true
3. **a.** $<$ **b.** $<$ **c.** $>$ **d.** $>$
4. **a.** $^-2 <\ ^-1$ **b.** $7 > 5$ **c.** $8 \leqslant 8$ **d.** $5 \not\geqslant 9$
5. $3,\ 0,\ ^-2,\ 5,\ ^-1\ \left(^-4 \not\geqslant\ ^-2,\ ^-2\frac{1}{2} \not\geqslant\ ^-2\right)$
6. **a.** true   **b.** false   **c.** true   **d.** true   **e.** false   **f.** false
7. **a.**   **b.**

 c.   **d.**

8. a.

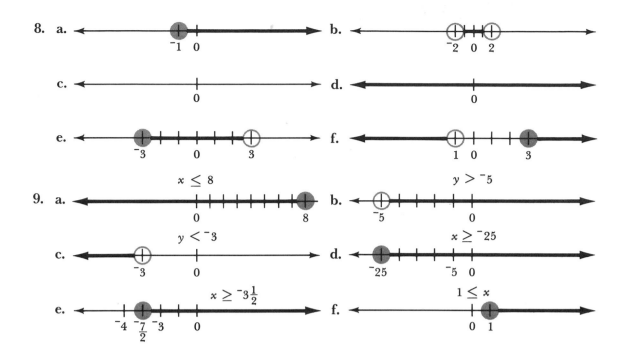

b.

c.

d.

e.

f.

$x \leq 8$

9. a.

$y > {}^-5$

b.

$y < {}^-3$

c.

$x \geq {}^-25$

d.

$x \geq {}^-3\frac{1}{2}$

e.

$1 \leq x$

f.

Chapter 9

Graphing Linear Equations and Inequalities

Some mathematical questions are more easily answered after they have been translated into geometric terms. To do this a system is needed that relates algebraic statements to the geometric plane. This chapter will develop the fundamental ideas of a system that allows geometric interpretations of algebraic statements.

9.1 Rectangular Coordinate System

---

1   A number line is a line with points corresponding to each integer. Rational numbers can also be located on a number line. The following section of a number line has a few integers shown. Of course, the line extends forever in each direction just as the integers continue in both the positive and negative directions.

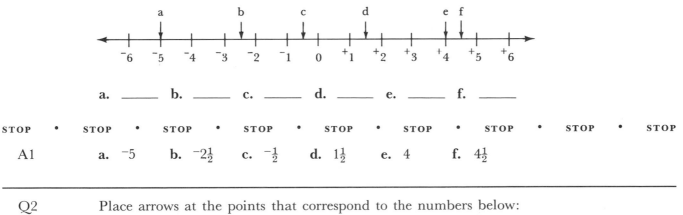

Q1   What numbers correspond to the points indicated by the arrows?

a. _____   b. _____   c. _____   d. _____   e. _____   f. _____

STOP • STOP • STOP • STOP • STOP • STOP • STOP • STOP • STOP

A1   a. $^-5$ b. $^-2\frac{1}{2}$ c. $^-\frac{1}{2}$ d. $1\frac{1}{2}$ e. 4 f. $4\frac{1}{2}$

Q2   Place arrows at the points that correspond to the numbers below:

     a. 3   b. $^-4$ c. $\frac{1}{2}$ d. $^-1\frac{1}{2}$ e. 4   f. $^-3$

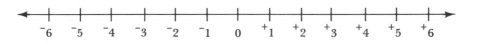

STOP • STOP • STOP • STOP • STOP • STOP • STOP • STOP • STOP

A2

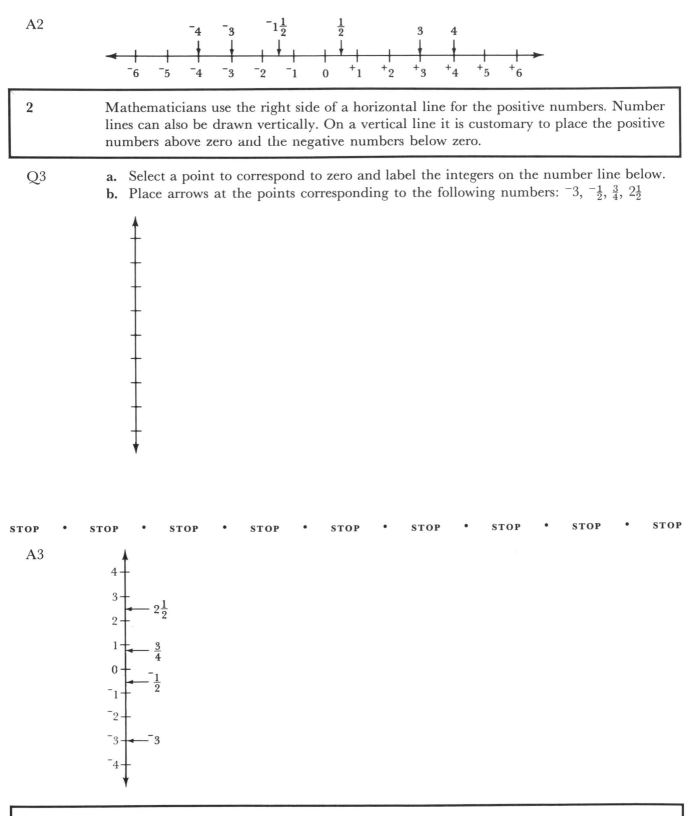

| 2 | Mathematicians use the right side of a horizontal line for the positive numbers. Number lines can also be drawn vertically. On a vertical line it is customary to place the positive numbers above zero and the negative numbers below zero. |

Q3   **a.** Select a point to correspond to zero and label the integers on the number line below.
      b. Place arrows at the points corresponding to the following numbers: $^-3$, $^-\frac{1}{2}$, $\frac{3}{4}$, $2\frac{1}{2}$

STOP • STOP • STOP • STOP • STOP • STOP • STOP • STOP • STOP

A3

| 3 | A *rectangular coordinate system* is formed by crossing one number line in a vertical position over another number line in a horizontal position. The horizontal line is called the *x axis* and the vertical line is the *y axis*. Together they are called the *coordinate axes*. The two lines intersect at the zero point on each axis. This point is called the *origin*. A rectangular coordinate system looks as follows: |

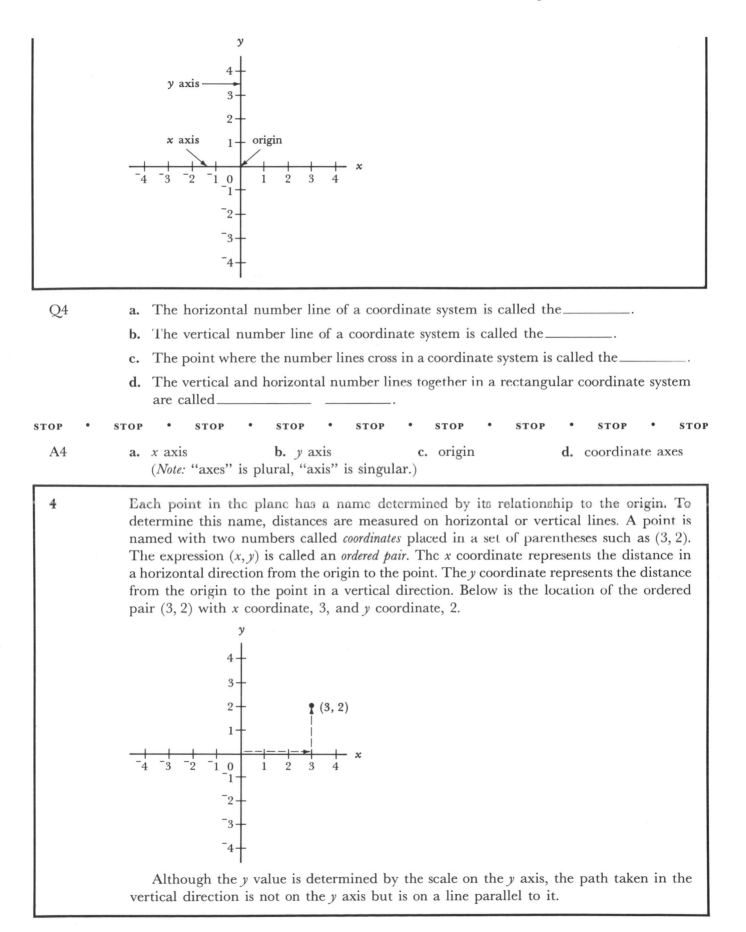

Q4   **a.** The horizontal number line of a coordinate system is called the_____.

b. The vertical number line of a coordinate system is called the_____.

c. The point where the number lines cross in a coordinate system is called the_____.

d. The vertical and horizontal number lines together in a rectangular coordinate system are called_____ _____.

STOP • **STOP** • **STOP** • **STOP** • **STOP** • **STOP** • **STOP** • **STOP** • **STOP**

A4   **a.** *x* axis   **b.** *y* axis   **c.** origin   **d.** coordinate axes
       (*Note:* "axes" is plural, "axis" is singular.)

4   Each point in the plane has a name determined by its relationship to the origin. To determine this name, distances are measured on horizontal or vertical lines. A point is named with two numbers called *coordinates* placed in a set of parentheses such as (3, 2). The expression (x, y) is called an *ordered pair*. The *x* coordinate represents the distance in a horizontal direction from the origin to the point. The *y* coordinate represents the distance from the origin to the point in a vertical direction. Below is the location of the ordered pair (3, 2) with *x* coordinate, 3, and *y* coordinate, 2.

Although the *y* value is determined by the scale on the *y* axis, the path taken in the vertical direction is not on the *y* axis but is on a line parallel to it.

Q5 Use the paths shown to name the points on the coordinate system:

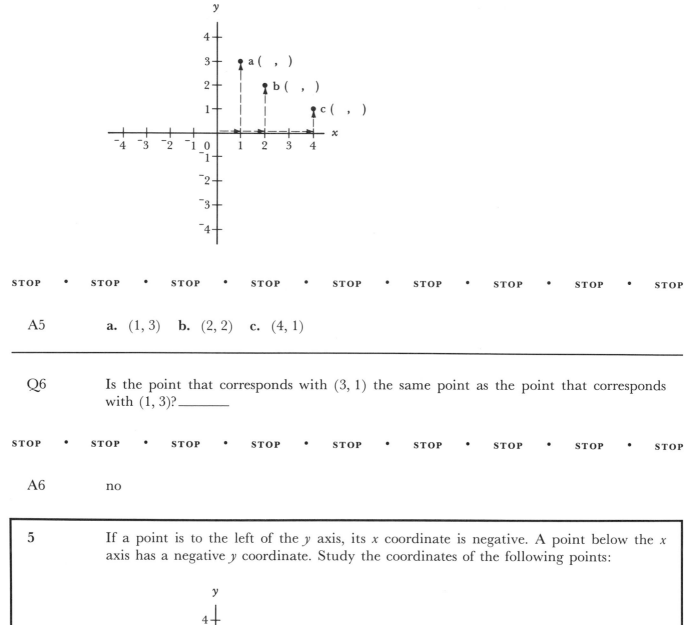

STOP • STOP • STOP • STOP • STOP • STOP • STOP • STOP • STOP

A5 **a.** $(1, 3)$ **b.** $(2, 2)$ **c.** $(4, 1)$

Q6 Is the point that corresponds with $(3, 1)$ the same point as the point that corresponds with $(1, 3)$?_____

STOP • STOP • STOP • STOP • STOP • STOP • STOP • STOP • STOP

A6 no

5 If a point is to the left of the y axis, its x coordinate is negative. A point below the x axis has a negative y coordinate. Study the coordinates of the following points:

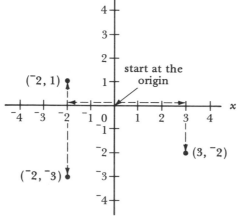

Q7   Name the points on the coordinate system:

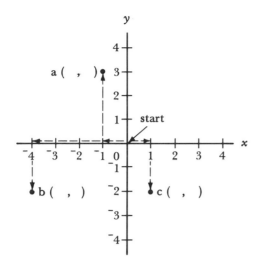

STOP • STOP • STOP • STOP • STOP • STOP • STOP • STOP • STOP

A7   **a.** $(^-1, 3)$ **b.** $(^-4, ^-2)$ **c.** $(1, ^-2)$

Q8   Name the coordinates of each of the following points:

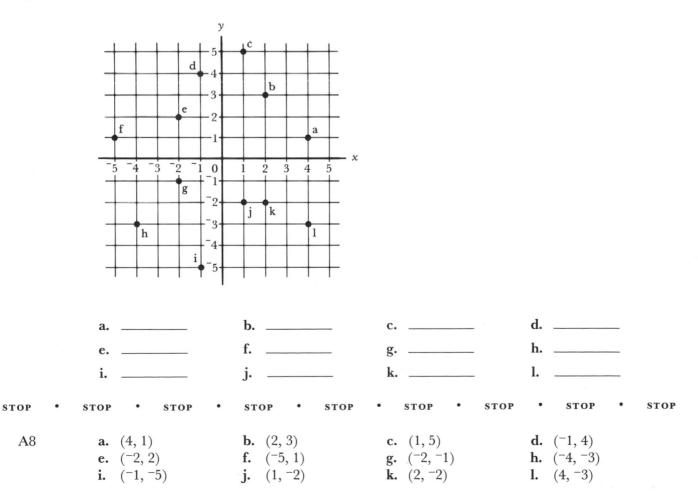

a. _____   b. _____   c. _____   d. _____

e. _____   f. _____   g. _____   h. _____

i. _____   j. _____   k. _____   l. _____

STOP • STOP • STOP • STOP • STOP • STOP • STOP • STOP • STOP

A8   **a.** $(4, 1)$ **b.** $(2, 3)$ **c.** $(1, 5)$ **d.** $(^-1, 4)$
            e. $(^-2, 2)$ **f.** $(^-5, 1)$ **g.** $(^-2, ^-1)$ **h.** $(^-4, ^-3)$
            i. $(^-1, ^-5)$ **j.** $(1, ^-2)$ **k.** $(2, ^-2)$ **l.** $(4, ^-3)$

6   A point is always named with an ordered pair of numbers. If a point is on the axes, one of its coordinates will be zero. For example, point a on the following coordinate system is 3 units from the origin in the horizontal direction but it is zero units from the origin in the vertical direction. It is named (3, 0). Point b is zero units from the origin in the horizontal direction but 2 units up. It is named (0, 2). Study the names of each of the points.

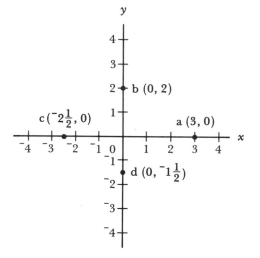

When graphing, numbers are more easily located if written in mixed-number form.

Q9   Name each point on the coordinate axes:

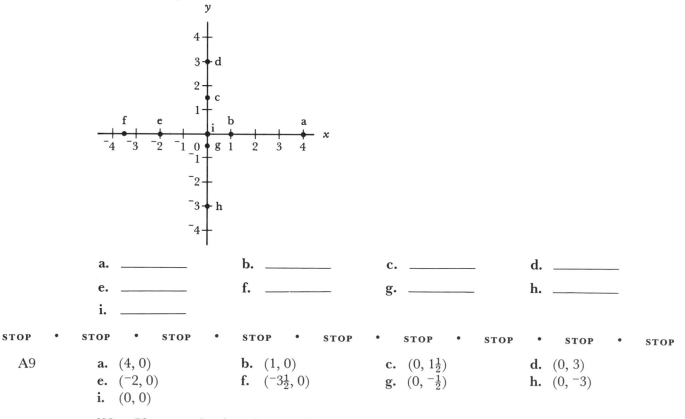

a. _____   b. _____   c. _____   d. _____

e. _____   f. _____   g. _____   h. _____

i. _____

STOP • STOP • STOP • STOP • STOP • STOP • STOP • STOP • STOP

A9   a. $(4, 0)$ b. $(1, 0)$ c. $(0, 1\frac{1}{2})$ d. $(0, 3)$
      e. $(-2, 0)$ f. $(-3\frac{1}{2}, 0)$ g. $(0, -\frac{1}{2})$ h. $(0, -3)$
      i. $(0, 0)$

[*Note:* If your point has the coordinates in the opposite order from the answers given, it is incorrect. For example, $(4, 0) \neq (0, 4)$.]

Q10   Name the points on the coordinate system:

a. _____   b. _____   c. _____   d. _____

e. _____   f. _____   g. _____   h. _____

i. _____   j. _____   k. _____   l. _____

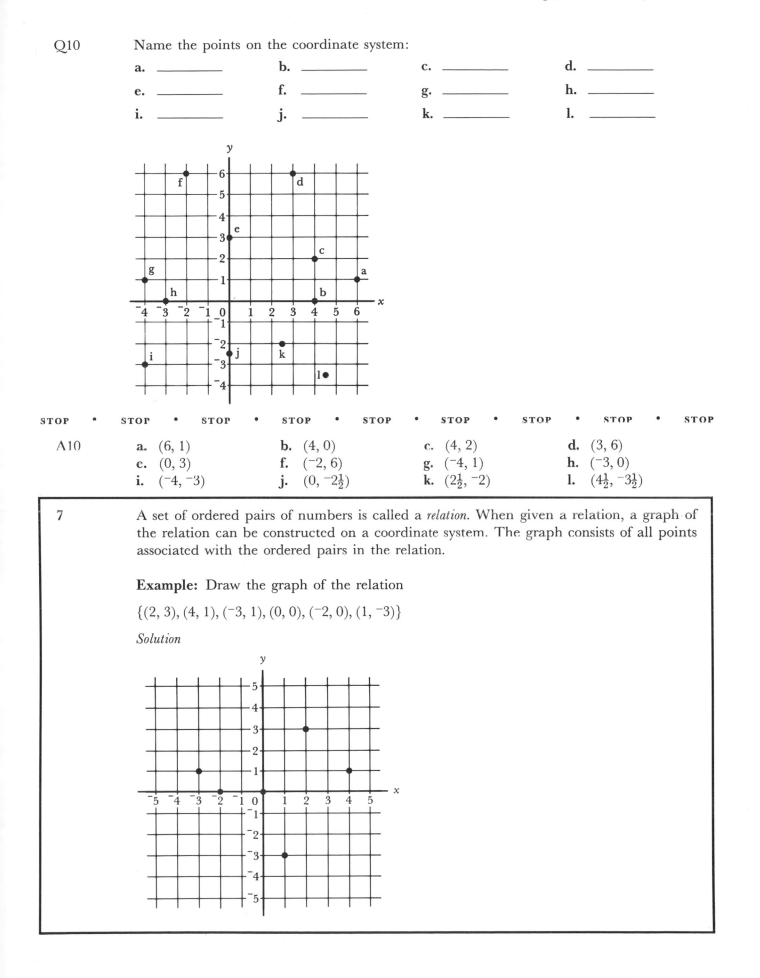

STOP • STOP • STOP • STOP • STOP • STOP • STOP • STOP • STOP

A10   a. $(6, 1)$ b. $(4, 0)$ c. $(4, 2)$ d. $(3, 6)$
        e. $(0, 3)$ f. $(^-2, 6)$ g. $(^-4, 1)$ h. $(^-3, 0)$
        i. $(^-4, ^-3)$ j. $(0, ^-2\frac{1}{2})$ k. $(2\frac{1}{2}, ^-2)$ l. $(4\frac{1}{2}, ^-3\frac{1}{2})$

---

7   A set of ordered pairs of numbers is called a *relation*. When given a relation, a graph of the relation can be constructed on a coordinate system. The graph consists of all points associated with the ordered pairs in the relation.

Example: Draw the graph of the relation

$\{(2, 3), (4, 1), (^-3, 1), (0, 0), (^-2, 0), (1, ^-3)\}$

Solution

Q11   **a.** Construct the graph of the relation

$$\{(3, 1), (^-4, 2), (^-1, 3), (1, 3), (^-2, ^-3), (1, ^-2), (3, ^-3)\}$$

b. Construct the graph of the relation

$$\{(5, 0), (4, 3), (^-4, ^-3), (3, 4), (^-3, ^-4), (3, ^-4), (^-3, 4), (^-5, 0), (0, ^-5), (0, 5),$$
$$(4, ^-3), (^-4, 3)\}$$

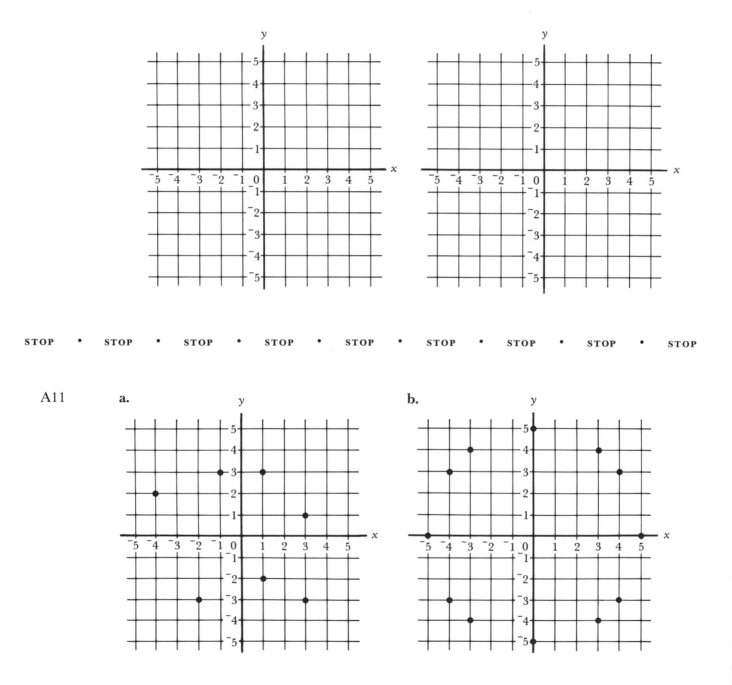

STOP • **STOP** • **STOP** • **STOP** • **STOP** • **STOP** • **STOP** • **STOP** • **STOP**

A11   **a.**   **b.**

In A11b, notice that all the points lie on a circle with center at the origin and radius 5.

Q12   All the points in the following relation are on the same line except two of them. Find the two points by graphing.

$$\{(1, 0), (0, ^-2), (3, 4), (5, 2), (^-1, ^-4), (^-1, 3), (2, 2)\}$$

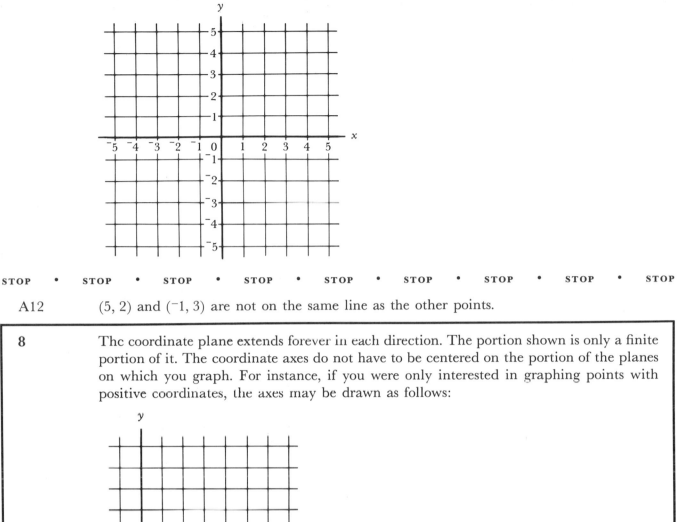

STOP • STOP • STOP • STOP • STOP • STOP • STOP • STOP • STOP

A12   (5, 2) and (⁻1, 3) are not on the same line as the other points.

8   The coordinate plane extends forever in each direction. The portion shown is only a finite portion of it. The coordinate axes do not have to be centered on the portion of the planes on which you graph. For instance, if you were only interested in graphing points with positive coordinates, the axes may be drawn as follows:

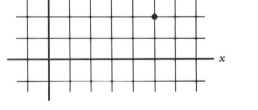

If numbers are not written along the axes, each space is assumed to represent 1 unit. Therefore, the point shown above is (5, 2).

The plane is separated (by the coordinate axes) into four parts called *quadrants*. The quadrants are numbered counterclockwise:

Quadrant II	Quadrant I
Quadrant III	Quadrant IV

A point on an axis is not in any quadrant.

Q13   Draw the x and y axes on the plane and graph the points. Write the points that are in each of the four quadrants.

(5, 2), (3, ⁻3), (3, 1), (⁻2, 1), (⁻1, 4), (1, 5), (⁻2, ⁻1), (3, ⁻4), (⁻4, 1), (⁻4, ⁻1), (⁻1, ⁻4), (1, ⁻4)

a. quadrant I   _____

b. quadrant II   _____

c. quadrant III   _____

d. quadrant IV   _____

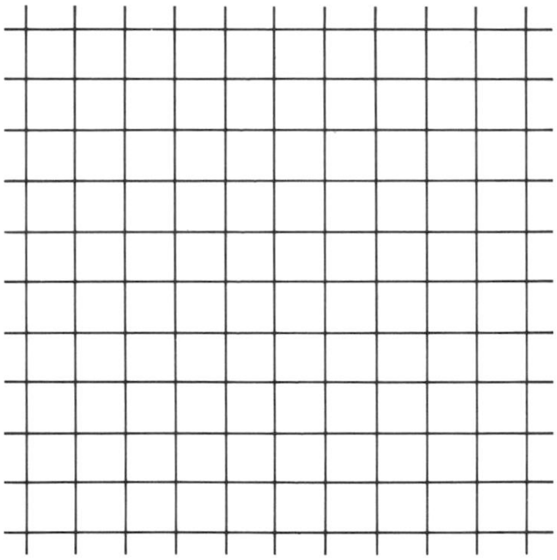

STOP • STOP • STOP • STOP • STOP • STOP • STOP • STOP • STOP

A13   a. (5, 2), (3, 1), (1, 5)   b. (⁻2, 1), (⁻1, 4), (⁻4, 1)
        c. (⁻2, ⁻1), (⁻4, ⁻1), (⁻1, ⁻4)   d. (3, ⁻3), (3, ⁻4), (1, ⁻4)

Q14   Describe the coordinates of a point if the point

a. lies in quadrant I.   _____

b. lies in quadrant II.   _____

c. lies in quadrant III.   _____

d. lies in quadrant IV.   _____

e. lies on the x axis.   _____

f. lies on the y axis.   _____

STOP • STOP • STOP • STOP • STOP • STOP • STOP • STOP • STOP

A14   a. Both x and y are positive.
        b. x is negative and y is positive.
        c. Both x and y are negative.
        d. x is positive and y is negative.
        e. x might be any number, y is zero.
        f. y might be any number, x is zero.

This completes the instruction for this section.

9.1   Exercises

1. Write the coordinates of the points shown on the coordinate system:

   a. _____   b. _____   c. _____   d. _____

   e. _____   f. _____   g. _____   h. _____

   i. _____   j. _____   k. _____   l. _____

   m. _____   n. _____   p. _____   q. _____

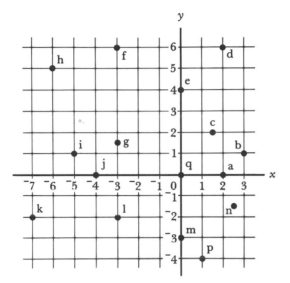

2. Name the letters of the points on the coordinate system in problem 1 in each of the following:

   a. quadrant I   b. quadrant II   c. quadrant III   d. quadrant IV

3. Graph the relation $\{(^-5, 2), (^-2, 2), (2, ^-4), (^-2, ^-3), (^-4, 0), (0, 3), (3, 2), (^-5, ^-5), (0, ^-2), (1, 2), (1, 0), (4, 0), (4, -2)\}$.

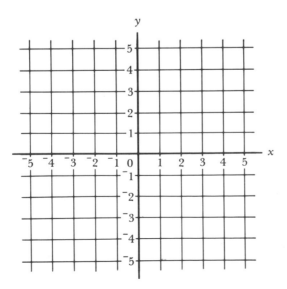

4. Use the grid to draw a coordinate system and determine if each set of three points is on a straight line:
    a. $(3, 2), (1, 0), (^-1, ^-2)$
    b. $(0, 0), (^-1, ^-2), (^-1, ^-4)$
    c. $(0, ^-5), (2, ^-2), (4, 1)$
    d. $(^-4, 2), (^-2, 1), (4, ^-2)$

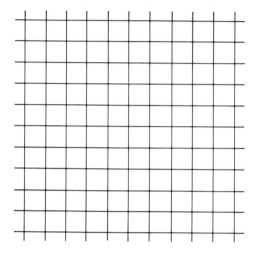

5. a. A vertical line segment is drawn from the point $(25, 14)$ to the x axis. How long is the segment?
    b. A horizontal line segment is drawn from the point $(17, 31)$ to the y axis. How long is the segment?

9.1   Exercise Answers

1. a. $(2, 0)$ b. $(3, 1)$ c. $(1\frac{1}{2}, 2)$ d. $(2, 6)$
    e. $(0, 4)$ f. $(^-3, 6)$ g. $(^-3, 1\frac{1}{2})$ h. $(^-6, 5)$
    i. $(^-5, 1)$ j. $(^-4, 0)$ k. $(^-7, ^-2)$ l. $(^-3, ^-2)$
    m. $(0, ^-3)$ n. $(2\frac{1}{2}, ^-1\frac{1}{2})$ p. $(1, ^-4)$ q. $(0, 0)$

2. a. b, c, and d   b. f, g, h, and i   c. k and 1   d. n and p
3.

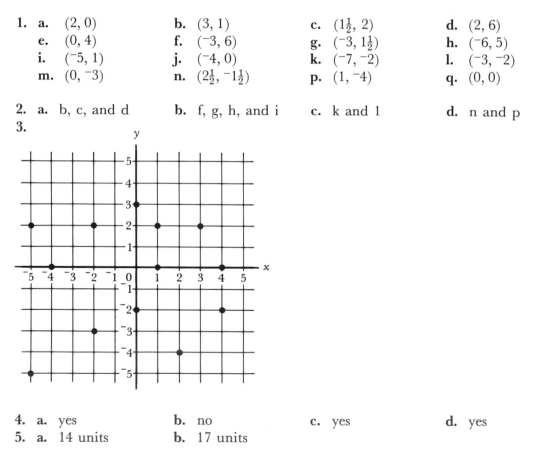

4. a. yes   b. no   c. yes   d. yes
5. a. 14 units   b. 17 units

9.2   Graphing Linear Equations

| 1 | Sets are sometimes defined by listing the elements. When this is inconvenient or impossible, a method using set builder notation is used. The set is described by first listing the type of element that is in the set and then listing any restriction that the elements must satisfy.

Examples:

{people|the person is a male over 6 feet tall}

This set defined is a set consisting of male people over 6 feet tall. The notation is read: "The set of all people such that the person is a male over 6 feet tall." The vertical line means "such that."

$\{x|x$ is a natural number less than 7$\}$

This set is the set $\{1, 2, 3, 4, 5, 6\}$ because only these numbers satisfy the restriction. The notation is read: "The set of all x such that x is a natural number less than 7." |

Q1   Write the words that would be used to read the following: $\{x|x$ is a natural number less than 3$\}$. _____

STOP • STOP • STOP • STOP • STOP • STOP • STOP • STOP • STOP

A1   The set of all x such that x is a natural number less than 3.

Q2   Consider the set $C = \{$animals$|$the animal has 4 legs$\}$. Answer true or false:

a. goat $\in C$ _____   **b.** fish $\in C$ _____

c. chicken $\notin C$ _____   **d.** dog $\in C$ _____

STOP • STOP • STOP • STOP • STOP • STOP • STOP • STOP • STOP

A2   **a.** true   **b.** false   **c.** true   **d.** true

| 2 | In Section 9.1 a relation was defined as a set of ordered pairs. An example is the relation $\{(2, 5), (4, 7), (3, {}^-1)\}$. If a relation has an infinite number of ordered pairs, set builder notation can be used to describe it. For example,

$\{(x,y)|x + y = 5\}$

The notation is read: "the set of all ordered pairs x,y such that $x + y = 5$." In order for an element to be in the set, it must be an ordered pair and the x and y values must sum to 5.
Let us name the set in this example f. Therefore,

$f = \{(x,y)|x + y = 5\}$

Verify that the following are true statements:

1. $(3, 2) \in f$ $3 + 2 = 5$
2. $(0, 5) \in f$ $0 + 5 = 5$
3. $({}^-2, 7) \in f$ ${}^-2 + 7 = 5$
4. $({}^-3, {}^-2) \notin f$ ${}^-3 + {}^-2 \neq 5$
5. $({}^-105, 110) \in f$ ${}^-105 + 110 = 5$

Since there are an infinite number of numbers whose sum is 5, this list could go on and on. |

Q3 Consider the set $h = \{(x, y) \mid x + y = 10\}$. Insert $\in$ or $\notin$ to form a true statement:

   a. $(8, 2)$_____h **b.** $(5, 4)$_____h **c.** $(0, 10)$_____h

   d. $(12, {}^-2)$_____h **e.** $(2, {}^-12)$_____h **f.** $(17, {}^-7)$_____h

STOP • **STOP** • **STOP** • **STOP** • **STOP** • **STOP** • **STOP** • **STOP** • **STOP**

A3   **a.** $\in$ **b.** $\notin$: because $5 + 4 \neq 10$ **c.** $\in$

      d. $\in$ **e.** $\notin$: because $2 + {}^-12 \neq 10$ **f.** $\in$

Q4 Consider the relation $f = \{(x, y) \mid x + 2y = 9\}$. Insert $\in$ or $\notin$ to form a true statement:

   a. $(5, 2)$_____f **b.** $(4, 5)$_____f

   c. $(7, 1)$_____f **d.** $(1, 7)$_____f

   e. $(13, {}^-2)$_____f **f.** $({}^-11, {}^-1)$_____f

STOP • **STOP** • **STOP** • **STOP** • **STOP** • **STOP** • **STOP** • **STOP** • **STOP**

A4   **a.** $\in$: $5 + 2(2) = 9$ **b.** $\notin$: $4 + 2(5) \neq 9$

      c. $\in$: $7 + 2(1) = 9$ **d.** $\notin$: $1 + 2(7) \neq 9$

      e. $\in$: $13 + 2({}^-2) = 9$ **f.** $\notin$: ${}^-11 + 2({}^-1) \neq 9$

---

3 The variables in a set of ordered pairs are assumed to represent real numbers (see p. 516) unless a restriction is expressed within the set. If the restriction that defines the set mentions only one variable, then the variable that is not restricted may take on any value. For example,

$$f = \{(x, y) \mid x = 4\}$$

Each ordered pair in the relation must have a 4 in the first position. Since there is no restriction on y, it may be any rational number. The following are elements of f:

$(4, 6)$, $(4, 0)$, $(4, 4)$, $(4, {}^-3)$, $(4, \frac{3}{2})$, $(4, \frac{{}^-2}{3})$, etc.

The following are not elements of f:

$(0, 4)$, $(3, 2)$, $({}^-10, 6)$, $({}^-12, 4)$, $(5, 7)$, etc.

Q5 Let $h = \{(x, y) \mid x = 6\}$. Answer true or false:

   a. $(3, 6) \in h$ _____   **b.** $(6, 4) \in h$ _____

   c. $(6, 0) \in h$ _____   **d.** $(6, 6) \in h$ _____

   e. $(5, 3) \in h$ _____   **f.** $({}^-6, 6) \in h$ _____

STOP • **STOP** • **STOP** • **STOP** • **STOP** • **STOP** • **STOP** • **STOP** • **STOP**

A5   **a.** false   **b.** true   **c.** true   **d.** true   **e.** false   **f.** false

---

Q6 Let $f = \{(x, y) \mid y = {}^-3\}$. Answer true or false:

   a. $({}^-3, 4) \in f$ _____   **b.** $(4, {}^-3) \in f$ _____

   c. $(0, {}^-3) \in f$ _____   **d.** $({}^-3, {}^-3) \in f$ _____

   e. $(\frac{1}{3}, {}^-3) \in f$ _____   **f.** $(\frac{{}^-7}{4}, \frac{{}^-9}{3}) \in f$ _____

STOP • **STOP** • **STOP** • **STOP** • **STOP** • **STOP** • **STOP** • **STOP** • **STOP**

A6   **a.** false   **b.** true   **c.** true   **d.** true   **e.** true   **f.** true: $\frac{-9}{3} = {}^-3$

---

4   The graph of a relation consisting of an infinite number of ordered pairs is an infinite number of points on a coordinate system. It would be impossible to locate them all individually. Fortunately, the graphs of linear equations,* which this section discusses, consist of points that all fall on the same straight line. To graph a relation, a few representative ordered pairs must be obtained.

Graph the relation $f = \{(x, y) \mid x + y = 4\}$. First obtain a few representative ordered pairs which are in the set. Some that could be used are:

$(4, 0)$ because $4 + 0 = 4$
$(3, 1)$ because $3 + 1 = 4$
$(2, 2)$ because $2 + 2 = 4$
$(1, 3)$ because $1 + 3 = 4$
$(0, 4)$ because $0 + 4 = 4$
$({}^-1, 5)$ because ${}^-1 + 5 = 4$
$(5, {}^-1)$ because $5 + {}^-1 = 4$

These ordered pairs are located on a coordinate system.

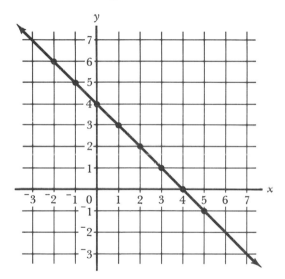

The points are connected by means of a straight line with arrows on each end to indicate that it extends forever. This corresponds with the fact that the set of ordered pairs is infinite.

*A linear equation is any equation that can be written in the form $ax + by = c$ (a and b not both zero), where a, b, and c are rational numbers. The values a, b, and c can also be real numbers. Real numbers will be discussed in Chapter 13.

Q7   **a.** Determine six ordered pairs that are in the relation $\{(x, y) \mid x + y = 6\}$. (Some possible choices for x or y are given. You may select others.)

$(5, \underline{\quad})$

$(\underline{\quad}, 3)$

$(\underline{\quad}, 0)$

$({}^-1, \underline{\quad})$

$(\underline{\quad}, \underline{\quad})$

$(\underline{\quad}, \underline{\quad})$

b. Graph the points and connect them with a line.

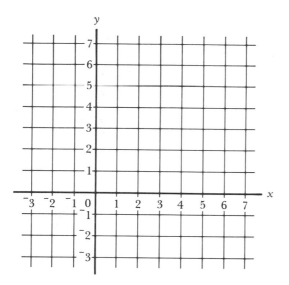

STOP • STOP • STOP • STOP • STOP • STOP • STOP • STOP • STOP

A7   **a.** Many ordered pairs could be used. The sum of the coordinates of each must be six. Some examples are: (5, 1), (3, 3), (6, 0), (⁻1, 7), (4, 2), (0, 6), (7, ⁻1). After these points have been graphed, other points can be seen to be on the graph. For example, (1, 5), (2, 4), ($1\frac{1}{2}$, $4\frac{1}{2}$), ($2\frac{1}{2}$, $3\frac{1}{2}$), ($6\frac{1}{2}$, $\frac{-1}{2}$), etc.

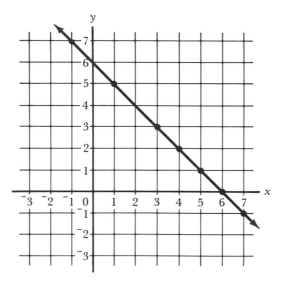

---

5   The equation is often more complicated so that ordered pairs which make the equation true are not easily found. A more systematic approach is needed.

To find an ordered pair that is in the relation, any value of one of the variables may be chosen. It is completely arbitrary, although it is good to choose numbers close to the origin so the point may be easily graphed. After the first number is chosen, the value of the second variable is determined by solving the equation.

Example: Find four ordered pairs that are in the relation $f = \{(x, y) \mid 2x + y = 3\}$ and graph the relation.

Solution

Any numbers may be used for x. Suppose that $x = 2$, $x = {}^-1$, $x = 0$, and $x = 4$ are selected. The values of y are found by substituting each value of x in the equation and solving the resulting equation for y.

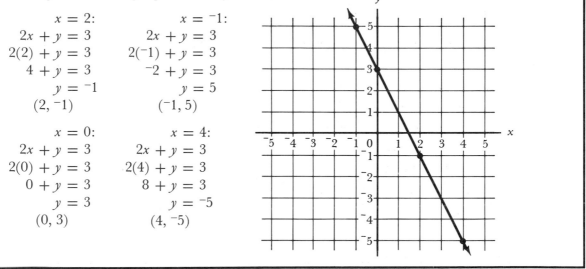

$$x = 2:$$
$$2x + y = 3$$
$$2(2) + y = 3$$
$$4 + y = 3$$
$$y = {}^-1$$
$$(2, {}^-1)$$

$$x = {}^-1:$$
$$2x + y = 3$$
$$2({}^-1) + y = 3$$
$${}^-2 + y = 3$$
$$y = 5$$
$$({}^-1, 5)$$

$$x = 0:$$
$$2x + y = 3$$
$$2(0) + y = 3$$
$$0 + y = 3$$
$$y = 3$$
$$(0, 3)$$

$$x = 4:$$
$$2x + y = 3$$
$$2(4) + y = 3$$
$$8 + y = 3$$
$$y = {}^-5$$
$$(4, {}^-5)$$

Q8   **a.** By reading the graph in Frame 5, find the y values that would be matched with (1) $x = 1$ and (2) $x = 3$.

b. Substitute $x = 1$ and $x = 3$ in the equation $2x + y = 3$ and verify the values of y that were obtained in part a.

STOP   •   **STOP**   •   **STOP**   •   **STOP**   •   **STOP**   •   **STOP**   •   **STOP**   •   **STOP**   •   **STOP**

A8   **a.** (1) $y = 1$: $(1, 1) \in f$ (2) $y = {}^-3$: $(3, -3) \in f$

b. $2(1) + y = 3$ $\qquad$ $2(3) + y = 3$

$\qquad 2 + y = 3$ $\qquad\qquad$ $6 + y = 3$

$\qquad\qquad y = 1$ $\qquad\qquad\qquad$ $y = {}^-3$

Q9   Consider the relation $f = \{(x, y) \mid x + y = {}^-5\}$.

a. Find the y values for the following elements of the relation: $({}^-6, \text{\_\_})$, $({}^-3, \text{\_\_})$, $(0, \text{\_\_})$, $(2, \text{\_\_})$

b. Choose two more values of x between ${}^-7$ and 2 and find the values of y associated with them.

c. Graph the relation f by using the six points found in parts a and b.

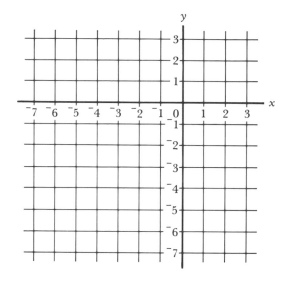

STOP • **STOP** • **STOP** • **STOP** • **STOP** • **STOP** • **STOP** • **STOP** • **STOP**

A9   **a.** $(^-6, 1), (^-3, ^-2), (0, ^-5), (2, ^-7)$: these are obtained by substituting the value of x into the equation $x + y = ^-5$ and solving for y.

$$^-6 + y = ^-5 \quad ^-3 + y = ^-5 \quad 0 + y = ^-5 \quad 2 + y = ^-5$$
$$y = 1 \qquad\quad y = ^-2 \qquad\quad y = ^-5 \qquad\quad y = ^-7$$
$$(^-6, 1) \qquad\quad (^-3, ^-2) \qquad\quad (0, ^-5) \qquad\quad (2, ^-7)$$

b. Many points are possible. Some examples are: $(^-5, 0), (^-4, ^-1), (^-2, ^-3), (^-1, ^-4), (1, ^-6)$, $(\frac{^-1}{2}, ^-4\frac{1}{2})$, etc. When plotted in part c, they should all fall on the same straight line.

c.

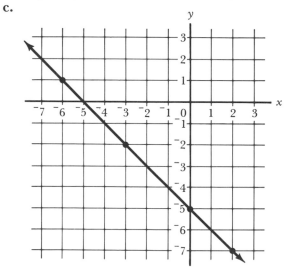

Q10   Consider the relation $g = \{(x, y) \mid x - y = 3\}$.

     a. Use the x values of $^-2$, 0, 3, and 5 along with two values of your choice to find six ordered pairs in g.

b. Graph g by using the six points from part a.

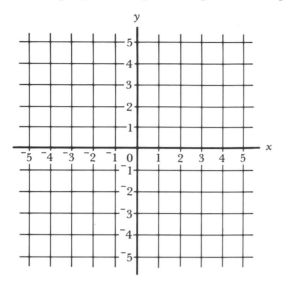

STOP • STOP • STOP • STOP • STOP • STOP • STOP • STOP • STOP

A10   **a.** $(^-2, ^-5)$, $(0, ^-3)$, $(3, 0)$, $(5, 2)$: these ordered pairs result from substituting the value of x into $x - y = 3$.

$^-2 - y = 3$	$0 - y = 3$	**b.**
$^-y = 5$	$^-y = 3$	
$y = ^-5$	$y = ^-3$	
$(^-2, ^-5)$	$(0, ^-3)$	
$3 - y = 3$	$5 - y = 3$	
$^-y = 0$	$^-y = ^-2$	
$y = 0$	$y = 2$	
$(3, 0)$	$(5, 2)$	

Other ordered pairs resulting from your choices of x might be $(^-1, ^-4)$, $(1, ^-2)$, $(2, ^-1)$, $(4, 1)$, $(1\frac{1}{2}, ^-1\frac{1}{2})$, etc.

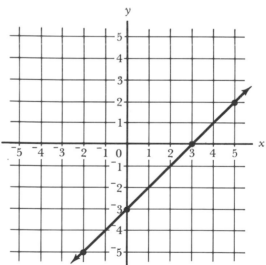

Q11   Consider the relation $h = \{(x, y)\,|\,y = \frac{1}{2}x - 1\}$.
             a. Use the x values of $^-2$ and 4 along with four choices of your own to find six elements of h.

b. Graph *h*.

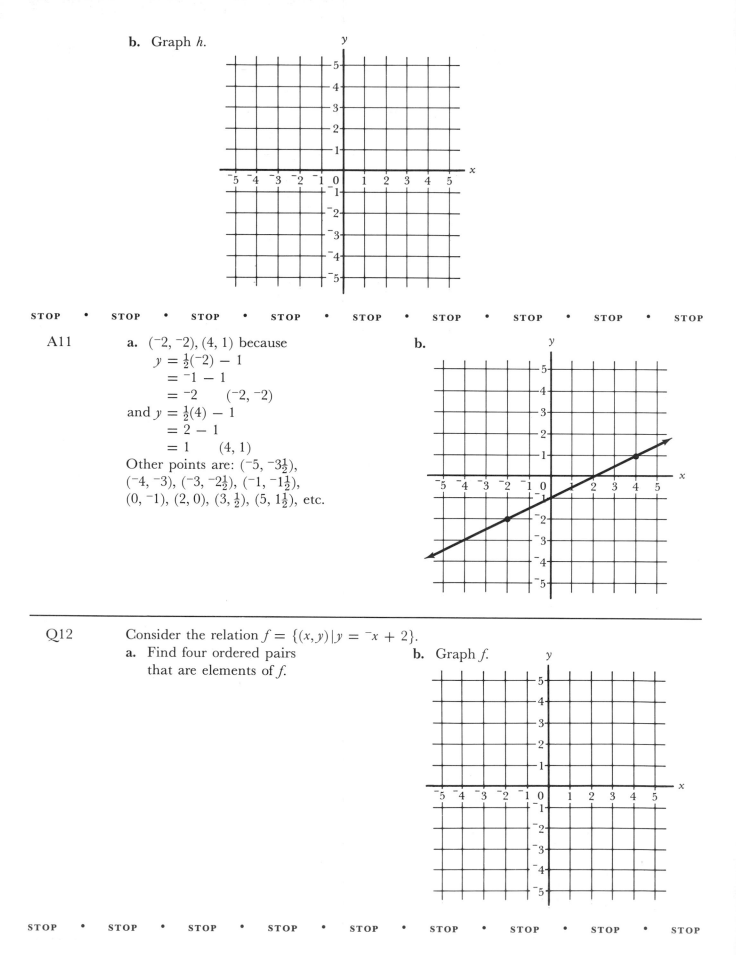

STOP • **STOP** • **STOP** • **STOP** • **STOP** • **STOP** • **STOP** • **STOP** • **STOP**

A11   **a.** $(^-2, ^-2), (4, 1)$ because

$$y = \tfrac{1}{2}(^-2) - 1$$
$$= ^-1 - 1$$
$$= ^-2 \qquad (^-2, ^-2)$$

and $y = \tfrac{1}{2}(4) - 1$
$$= 2 - 1$$
$$= 1 \qquad (4, 1)$$

Other points are: $(^-5, ^-3\tfrac{1}{2})$, $(^-4, ^-3)$, $(^-3, ^-2\tfrac{1}{2})$, $(^-1, ^-1\tfrac{1}{2})$, $(0, ^-1)$, $(2, 0)$, $(3, \tfrac{1}{2})$, $(5, 1\tfrac{1}{2})$, etc.

b.

Q12   Consider the relation $f = \{(x, y) \mid y = \,^-x + 2\}$.

a. Find four ordered pairs that are elements of *f*.

b. Graph *f*.

STOP • **STOP** • **STOP** • **STOP** • **STOP** • **STOP** • **STOP** • **STOP** • **STOP**

A12

a. Some ordered pairs that could be used are $(^-3, 5)$, $(^-2, 4)$, $(^-1, 3)$, $(0, 2)$, $(1, 1)$, $(2, 0)$, $(3, ^-1)$, $(4, ^-2)$, $(5, ^-3)$.

b.

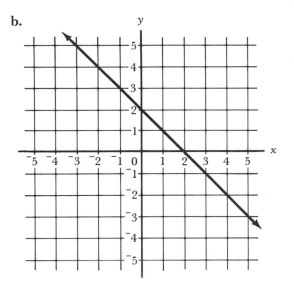

6

Sometimes the instruction to "find the graph of the relation $f = \{(x, y) \mid y = 3x - 1\}$" is shortened to "graph the line $y = 3x - 1$," or simply, "graph $y = 3x - 1$." When you read the expression, "Find the graph of $3x - y = 6$," it means that you should graph the set of all ordered pairs of numbers which make the equation true. This is done by finding some representative ordered pairs and connecting them with a straight line.

Example: Graph $y = 3x - 1$.

Solution

Ordered pairs can be found mentally by letting x equal several numbers and finding y: $(1, 2)$, $(0, ^-1)$, $(2, 5)$, $(^-1, ^-4)$. Notice that if too-large values of x are used, the point will not be located on the section of the coordinate space provided.

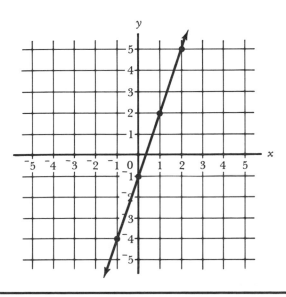

Q13   **a.** Graph $x - y = 4$.   **b.** Graph $2x - y = 0$.

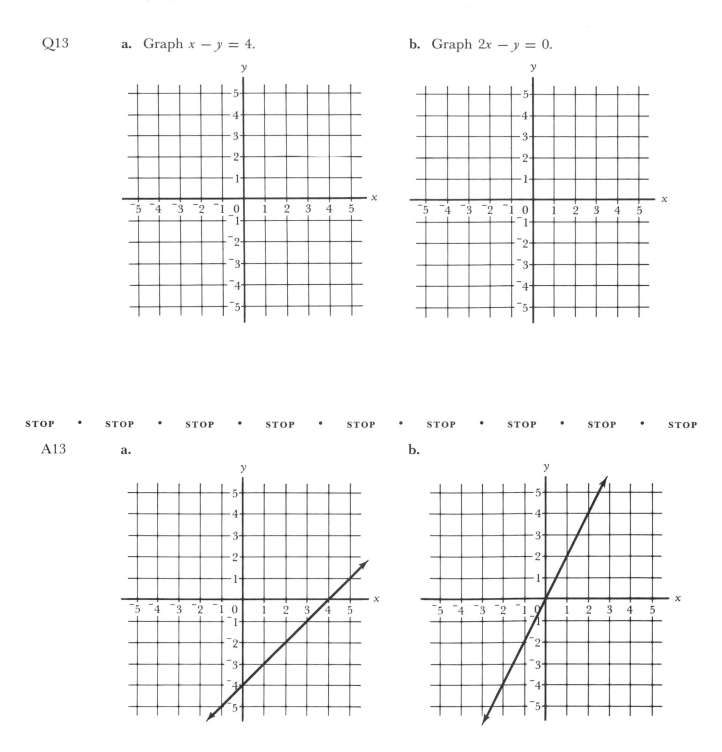

STOP • **STOP** • **STOP** • **STOP** • **STOP** • **STOP** • **STOP** • **STOP** • **STOP**

A13   **a.**   **b.**

7   The graph of the relation $f = \{(x, y) \mid x = 3\}$ is also a straight line. The graph may be found by writing a few ordered pairs that satisfy the restriction that x must be 3 and locating them on a coordinate system.

Example: Graph $\{(x, y) \mid x = 3\}$.

Solution

The following ordered
pairs are in the relation:
(3, 2), (3, 0), (3, ⁻3), (3, 5).
These points are located on the
graph to the right. Another way
of describing the graph is to
say that it is the graph of the line
$x = 3$.

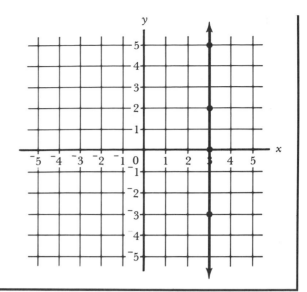

Q14   **a.** Graph the line $x = {}^-2$.   **b.** Graph the line $y = 4$.

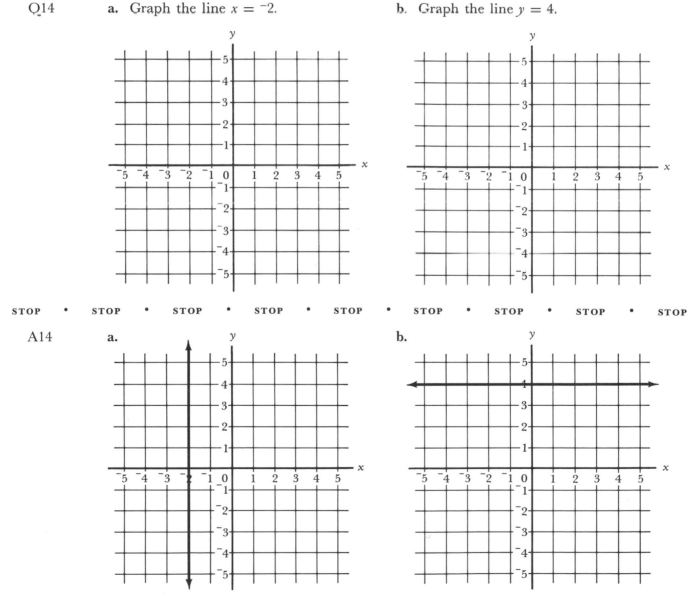

STOP   •   STOP   •   STOP   •   STOP   •   STOP   •   STOP   •   STOP   •   STOP   •   STOP

A14   **a.**   **b.**

8   Frequently graphs are drawn on graph paper that does not have the axes included. When graphing a line, find several ordered pairs on the line first before the axes are drawn. By considering the ordered pairs that must be located, the axes can sometimes be adjusted horizontally or vertically to make it easier to locate the points.

Example: Graph $x + y = 7$.

Solution

A few ordered pairs that make the equation true are: $(7, 0)$, $(3, 4)$, $(5, 2)$, $(0, 7)$, $(1, 6)$. These are all in the first quadrant, so the axes are drawn so more of the first quadrant is visible. With the axes moved to the left and down, the intersections of the line with the axes become visible.

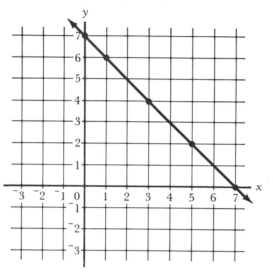

Q15   Draw coordinate axes and graph each line:
a. $x = 6$ **b.** $x + y = {}^-5$

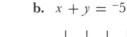

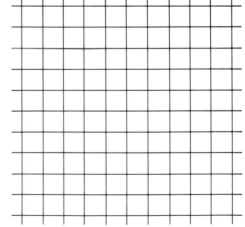

c. $y = {}^-6$

d. $x - 3y = 9$

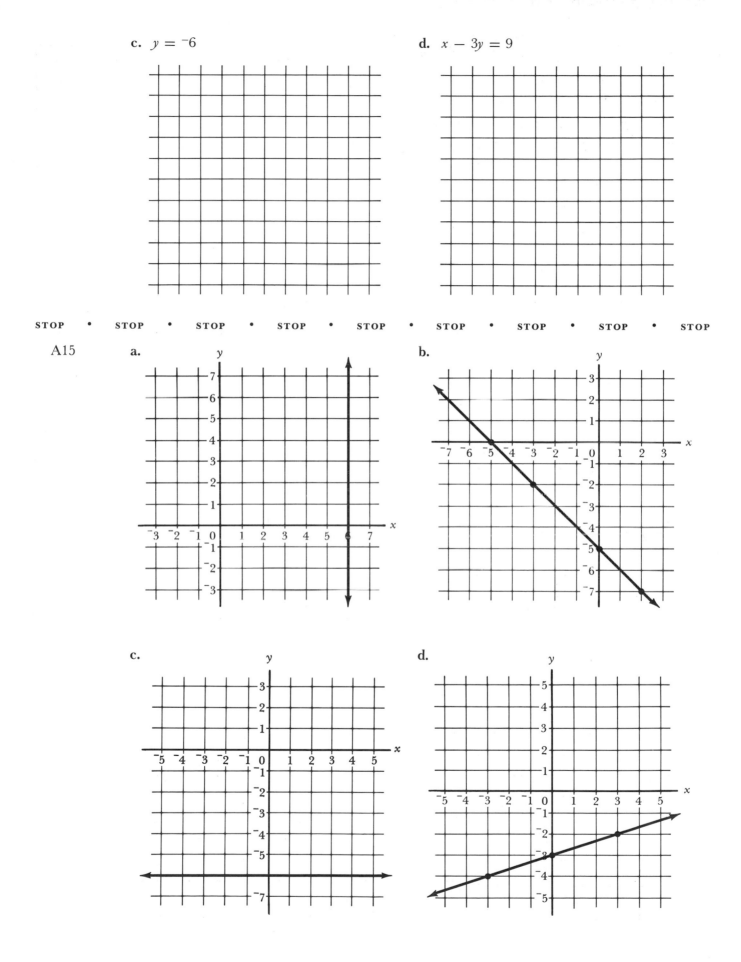

STOP • STOP • STOP • STOP • STOP • STOP • STOP • STOP • STOP

A15

a.

b.

c.

d.

9   In the problems of this section several points were graphed before the line was drawn through the points. The question arises: "How many points are necessary before the line is determined?" Since two points determine a unique line, the answer is "two." However, a good policy is to always graph at least three points because the third point is a good check. If the three points are not on a line, a fourth point will probably tell which of the previous three was in error because three of the four points would likely fall on a line. When graphing an equation in the form $ax + by = c$, two easy points to obtain are the point with $x = 0$ and the point with $y = 0$. The ordered pairs can be listed in a chart.

Example: Graph $3x + 4y = 12$.

Solution

Let $x = 0$,

$3(0) + 4y = 12$
$\qquad 4y = 12$
$\qquad\ y = 3$

x	y
0	3
4	0
2	$1\frac{1}{2}$

Let $y = 0$,

$3x + 4(0) = 12$
$\qquad 3x = 12$
$\qquad\ x = 4$

For a check let $x = 2$,

$3(2) + 4y = 12$
$\quad 6 + 4y = 12$
$\qquad\ 4y = 6$

$\qquad\ y = \dfrac{3}{2}$

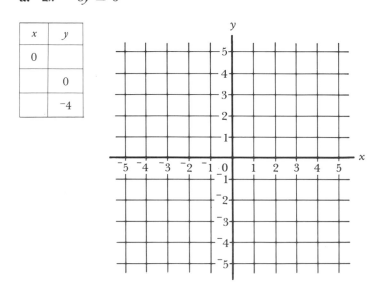

Notice that the point where $x = 0$ is on the y axis and that the point where $y = 0$ is on the x axis.

Q16   Graph the equation by finding three points. In each case let $x = 0$ for one point and $y = 0$ for another point.

a. $2x - 3y = 6$

x	y
0	
	0
	−4

b. $^-2x + 3y = 9$

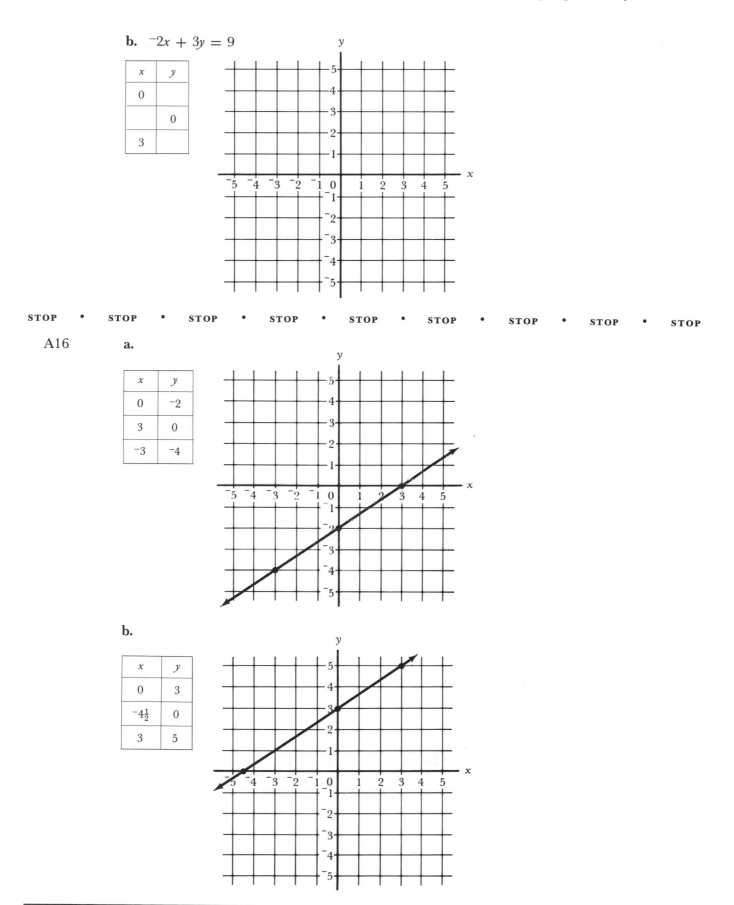

x	y
0	
	0
3	

STOP • STOP • STOP • STOP • STOP • STOP • STOP • STOP • STOP

A16   **a.**

x	y
0	$^-2$
3	0
$^-3$	$^-4$

b.

x	y
0	3
$^-4\frac{1}{2}$	0
3	5

This completes the instruction for this section.

9.2   Exercises

1. Use the coordinate systems to graph each relation:

   a. $\{(x,y) \mid x + y = 1\}$

   b. $\{(x,y) \mid y = \frac{1}{2}x + 1\}$

   c. $\{(x,y) \mid y = {}^-2x\}$

   d. $\{(x,y) \mid x - y = 3\}$

   e. $\{(x,y) \mid y = {}^-2x - 6\}$

   f. $\{(x,y) \mid y = 6\frac{1}{2}\}$

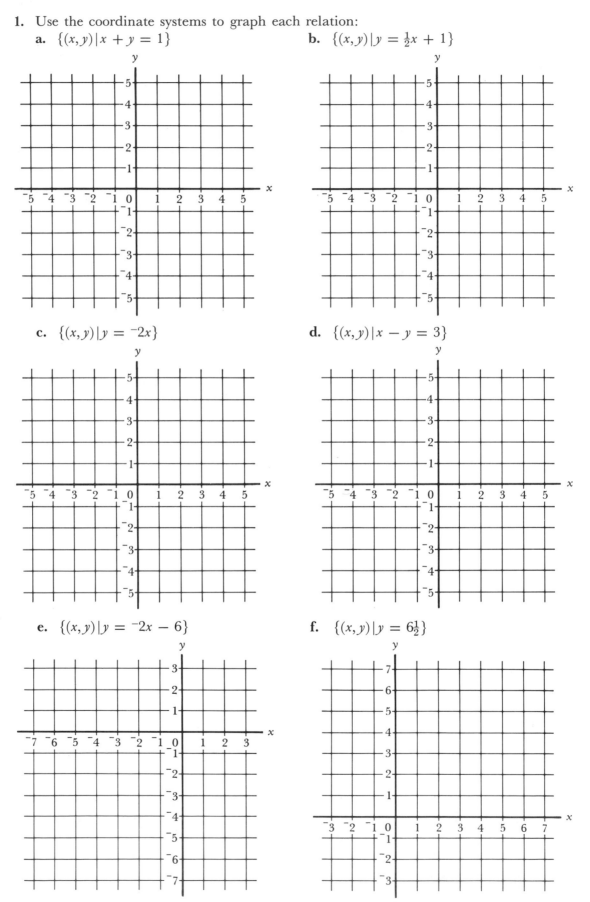

2. Graph the lines on the grids provided:

a. $y + 2x = 0$

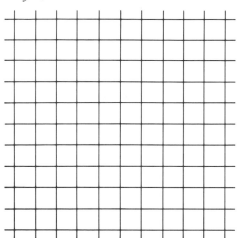

b. $3y = x + 4$

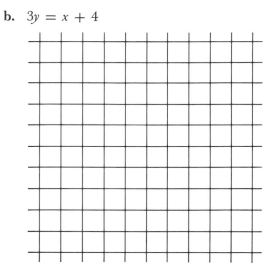

c. $x + 2y = {}^-6$

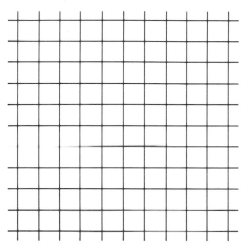

d. $x = 5\frac{1}{2}$

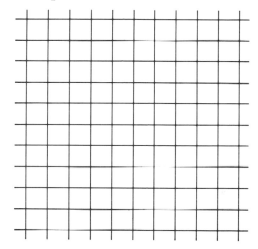

9.2   Exercise Answers

1. a.

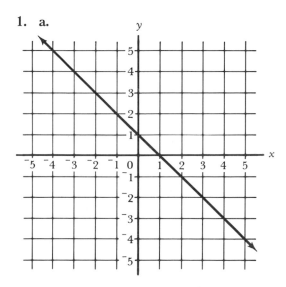

b.

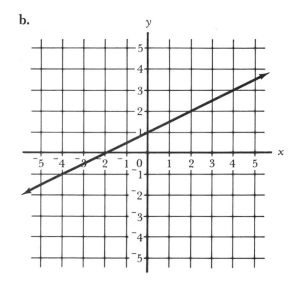

c.

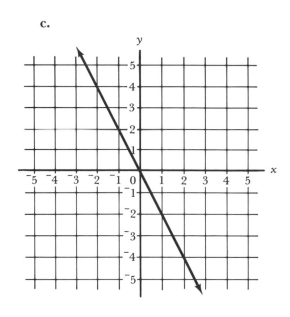

d.

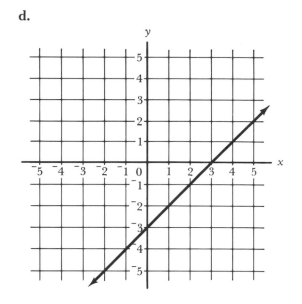

e.

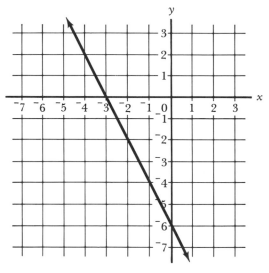

f.

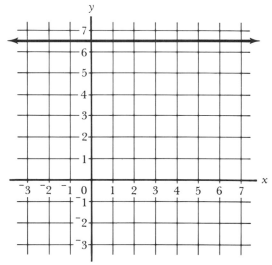

2. a.

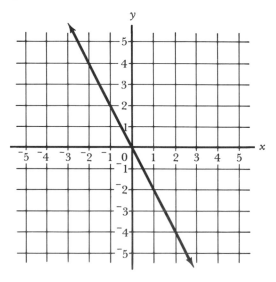

b.

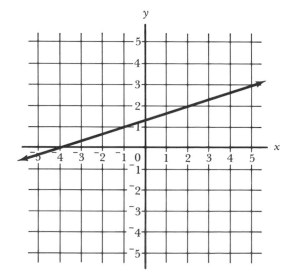

c.   **d.**

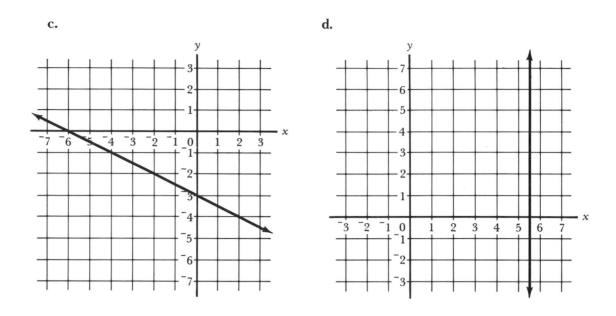

9.3   Graphing Inequalities in the Plane

1

Sets described by statements on inequality have been graphed on a line.

Example 1: Graph $\{x \mid x \geqslant 2\}$.

Solution

Sets of ordered pairs are sometimes described by a restriction involving an inequality. Any ordered pair that satisfies the restriction is a part of the set.

Example 2: Graph $f = \{(x, y) \mid x \geqslant 2\}$.

Solution

To find the points that are a part of the graph, examine all points (1) on the line $x = 2$, (2) on the left of the line $x = 2$, and (3) on the right of the line $x = 2$ to see if they satisfy the restriction that $x \geqslant 2$. Some examples are shown on the graph.

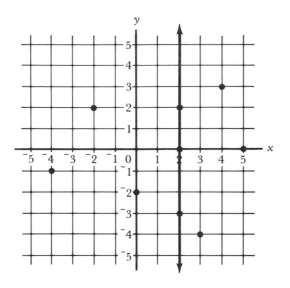

1. On the line $x = 2$:

$(2, {}^-3) \in f$ because $2 \geqslant 2$ is true.
$(2, 0) \in f$ because $2 \geqslant 2$ is true.
$(2, 2) \in f$ because $2 \geqslant 2$ is true.

2. On the left of $x = 2$:

$({}^-2, 2) \notin f$ because ${}^-2 \geqslant 2$ is false.
$({}^-4, {}^-1) \notin f$ because ${}^-4 \geqslant 2$ is false.
$(0, {}^-2) \notin f$ because $0 \geqslant 2$ is false.

3. On the right of the line $x = 2$:

$(4, 3) \in f$ because $4 \geqslant 2$ is true.
$(5, 0) \in f$ because $5 \geqslant 2$ is true.
$(3, {}^-4) \in f$ because $3 \geqslant 2$ is true.

The points that fall on the line $x = 2$ or to its right all satisfy the restriction that $x \geqslant 2$. All other points on the line or to its right will also satisfy the restriction. The portion of the plane on the right of the line $x = 2$ is shaded to indicate that all those points are a part of the graph.

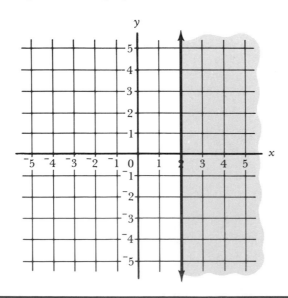

Q1   Graph the relation $g = \{(x,y)\,|\,x \geqslant {}^{-}1\}$ by examining points on the line $x = {}^{-}1$, on the right of $x = {}^{-}1$, and on the left of $x = {}^{-}1$.

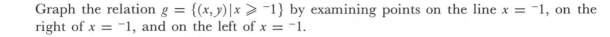

STOP • **STOP** • **STOP** • **STOP** • **STOP** • **STOP** • **STOP** • **STOP** • **STOP**

A1   All points on the line $x = {}^{-}1$ and to its right satisfy the restriction $x \geqslant {}^{-}1$. All points on the left of $x = {}^{-}1$ do not satisfy $x \geqslant {}^{-}1$.

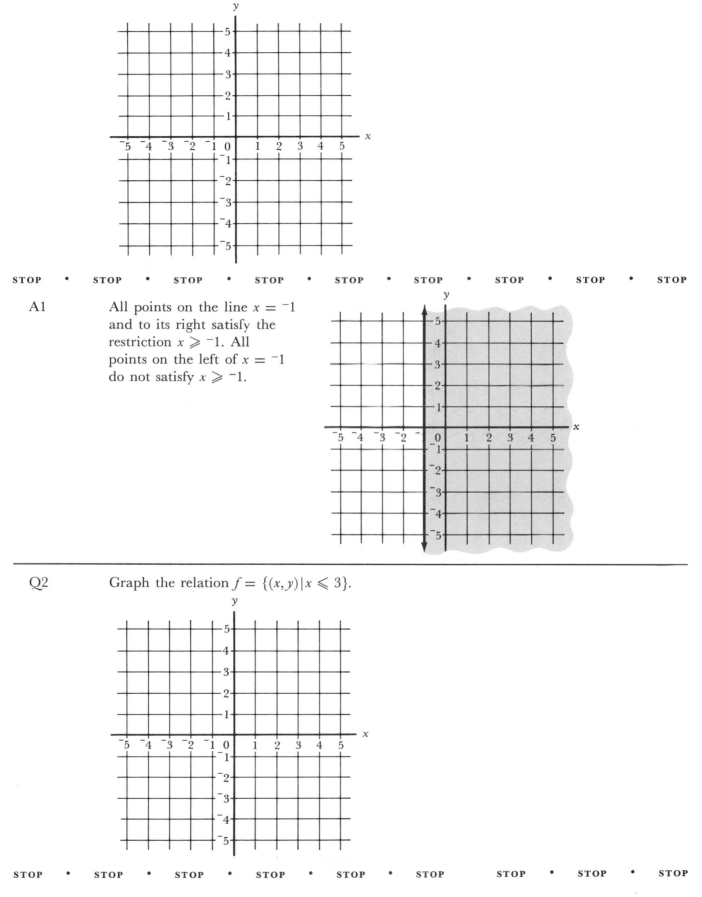

Q2   Graph the relation $f = \{(x,y)\,|\,x \leqslant 3\}$.

STOP • **STOP** • **STOP** • **STOP** • **STOP** • **STOP** • **STOP** • **STOP** • **STOP**

A2

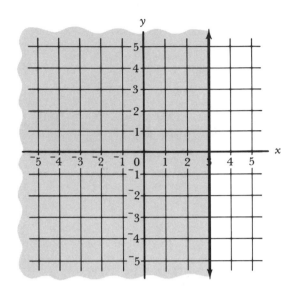

2   If the restriction is placed only on the variable y in the description of the relation, points on each side of a horizontal line must be examined.

Example: Graph $h = \{(x, y) \mid y < 1\}$.

Solution

Examine the points (1) on the line $y = 1$, (2) above the line $y = 1$, and (3) below the line $y = 1$.

1. On the line $y = 1$: $(^-3, 1)$, $(0, 1)$, and $(4, 1)$ all make the restriction $y < 1$ false.
2. Above the line $y = 1$: $(^-2, 3)$, $(0, 2)$, and $(3, 4)$ all make the restriction $y < 1$ false.
3. Below the line $y = 1$: $(^-2, 0)$, $(0, ^-4)$, and $(3, ^-2)$ all make the restriction $y < 1$ true.

Therefore the graph includes all points below the line $y = 1$. To show that the line $y = 1$ is not a part of the graph, it is drawn with a dashed line.

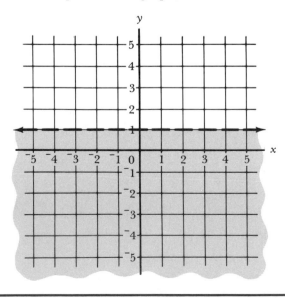

Q3   Graph $\{(x, y) | y > 2\}$.

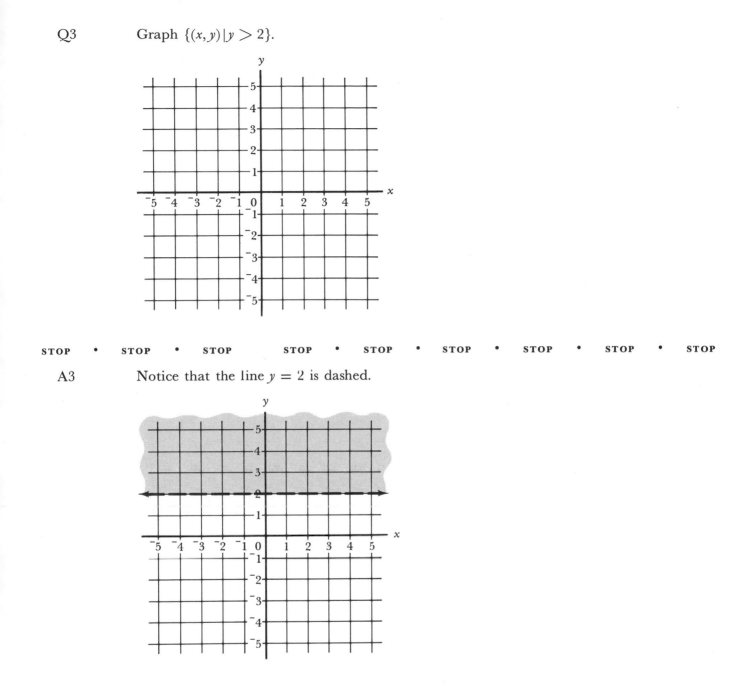

STOP • STOP • STOP • STOP • STOP • STOP • STOP • STOP • STOP

A3   Notice that the line $y = 2$ is dashed.

---

3   A relation is sometimes described with a restriction on both x and y: for example, $x + y \leqslant 3$. The graph of the relation would certainly include the line $x + y = 3$, because every point on that line would also make the inequality $x + y \leqslant 3$ true. In addition to those points, all the points on one side of the line make the restriction true while all points on the other side make it false. To discover which side makes it true, only one point needs to be tested.

Example: Graph the relation $f = \{(x, y) | x + y \leqslant 3\}$.

Solution

Step 1: The line $x + y = 3$ is graphed. Points on the line are: $(1, 2)$, $(3, 0)$, $(^-1, 4)$.

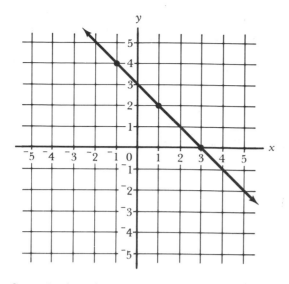

Step 2: A point not on the line is tested in the restriction. $(0, 0)$ is probably the easiest one to test. Substituting the coordinates of $(0, 0)$ in $x + y \leqslant 3$, the statement $0 + 0 \leqslant 3$ is obtained. Since this is a true statement, $(0, 0)$ and all other points below the line are a part of the graph.

Step 3: The final graph is shown below.

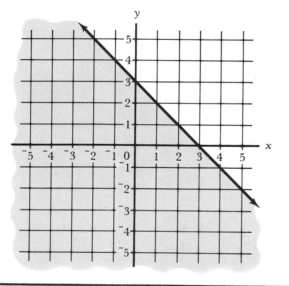

Q4   Consider the relation $\{(x, y) \mid x + y \leqslant 3\}$ graphed in Frame 3 to answer the following:
  a. Substitute the coordinates of $(^-2, 3)$ and $(1, 1)$ along with two other points of your choice from below the line $x + y = 3$ into the restriction $x + y \leqslant 3$ and show that a true statement results.

  b. Substitute $(1, 4)$ and $(5, ^-1)$ along with two other points of your choice from above the line $x + y = 3$ into the restriction and show that a false statement results.

c. Why is it always sufficient to test only one point in the restriction to find which side of the line is a part of the graph?

STOP • STOP • STOP • STOP • STOP • STOP • STOP • STOP • STOP

A4

a. $^-2 + 3 \leqslant 3$ $1 + 1 \leqslant 3$ Any other points below
  $1 \leqslant 3$ $2 \leqslant 3$ the line also make it true.
  true   true

b. $1 + 4 \leqslant 3$ $5 + {}^-1 \leqslant 3$ Any other points above the
  $5 \leqslant 3$ $4 \leqslant 3$ line also make it false.
  false   false

c. If a point satisfies the restriction, all points on the same side of the line as the tested point are on the graph. If the point makes the restriction false, all points on the other side of the line are on the graph.

Q5   Graph $\{(x,y) \mid 2x + y \leqslant 4\}$.

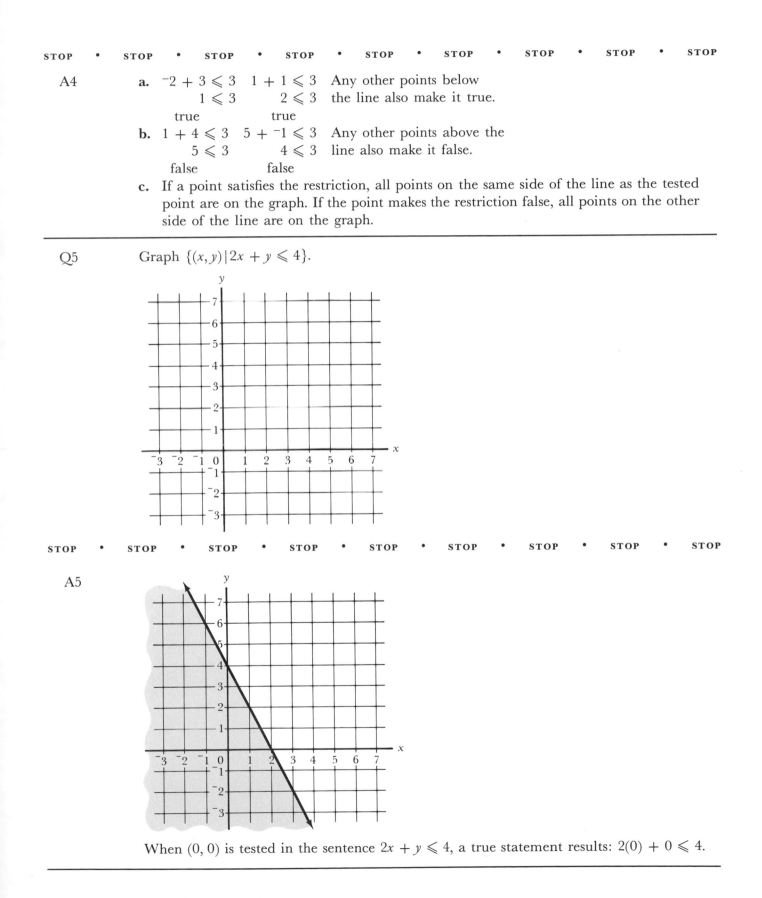

STOP • STOP • STOP • STOP • STOP • STOP • STOP • STOP • STOP

A5

When $(0, 0)$ is tested in the sentence $2x + y \leqslant 4$, a true statement results: $2(0) + 0 \leqslant 4$.

Q6   **a.** Graph $y \geqslant x + 1$.   **b.** Graph $x - y \leqslant 3$.

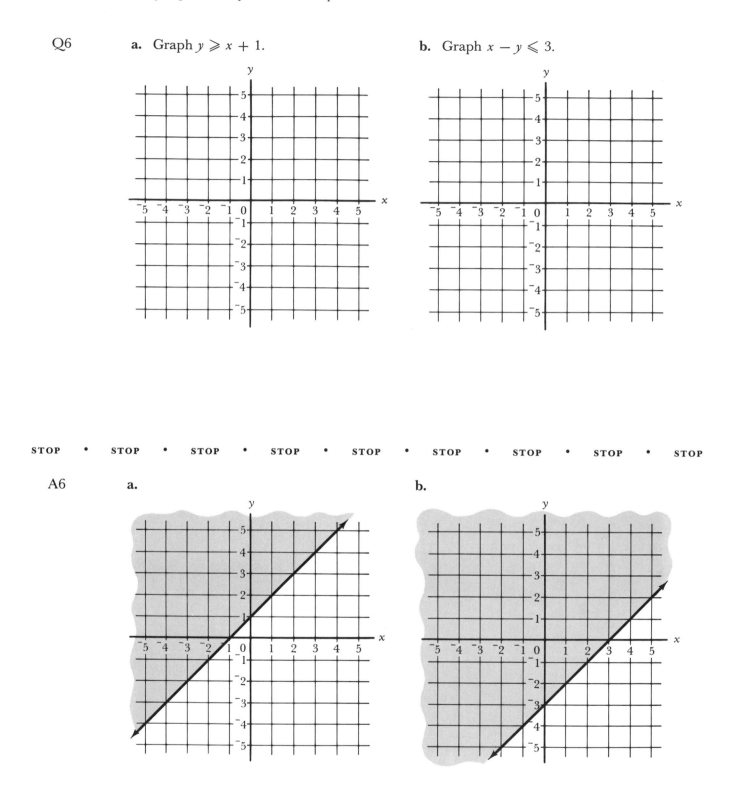

STOP • **STOP** • **STOP** • **STOP** • **STOP** • **STOP** • **STOP** • **STOP** • **STOP**

A6   **a.**   **b.**

Notice that the point $(0, 0)$ does not satisfy the restriction $y \geqslant x + 1$ ($0 \geqslant 0 + 1$ is false); therefore, the graph consists of the points on the line and above the line $y = x + 1$.

Q7   **a.** Graph $y \leqslant 2x - 1$.   **b.** Graph $2y \leqslant x + 1$.

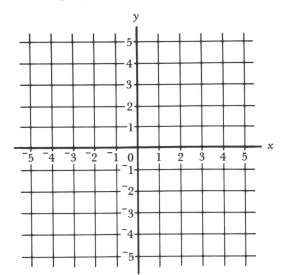   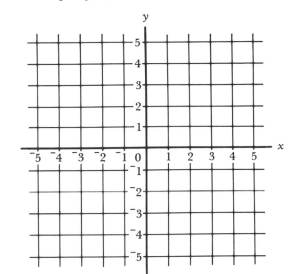

STOP • STOP • STOP • STOP • STOP • STOP • STOP • STOP • STOP

A7   **a.**   **b.**

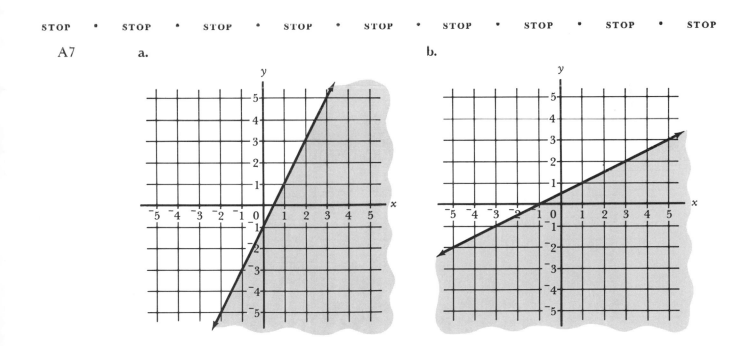

Q8   **a.** Graph $y < x$.   **b.** Graph $3x - y > 6$.

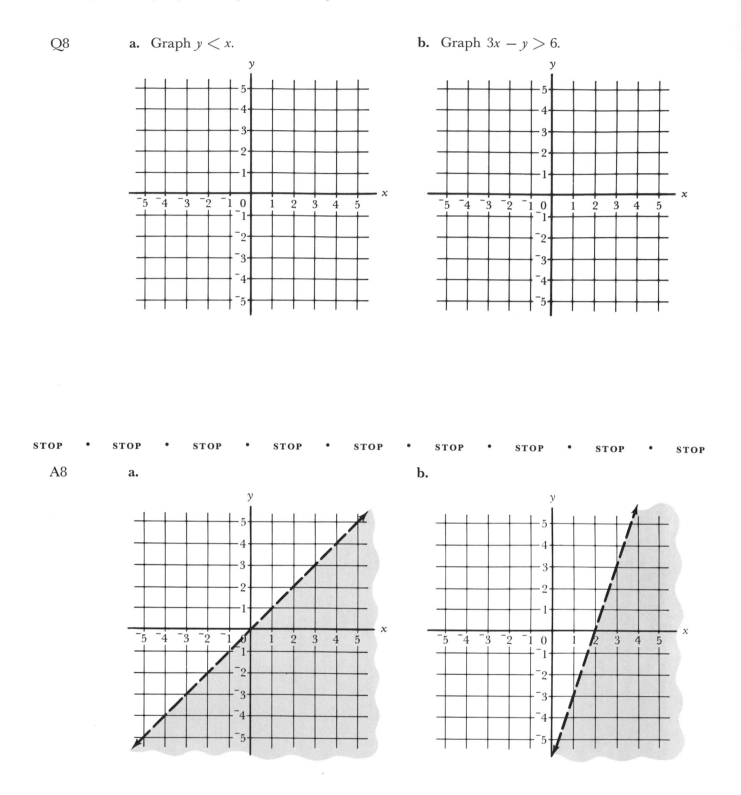

STOP • **STOP** • **STOP** • **STOP** • **STOP** • **STOP** • **STOP** • **STOP** • **STOP**

A8   **a.**   **b.**

In part a some point other than $(0, 0)$ must be used to test the restriction $y < x$.

This completes the instruction for this section.

9.3   Exercises

Use the coordinate systems provided to graph the relations.

1. $\{(x,y)\,|\,x > {}^+3\}$

2. $\{(x,y)\,|\,y \leqslant {}^-1\}$

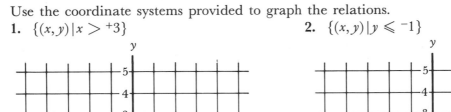

3. $\{(x,y)\,|\,x + y \leqslant {}^-2\}$

4. $\{(x,y)\,|\,x - 3y < 6\}$

5. $\{(x,y)\,|\,y \leqslant 2x\}$

6. $y > \tfrac{1}{3}x$

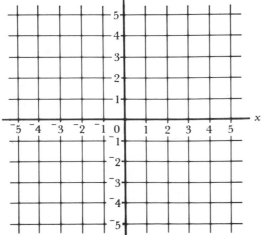

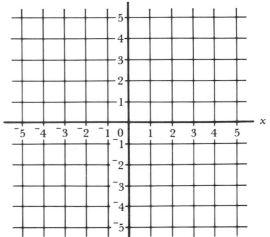

7. $y \leqslant {}^-2x - 6$

8. $4x + y < 6$

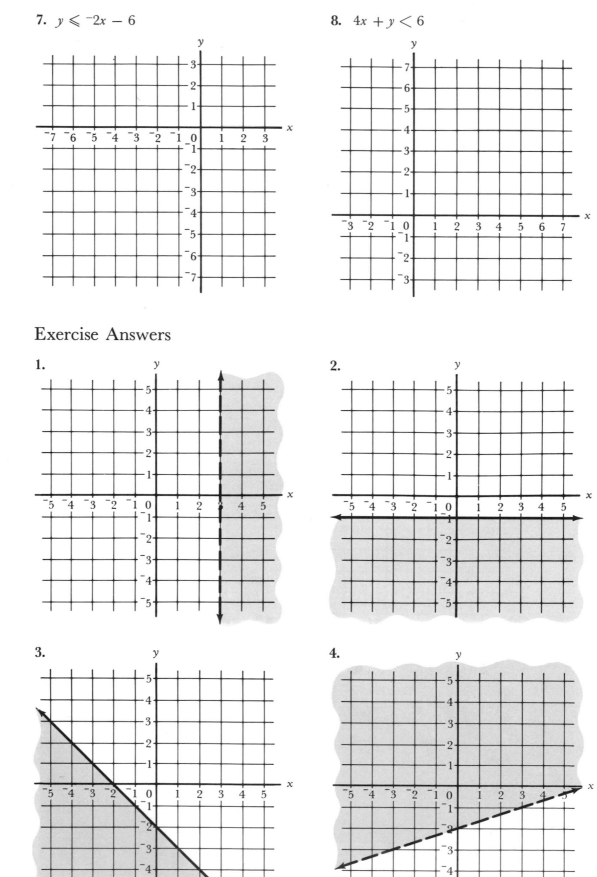

9.3   Exercise Answers

1.

2.

3.

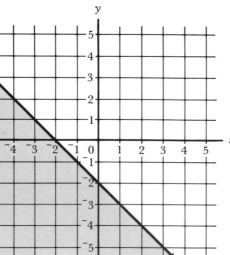

4.

5.

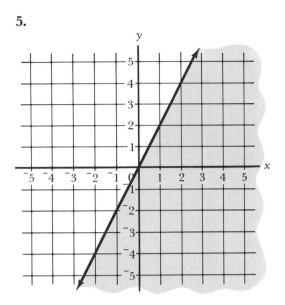

6.

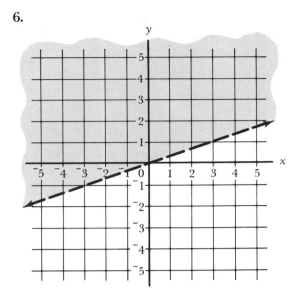

7.

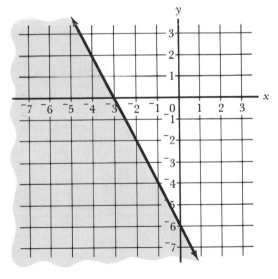

8.

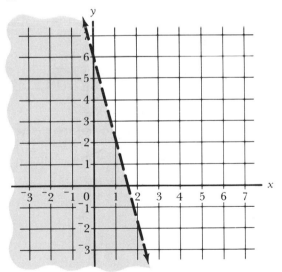

Chapter 9 Sample Test

At the completion of Chapter 9 it is expected that you will be able to work the following problems.

9.1   Rectangular Coordinate System

1. Match the letter from the figure to the right with the proper word.

   a. second quadrant _____
   b. x axis _____
   c. x coordinate _____
   d. third quadrant _____
   e. fourth quadrant _____
   f. origin _____
   g. y axis _____
   h. y coordinate _____

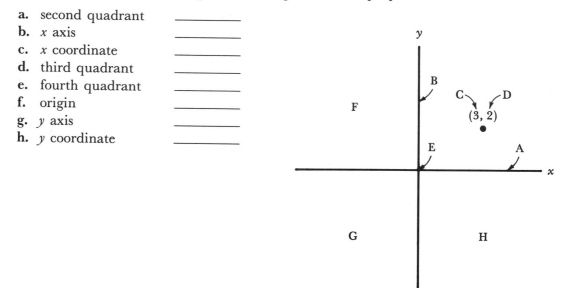

2. Locate the points of the following relation on the coordinate system shown:
   $\{(4, 0), (^-4, ^-2), (2, 1), (0, 2),$
   $(0, 0), (^-2, ^-3), (4, ^-3), (0, ^-3),$
   $(4, 5), (^-1, 0)\}$

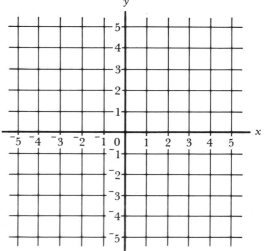

3. Name the coordinates of each of the points on the coordinate system shown:

a. _____   b. _____
c. _____   d. _____
e. _____   f. _____
g. _____   h. _____

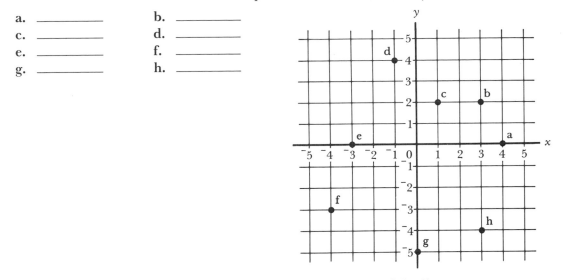

4. Use the grid to determine if each set of points is on a straight line.

a. $(^-2, ^-3), (1, ^-1), (4, 1)$ _____
b. $(^-2, ^-2), (2, 0), (1, 4)$ _____
c. $(^-3, 1), (1, ^-2), (4, ^-5)$ _____

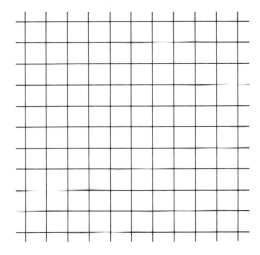

9.2   Graphing Linear Equations

5. Graph the relation
$f = \{(x,y)\,|\,x - y = 4\}$.

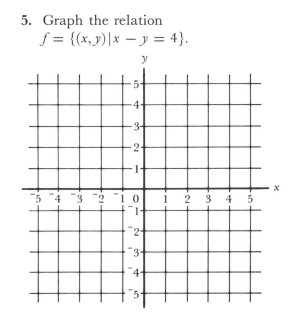

6. Graph the relation
$h = \{(x,y)\,|\,x = 3\}$.

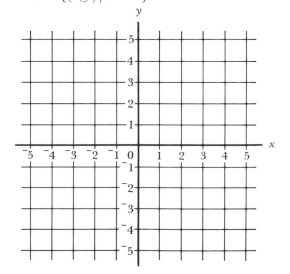

7. Graph $2x + y = 4$.

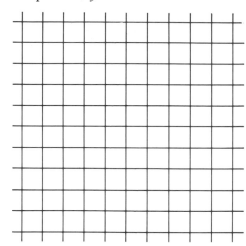

8. Graph $y = 3x$.

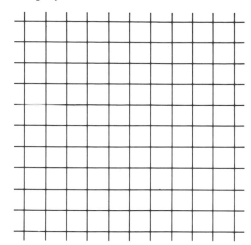

9.3   Graphing Inequalities in the Plane

9. Graph the relation
$\{(x,y)\,|\,x + y < 5\}$.

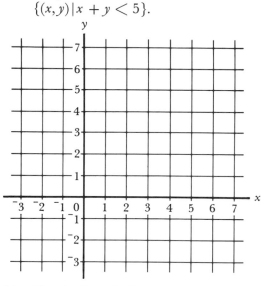

10. Graph the relation
$\{(x,y)\,|\,2x + y \geqslant 4\}$.

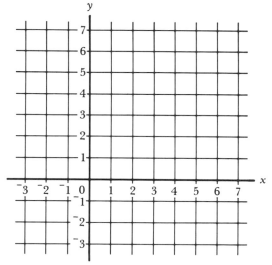

11. Graph the relation
$\{(x,y)\,|\,x < 2\}$.

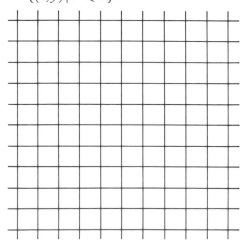

12. Graph the relation
$\{(x,y)\,|\,y \leqslant {}^-2x\}$.

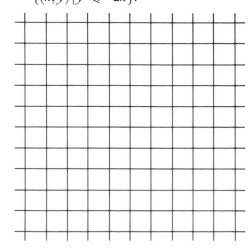

Chapter 9 Sample Test Answers

1. **a.** F   **b.** A   **c.** C   **d.** G   **e.** H   **f.** E   **g.** B
   h. D

2.

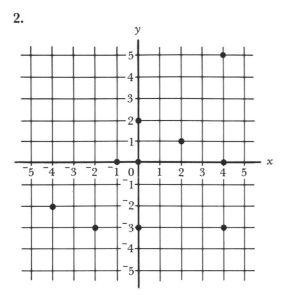

3. **a.** $(4, 0)$ **b.** $(3, 2)$ **c.** $(1, 2)$ **d.** $(^-1, 4)$
   e. $(^-3, 0)$ **f.** $(^-4, ^-3)$ **g.** $(0, ^-5)$ **h.** $(3, ^-4)$

4. **a.** yes   **b.** no   **c.** no

5.

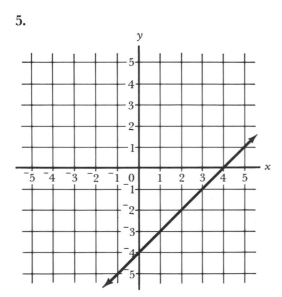

6.

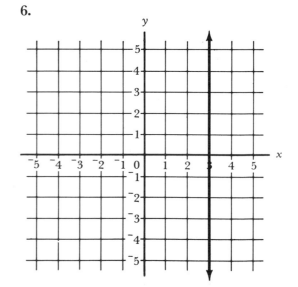

7.

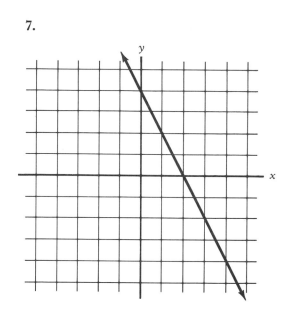

8.

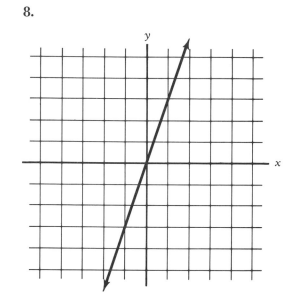

9.

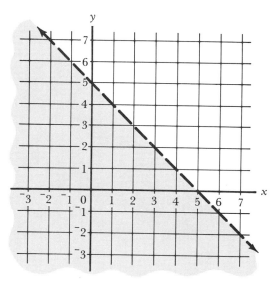

10.

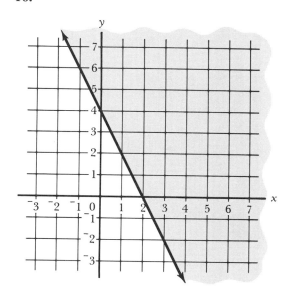

11.

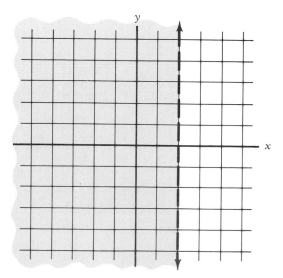

12.

Chapter 10

Systems of Equations

10.1   Solution by Graphing

<table>
<tr>
<td>1</td>
<td>

The general form for a linear equation in two variables is $ax + by = c$, where a, b, and c are rational numbers. A linear equation in two variables has an infinite number of solutions where a *solution* is any ordered pair of numbers (x, y) that converts the open sentence into a true statement.

Examples:

1. $x + y = 5$

The ordered pair $(3, 2)$ is a solution to $x + y = 5$, because $3 + 2 = 5$. The ordered pair $(^-2, 9)$ is not a solution, because $^-2 + 9 \neq 5$.

2. $3x = 4y - 11$

The ordered pair $(^-1, 2)$ is a solution to $3x = 4y - 11$, because

$$3(^-1) \overset{?}{=} 4(2) - 11$$
$$^-3 \overset{?}{=} 8 - 11$$
$$^-3 = ^-3 \qquad \text{(true statement)}$$

The ordered pair $(0, 5)$ is not a solution, because

$$3(0) \overset{?}{=} 4(5) - 11$$
$$0 \overset{?}{=} 20 - 11 \qquad \text{(false statement)}$$
$$0 \neq 9 \qquad \text{(true statement)}$$

</td>
</tr>
</table>

Q1   Choose the ordered pairs below which are solutions of $x - 2y = 10$:
    a. $(0, ^-5)$ **b.** $(^-3, ^-1)$ **c.** $(2, ^-4)$ **d.** $(10, 0)$
    e. $(4, 7)$ **f.** $(^-6, ^-8)$

STOP • STOP • STOP • STOP • STOP • STOP • STOP • STOP • STOP

A1   a, c, d, and f

<table>
<tr>
<td>2</td>
<td>A set of two linear equations in two variables is called a *system of two equations in two unknowns*. The general form for a system of equations of this type is as follows:</td>
</tr>
</table>

$$ax + by = c$$
$$dx + ey = f \qquad (a, b, c, d, e, \text{ and } f \text{ are rational numbers})$$

A *solution* of a system of two equations in two unknowns is any ordered pair of numbers (x, y) which converts *both* equations into true statements.

Examples:

1. $x + y = 7$
      $2x - y = 5$

The ordered pair $(4, 3)$ is a solution of the system, because

$$x + y = 7 \qquad \text{and} \qquad 2x - y = 5$$
$$4 + 3 \overset{?}{=} 7 \qquad\qquad\qquad 2(4) - 3 \overset{?}{=} 5$$
$$7 = 7 \qquad\qquad\qquad\qquad 5 = 5$$

2. $5x - 2y = {}^{-}12$
    $7x = {}^{-}2y - 12$

The ordered pair $({}^{-}2, 1)$ is a solution of the system, because

$$5x - 2y = {}^{-}12 \qquad \text{and} \qquad 7x = {}^{-}2y - 12$$
$$5({}^{-}2) - 2(1) \overset{?}{=} {}^{-}12 \qquad\qquad 7({}^{-}2) \overset{?}{=} {}^{-}2(1) - 12$$
$${}^{-}10 - 2 \overset{?}{=} {}^{-}12 \qquad\qquad\qquad {}^{-}14 \overset{?}{=} {}^{-}2 - 12$$
$${}^{-}12 = {}^{-}12 \qquad\qquad\qquad\qquad {}^{-}14 = {}^{-}14$$

Q2   Verify that $({}^{-}8, 2)$ is a solution of the system below by showing that it is a solution of both equations.
$$x - y = {}^{-}10$$
$$x + 2y = {}^{-}4$$

STOP • STOP • STOP • STOP • STOP • STOP • STOP • STOP • STOP

A2 $x - y = {}^{-}10 \qquad \text{and} \qquad x + 2y = {}^{-}4$
$${}^{-}8 - 2 \overset{?}{=} {}^{-}10 \qquad\qquad {}^{-}8 + 2(2) \overset{?}{=} {}^{-}4$$
$${}^{-}10 = {}^{-}10 \qquad\qquad\qquad {}^{-}8 + 4 \overset{?}{=} {}^{-}4$$
$${}^{-}4 = {}^{-}4$$

Q3   Verify that $(0, {}^{-}3)$ is not a solution of the system below by showing that it is not a solution of *both* equations.
$$7x - y = 3$$
$$y = x + 2$$

STOP • STOP • STOP • STOP • STOP • STOP • STOP • STOP • STOP

A3 $7x - y = 3 \qquad \text{but} \qquad y = x + 2$
$$7(0) - ({}^{-}3) \overset{?}{=} 3 \qquad\qquad {}^{-}3 \overset{?}{=} 0 + 2$$
$$0 + 3 \overset{?}{=} 3 \qquad\qquad\qquad {}^{-}3 \neq 2$$
$$3 = 3$$

Q4   Indicate whether the given ordered pair is or is not a solution to the given system of equations by responding yes or no:

a. $y = x + 2$; $(7, 9)$ 　　　　　**b.** $x = 5 + y$; $(1, 5)$
　　$x - y = {}^{-}2$ 　　　　　　　　　$3x - 2y = {}^{-}10$

c. $3x - 2y = {}^{-}9$; $({}^{-}1, 3)$ 　　**d.** ${}^{-}3 = x + 2y$; $({}^{-}1, {}^{-}1)$
　　$x + 2y = 5$ 　　　　　　　　　　$2x = {}^{-}5 - 3y$

STOP　•　STOP　•　STOP　•　STOP　•　STOP　•　STOP　•　STOP　•　STOP　•　STOP

A4 　　　**a.** yes 　　**b.** no 　　**c.** yes 　　**d.** yes

3 　　In Chapter 9, many of the relations graphed were linear equations. Each ordered pair in a relation is a solution to the equation. The set of solutions for a linear equation can be represented in picture form by the graph of the equation. It is likewise possible to picture the solution set for a system of equations in two unknowns by the method of graphing. Since any solution for a system of equations must be a solution common to each of the equations, the solution set for a system of equations in two unknowns is the *intersection* of the graphs of the two equations.

Examples:

1. ${}^{-}x + 2y = 2$
　　$x + y = 4$

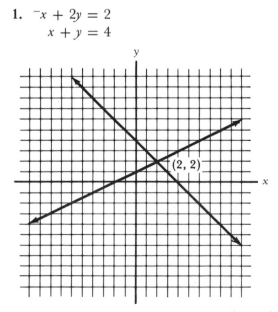

The solution set for the system is $\{(2, 2)\}$.

2. $x - 3y = ^-9$
$2x - 6y = 12$

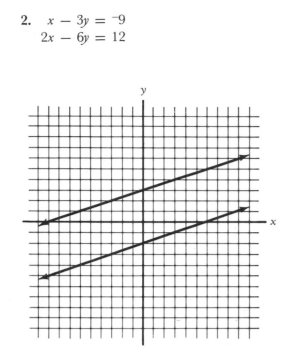

Since the lines are parallel and do not intersect, there is *no solution* for the system. Thus, the solution set is { } or ∅.

Q5   **a.** Determine the solution to the system $\begin{cases} 2x - y = 5 \\ x + y = ^-2 \end{cases}$ from the graph below. _____

         b. What is the solution set? _____

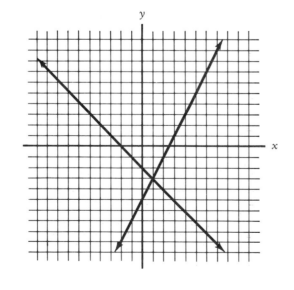

STOP • **STOP** • **STOP** • **STOP** • **STOP** • **STOP** • **STOP** • **STOP** • **STOP**

A5   **a.** $(1, ^-3)$ **b.** $\{(1, ^-3)\}$

Q6   Determine the solution set of each of the following systems from the graphs:

a. $2x + y = 4$ **b.** $x = y - 3$
   $3x - y = 6$ $y - x = 1$

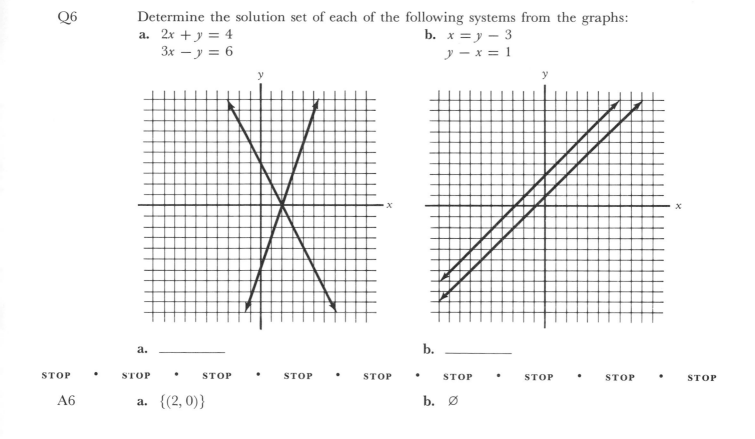

a. _____   b. _____

STOP • STOP • STOP • STOP • STOP • STOP • STOP • STOP • STOP

A6   **a.** $\{(2, 0)\}$ **b.** $\varnothing$

---

4   To find the solution of a system of two equations in two unknowns by graphing, graph each of the equations and determine the intersection of the two graphs. If the graphs are parallel, there is no solution for the system. For example,

$x - y = {}^-7$
$2x + y = 1$

$x - y = {}^-7$ $2x + y = 1$

x	y
0	7
$^-7$	0

x	y
0	1
$\frac{1}{2}$	0

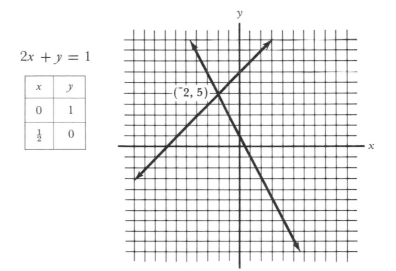

The solution is $({}^-2, 5)$ and is checked by substitution into each equation.

$x - y = {}^-7$ and $2x + y = 1$
$({}^-2) - (5) \overset{?}{=} {}^-7$ $2({}^-2) + 5 \overset{?}{=} 1$
$\qquad {}^-7 = {}^-7$ $\quad {}^-4 + 5 \overset{?}{=} 1$
$\qquad\qquad\qquad\qquad\qquad\qquad\quad 1 = 1$

Q7   **a.** Solve the following system by graphing.

$$x + y = {}^-1$$
$$2x - y = 4$$

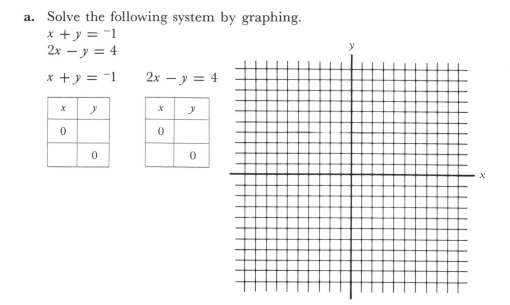

x	y
0	
	0

x	y
0	
	0

b. Check the solution by substitution.

STOP • **STOP** • **STOP** • **STOP** • **STOP** • **STOP** • **STOP** • **STOP** • **STOP**

A7   **a.** $(1, {}^-2)$:   the solution set would be $\{(1, {}^-2)\}$.

$$x + y = {}^-1 \qquad 2x - y = 4$$

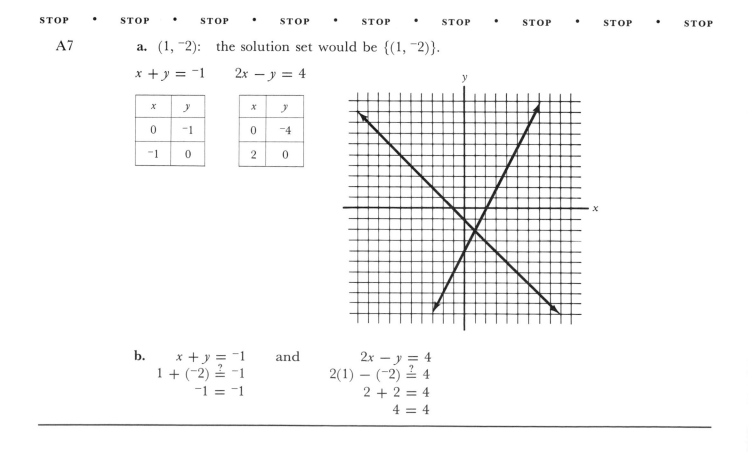

x	y
0	$^-1$
$^-1$	0

x	y
0	$^-4$
2	0

b. $x + y = {}^-1$ and $2x - y = 4$
$$1 + ({}^-2) \overset{?}{=} {}^-1 \qquad\qquad 2(1) - ({}^-2) \overset{?}{=} 4$$
$${}^-1 = {}^-1 \qquad\qquad\qquad 2 + 2 = 4$$
$$4 = 4$$

Q8   **a.** Solve the following system by graphing.

$$4x + y = {}^-1$$
$$x + 3y = 8$$

$4x + y = {}^-1$ $x + 3y = 8$

x	y
0	
	0

x	y
0	
	0

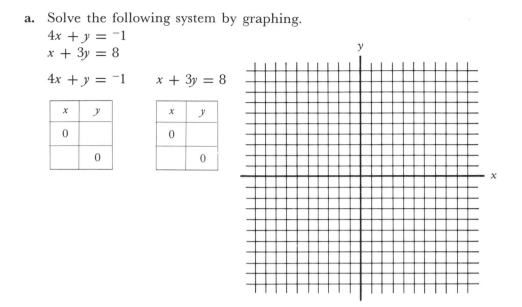

b. Check the solution by substitution.

STOP   •   STOP   •   STOP   •   STOP   •   STOP   •   STOP   •   STOP   •   STOP   •   STOP

A8   **a.** $({}^-1, 3)$:   the solution set would be $\{({}^-1, 3)\}$.

$4x + y = 1$ $x + 3y = 8$

x	y
0	$^-1$
$^-\frac{1}{4}$	0

x	y
0	$2\frac{2}{3}$
8	0

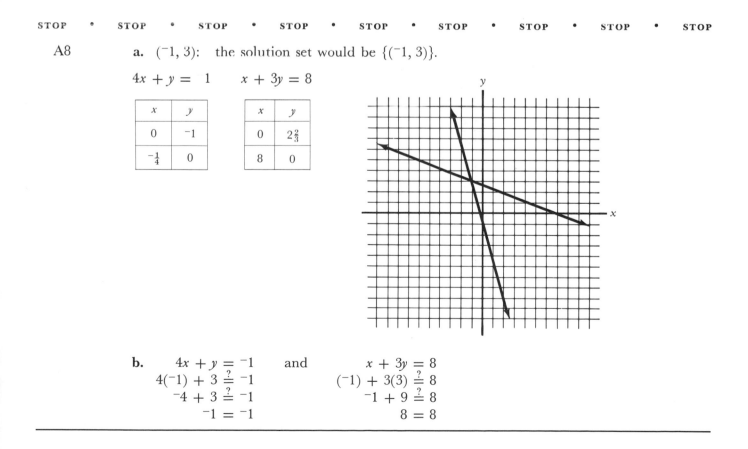

b.
$$
\begin{array}{ll}
4x + y = {}^-1 & \qquad \text{and} \\
4({}^-1) + 3 \overset{?}{=} {}^-1 \\
{}^-4 + 3 \overset{?}{=} {}^-1 \\
{}^-1 = {}^-1
\end{array}
\qquad
\begin{array}{l}
x + 3y = 8 \\
({}^-1) + 3(3) \overset{?}{=} 8 \\
{}^-1 + 9 \overset{?}{=} 8 \\
8 = 8
\end{array}
$$

Q9   Find the solution set for each of the following systems by graphing:

a. $x - y = 7$ **b.** $2x + y = 4$
　　$2y = 2x + 4$ $2x = y - 4$

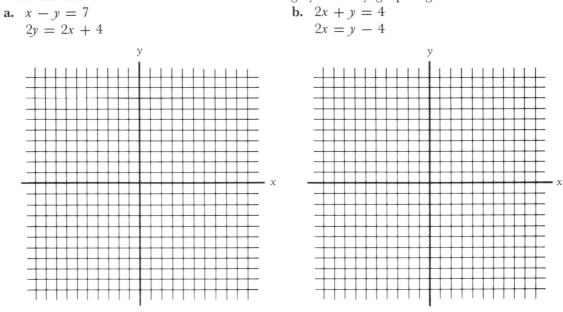

c. $^-2x + y = 6$ **d.** $y = ^-2x + 7$
　　$x + y = ^-9$ 　$4y + 8x = ^-4$

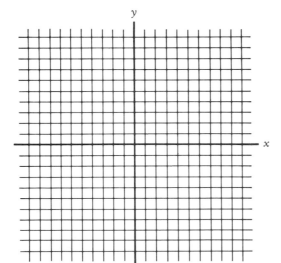
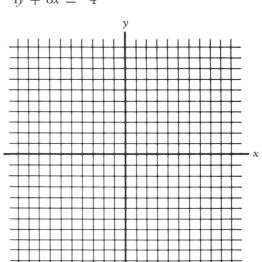

STOP • **STOP** • **STOP** • **STOP** • **STOP** • **STOP** • **STOP** • **STOP** • **STOP**

A9   **a.** $\varnothing$ **b.** $\{(0, 4)\}$ **c.** $\{(^-5, ^-4)\}$ **d.** $\varnothing$

5   **Example:** Consider the solution of the system $\begin{cases} x + 2y = {}^-1 \\ 4x + 8y = {}^-4 \end{cases}$ by graphing.

Solution

$x + 2y = {}^-1$ $4x + 8y = {}^-4$

x	y
0	$-\frac{1}{2}$
$^-1$	0

x	y
0	$-\frac{1}{2}$
$^-1$	0

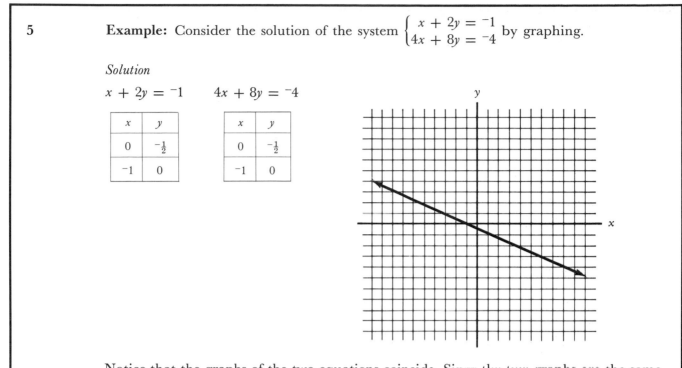

Notice that the graphs of the two equations coincide. Since the two graphs are the same, every ordered pair of values that is a solution of one equation is also a solution of the second equation. As *each* equation has an infinite number of solutions, the system of equations also has an infinite number of solutions: namely, the set of ordered pairs that satisfy either of the equations.

Q10   **a.** Graph each of the equations in the system $\begin{cases} 4x - y = 3 \\ 8x - 2y = 6 \end{cases}$.

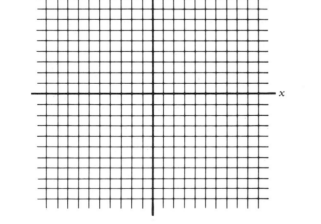

b. Does the system have one solution, no solution, or an infinite number of solutions?

_____

STOP • STOP • STOP • STOP • STOP • STOP • STOP • STOP • STOP

A10

a. $4x - y = 3$ $8x - 2y = 6$

x	y
0	-3
$\frac{3}{4}$	0

x	y
0	-3
$\frac{3}{4}$	0

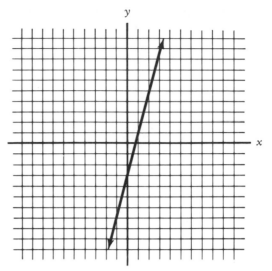

b. An infinite number of solutions.

---

6   If two equations in two unknowns have only one solution, they are called *consistent equations*. If two equations in two unknowns have no common solution, they are called *inconsistent equations*. Two equations in two unknowns with an infinite number of solutions are called *dependent equations*. An example of each of the three system types is graphed below:

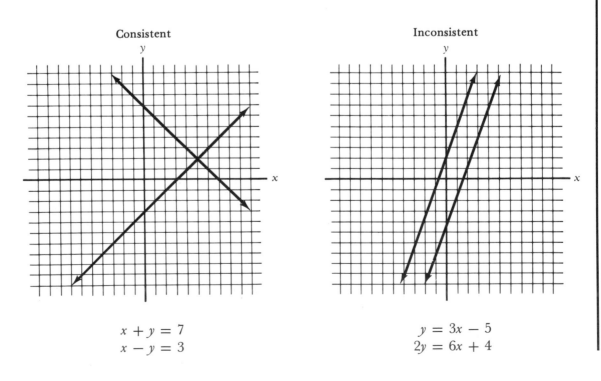

Consistent

$x + y = 7$
$x - y = 3$

Inconsistent

$y = 3x - 5$
$2y = 6x + 4$

Dependent

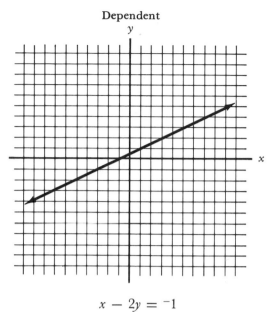

$$x - 2y = {}^-1$$
$$2x - 4y = {}^-2$$

The following chart summarizes the above information:

Type of system	Graphs	Solution
1. Consistent	Intersect in one point	One solution
2. Inconsistent	Are parallel	No solution
3. Dependent	Coincide	Infinite number of solutions

Q11   **a.** Solve the following system by graphing.

$$3x - y = 5$$
$$x + 2y = {}^-3$$

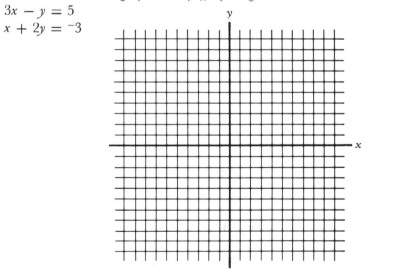

   b. Is the system consistent, inconsistent, or dependent? _____

STOP • STOP • STOP • STOP • STOP • STOP • STOP • STOP • STOP

A11   **a.** $(1, {}^-2)$ **b.** consistent

Q12   **a.** Solve the following system by graphing.

$2x + y = 7$

$^-2y = 4x - 14$

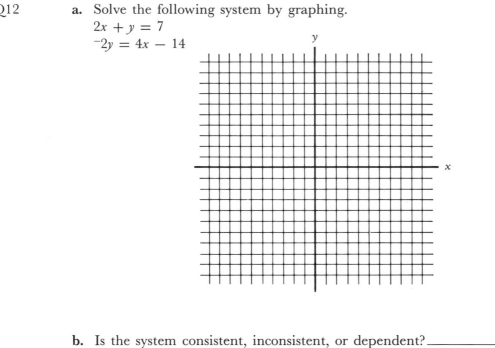

b. Is the system consistent, inconsistent, or dependent?_____

STOP • **STOP** • **STOP** • **STOP** • **STOP** • **STOP** • **STOP** • **STOP** • **STOP**

A12   **a.**   **b.** dependent

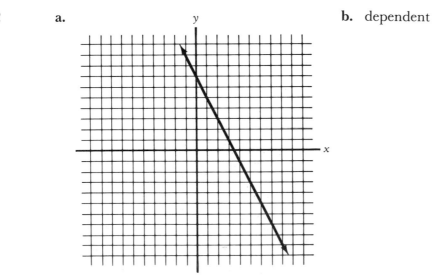

Q13   (1) Solve each of the following systems by graphing and (2) label each as consistent, inconsistent, or dependent:

a. $2x + 3y = 6$
     $2x + 3y = {}^-6$

b. $x + y = 1$
     $x - 2y = 10$

(1) _____   (1) _____

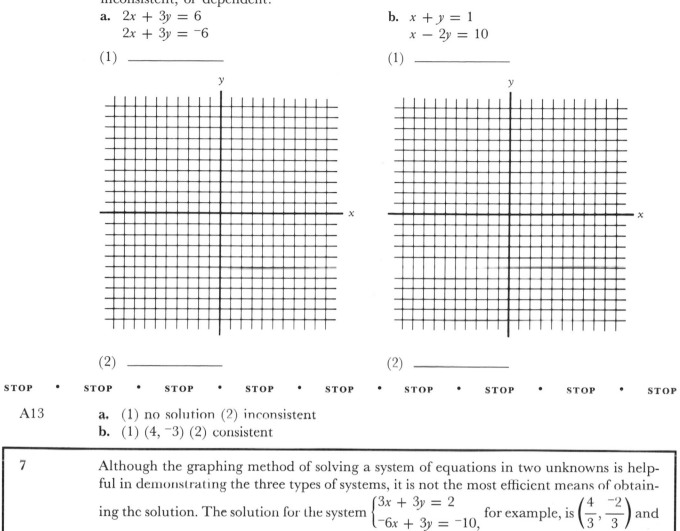

(2) _____   (2) _____

STOP • **STOP** • **STOP** • **STOP** • **STOP** • **STOP** • **STOP** • **STOP** • **STOP**

A13   a.   (1) no solution (2) inconsistent
      b.   (1) $(4, {}^-3)$ (2) consistent

---

7   Although the graphing method of solving a system of equations in two unknowns is helpful in demonstrating the three types of systems, it is not the most efficient means of obtaining the solution. The solution for the system $\begin{cases} 3x + 3y = 2 \\ {}^-6x + 3y = {}^-10, \end{cases}$ for example, is $\left(\dfrac{4}{3}, \dfrac{{}^-2}{3}\right)$ and would be difficult to determine by observing the intersection point of the two graphs.

The remaining sections of this chapter will develop algebraic means of solving systems of equations in two unknowns.

This completes the instruction for this section.

10.1   Exercises

1. Which of the following ordered pairs are solutions of the equation $4x - 3y = {}^-12$?

   $(3, 0), (0, 4), (1, 3), ({}^-3, 0), ({}^-6, {}^-4), \left({}^-2, \dfrac{4}{3}\right)$

2. Find the solution for each of the following systems from the list of ordered pairs below:
   $({}^-3, 2), (0, {}^-1), (1, 7), (4, 4)$

   a. $2x + y = {}^-1$
        $x - y = 1$

   b. $4x - y = {}^-3$
        ${}^-2y + 9 = {}^-5x$

   c. $x + 2y = 1$
        $2x + 5y = 4$

   d. $2x - y = 4$
        $x - y = 0$

3. The set of infinite solutions for a linear equation can be represented in picture form by a _____.

4. Two equations in two unknowns are called either _____ equations, _____ equations, or _____ equations.

5. **a.** Two equations with no common solution are called _____ equations.

   b. Two equations with only one common solution are called _____ equations.

   c. Two equations with an infinite number of common solutions are called _____ equations.

6. Label each of the graphs below as consistent, inconsistent, or dependent.

   a. $y = 2x + 7$
    $y - 2x = {}^-1$

   b. $4x - 2y = 8$
    $y - 2x = {}^-4$

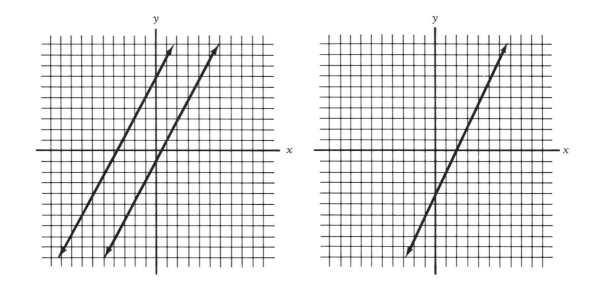

   c. $2x + y = 7$
    $x - y = 2$

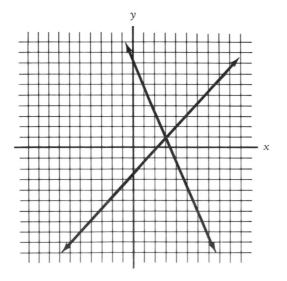

7. Solve each of the following systems by graphing and label each as consistent, inconsistent, or dependent:

a. $y = {}^-2x + 5$
      $y = {}^-x + 3$

b. $y - 3x = 3$
      $y = 3x - 6$

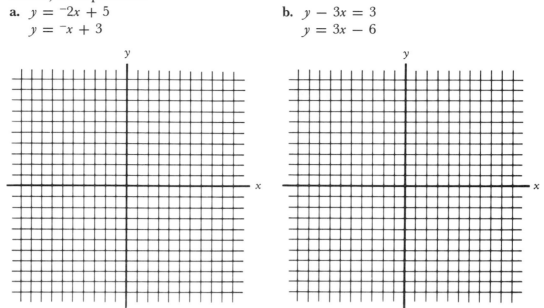

c. $2x - 3y = 6$
      $4x - 6y = 12$

d. $2x - 5y = {}^-10$
      $x + y = {}^-5$

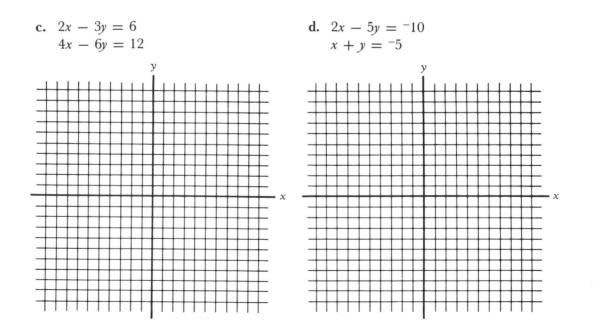

e. $x - 3y = 6$
    $y = 4x + 9$

f. $3y = x - 6$
    $^-6y + 2x = 12$

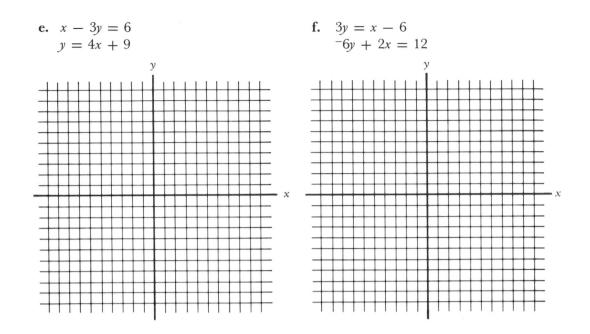

g. $2y - x = {}^-10$
    $x - 2y = {}^-4$

h. $4x - 3y = 12$
    $x + y = {}^-4$

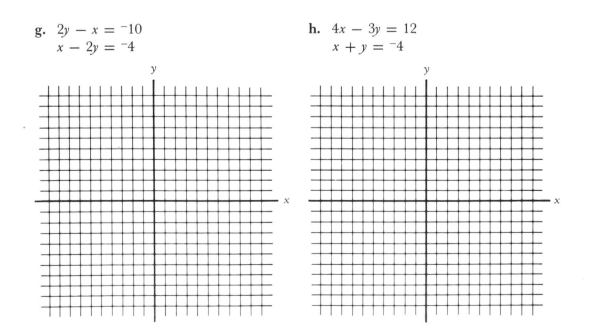

i. $x - 3y = 9$
$x + y = 1$

j. $x + y = 7$
$x + y = {}^-2$

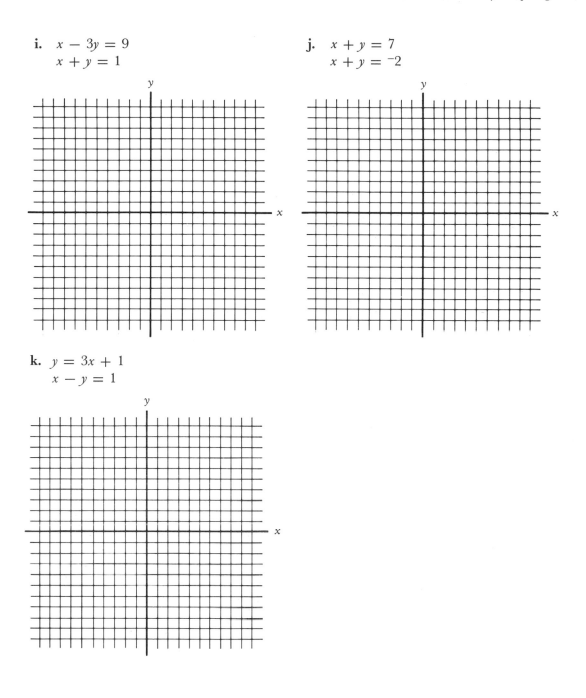

k. $y = 3x + 1$
$x - y = 1$

10.1   Exercise Answers

1. $(0, 4), ({}^-3, 0), ({}^-6, {}^-4), \left({}^-2, \dfrac{4}{3}\right)$

2. a. $(0, {}^-1)$ **b.** $(1, 7)$ **c.** $({}^-3, 2)$ **d.** $(4, 4)$
3. graph (straight line)
4. consistent, inconsistent, dependent (the order of your answers may vary)
5. a. inconsistent   **b.** consistent   **d.** dependent
6. a. inconsistent   **b.** dependent   **c.** consistent

7. a. (2, 1), consistent

b. no solution, inconsistent

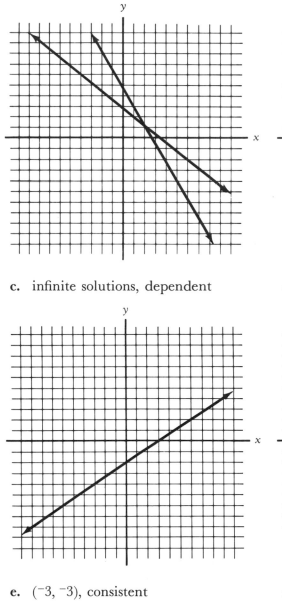

c. infinite solutions, dependent

d. (⁻5, 0), consistent

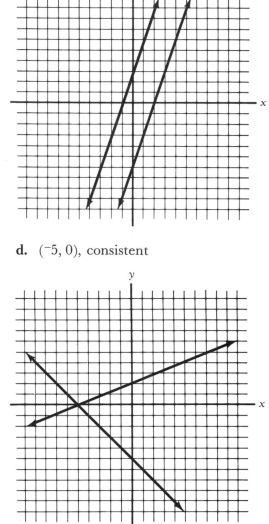

e. (⁻3, ⁻3), consistent

f. infinite solutions, dependent

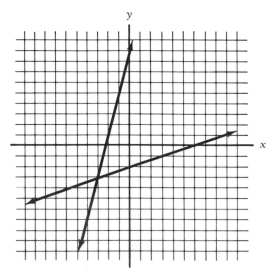

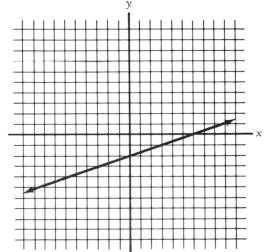

g. no solution, inconsistent

h. (0, ⁻4), consistent

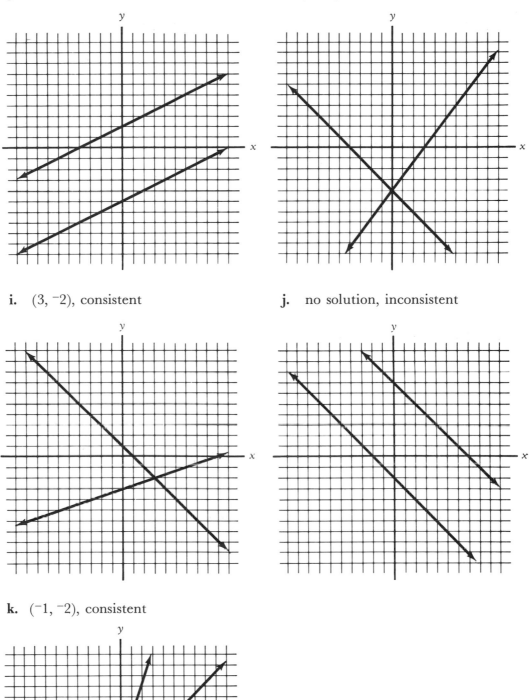

i. (3, ⁻2), consistent

j. no solution, inconsistent

k. (⁻1, ⁻2), consistent

10.2   Solution by Addition

---

1   There are two methods that can be used in finding the solution of a system of two linear equations in two unknowns which are more efficient, in general, than the graphing method. These are called the *method of elimination by addition* and the *substitution method*. This section will be concerned with the method of elimination by addition.

Consider the effect of adding like terms in the following system of equations:

$$\begin{array}{l} x + y = 8 \\ \underline{x - y = 2} \\ 2x \quad\; = 10 \end{array}$$

Since the coefficients of the y terms are opposites ($^+1$ and $^-1$), their sum is zero, and adding the two equations results in the elimination of the y terms. The resulting equation, $2x = 10$, can be solved for the value of x.

$$\begin{array}{l} 2x = 10 \\ \;\; x = 5 \end{array}$$

The value of y is found by substituting the value of x into either one of the original equations.

$$\begin{array}{lcl} x + y = 8 & \quad\text{or}\quad & x - y = 2 \\ 5 + y = 8 & & 5 - y = 2 \\ \quad\;\; y = 3 & & \quad\; ^-y = ^-3 \\ & & \quad\;\; y = 3 \end{array}$$

Therefore, the solution for the system of equations is the ordered pair whose x coordinate is 5 and whose y coordinate is 3 or (5, 3). The solution is checked by showing that the ordered pair converts each equation into a true statement.

$$\begin{array}{lcl} x + y = 8 & & x - y = 2 \\ 5 + 3 = 8 \quad \text{(true)} & & 5 - 3 = 2 \quad \text{(true)} \end{array}$$

Q1

$$\begin{array}{l} x - 2y = 3 \\ 2x + 2y = 6 \end{array}$$

a. Which term will be eliminated, the x term or the y term, if the like terms in the system are added? _____

b. Add the like terms in the system and write the resulting equation.

c. Solve the equation for x.

     d. Substitute the value for x into one of the original equations and solve for y.

     e. Write the solution to the system. _____

STOP • STOP • STOP • STOP • STOP • STOP • STOP • STOP • STOP

A1

a. y term

b. $3x = 9$:
$$x - 2y = 3$$
$$2x + 2y = 6$$
$$\overline{3x \qquad = 9}$$

c. $x = 3$

d. $y = 0$:
$$\begin{aligned} x - 2y &= 3 \\ 3 - 2y &= 3 \\ {}^-2y &= 0 \\ y &= 0 \end{aligned}$$
or
$$\begin{aligned} 2x + 2y &= 6 \\ 2(3) + 2y &= 6 \\ 6 + 2y &= 6 \\ 2y &= 0 \\ y &= 0 \end{aligned}$$

e. $(3, 0)$

Q2

Solve by the addition method:
$$^-x + y = 2$$
$$2x - y = {}^-3$$

STOP • STOP • STOP • STOP • STOP • STOP • STOP • STOP • STOP

A2

$({}^-1, 1)$:
$$^-x + y = 2$$
$$2x - y = {}^-3$$
$$\overline{x \qquad = {}^-1}$$
Recall: $^-x = {}^-1x$

To find y, substitute the value $x = {}^-1$ into either of the original equations.
$$\begin{aligned} ^-x + y &= 2 \\ {}^-({}^-1) + y &= 2 \\ 1 + y &= 2 \\ y &= 1 \end{aligned}$$
or
$$\begin{aligned} 2x - y &= {}^-3 \\ 2({}^-1) - y &= {}^-3 \\ {}^-2 - y &= {}^-3 \\ {}^-y &= {}^-1 \\ y &= 1 \end{aligned}$$

Thus, the solution is $({}^-1, 1)$.

Q3 Solve by the addition method:

$$x + 7y = 2$$
$$5x - 7y = ^-32$$

STOP • **STOP** • **STOP** • **STOP** • **STOP** • **STOP** • **STOP** • **STOP** • **STOP**

A3 $(^-5, 1)$:
$$\begin{array}{r} x + 7y = 2 \\ 5x - 7y = ^-32 \\ \hline 6x \quad\quad = ^-30 \\ x = ^-5 \end{array}$$

To find y, substitute the value $x = ^-5$ into either of the original equations.

$$\begin{array}{ccc}
5x - 7y = ^-32 & \text{or} & x + 7y = 2 \\
5(^-5) - 7y = ^-32 & & ^-5 + 7y = 2 \\
^-25 - 7y = ^-32 & & 7y = 7 \\
^-7y = ^-7 & & y = 1 \\
y = 1 & &
\end{array}$$

Thus, the solution is $(^-5, 1)$.

Q4 Solve each of the following by the addition method:

 a. $2x + 3y = ^-8$ **b.** $7x + 2y = ^-4$ **c.** $2x - 4y = 6$
   $^-5x - 3y = 2$ $13x - 2y = 4$ $^-3x + 4y = ^-10$

STOP • **STOP** • **STOP** • **STOP** • **STOP** • **STOP** • **STOP** • **STOP** • **STOP**

A4 **a.** $(2, ^-4)$ **b.** $(0, ^-2)$ **c.** $\left(4, \dfrac{1}{2}\right)$

---

2 It is sometimes possible to eliminate the x terms when the like terms of the equations of a system are added. For example, consider the system $\begin{cases} ^-2x + y = 7 \\ 2x + 3y = 5. \end{cases}$ Since the coefficients of the x terms are opposites ($^-2$ and $^+2$), their sum is zero, and the result of adding the two equations is the elimination of the x terms.

$$\begin{array}{r} ^-2x + y = 7 \\ 2x + 3y = 5 \\ \hline 4y = 12 \end{array}$$

The resulting equation, $4y = 12$, can be solved for the value of y.

$$4y = 12$$
$$y = 3$$

To find the value of x, substitute the value $y = 3$ into either one of the original equations.

$$\begin{array}{lcl} {}^-2x + y = 7 & \text{or} & 2x + 3y = 5 \\ {}^-2x + 3 = 7 & & 2x + 3(3) = 5 \\ {}^-2x = 4 & & 2x + 9 = 5 \\ x = {}^-2 & & 2x = {}^-4 \\ & & x = {}^-2 \end{array}$$

Thus, the solution for the system is $({}^-2, 3)$. The solution is checked by showing it converts each equation into a truc statement.

$$\begin{array}{lcl} {}^-2x + y = 7 & \text{and} & 2x + 3y = 5 \\ {}^-2({}^-2) + 3 \overset{?}{=} 7 & & 2({}^-2) + 3(3) \overset{?}{=} 5 \\ 4 + 3 = 7 \quad \text{(true)} & & {}^-4 + 9 = 5 \quad \text{(true)} \end{array}$$

Q5   Solve by the addition method:
$$x - 3y = {}^-4$$
$${}^-x - 2y = {}^-1$$

STOP • **STOP** • **STOP** • **STOP** • **STOP** • **STOP** • **STOP** • **STOP** • **STOP**

A5 $({}^-1, 1)$:
$$\begin{array}{l} x - 3y = {}^-4 \\ \underline{{}^-x - 2y = {}^-1} \\ \quad {}^-5y = {}^-5 \\ \quad\quad y = 1 \end{array}$$

To find x, substitute the value $y = 1$ into either one of the original equations.

$$\begin{array}{lcl} x - 3y = {}^-4 & \text{or} & {}^-x - 2y = {}^-1 \\ x - 3(1) = {}^-4 & & {}^-x - 2(1) = {}^-1 \\ x - 3 = {}^-4 & & {}^-x - 2 = {}^-1 \\ x = {}^-1 & & {}^-x = 1 \\ & & x = {}^-1 \end{array}$$

Thus, the solution for the system is $({}^-1, 1)$.

Q6   Check the solution to Q5 by showing that it converts each of the two equations into a true statement.

STOP • **STOP** • **STOP** • **STOP** • **STOP** • **STOP** • **STOP** • **STOP** • **STOP**

A6

$$x - 3y = {}^-4$$
$$({}^-1) - 3(1) \stackrel{?}{=} {}^-4$$
$${}^-1 - 3 \stackrel{?}{=} {}^-4$$
$${}^-4 = {}^-4 \quad \text{(true)}$$

$${}^-x - 2y = {}^-1$$
$${}^-({}^-1) - 2(1) \stackrel{?}{=} {}^-1$$
$$1 - 2 \stackrel{?}{=} {}^-1$$
$${}^-1 = {}^-1 \quad \text{(true)}$$

Q7   Solve by the addition method:
$$5x - y = 8$$
$${}^-5x + 7y = 4$$

STOP • **STOP** • **STOP** • **STOP** • **STOP** • **STOP** • **STOP** • **STOP** • **STOP**

A7 $(2, 2)$:
$$5x - y = 8$$
$$\underline{{}^-5x + 7y = 4}$$
$$6y = 12$$
$$y = 2$$

To find x, substitute $y = 2$ into either one of the original equations.
$$5x - y = 8 \qquad \text{(other choice not shown)}$$
$$5x - 2 = 8$$
$$5x = 10$$
$$x = 2$$

Q8   Solve by the addition method:
$$3x - y = {}^-5$$
$${}^-2x + y = 11$$

STOP • **STOP** • **STOP** • **STOP** • **STOP** • **STOP** • **STOP** • **STOP** • **STOP**

A8 $(6, 23)$:
$$3x - y = {}^-5$$
$$\underline{{}^-2x + y = 11}$$
$$x = 6$$

To find y, substitute $x = 6$ into either one of the original equations.
$$3x - y = {}^-5 \qquad \text{(other choice not shown)}$$
$$3(6) - y = {}^-5$$
$$18 - y = {}^-5$$
$${}^-y = {}^-23$$
$$y = 23$$

Q9   Solve each of the following by the addition method:

a. $x - y = 12$ b. $3x - 3y = {}^-12$ c. $x + 2y = 1$
   ${}^-x - y = 6$ $x + 3y = 12$ ${}^-x + 2y = 1$

STOP • **STOP** • **STOP** • **STOP** • **STOP** • **STOP** • **STOP** • **STOP** • **STOP**

A9   a. $(3, {}^-9)$ b. $(0, 4)$ c. $\left(0, \dfrac{1}{2}\right)$

---

3   If the coefficients of the x terms or the y terms are not opposites, the sum of the two equations will *not* result in the elimination of one of the terms. In this situation it is impossible to solve for the numerical value of one of the variables by the addition method presented so far.

Examples:

1. $3x + y = 7$
   ${}^-x + 2y = 3$
   ─────────────
   $2x + 3y = 10$

Since the coefficients of neither the x nor the y terms are opposites, the sum of the two equations fails to eliminate one of the variables.

2. $5x + 2y = 4$
   $3x - 2y = 12$
   ─────────────
   $8x\ \ \ \ \ = 16$

Since the coefficients of the y terms are opposites (${}^+2$ and ${}^-2$), the sum of the two equations eliminates the y term.

Q10   In which of the following systems will addition of the two equations *not* result in the elimination of one of the terms? _____

a. $2x - y = {}^-5$ b. $2x - 4y = 7$ c. $6x + y = 2$
   ${}^-3x + y = 5$ $x + 3y = 0$ $6x + 4y = 8$

STOP • **STOP** • **STOP** • **STOP** • **STOP** • **STOP** • **STOP** • **STOP** • **STOP**

A10   b and c

---

4   If the sum of the equations of a system will not result in the elimination of one of the terms, it is often possible to multiply both sides of one of the equations by the same integer to produce opposite coefficients on one of the terms. (Recall that multiplying both sides of an equation by the same nonzero number produces an equivalent equation and does not affect the solution set.)

Examples:

1. $^-2x + 3y = 11$ $\,-\,-\,-\,-\,-\,-\,-\,\rightarrow$ $^-2x + 3y = 11$

    $x - y = ^-2$ $\,-\,\underset{\text{multiply by 3}}{-\,-\,-\,-\,-\,}\,\rightarrow$ $\underline{3x - 3y = ^-6}$

    $\qquad\qquad\qquad\qquad\qquad\qquad\qquad x \qquad\; = 5$

The value of y is found by substituting the value $x = 5$ into either of the original equations.

$x - y = ^-2 \qquad$ (other choice not shown)

$5 - y = ^-2$

$\quad ^-y = ^-7$

$\quad\; y = 7 \qquad$ The solution is $(5, 7)$.

2. $3x + 5y = 8$ $\,-\,\underset{\text{multiply by } ^-5}{-\,-\,-\,-\,-\,}\,\rightarrow$ $^-15x - 25y = ^-40$

    $15x + 2y = ^-6$ $\,-\,-\,-\,-\,-\,-\,-\,\rightarrow$ $\underline{15x + 2y = ^-6}$

    $\qquad\qquad\qquad\qquad\qquad\qquad\qquad\qquad ^-23y = ^-46$

    $\qquad\qquad\qquad\qquad\qquad\qquad\qquad\qquad\quad y = 2$

The value of x is found by substituting the value $y = 2$ into either of the original equations.

$3x + 5y = 8 \qquad$ (other choice not shown)

$3x + 5(2) = 8$

$3x + 10 = 8$

$\quad 3x = ^-2$

$\quad\; x = \dfrac{^-2}{3} \qquad$ The solution is $\left(\dfrac{^-2}{3}, 2\right)$.

Q11 $\quad ^-3x - 2y = 7$

$\qquad\; 2x + y = ^-2$

a. What integer could be multiplied times the second equation to make the y terms opposites? _____

b. Solve the system using the answer to part a.

STOP • **STOP** • **STOP** • **STOP** • **STOP** • **STOP** • **STOP** • **STOP** • **STOP**

A11 $\qquad$ **a.** 2 $\qquad$ **b.** $(3, ^-8)$.

$\qquad\qquad\qquad\qquad ^-3x - 2y = 7$ $\,-\,-\,-\,-\,-\,-\,-\,\rightarrow$ $^-3x - 2y = 7$

$\qquad\qquad\qquad\qquad\; 2x + y = ^-2$ $\,-\,\underset{\text{multiply by 2}}{-\,-\,-\,-\,-\,}\,\rightarrow$ $\underline{4x + 2y = ^-4}$

$\qquad\qquad\qquad\qquad\qquad\qquad\qquad\qquad\qquad\qquad x \qquad\; = 3$

$\qquad\qquad\qquad\qquad\; 2x + y = ^-2 \qquad$ (other choice not shown)

$\qquad\qquad\qquad\qquad\; 2(3) + y = ^-2$

$\qquad\qquad\qquad\qquad\quad 6 + y = ^-2$

$\qquad\qquad\qquad\qquad\qquad\quad y = ^-8 \qquad$ The solution is $(3, ^-8)$.

Q12 $x - 7y = 4$
        $3x + 5y = 12$

  a. What integer could be multiplied times the first equation to make the x terms opposites? _____

  b. Solve the system using the answer to part a.

STOP • STOP • STOP • STOP • STOP • STOP • STOP • STOP • STOP

A12   **a.** $^-3$ **b.** $(4, 0)$:

$$\begin{array}{ll} x - 7y = 4 & \xrightarrow{\text{multiply by } ^-3} \quad ^-3x + 21y = ^-12 \\ 3x + 5y = 12 & \dashrightarrow \qquad\qquad\quad \underline{3x + 5y = 12} \\ & \qquad\qquad\qquad\qquad\quad 26y = 0 \\ & \qquad\qquad\qquad\qquad\qquad y = 0 \end{array}$$

$\quad x - 7y = 4$ (other choice not shown)
$\quad x - 7(0) = 4$
$\quad x - 0 = 4$
$\quad x = 4$ The solution is $(4, 0)$.

Q13   Solve the system $\begin{cases} ^-3x - 5y = 7 \\ 2x - 5y = ^-3 \end{cases}$ by multiplying the first equation by $^-1$.

STOP • STOP • STOP • STOP • STOP • STOP • STOP • STOP • STOP

A13 $\left(^-2, \dfrac{^-1}{5}\right)$: $^-3x - 5y = 7 \quad \xrightarrow{\text{multiply by } ^-1} \quad 3x + 5y = ^-7$

$$\begin{array}{ll} 2x - 5y = ^-3 & \dashrightarrow \quad \underline{2x - 5y = ^-3} \\ 2x - 5y = ^-3 & \qquad\quad 5x \qquad = ^-10 \\ 2(^-2) - 5y = ^-3 & \qquad\quad x \qquad = ^-2 \\ ^-4 - 5y = ^-3 \\ ^-5y = 1 \end{array}$$

$$y = \frac{^-1}{5} \quad \text{The solution is } \left(^-2, \frac{^-1}{5}\right).$$

Q14   Solve each of the following systems:

a. $x - y = 3$ b. $2x + 3y = 7$ c. $2x + 7y = ^-14$
   $2x + 3y = 16$ $3x + y = 7$ $3x - 7y = 14$

STOP • STOP • STOP • STOP • STOP • STOP • STOP • STOP • STOP

A14   a. $(5, 2)$ b. $(2, 1)$ c. $(0, ^-2)$

---

5   It is sometimes necessary to multiply both equations by two different integers to produce opposite coefficients on the x or y terms. For example,

1. $2x - 7y = 2$ $\xrightarrow{\text{multiply by 2}}$ $4x - 14y = 4$

    $3x + 2y = 3$ $\xrightarrow{\text{multiply by 7}}$ $\underline{21x + 14y = 21}$

$$25x \qquad = 25$$
$$x \qquad = 1$$

(*Note:* We eliminate the y terms because they already have opposite signs.)

$$2x - 7y = 2$$
$$2(1) - 7y = 2$$
$$2 - 7y = 2$$
$$^-7y = 0$$
$$y = 0$$

Therefore, the solution is $(1, 0)$.

2. $5x - 3y = 4$ $\xrightarrow{\text{multiply by 3}}$ $15x - 9y = 12$

    $3x - 4y = ^-2$ $\xrightarrow{\text{multiply by } ^-5}$ $\underline{^-15x + 20y = 10}$

$$11y = 22$$
$$y = 2$$

(*Note:* Either the x or the y term could be eliminated because both terms have the same sign. Here the x terms were eliminated.)

$$5x - 3y = 4$$
$$5x - 3(2) = 4$$
$$5x - 6 = 4$$
$$5x = 10$$
$$x = 2$$

Therefore, the solution is $(2, 2)$.

To eliminate the y terms first instead of the x terms as above, multiply the first equation by 4 and the second equation by $^-3$, or the first by $^-4$ and the second by 3.

Q15 $3x - 2y = 7$
$4x + 5y = {}^-6$

    a. To eliminate the y terms, it would be necessary to multiply the first equation by _____ and the second equation by _____.

    b. To eliminate the x terms, it would be necessary to multiply the first equation by _____ and the second equation by _____.

STOP • **STOP** • **STOP** • **STOP** • **STOP** • **STOP** • **STOP** • **STOP** • **STOP**

A15   **a.** 5, 2   **b.** 4, ⁻3 or ⁻4, 3

Q16   Solve the system of Q15.

STOP • **STOP** • **STOP** • **STOP** • **STOP** • **STOP** • **STOP** • **STOP** • **STOP**

A16 $(1, {}^-2)$: using the procedure of Q15a, the solution is as follows:

$$3x - 2y = 7 \quad \xrightarrow{\text{multiply by 5}} \quad 15x - 10y = 35$$
$$4x + 5y = {}^-6 \quad \xrightarrow{\text{multiply by 2}} \quad \underline{8x + 10y = {}^-12}$$
$$23x \qquad = 23$$
$$x \qquad = 1$$

$$3x - 2y = 7$$
$$3(1) - 2y = 7$$
$$3 - 2y = 7$$
$${}^-2y = 4$$
$$y = {}^-2 \qquad \text{The solution is } (1, {}^-2).$$

Q17 $11x + 6y = 1$
$^-7x + 5y = 17$

    a. To eliminate the y terms, it would be necessary to multiply the first equation by _____ and the second equation by _____.

    b. To eliminate the x terms, it would be necessary to eliminate the first equation by _____ and the second equation by _____.

STOP • **STOP** • **STOP** • **STOP** • **STOP** • **STOP** • **STOP** • **STOP** • **STOP**

A17   **a.** ⁻5, 6 or 5, ⁻6   **b.** 7, 11

Q18   Solve the system of Q17.

STOP   •   STOP   •   STOP   •   STOP   •   STOP   •   STOP   •   STOP   •   STOP   •   STOP

A18 $(^-1, 2)$: using the procedure of Q17a, the solution is as follows:

$$11x + 6y = 1 \quad \xrightarrow{\text{multiply by } ^-5} \quad ^-55x - 30y = ^-5$$

$$^-7x + 5y = 17 \quad \xrightarrow{\text{multiply by } 6} \quad \underline{^-42x + 30y = 102}$$

$$^-97x \qquad = 97$$

$$x \qquad = ^-1$$

$$11x + 6y = 1$$
$$11(^-1) + 6y = 1$$
$$^-11 + 6y = 1$$
$$6y = 12$$
$$y = 2$$

Q19   Solve each of the following systems:

a. $3x - 5y = 22$ **b.** $2x + 5y = ^-3$ **c.** $13x - 5y = ^-40$

$7x - 4y = 13$ $8x - 7y = ^-12$ $9x - 3y = ^-24$

STOP   •   STOP   •   STOP   •   STOP   •   STOP   •   STOP   •   STOP   •   STOP   •   STOP

A19   **a.** $(^-1, ^-5)$ **b.** $\left(-\dfrac{3}{2}, 0\right)$ or $\left(-1\dfrac{1}{2}, 0\right)$ **c.** $(0, 8)$

---

6 | Recall from Section 10.1 the names given to the various types of systems. If two equations in two unknowns have only one solution, they are called *consistent equations*. If two equations in two unknowns have no common solution, they are called *inconsistent equations*. Two equations in two unknowns with an infinite number of solutions are called *dependent equations*.

Q20   **a.** Each system of equations solved in this section so far had _____ solution.

          b. Are the systems solved in this section so far correctly labeled consistent, inconsistent,

          or dependent? _____

STOP   •   **STOP**   •   **STOP**   •   **STOP**   •   **STOP**   •   **STOP**   •   **STOP**   •   **STOP**   •   **STOP**

A20   **a.** one   **b.** consistent

---

7   Consider the solution of the following system:

$$2x - 3y = 6 \quad \xrightarrow{\text{multiply by 2}} \quad 4x - 6y = 12$$
$$^-4x + 6y = {}^-12 \quad \text{- - - - - - - -} \rightarrow \quad \underline{{}^-4x + 6y = {}^-12}$$
$$0 + 0 = 0$$
$$0 = 0$$

The sum of the two equations produces the true statement $0 = 0$, which indicates that the two equations differ only by a factor of negative one. The equations are, therefore, equivalent and the solutions for each are the same. Since any linear equation has an infinite number of solutions, the solution set for each of the two equations and the system is infinite. That the solutions for the two equations are the same is seen in the following graph:

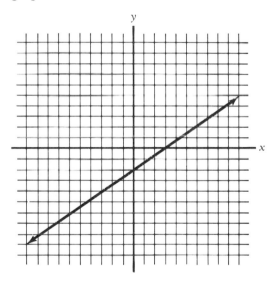

When the sum of a system of two equations in two unknowns produces a statement that is always true such as $4 = 4$ or $0 = 0$, the equations are *dependent* and *the system has an infinite number of solutions.*

Q21   **a.** Solve the system:
          $$2x - 3y = 6$$
          $$^-4x + 6y = {}^-12$$

b. Is the system consistent, inconsistent, or dependent? _____

c. Does the system have one solution, no solution, or an infinite number of solutions?

_____

STOP • STOP • STOP • STOP • STOP • STOP • STOP • STOP • STOP

A21   **a.** The equations are equivalent and the solutions for each are the same.
         b. dependent   **c.** an infinite number

---

8   Consider the solution of the following system:

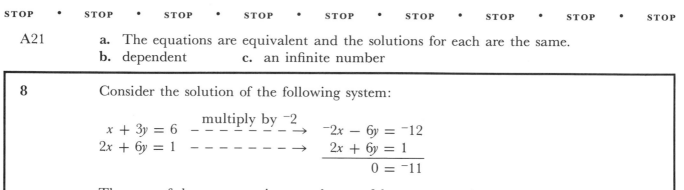

$$x + 3y = 6 \quad \xrightarrow{\text{multiply by } ^-2} \quad ^-2x - 6y = ^-12$$
$$2x + 6y = 1 \quad \text{-------}\rightarrow \quad \underline{2x + 6y = 1}$$
$$0 = ^-11$$

The sum of the two equations produces a false statement $0 = ^-11$, which indicates that there is no ordered pair of number (x, y) which solves the system of equations. The fact that the system has no solution is seen in the graphs of the two equations.

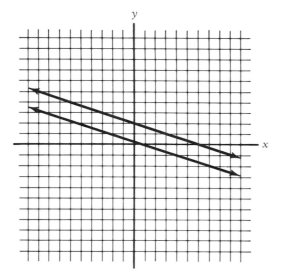

The graphs of the two equations are parallel and are, therefore, inconsistent. When the sum of a system of two equations in two unknowns produces a statement that is always false such as $0 = 8$ or $4 = 6$, the equations are *inconsistent* and *the system has no solution.*

Q22   **a.** Solve the system:
$$5x - 2y = 7$$
$$^-10x + 4y = 4$$

b. Is the system consistent, inconsistent, or dependent? _____

STOP • STOP • STOP • STOP • STOP • STOP • STOP • STOP • STOP

A22   **a.** The system has no solution.   **b.** inconsistent

Q23 Solve each of the following systems and classify each as consistent, inconsistent, or dependent:

a. $3x + y = 2$
   $x + 2y = 9$

b. $^-x + 3y = ^-3$
   $2x - 6y = 5$

c. $4x - 2y = 10$
   $2x - y = 5$

STOP • STOP • STOP • STOP • STOP • STOP • STOP • STOP • STOP

A23 a. $(^-1, 5)$, consistent   b. no solution, inconsistent
    c. infinite number of solutions, dependent

---

9 The general form for two equations in two unknowns is $\begin{cases} ax + by = c \\ dx + ey = f \end{cases}$ where $a, b, c, d, e,$ and f are rational numbers.

If the system is not already in the general form, it is necessary to place it in this form by aligning the like terms. For example,

1. $x - 5 = 2y$ The system is not in the general form.
   $y + x = 7$ Each of the equations can be rearranged as follows:

$$x - 5 = 2y \qquad\qquad y + x = 7$$
$$x - 5 + 5 = 2y + 5 \qquad x + y = 7$$
$$x = 2y + 5$$
$$x - 2y = 2y + 5 - 2y$$
$$x - 2y = 5$$

The system in general form is $\begin{cases} x - 2y = 5 \\ x + y = 7 \end{cases}$.

2. $4 - y = 5x$ The system is not in the general form.
   $8 - x = y$ Each of the equations can be rearranged as follows:

$$4 - y = 5x \qquad\qquad 8 - x = y$$
$$4 - y + y = 5x + y \qquad 8 - x + x = y + x$$
$$4 = 5x + y \qquad\qquad 8 = y + x$$
$$\qquad\qquad\qquad\qquad 8 = x + y$$

The system in general form is $\begin{cases} 5x + y = 4 \\ x + y = 8 \end{cases}$.

Q24 Write in general form:
   $x + 7 = 3y$
   $^-4 - x = 3y$

STOP • STOP • STOP • STOP • STOP • STOP • STOP • STOP • STOP

A24

$$x - 3y = {}^-7 \qquad x + 7 = 3y \qquad\qquad {}^-4 - x = 3y$$
$${}^-x - 3y = 4: \quad x + 7 - 7 = 3y - 7 \qquad {}^-4 - x + 4 = 3y + 4$$
$$x = 3y - 7 \qquad\qquad {}^-x = 3y + 4$$
$$x - 3y = 3y - 7 - 3y \qquad {}^-x - 3y = 3y + 4 - 3y$$
$$x - 3y = {}^-7 \qquad\qquad {}^-x - 3y = 4$$

Q25   **a.** Write in general form:

$$y = 8 - 3x$$
$$2x = y + 2$$

b. Solve the system of part a.

STOP • **STOP** • **STOP** • **STOP** • **STOP** • **STOP** • **STOP** • **STOP** • **STOP**

A25   **a.** $3x + y = 8 \qquad y = 8 - 3x \qquad\qquad 2x = y + 2$
$\qquad\qquad 2x - y = 2: \quad y + 3x = 8 - 3x + 3x \quad 2x - y = y + 2 - y$
$\qquad\qquad\qquad\qquad\qquad y + 3x = 8 \qquad\qquad 2x - y = 2$
$\qquad\qquad\qquad\qquad\qquad 3x + y = 8$

b. $(2, 2)$

Q26   Solve the system:
$$3x - 5y = 2$$
$$15y - 9x = {}^-2$$

STOP • **STOP** • **STOP** • **STOP** • **STOP** • **STOP** • **STOP** • **STOP** • **STOP**

A26   There is no solution, the equations are inconsistent.

---

10   To solve a system of two equations in two unknowns in which one or both of the equations has fractional coefficients, proceed as follows:

1. Clear each equation of fractions by multiplying each term of the equation by the least common denominator (LCD).
2. Write each equation in general form.
3. Solve using the addition method.

Examples:

1. $2x - 3y = 12$ $- - - - - - - - \rightarrow$ $2x - 3y = 12$

   $\dfrac{1}{3}x = {}^-y$ $\underset{\text{multiply by 3}}{- - - - - - - \rightarrow}$ $x = {}^-3y$

The system can now be placed in general form and solved.

$$\begin{array}{r} 2x - 3y = 12 \\ x + 3y = 0 \\ \hline 3x = 12 \\ x = 4 \end{array}$$

To find y, substitute $x = 4$ into either of the original equations.

$$\begin{array}{r} 2x - 3y = 12 \\ 2(4) - 3y = 12 \\ {}^-3y = 4 \\ y = \dfrac{4}{3} \end{array}$$ The solution is $\left(4, \dfrac{{}^-4}{3}\right)$ or $\left(4, {}^-1\dfrac{1}{3}\right)$.

2. $\dfrac{2}{3}x + y = {}^-1$ (the LCD is 3)

   $\dfrac{1}{2}x + \dfrac{4}{5}y = 5$ (the LCD is 10)

$\dfrac{2}{3}x + y = {}^-1$ $\underset{\text{multiply by 3}}{- - - - - - - \rightarrow}$ $2x + 3y = {}^-3$ $\underset{\text{multiply by }{}^-5}{- - - - - - - \rightarrow}$

${}^-10y - 15y = 15$

$\dfrac{1}{2}x + \dfrac{4}{5}y = 5$ $\underset{\text{multiply by 10}}{- - - - - - - \rightarrow}$ $5x + 8y = 50$ $\underset{\text{multiply by 2}}{- - - - - - - \rightarrow}$

$$\begin{array}{r} 10x + 16y = 100 \\ \hline y = 115 \end{array}$$

To find x, substitute $y = 115$ into either of the original equations (choose the least complicated equation).

$\dfrac{2}{3}x + y = {}^-1$

$\dfrac{2}{3}x + 115 = {}^-1$

$\dfrac{2}{3}x = {}^-116$

$\dfrac{3}{2}\left(\dfrac{2}{3}x\right) = \dfrac{3}{2}({}^-116)$

$x = {}^-174$

Therefore, the solution is $({}^-174, 115)$.

Q27 Solve the system:

$$x + \frac{2}{5}y = {}^-1$$
$$3x - 2y = 5$$

STOP • STOP • STOP • STOP • STOP • STOP • STOP • STOP • STOP

A27 $\left(0, \dfrac{^-5}{2}\right)$ or $\left(0, {}^-2\dfrac{1}{2}\right)$: $x + \dfrac{2}{5}y = {}^-1$ $\xrightarrow{\text{multiply by 5}}$ $5x + 2y = {}^-5$

$\qquad\qquad\qquad\qquad\qquad\qquad\qquad 3x - 2y = 5$ $\dashrightarrow$ $3x - 2y = 5$

$\qquad\qquad\qquad\qquad\qquad\qquad\qquad\qquad\qquad\qquad\qquad\qquad\overline{}$

$\qquad\qquad\qquad\qquad\qquad\qquad\qquad\qquad\qquad\qquad\qquad\qquad 8x \qquad\ = 0$

$\qquad\qquad\qquad\qquad\qquad\qquad\qquad\qquad\qquad\qquad\qquad\qquad\ x \qquad\ = 0$

$$3x - 2y = 5$$
$$3(0) - 2y = 5$$
$$^-2y = 5$$
$$y = \frac{^-5}{2}$$

Q28 Solve the system:

$$\frac{1}{4}y - 2 = x$$
$$x + \frac{2}{3}y = \frac{5}{3}$$

STOP • STOP • STOP • STOP • STOP • STOP • STOP • STOP • STOP

A28 $({}^-1, 4)$: $\dfrac{1}{4}y - 2 = x$ $\xrightarrow{\text{multiply by 4}}$ $y - 8 = 4x$

$\qquad\qquad\qquad x + \dfrac{2}{3}y = \dfrac{5}{3}$ $\xrightarrow{\text{multiply by 3}}$ $3x + 2y = 5$

The system in general form is: $4x - y = {}^-8$

$\qquad\qquad\qquad\qquad\qquad\qquad\qquad\qquad\quad 3x + 2y = 5$

The solution by the addition method is as follows:

$4x - y = {}^-8$ $\dashrightarrow$ $8x - 2y = {}^-16$

$3x + 2y = 5$ $\dashrightarrow$ $3x + 2y = 5$

$\qquad\qquad\qquad\qquad\qquad\qquad\quad\overline{}$

$\qquad\qquad\qquad\qquad\qquad\qquad\quad 11x \qquad\ = {}^-11$

$\qquad\qquad\qquad\qquad\qquad\qquad\qquad x \qquad\ = {}^-1$

$$\frac{1}{4}y - 2 = x$$

$$\frac{1}{4}y - 2 = ^-1$$

$$\frac{1}{4}y = 1$$

$$y = 4$$

This completes the instruction for this section.

10.2   Exercises

1. Solve each of the following systems by addition:

   a. $x - y = 5$
      $x + y = 7$

   b. $2x - 3y = 4$
      $^-2x - 3y = 8$

   c. $2x + y = 3$
      $^-3x + 3y = 9$

   d. $3x - 4y = 5$
      $^-2x + 5y = ^-6$

2. Solve each of the following systems by addition and classify each as consistent, inconsistent or dependent:

   a. $2x - 3y = 7$
      $^-9 - y = 2x$

   b. $x = y + 2$
      $x - y = ^-5$

   c. $5x = 2y$
      $^-x = y + 7$

   d. $6 = 12x - 3y$
      $y - 4x = ^-2$

3. Solve each of the following systems:

   a. $3x - 7 = 2y$
      $5y + 4x = ^-6$

   b. $\frac{2}{3}x + y = 0$
      $x - \frac{1}{4}y = 7$

   c. $^-2x - 3 = ^-5y$
      $6x - 15y = ^-9$

   d. $\frac{3}{7}x - \frac{2}{5}y = ^-2$
      $\frac{5}{8}x + y = 5$

   e. $y = 4x + 11$
      $8x = 3 + 2y$

   f. $\frac{3}{4}x + \frac{1}{7}y = ^-8$
      $\frac{^-1}{3}x = \frac{4}{7}y$

4. Solve each of the following systems by graphing and check the answer using the addition method:

   a. $2x - 3y = 1$
      $x + y = 8$

   b. $4x + 5y = ^-2$
      $^-2x - 7y = 1$

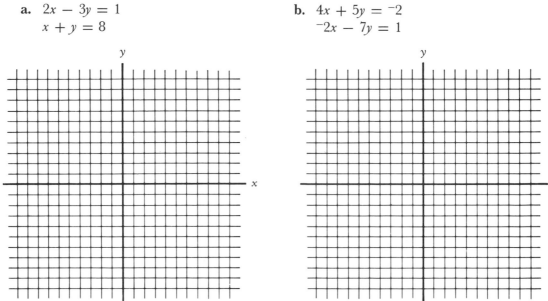

10.2   Exercise Answers

1. **a.** $(6, 1)$ **b.** $(^-1, ^-2)$ **c.** $(0, 3)$ **d.** $\left(\dfrac{1}{7}, \dfrac{^-8}{7}\right)$

2. **a.** $\left(\dfrac{^-5}{2}, ^-4\right)$, consistent   **b.** no solution, inconsistent

    c. $(^-2, ^-5)$, consistent   **d.** infinite solutions, dependent

3. **a.** $(1, ^-2)$ **b.** $(6, ^-4)$
    c. infinite solutions (dependent equations)   **d.** $(0, 5)$
    e. no solution (inconsistent equations)   **f.** $(^-12, 7)$

4. **a.** $(5, 3)$ **b.** $\left(\dfrac{^-1}{2}, 0\right)$

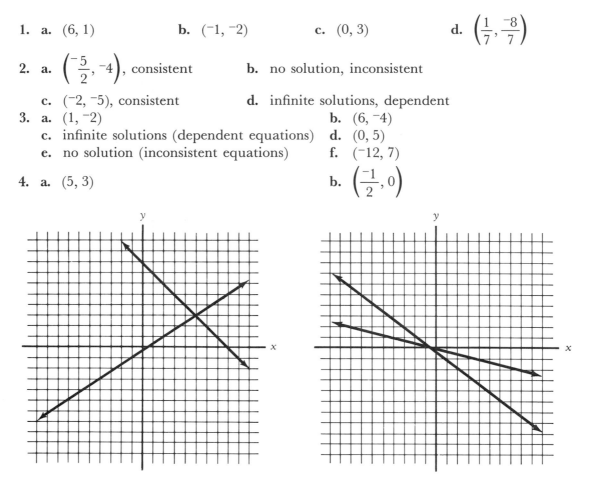

10.3   Solution by Substitution

1 | The substitution method is a second procedure for solving a system of two equations in two unknowns which is more efficient than the graphing method. The substitution method is most conveniently used when there is a coefficient of 1 or $^-1$ on either the x or y term in one of the equations.

$3x + y = 12$
$2x + 5y = 8$

The substitution method is convenient, because the coefficient of the y term in the first equation is 1.

$^-2x + 7y = 17$
$5x - 3y = ^-5$

The substitution method is *not* convenient, because neither equation has a coefficient of 1 or $^-1$ on the x or y terms.

Q1   Which of the systems are appropriate for solution by the substitution method? _____

    a. $x - y = 5$ **b.** $7x - 5y = {}^-10$ **c.** $9x - 4 = 2y$

        $3x + 5y = 0$ $2x + 6y = 9$ $x = 7y - 3$

STOP • STOP • STOP • STOP • STOP • STOP • STOP • STOP • STOP

A1   a and c

---

2   The system $\begin{cases} 3x + y = 12 \\ 2x + 5y = 8 \end{cases}$ has a term whose coefficient is 1 (the y term) and is, therefore, convenient for solution by the substitution method. Since the y term appears in the first equation, solve this equation for y in terms of x.

$$3x + y = 12$$
$$3x + y - 3x = 12 - 3x$$
$$y = 12 - 3x$$

The value of y is now substituted into the second equation.

$$2x + 5y = 8$$
$$2x + 5(12 - 3x) = 8$$

The resulting equation can now be solved for x.

$$2x + 60 - 15x = 8$$
$${}^-13x + 60 = 8$$
$${}^-13x = {}^-52$$
$$x = 4$$

The value of y is found by substituting $x = 4$ into the equation $y = 12 - 3x$.

$$y = 12 - 3x$$
$$= 12 - 3(4)$$
$$= 12 - 12$$
$$= 0$$

Therefore, the solution of the system is $(4, 0)$.

Q2 $3x + 2y = 0$

    $x - 4y = 14$

    a. Which variable has a coefficient of 1 or ${}^-1$? _____

    b. Solve the equation for this variable in terms of the other variable.

    c. Substitute the value of x into the remaining equation of the system and solve for y.

    d. Substitute the value of y into the equation of part b.

    e. The solution of the system is ＿＿＿＿＿.

STOP • **STOP** • **STOP** • **STOP** • **STOP** • **STOP** • **STOP** • **STOP** • **STOP**

A2

    a. The x term in the second equation.

    b. $x = 14 + 4y$:
$$x - 4y = 14$$
$$x - 4y + 4y = 14 + 4y$$
$$x = 14 + 4y$$

    c. $y = {}^-3$:
$$3x + 2y = 0$$
$$3(14 + 4y) + 2y = 0$$
$$42 + 12y + 2y = 0$$
$$42 + 14y = 0$$
$$14y = {}^-42$$
$$y = {}^-3$$

    d. $x = 2$:
$$x = 14 + 4y$$
$$= 14 + 4({}^-3)$$
$$= 14 + {}^-12$$
$$= 2$$

    e. $(2, {}^-3)$

Q3

    Solve by the substitution method:
$$2x + y = {}^-4$$
$$3x = {}^-2y - 3$$

STOP • **STOP** • **STOP** • **STOP** • **STOP** • **STOP** • **STOP** • **STOP** • **STOP**

A3

    $({}^-5, 6)$:   Solve the first equation for y.
$$2x + y = {}^-4$$
$$2x + y - 2x = {}^-4 - 2x$$
$$y = {}^-4 - 2x$$

    Substitute $4 - 2x$ for y into the second equation and solve for x.
$$3x = {}^-2y - 3$$
$$3x = {}^-2({}^-4 - 2x) - 3$$
$$3x = 8 + 4x - 3$$
$$3x = 5 + 4x$$
$$3x - 4x = 5 + 4x - 4x$$
$${}^-x = 5$$
$$x = {}^-5$$

Substitute $^-5$ for x into $y = ^-4 - 2x$ to find y.

$$y = ^-4 - 2x$$
$$= ^-4 - 2(^-5)$$
$$= ^-4 + 10$$
$$= 6$$

Q4   Solve by the substitution method:

$$2x - 3y = ^-14$$
$$y + 4x = 14$$

STOP • **STOP** • **STOP** • **STOP** • **STOP** • **STOP** • **STOP** • **STOP** • **STOP**

A4 $(2, 6)$:   Solve the second equation for y.

$$y + 4x = 14$$
$$y + 4x - 4x = 14 - 4x$$
$$y = 14 - 4x$$

Substitute $14 - 4x$ for y into the first equation and solve for x.

$$2x - 3y = ^-14$$
$$2x - 3(14 - 4x) = ^-14$$
$$2x - 42 + 12x = ^-14$$
$$14x - 42 + 42 = ^-14 + 42$$
$$14x = 28$$
$$x = 2$$

Substitute $x = 2$ into $y = 14 - 4x$ to find y.

$$y = 14 - 4x$$
$$= 14 - 4(2)$$
$$= 14 - 8$$
$$= 6$$

---

3   The system $\begin{cases} 3x + 7y = ^-15 \\ 2y - x = 5 \end{cases}$ has a term whose coefficient is $^-1$ (the x term) and, therefore, can be solved conveniently by the substitution method. Since the x term appears in the second equation, solve this equation for x in terms of y.

$$2y - x = 5$$
$$2y - x - 2y = 5 - 2y$$
$$^-x = 5 - 2y$$
$$(^-1)(^-x) = ^-1(5 - 2y)$$
$$x = ^-5 + 2y$$

The value of x is now substituted into the first equation.

$$3x + 7y = {}^-15$$
$$3({}^-5 + 2y) + 7y = {}^-15$$

The resulting equation can now be solved for y.

$$^-15 + 6y + 7y = {}^-15$$
$$^-15 + 13y = {}^-15$$
$$13y = 0$$
$$y = 0$$

The value of x is found by substituting $y = 0$ into the equation $x = {}^-5 + 2y$.

$$x = {}^-5 + 2y$$
$$= {}^-5 + 2(0)$$
$$= {}^-5 + 0$$
$$= {}^-5$$

Therefore, the solution of the system is $({}^-5, 0)$.

Q5 $2x - y = 1$
      $3x + 5y = {}^-44$

a. Which variable has a coefficient of 1 or $^-1$? _____

b. Solve the equation for this variable in terms of the other variable.

c. Substitute the value from part b into the other equation of the system and solve for the variable x.

d. Substitute the value from part c into the equation in part b and solve for y.

e. The solution of the system is _____.

STOP • STOP • STOP • STOP • STOP • STOP • STOP • STOP • STOP

A5   a. The y term in the first equation.
      b. $y = {}^-1 + 2x$: $2x - y = 1$
                              $2x - y - 2x = 1 - 2x$
                                    $^-y = 1 - 2x$
                              $^-1({}^-y) = {}^-1(1 - 2x)$
                                     $y = {}^-1 + 2x$

c. $x = {}^-3$:
$$3x + 5y = {}^-44$$
$$3x + 5({}^-1 + 2x) = {}^-44$$
$$13x - 5 = {}^-44$$
$$13x = {}^-39$$
$$x = {}^-3$$

d. $y = {}^-7$: $y = {}^-1 + 2x$
$$= {}^-1 + 2({}^-3)$$
$$= {}^-1 - 6$$
$$= {}^-7$$

e. $({}^-3, {}^-7)$

Q6   Solve by the substitution method:
$${}^-y + 3x = 6$$
$$4x - 3y = 13$$

STOP   •   STOP   •   STOP   •   STOP   •   STOP   •   STOP   •   STOP   •   STOP   •   STOP

A6 $(1, {}^-3)$:   Solve the first equation for y.
$${}^-y + 3x = 6$$
$${}^-y + 3x - 3x = 6 - 3x$$
$${}^-y = 6 - 3x$$
$${}^-1({}^-y) = {}^-1(6 - 3x)$$
$$y = {}^-6 + 3x$$

Substitute ${}^-6 + 3x$ for y into the second equation and solve for x.
$$4x - 3y = 13$$
$$4x - 3({}^-6 + 3x) = 13$$
$$4x + 18 - 9x = 13$$
$${}^-5x + 18 = 13$$
$${}^-5x = {}^-5$$
$$x = 1$$

Substitute 1 for x into $y = {}^-6 + 3x$ to find y.
$$y = {}^-6 + 3x$$
$$= {}^-6 + 3(1)$$
$$= {}^-6 + 3$$
$$= {}^-3$$

Therefore, the solution is $(1, {}^-3)$.

---

4   The steps used in solving a system of two equations in two unknowns by the substitution method are as follows:

1. Solve one of the equations for y in terms of x (or x in terms of y).
2. Substitute the value of y (or x) from step 1 into the other equation and solve for x (or y).
3. Substitute the value of x (or y) into the equation of step 1 and find the value of y (or x).
4. State the solution as an ordered pair of the form (x, y).
5. Check the solution by substituting the values of x and y into the equation not used in step 1.

Q7   Solve each of the systems by the substitution method:

     a. $5x + 3y = {}^-8$ **b.** $2x + y - 13 = 0$ **c.** $5x - 2 = y$

         $x - 5y = {}^-24$ $5y = 3x + 26$ $7x + 2y = {}^-4$

STOP • **STOP** • **STOP** • **STOP** • **STOP** • **STOP** • **STOP** • **STOP** • **STOP**

A7   **a.** $({}^-4, 4)$ **b.** $(3, 7)$ **c.** $(0, {}^-2)$

---

5   Consider the solution by substitution of the system $\begin{cases} 2y - 4x = {}^-3 \\ 2x - y = 10 \end{cases}$. Solve the second equation for y.

$$2x - y = 10$$
$$2x - y - 2x = 10 - 2x$$
$${}^-y = 10 - 2x$$
$${}^-1({}^-y) = {}^-1(10 - 2x)$$
$$y = {}^-10 + 2x$$

Substitute $^-10 + 2x$ for y into the first equation.

$$2y - 4x = {}^-3$$
$$2({}^-10 + 2x) - 4x = {}^-3$$
$${}^-20 + 4x - 4x = {}^-3$$
$${}^-20 = {}^-3$$

The resulting equality $^-20 = {}^-3$ is a false statement, indicating that the equations are inconsistent. Thus, the system has no solution.

Q8   Solve by the substitution method:
$$x + 7y = 3$$
$$^-2x = 14y$$

STOP • STOP • STOP • STOP • STOP • STOP • STOP • STOP • STOP

A8   no solution: $x + 7y = 3$
$$x + 7y - y = 3 - 7y$$
$$x = 3 - 7y$$

Substitute $3 - 7y$ for x into the second equation.
$$^-2x = 14y$$
$$^-2(3 - 7y) = 14y$$
$$^-6 + 14y = 14y$$
$$^-6 = 0 \quad \text{(false)}$$

Therefore, the equations are inconsistent and the system has no solution.

---

6   Consider the solution by substitution of the system $\begin{cases} x = y + 10 \\ ^-2x + 2y = ^-20 \end{cases}$. Since the first equation is already solved for x, substitute this value into the second equation.

$$^-2x + 2y = ^-20$$
$$^-2(y + 10) + 2y = ^-20$$
$$^-2y - 20 + 2y = ^-20$$
$$^-20 = ^-20$$

The resulting equality is a true statement, indicating that the equations are dependent. Thus, the system has an infinite number of solutions.

Q9   Solve by the substitution method:
$$3x + 6y = 12$$
$$x + 2y = 4$$

STOP • STOP • STOP • STOP • STOP • STOP • STOP • STOP • STOP

A9   The equations are dependent and there are an infinite number of solutions. $x = 4 - 2y$.

$$3x + 6y = 12$$
$$3(4 - 2y) + 6y = 12$$
$$12 - 6y + 6y = 12$$
$$12 = 12 \quad \text{(true)}$$

The true statement indicates that the equations are dependent and that the system has an infinite number of solutions.

---

7   To solve the system $\begin{cases} \dfrac{1}{3}x + 2y = 14 \\[2mm] \dfrac{2}{3}x - \dfrac{1}{4}y = {}^-6 \end{cases}$ by the method of substitution, each equation is first cleared of fractions using the procedure of Section 10.2.

$$\frac{1}{3}x + 2y = 14 \quad \xrightarrow{\text{multiply by 3}} \quad x + 6y = 42$$

$$\frac{2}{3}x - \frac{1}{4}y = {}^-6 \quad \xrightarrow{\text{multiply by 12}} \quad 8x - 3y = {}^-72$$

(Recall that each equation is cleared of fractions by multiplying each term of the equation by the LCD.)

The solution is now completed using the procedures already developed.

$$x + 6y = 42$$
$$x = 42 - 6y$$

Substitute $42 - 6y$ for x into the second equation.

$$8x - 3y = {}^-72$$
$$8(42 - 6y) - 3y = {}^-72$$
$$336 - 48y - 3y = {}^-72$$
$$336 - 51y = {}^-72$$
$${}^-51y = {}^-408$$
$$y = 8$$

Substitute 8 for y into $x = 42 - 6y$ to find x.

$$x = 42 - 6y$$
$$= 42 - 6(8)$$
$$= 42 - 8$$
$$= {}^-6$$

The solution $({}^-6, 8)$ should now be checked in both original equations.

$$\frac{1}{3}x + 2y = 14 \qquad\qquad \frac{2}{3}x - \frac{1}{4}y = {}^-6$$

$$\frac{1}{3}({}^-6) + 2(8) \stackrel{?}{=} 14 \qquad\qquad \frac{2}{3}({}^-6) - \frac{1}{4}(8) \stackrel{?}{=} {}^-6$$

$${}^-2 + 16 \stackrel{?}{=} 14 \qquad\qquad {}^-4 - 2 \stackrel{?}{=} {}^-6$$

$$14 = 14 \quad \text{(true)} \qquad\qquad {}^-6 = {}^-6 \quad \text{(true)}$$

Q10   Solve by the substitution method:

$$\frac{1}{2}x - \frac{5}{7}y = {}^-1$$

$$x + 5y = {}^-2$$

STOP • **STOP** • **STOP** • **STOP** • **STOP** • **STOP** • **STOP** • **STOP** • **STOP**

A10 $({}^-2, 0)$: $\dfrac{1}{2}x - \dfrac{5}{7}y = {}^-1$ $\xrightarrow{\text{multiply by 14}}$ $7x - 10y = {}^-14$

                $x + 5y = {}^-2$ $----\rightarrow$ $x + 5y = {}^-2$

                                     $x + 5y = {}^-2$

                                        $x = {}^-2 - 5y$

Substitute ${}^-2 - 5y$ for x into the first equation.

$$7x - 10y = {}^-14$$
$$7({}^-2 - 5y) - 10y = {}^-14$$
$${}^-14 - 35y - 10y = {}^-14$$
$${}^-14 - 45y = {}^-14$$
$${}^-45y = 0$$
$$y = 0$$

Substitute 0 for y into $x = {}^-2 - 5y$ to find x.

$$x = {}^-2 - 5y$$
$$= {}^-2 - 5(0)$$
$$= {}^-2 - 0$$
$$= {}^-2$$

Check the solution $({}^-2, 0)$.

       $\dfrac{1}{2}x - \dfrac{5}{7} = {}^-1$ $x + 5y = {}^-2$

$\dfrac{1}{2}({}^-2) - \dfrac{5}{7}(0) \overset{?}{=} {}^-1$ ${}^-2 + 5(0) \overset{?}{=} {}^-2$

          ${}^-1 - 0 = {}^-1$ (true) ${}^-2 + 0 = {}^-2$ (true)

Q11　　　Solve by the substitution method:

$$3x - \frac{1}{5}y = {}^-5$$

$$5x + \frac{3}{10}y = {}^-2$$

STOP • **STOP** • **STOP** • **STOP** • **STOP** • **STOP** • **STOP** • **STOP** • **STOP**

A11　　　$({}^-1, 10)$

This completes the instruction for this section.

10.3　Exercises

1. Solve each of the following systems by the method of substitution:
   a. $x - y = 0$　　　　　　　b. $3x - 4y = 12$　　　　　c. $2x + y = {}^-8$
   　　$2x + 3y = 5$　　　　　　　$y = 7x - 3$　　　　　　　${}^-3x - 4y = 7$
   d. $5x + 2y = 12$
   　　$3y = x - 33$

2. Solve each of the following systems by the method of substitution and check each of the solutions that exist.

   a. $\frac{1}{4}x + y = {}^-1$　　　　　b. $2x - 6y = {}^-8$　　　　c. $\frac{{}^-1}{2}x = \frac{2}{3}y$
   　　　　　　　　　　　　　　　　$3 - x = {}^-3y$
   　　$\frac{3}{4}x - \frac{5}{7}y = {}^-3$　　　　　　　　　　　　　　　　$x = {}^-y - 3$

   d. $\frac{1}{3}x + 2y = {}^-6$

   　　$2x + 12y = {}^-5$

3. Solve each of the following by all three methods for solving systems of equations: (1) solution by graphing, (2) the addition method, and (3) the method of substitution.

a. $2x - y = 15$
$x + 3y = {}^{-}17$

b. $3x - 5y = {}^{-}5$
$y + 7x = 1$

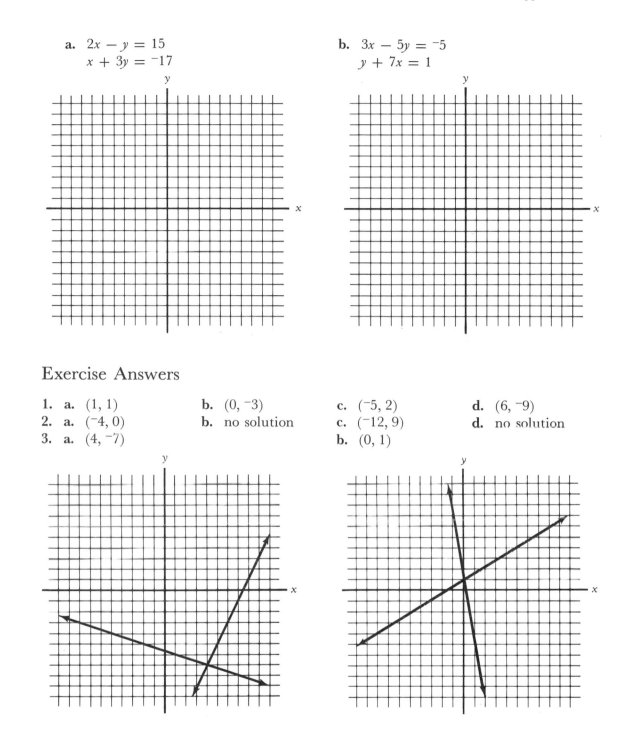

10.3   Exercise Answers

1. **a.** $(1, 1)$ **b.** $(0, {}^{-}3)$ **c.** $({}^{-}5, 2)$ **d.** $(6, {}^{-}9)$
2. **a.** $({}^{-}4, 0)$ **b.** no solution   **c.** $({}^{-}12, 9)$ **d.** no solution
3. **a.** $(4, {}^{-}7)$ **b.** $(0, 1)$

10.4   **Applications**

1   Chapter 7 presented techniques for solving verbal problems by use of one variable. If two unknowns were present, it was necessary to use one of the facts given to express each unknown in terms of the variable selected. The other fact could then be used to write an equation.

Example: The sum of two numbers is 21 and their difference is 9. Find the two numbers.

Solution

Let $n =$ one of the numbers. Since the sum of the two numbers is 21, $21 - n =$ the other number. The equation can now be written based upon the fact that their difference is 9.

$$n - (21 - n) = 9$$
$$n - 21 + n = 9$$
$$2n - 21 = 9$$
$$2n = 30$$
$$n = 15$$

Hence, one number is 15 and the other is $21 - 15$, or 6. These numbers satisfy the facts of the problem, because it is also true that $15 - 6 = 9$.

Q1 Use one variable (n) to solve these problems:
  a. The sum of two numbers is 63 and the difference between the larger and one half of the smaller number is 30. What are the two numbers?

  b. George Tripp made two investments, the first at 5 percent and the second at 6 percent. He received a yearly income from them of $275. If the total investment was $5,000, how much did he invest at each rate?

STOP • STOP • STOP • STOP • STOP • STOP • STOP • STOP • STOP

A1 **a.** 41 and 22: Let $n =$ the larger number and $63 - n =$ the smaller number.

$$n - \frac{1}{2}(63 - n) = 30$$
$$2n - (63 - n) = 60 \quad \text{(multiplying both sides by 2)}$$
$$2n - 63 + n = 60$$
$$3n - 63 = 60$$
$$3n = 123$$
$$n = 41$$
$$63 - n = 63 - 41 = 22$$

Hence, the larger number is 41 and the smaller number is 22. Check: $41 - \frac{1}{2}(22) = 30$.

  b. $2,500 at 5% and $2,500 at 6%: Let $n =$ amount invested at 5% and $5,000 - n =$ amount invested at 6%.

$$0.05n + 0.06(5,000 - n) = 275$$
$$5n + 6(5,000 - n) = 27,500 \quad \text{(multiplying both sides by 100)}$$
$$5n + 30,000 - 6n = 27,500$$
$$^-n + 30,000 = 27,500$$
$$^-n = {}^-2,500$$
$$n = 2,500$$
$$5,000 - n = 2,500$$

Hence, $2,500 was invested at 5% and $2,500 was invested at 6%.

Check: 0.05 of $2,500 = $125
0.06 of $2,500 = $150
$275

2	Many problems can be solved by the use of two variables.

Example: The sum of two numbers is 21 and their difference is 9. Find the two numbers.

Solution

Let $x =$ the larger number and $y =$ the smaller number. (*Note:* Any two variables can be used.) Use these variables to write the facts in the form of a system of two equations.

$$x + y = 21$$
$$x - y = 9$$

Now solve the system by the most convenient method. By addition:

$$2x = 30$$
$$x = 15$$

Determine the value of y by substituting the value of x into either of the original equations.

$$x + y = 21$$
$$15 + y = 21$$
$$y = 6$$

Hence, the two numbers are 15 and 6. Check by comparing the results with the facts stated in the problem.

The sum of two numbers is 21: $15 + 6 = 21$ (true)
Their difference is 9: $15 - 6 = 9$ (true)

Q2   Write the facts in the form of a system of two equations. Do not solve the equations.
a.   The sum of two numbers is 63 and the difference between the larger and one half of the smaller is 30. Let $x =$ the larger number and $y =$ the smaller number.

b.   George Tripp made two investments, the first at 5 percent and the second at 6 percent. He received a yearly income from them of $275 by investing a total of $5,000. Let $x =$ the amount invested at 5 percent and $y =$ the amount invested at 6 percent.

STOP   •   **STOP**   •   **STOP**   •   **STOP**   •   **STOP**   •   **STOP**   •   **STOP**   •   **STOP**   •   **STOP**

A2   **a.** $x + y = 63$ **b.** $0.05x + 0.06y = 275$

$x - \dfrac{1}{2}y = 30$ $x + y = 5{,}000$

Q3   Write the facts in the form of a system of two equations and solve the system:
a. If the larger of two numbers is diminished by one-half of the smaller number, the result is 67. The difference between the two numbers is 45. Find the two numbers. Let $x =$ the larger number and $y =$ the smaller number.

b. The sum of two numbers is 49. The larger is 7 more than the smaller. Find the two numbers. Let $x =$ the larger number and $y =$ the smaller number.

STOP • **STOP** • **STOP** • **STOP** • **STOP** • **STOP** • **STOP** • **STOP** • **STOP**

A3   **a.** 89 and 44: $x - \dfrac{1}{2}y = 67$ **b.** 28 and 21: $x + y = 49$

$x - y = 45$ $x = y + 7$

---

3   Consecutive integers are integers that increase by 1. For example, 4, 5, 6, 7, ... or $^-5$, $^-4$, $^-3$, $^-2$, ... By letting the first number be n, the next two consecutive integers would be $n + 1$ and $n + 2$.

On the other hand, two consecutive integers differ by 1. Hence, if the first integer is x and the second integer is y, from the fact that the integers are consecutive any one of the following equations can be written:

1. $y - x = 1$, or
2. $x = y - 1$, or
3. $y = x + 1$

Q4 Use x for the first integer and y for the second integer and find two consecutive integers whose sum is 115.

STOP • STOP • STOP • STOP • STOP • STOP • STOP • STOP • STOP

A4 57 and 58: possible systems:

$$y - x = 1 \quad \text{or} \quad x = y - 1 \quad \text{or} \quad y = x + 1$$
$$x + y = 115 \qquad x + y = 115 \qquad x + y = 115$$

Q5 Find two consecutive even integers so that twice the first increased by three times the second is 56. Let $x = $ the first integer and $y = $ the second integer.
[*Hint:* Even (or odd) integers differ by 2.]

STOP • STOP • STOP • STOP • STOP • STOP • STOP • STOP • STOP

A5 10 and 12: possible systems:

$$y - x = 2 \quad \text{or} \quad x = y - 2 \quad \text{or} \quad y = x + 2$$
$$2x + 3y = 56 \qquad 2x + 3y = 56 \qquad 2x + 3y = 56$$

4 It is important to note that in order to solve a problem by the use of two variables, a system of two equations must be obtained from the facts. This requires that two unrelated facts be given about the two unknowns.

Q6 Mr. Rose is four times as old as his son. Eighteen years from now he will be only twice as old as his son. Find their present ages. Let $x = $ Mr. Rose's age and $y = $ his son's age.

STOP • STOP • STOP • STOP • STOP • STOP • STOP • STOP • STOP

A6 Mr. Rose is 36 years old and his son is 9 years old:

$$x = 4y$$
$$x + 18 = 2(y + 18)$$

Q7   The acute angles of any right triangle are complementary; that is, their sum equals 90°. If the larger acute angle is 6 degrees less than three times the smaller angle, how many degrees has each acute angle? Let x = number of degrees in the larger acute angle, and y = number of degrees in the smaller acute angle.

STOP • **STOP** • **STOP** • **STOP** • **STOP** • **STOP** • **STOP** • **STOP** • **STOP**

A7   24° and 66°: $x + y = 90$
$$x = 3y - 6$$

Q8   During a two-day motor trip, Marcia Joiner drove 1,240 miles. If during the first day she drove 100 miles more than during the second day, how many miles did she drive each day? Let x = number of miles driven the first day, and y = number of miles driven the second day.

STOP • **STOP** • **STOP** • **STOP** • **STOP** • **STOP** • **STOP** • **STOP** • **STOP**

A8   670 miles the first day and
570 miles the second day: $x + y = 1,240$
$$x = y + 100$$

Q9   John canoed 20 miles downstream in 2 hours. The trip upstream took 5 hours. Find his speed in still water and the speed of the current. Let x = his rate (speed) in still water, and y = rate (speed) of the current.
[*Hint:* rate with the current: $x + y$
rate against the current: $x - y$
distance = rate times time ($d = rt$)]

STOP • **STOP** • **STOP** • **STOP** • **STOP** • **STOP** • **STOP** • **STOP** • **STOP**

A9   rate in still water: 7 mph
rate of the current: 3 mph: $2(x + y) = 20$ and $5(x - y) = 20$

Q10   Given two grades of zinc ore, the first containing 45 percent zinc and the second 25 percent zinc, find how many pounds of each grade must be used to make a mixture of 2,000 pounds containing 40 percent zinc. Let x = number of pounds of ore that is 45 percent zinc and y = number of pounds of ore that is 25 percent zinc.

[*Hint:* $x + y = $ _____

$(\ \ )x + (\ \ )y = (\ \ ) \cdot 2,000$]

STOP • **STOP** • **STOP** • **STOP** • **STOP** • **STOP** • **STOP** • **STOP** • **STOP**

A10   1,500 pounds of 45 percent zinc ore
500 pounds of 25 percent zinc ore:

$$x + y = 2,000$$
$$(0.45)x + (0.25)y = (0.40) \cdot 2,000$$
$$x + y = 2,000$$
$$0.45x + 0.25y = 800$$
$$x + y = 2,000$$
$$45x + 25y = 80,000$$

This completes the instruction for this section.

10.4   Exercises

Determine the answers to the following problems by writing and solving a system of equations:

1. The sum of two numbers is 33. The larger is twice the smaller. What are the numbers?
2. The larger of two numbers exceeds the smaller by 15. One half the smaller number tripled is equal to the larger number. What are the two numbers?
3. The sum of two consecutive integers is 217. What are the integers?
4. Find two consecutive odd integers whose sum is 232.
5. The pilot of a light plane finds that it takes him 16 minutes to go 48 miles when traveling with the wind, whereas it takes him 21 minutes to go the 48 miles on the return trip against the wind. Calculate his average airspeed and the speed of the wind; assume that it is constant during the flight.
6. Two hikers start at the same time from points that are 19 miles apart. One walks approximately 1 mile per hour faster than the other. They meet in 2 hours. What is the average rate of each hiker?
7. Mr. Clyde Wu invests part of $8,000 at 5 percent interest and the rest at $5\frac{1}{2}$ percent. His return on the investment at the end of one year is $430. How much has he invested at each rate?
8. George is twice as old as his sister Betty. Four years ago he was only three times as old. What are their ages now?
9. The perimeter of a rectangle is 34 centimeters. If the width is increased by 2 centimeters and the length is decreased by 1 centimeter, a square is formed. Find the dimensions of the original rectangle.
10. How many milligrams of uranium 75 percent pure and of uranium 85 percent pure must be mixed to give 9 milligrams of uranium 80 percent pure?

10.4   Exercise Answers

1. 11 and 22
2. 30 and 45
3. 108 and 109
4. 115 and 117
5. Airspeed is approximately 158.6 mph, wind is approximately 21.4 mph.
6. $4\frac{1}{4}$ mph and $5\frac{1}{4}$ mph
7. $6,000 at $5\frac{1}{2}$ percent and $2,000 at 5 percent
8. George is 16 years old and Betty is 8 years old.
9. Length is 10 centimeters and width is 7 centimeters.
10. 4.5 milligrams of uranium 75 percent pure and 4.5 milligrams of uranium 85 percent pure

Chapter 10 Sample Test

At the completion of Chapter 10 it is expected that you will be able to work the following problems.

10.1   **Solution by Graphing**

1. Find the solution for each of the following systems from the list of ordered pairs below:
   $(5, 0), (0, 5), (^-1, 3), (^-2, ^-2), (2, ^-2), (6, ^-3)$
   a. $3x - y = ^-5$ **b.** $^-x + y = 0$ **c.** $5x = ^-3y + 4$
   $x + 4y = 20$ $x + y = ^-4$ $2y = ^-6x$
2. The set of infinite solutions for a linear equation can be represented in picture form by a _____.
3. Determine the solution for each of the following systems from the graphs:
   a. $x + y = 4$ **b.** $3y = 2x - 6$
   $x - 3y = 0$ $^-2x + 3y = 3$

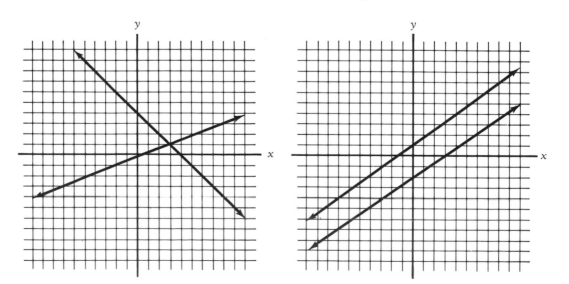

c. $y = \dfrac{1}{3}x - 2$

$\quad\;\; 3y - 4x = 12$

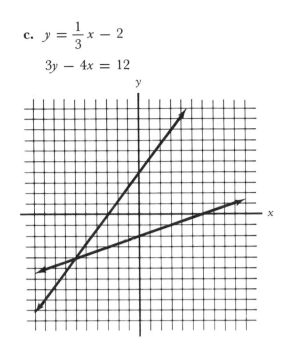

4. Label each of the graphs below as consistent, inconsistent, or dependent:

a.

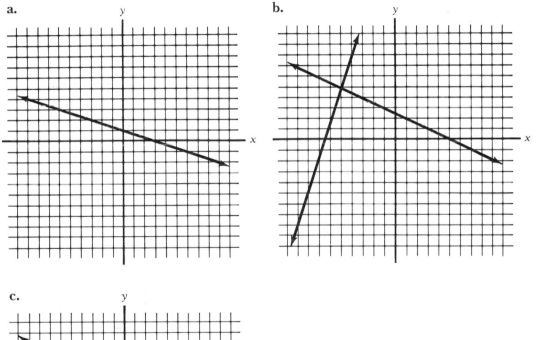

b.

c.

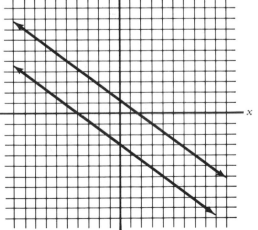

5. Label each of the graphs from problem 4 as having one solution, no solution, or an infinite number of solutions.

6. Solve each of the following systems by graphing and label each as consistent, inconsistent, or dependent:

   a. $2x - y = 6$
      $y = x - 3$

   b. $3x + 2y = 6$
      $2y - 4 = {}^-3x$

   c. $x + 2y = 5$
      $4x + 8y = 20$

   d. $x + 6y = {}^-2$
      ${}^-x - 2y = 6$

10.2   Solution by Addition

7. Solve each of the following systems by the method of elimination by addition:

   a. $3x + 2y = 4$
      $x - 2y = {}^-4$

   b. ${}^-x + 2y = {}^-1$
      ${}^-4y = 5 - x$

   c. $3x - 5y = {}^-23$

      $7x + 4y = 9$

   d. $\dfrac{1}{4}y - 2 = x$

      $x + \dfrac{2}{3}y = \dfrac{5}{3}$

10.3   Solution by Substitution

8. When is the substitution method convenient to use in solving a system of two equations in two unknowns?

9. Solve each of the following systems by the method of substitution:

   a. $2x + y = {}^-8$
      ${}^-3x - 4y = 7$

   b. $2y - 4x = {}^-3$
      $2x - y = 10$

   c. $\dfrac{{}^-1}{2}x = \dfrac{2}{3}y$
      $x = {}^-y + 3$

10.4   Applications

Find the answers by writing and solving a system of equations:

10. Paul is paid $5 more a week than Jim. In 25 weeks Jim earns as much as Paul does in 24 weeks. What is the weekly wage of each?

11. The sum of two consecutive even integers is 126. What are the two integers?

12. A freighter can make a trip of 60 miles upstream in 6 hours. The return trip with the current can be made in 5 hours. Assume that the ship travels at a constant rate. What would be the speed of the ship in still water, and what is the average rate of the current?

13. If part of $5,000 is invested at $5\dfrac{1}{2}$ percent interest and the other part is invested at 6 percent interest, the total annual interest is $282.50. What are the two amounts invested?

14. David is now 2 years younger than his brother Steve. In 3 years, six times David's age will equal four times Steve's age. What are their present ages?

15. The perimeter of a rectangular lot is 60 meters. If the length of the lot is six less than twice the width, what are the dimensions of the lot?

16. A chemist has two stock solutions of boric acid, one 10 percent boric acid and the other 15 percent boric acid. How much of each must he use to obtain 10 liters of a solution that is 12 percent boric acid?

Chapter 10 Sample Test Answers

1. **a.** $(0, 5)$ **b.** $(^-2, ^-2)$ **c.** $(^-1, 3)$
2. line
3. **a.** $(3, 1)$ **b.** no solution **c.** $(^-6, ^-4)$
4. **a.** dependent **b.** consistent **c.** inconsistent
5. **a.** infinite solutions **b.** one solution **c.** no solution
6. **a.** $(3, 0)$, consistent **b.** no solution, inconsistent **c.** infinite solutions, dependent **d.** $(^-8, 1)$, consistent
7. **a.** $(0, 2)$ **b.** $(^-3, ^-2)$ **c.** $(^-1, 4)$ **d.** $(^-1, 4)$
8. When there is a coefficient of 1 or $^-1$ on either the x or y term in one of the equations.
9. **a.** $(^-5, 2)$ **b.** no solution **c.** $(12, ^-9)$
10. Jim: \$120; Paul: \$125
11. 62 and 64
12. rate of the ship: 11 mph, rate of the current: 1 mph
13. \$1,500 at 6 percent and \$3,500 at $5\frac{1}{2}$ percent
14. David's age: 1 year old, Steve's age: 3 years old
15. length: 18 meters, width: 12 meters
16. 4 liters of 15 percent solution and 6 liters of 10 percent solution

Chapter 11

Polynomials

In this chapter you will study basic operations with algebraic expressions called polynomials.

11.1   Introduction

---

1   In the expression x^n, n is called an *exponent* and x is called the *base*. Symbols such as 2^3 and x^5 are called *powers*. The exponent of a variable such as x is understood to be 1. That is, $x = x^1$ and is read as "x" or "the first power of x." x^2 is read "x squared" or "the second power of x." x^3 is read "x cubed" or "the third power of x." Other powers are read according to the number used as an exponent: for example, x^6 is read "the sixth power of x."

Q1   Read (write) each of the following:

    a. x^5 _____

    b. y^3 _____

    c. $^-z^4$ _____

    d. x^2y _____

STOP • **STOP** • **STOP** • **STOP** • **STOP** • **STOP** • **STOP** • **STOP** • **STOP**

A1   **a.** the fifth power of x
    b. y cubed or the third power of y
    c. the opposite of the fourth power of z
    d. x squared times y
(*Note:* There are numerous other acceptable ways of reading powers. For example, x^2 may be read "x to the second power" or simply "x to the second." x^2 may also be read "the square of x." x^3 may be read "the cube of x," "x to the third power," or simply "x to the third." In general, x^n may be read "x to the nth power" or simply "x to the nth.")

---

2   Natural-number exponents indicate the number of times the base appears as a factor.

Examples:

1. x means x^1 or one factor of x.
2. x^2 means $x \cdot x$ or two factors of x.
3. x^3 means $x \cdot x \cdot x$ or three factors of x.

In general,

$$x^m = \underbrace{x \cdot x \cdot x \cdots x}_{m \text{ factors of } x} \qquad \text{where } m \in N$$

4. $x^5 y^2$ means $x \cdot x \cdot x \cdot x \cdot x \cdot y \cdot y$ or five factors of x times two factors of y.

Q2   Complete to form a true statement:

    a. y^3 means ＿＿＿＿＿ or ＿＿＿＿＿ factors of y.

    b. z^2 means ＿＿＿＿＿ or two ＿＿＿＿＿ of z.

    c. m^1 means ＿＿＿＿＿ or ＿＿＿＿＿ factor of m.

    d. $x^3 y^4$ means ＿＿＿＿＿＿＿＿＿＿＿.

STOP • **STOP** • **STOP** • **STOP** • **STOP** • **STOP** • **STOP** • **STOP** • **STOP**

A2   **a.** $y \cdot y \cdot y$, three   **b.** $z \cdot z$, factors   **c.** m, one   **d.** $x \cdot x \cdot x \cdot y \cdot y \cdot y \cdot y$

---

3   Special care must be exercised when evaluating expressions involving exponents.

Example 1: Evaluate x^4 when $x = {}^-3$.

Solution

$$\begin{aligned}
x^4 = ({}^-3)^4 &= \underbrace{({}^-3)({}^-3)}({}^-3)({}^-3) \\
&= \underbrace{({}^+9)({}^-3)}({}^-3) \\
&= ({}^-27)({}^-3) \\
&= {}^+81 \quad \text{or} \quad 81
\end{aligned}$$

A useful technique when evaluating a product that involves numerous factors of the same number is to pair the factors in the process of multiplying. For example,

$$\begin{aligned}
({}^-3)^4 &= \underbrace{({}^-3)({}^-3)}\,\underbrace{({}^-3)({}^-3)} \\
&= ({}^+9)({}^+9) \\
&= {}^+81 \quad \text{or} \quad 81
\end{aligned}$$

It is important to realize that the notation ${}^-x^2$ means ${}^-(x^2)$, which indicates the opposite of the square of x. $({}^-x)^2$ would indicate the square of the opposite of x. Similarly, ${}^-5^2$ means ${}^-(5^2)$ or ${}^-(25)$, which is ${}^-25$. $({}^-5)^2$ means $({}^-5)({}^-5)$, which is 25.

Example 2: Evaluate ${}^-x^2$ when $x = 5$.

Solution

$$\begin{aligned}
{}^-x^2 &= {}^-(5^2) &&\text{``the opposite of 5 squared''} \\
&= {}^-(25) &&\text{``the opposite of 25''} \\
&= {}^-25 &&\text{``negative 25''}
\end{aligned}$$

Example 3: Evaluate ${}^-x^3$ when $x = {}^-4$.

Solution

$$^-x^3 = ^-[(^-4)^3]$$
$$= ^-[(^-4)(^-4)(^-4)]$$
$$= ^-[(16)(^-4)]$$
$$= ^-(^-64)   \text{"the opposite of negative 64"}$$
$$= ^+64   \text{or}   64$$

Q3   Evaluate for the given values of the variables:

a. x^2 when $x = 4$ **b.** x^2 when $x = ^-3$ **c.** x^3 when $x = 3$

d. x^3 when $x = ^-2$ **e.** x^4 when $x = 2$ **f.** y^4 when $y = ^-2$

g. y^5 when $y = 2$ **h.** z^5 when $z = ^-2$ **i.** $^-m^2$ when $m = 4$

j. $^-z^2$ when $z = ^-4$ **k.** $^-x^3$ when $x = 3$ **l.** $^-x^3$ when $x = ^-2$

STOP • **STOP** • **STOP** • **STOP** • **STOP** • **STOP** • **STOP** • **STOP** • **STOP**

A3

a. 16: $4^2 = 4 \cdot 4$ **b.** 9: $(^-3)^2 = (^-3)(^-3)$
c. 27: $3^3 = 3 \cdot 3 \cdot 3$ **d.** $^-8$: $(^-2)^3 = (^-2)(^-2)(^-2)$
e. 16: $2^4 = 2 \cdot 2 \cdot 2 \cdot 2$ **f.** 16: $(^-2)^4 = (^-2)(^-2)(^-2)(^-2)$
g. 32: $2^5 = 2 \cdot 2 \cdot 2 \cdot 2 \cdot 2$ **h.** $^-32$: $(^-2)^5 = (^-2)^4(^-2)$
$$= (16)(^-2)$$
$$= ^-32$$

i. $^-16$: $^-(4^2) = ^-(16)$ **j.** $^-16$: $^-[(^-4)^2] = ^-[(^-4)(^-4)]$
$$= ^-(16)$$

k. $^-27$: $^-(3^3) = ^-(27)$ **l.** 8: $^-[(^-2)^3] = ^-[(^-2)(^-2)(^-2)]$
$$= ^-(^-8)$$

4   A *polynomial* is defined as an algebraic expression made up of sums, differences, and products of variables and numbers. Examples of polynomials are

$$2x - 3, 5y^2 - 3y + 7, ^-5, x^2 - xy - y^2, \text{ and } ^-5x^2y$$

where the variables represent rational numbers.* Expressions containing quotients where the variable is in the denominator are not polynomials. Examples of expressions that are

---

*For any rational-number replacement of the variables, a polynomial represents a rational number. Hence, all properties that are valid for the set of rational numbers can be used when performing operations with polynomials. Polynomials are also defined for real-number replacements of the variables. The set of real numbers will be discussed in Chapter 13.

not polynomials are

$$1 + \frac{3}{x} \qquad \frac{5}{3x - 7} \qquad 4y^2 + 3y + \frac{1}{y}$$

Operations with expressions of this type will be discussed in Chapter 12.

You should observe that all *constants* (numbers that do not vary in value), such as $^-5$, $\frac{1}{2}$, and $\frac{^-2}{3}$, are polynomials.

Q4   Write "yes" if the algebraic expression is a polynomial. Otherwise, write "no."

a. $x^2 + 2xy + y^2$ _____   b. ^-3x _____

c. $5y - 5$ _____   d. $\frac{4x^2}{10y^3}$ _____

e. $5x^2 + 3x - \frac{1}{x}$ _____   f. 0 _____

g. $2xy + 9$ _____   h. $4x^5 + 6x^2 - 2x + 7$ _____

i. $\frac{1}{7}x^3$ _____   j. $\frac{7x - 8}{2x + 3}$ _____

STOP • **STOP** • **STOP** • **STOP** • **STOP** • **STOP** • **STOP** • **STOP** • **STOP**

A4   **a.** yes   **b.** yes   **c.** yes   **d.** no   **e.** no   **f.** yes   **g.** yes
h. yes   **i.** yes   **j.** no

---

5   The definition of subtraction

$x - y = x + {}^-y$ where x and $y \in Q$

can be applied to any polynomial that contains a subtraction sign. The terms of a polynomial are the expressions separated by addition signs.

Examples:

	Terms
$5y - 5 = 5y + {}^-5$	$5y$ and $^-5$
$5x^2 - 3x + 7 = 5x^2 + {}^-3x + 7$	$5x^2$, ^-3x, and 7

Q5   Identify the terms of each polynomial:

a. $^-7 + x$ _____

b. $5x^2 - 3$ _____

c. $7x^3 - 8x^2 + x - 2$ _____

d. $12x$ _____

e. $^-15$ _____

STOP • **STOP** • **STOP** • **STOP** • **STOP** • **STOP** • **STOP** • **STOP** • **STOP**

A5   **a.** $^-7$ and x **b.** $5x^2$ and $^-3$ **c.** $7x^3$, $^-8x^2$, x, and $^-2$
d. $12x$ **e.** $^-15$

6

Polynomials that contain only *one* term are called *monomials*. Polynomials that have exactly *two* terms are called *binomials*. Polynomials that have exactly *three* terms are called *trinomials*. No special names are associated with polynomials with more than three terms. You should observe that a monomial, a binomial, and a trinomial are *all* polynomials.

Q6   Identify each polynomial as a monomial, binomial, or trinomial. Otherwise, write polynomial.

   a. $5x^2 - 2x - 3$ _____

   b. $x^2 - 9$ _____

   c. ^-5xyz _____

   d. $^-4m^3 + 7m^2 + 9m - 3$ _____

   e. 0 _____

STOP   •   STOP   •   STOP   •   STOP   •   STOP   •   STOP   •   STOP   •   STOP   •   STOP

A6   **a.** trinomial   **b.** binomial   **c.** monomial   **d.** polynomial
      e. monomial

7

Every term, except zero, has a *degree*. It is agreed that the degree of a constant is zero except for the constant 0, which is said to have no degree. The degree of a term with one variable is determined by the exponent on the variable.

Examples:

	Degree
y	1, because $y = y^1$
5	0
$^-3y^2$	2
$9x^5$	5
0	no degree

Q7   Indicate the degree of each term:

   a. ^-3t _____   **b.** $^-8$ _____

   c. $5m^2$ _____   **d.** 0 _____

   e. $14x^7$ _____   **f.** y^3 _____

STOP   •   STOP   •   STOP   •   STOP   •   STOP   •   STOP   •   STOP   •   STOP   •   STOP

A7   **a.** 1: $^-3t = ^-3t^1$ **b.** 0   **c.** 2   **d.** no degree
      e. 7   **f.** 3

8

The *degree of a polynomial* is the greatest degree of any of its terms.

Examples:

	Degree of the polynomial
$x^2 - x - 6$	2
$x + 5$	1
$3x^4 + 2x^2 - 5$	4
12	0
$7x^3$	3
0	no degree

Q8   Indicate the degree of each polynomial:

     a. $-3x^2 + x - 2$ _____   **b.** $x^2 - 4$ _____

     c. $-x + 9$ _____   **d.** 5 _____

     e. $3x^5$ _____   **f.** $8y^3 - y + 1$ _____

     g. 0 _____   **h.** $1 - x^2 - 3x^4$ _____

STOP • **STOP** • **STOP** • **STOP** • **STOP** • **STOP** • **STOP** • **STOP** • **STOP**

A8   **a.** 2   **b.** 2   **c.** 1   **d.** 0

        e. 5   **f.** 3   **g.** no degree   **h.** 4

---

9   The *numerical coefficient* (number factor) of $-3x^2$ is -3. The *literal coefficient* (letter factor) of $-3x^2$ is x^2. The numerical coefficient is generally called simply the "coefficient" of the term. By the multiplication property of one, the coefficient of a term with no coefficient shown is 1. That is, $7x^2 + x = 7x^2 + 1x$.

Q9   True or false:

     a. 2 is the numerical coefficient of $2x^3$.   _____

     b. x^3 is the literal coefficient of $2x^3$.   _____

     c. The coefficient of x^7 is 1.   _____

     d. The coefficient of $-x$ is -1.   _____

STOP • **STOP** • **STOP** • **STOP** • **STOP** • **STOP** • **STOP** • **STOP** • **STOP**

A9   **a.** true   **b.** true   **c.** true

        d. true: by the multiplication property of -1, $x = -1x$

Q10   Indicate (in order from left to right) the coefficient of each term. (*Note:* The constant term of a polynomial is considered to be a coefficient.)

     a. $x^3 - 3x^2 + x - 4$ _____

     b. $2x^3 - x^2 + 5x - 8$ _____

STOP • **STOP** • **STOP** • **STOP** • **STOP** • **STOP** • **STOP** • **STOP** • **STOP**

A10   **a.** $1, -3, 1, -4$ **b.** $2, -1, 5, -8$

---

10   It is common (and sometimes necessary) to rearrange the order of the terms of a polynomial. The commutative and associative properties of addition together with the definition of subtraction permit the following rule: *The terms of a polynomial can be rearranged in any order as long as the original sign of each term is left unchanged.*

     Examples:

$$-2 + x = x - 2$$
$$-3y - 5y^2 + 7 = 7 - 3y - 5y^2$$
$$5x^2 - 3 + 2x = 5x^2 + 2x - 3$$
$$15x + x^2 - 34 = x^2 + 15x - 34$$
$$4 - x = -x + 4$$

Q11   Complete the following correctly:

a. $5 - x = \underline{\hspace{1cm}} + 5$ b. $^{-}3 + 5x = 5x\underline{\hspace{1cm}}$

c. $2x^2 - 7 + 9x = 2x^2\underline{\hspace{1cm}} - 7$ d. $9 - 10y + y^2 = y^2 - 10y\underline{\hspace{1cm}}$

STOP • STOP • STOP • STOP • STOP • STOP • STOP • STOP • STOP

A11   a. ^{-}x b. -3 c. $+9x$ d. $+9$

---

11   When a polynomial in one variable is arranged so that the degree of each term decreases from left to right, the terms of the polynomial are said to be arranged in *descending order* with respect to the variable.

Examples:

Descending order

$2x^2 - 7 + 9x = 2x^2 + 9x - 7$
$5 - x \qquad = ^{-}x + 5$
$^{-}2 + 5y + y^2 = y^2 + 5y - 2$

Polynomials can be arranged so that the degree of each term increases from left to right. If so, the terms of the polynomial are said to be in *ascending order* with respect to the variable.

Q12   Arrange in descending order:

a. $3x^4 - x + 7x^3 - 3 + 2x^2 = $ _____

b. $9y^2 + 2 - 6y^3 + 5y = $ _____

c. $12 - 8r + r^2 = $ _____

d. $^{-}25 + m^2 = $ _____

STOP • STOP • STOP • STOP • STOP • STOP • STOP • STOP • STOP

A12   a. $3x^4 + 7x^3 + 2x^2 - x - 3$ b. $^{-}6y^3 + 9y^2 + 5y + 2$
        c. $r^2 - 8r + 12$ d. $m^2 - 25$

Q13   Arrange in ascending order:

a. $^{-}m^2 + 3m + 2 = $ _____

b. $15 + 9x^3 - 2x^2 + x = $ _____

c. $^{-}x^2 + 1 = $ _____

STOP • STOP • STOP • STOP • STOP • STOP • STOP • STOP • STOP

A13   a. $2 + 3m - m^2$ b. $15 + x - 2x^2 + 9x^3$ c. $1 - x^2$

This completes the instruction for this section.

11.1   Exercises

1. Read (write) each of the following:

a. x^4 b. y^3 c. m^2 d. z^8 e. ^{-}t f. x

2. **a.** What does x^2x^3 mean?
   b. Evaluate x^2x^3 for $x = ^-3$.
   c. How many factors of m are present in ^-3m?
   d. Evaluate $^-y^5$ for $y = ^-1$.
3. Indicate the terms of each polynomial:
   a. 23   **b.** $t^3 - 4t^2 - t + 8$ **c.** $x^2 + 7x + 6$
   d. $y^2 - 49$ **e.** $17n$ **f.** 0
4. Identify each polynomial in problem 3 as a monomial, binomial, or trinomial. Otherwise, write polynomial.
5. What is the degree of each polynomial in problem 3?
6. **a.** What is the literal coefficient of $^-4t^2$?
   b. What is the numerical coefficient of ^-t?
   c. In $x^2 - x - 6$, what is the coefficient of x^2? of x?
7. Arrange in descending order:
   a. $12x - 5x^2 + 4$ **b.** $7 + x$ **c.** $^-4x + 5 + 9x^2$ **d.** $^-6 - y + y^2$
8. In which polynomials of problem 7 are the terms arranged in ascending order?
9. Evaluate the polynomial for the given value of the variable:
   a. $x^2 - x - 6$ when $x = 0$ **b.** x^2y^5 when $x = 3$ and $y = 2$
   c. $y^2 - 49$ when $y = ^-7$ **d.** $t^3 - 4t^2 - t + 2$ when $t = 2$
   e. $^-m^3 + 5m^2 - m + 8$ when $m = ^-3$

11.1   Exercise Answers

1. Only one description from each part need be given.
   a. x to the fourth power, x to the fourth, the fourth power of x
   b. y cubed, y to the third power, y to the third, the cube of y, the third power of y
   c. m squared, m to the second power, m to the second, the square of m, the second power of m
   d. z to the eighth power, z to the eighth, the eighth power of z
   e. the opposite of t
   f. x, x to the first power, the first power of x, x to the first
2. **a.** $(x \cdot x)(x \cdot x \cdot x)$ **b.** $^-243$ **c.** one   **d.** 1
3. **a.** 23   **b.** $t^3, ^-4t^2, ^-t, 8$ **c.** $x^2, 7x, 6$ **d.** $y^2, ^-49$
   e. $17n$ **f.** 0
4. **a.** monomial   **b.** polynomial   **c.** trinomial   **d.** binomial
   e. monomial   **f.** monomial
5. **a.** 0   **b.** 3   **c.** 2   **d.** 2   **e.** 1   **f.** no degree
6. **a.** t^2 **b.** $^-1$ **c.** $1, ^-1$
7. **a.** $^-5x^2 + 12x + 4$ **b.** $x + 7$ **c.** $9x^2 - 4x + 5$
   d. $y^2 - y - 6$
8. b and d
9. **a.** $^-6$ **b.** 288   **c.** 0   **d.** $^-8$ **e.** 83

11.2   Addition and Subtraction

> **1**
>
> Terms of a polynomial are said to be *like terms* if they have the same literal coefficients (variables) as well as the same exponent on each variable: $2x$ and ^-3x are like terms; $2x$ and $^-3x^2$ are not like terms. Constant terms such as 5 and $^-8$ are considered to be like terms. The sum or difference of like terms may be combined or simplified.

Examples:

1. By the right distributive property of multiplication over addition,

   $$5x + 2x = (5 + 2)x = 7x$$

2. By the right distributive property of multiplication over subtraction and the multiplication property of $^-1$,

   $$3y - 4y = (3 - 4)y = {}^-1y = {}^-y$$

Q1   Combine like terms:
  a. $7x + 4x$ **b.** $12y - 5y$ **c.** $m - 3m$ **d.** $2x + 8x$

STOP • STOP • STOP • STOP • STOP • STOP • STOP • STOP • STOP

A1   **a.** $11x$ **b.** $7y$ **c.** ^-2m: $1m - 3m$ **d.** $10x$

Q2   Simplify (combine like terms):
  a. $9x - 8x$ **b.** $8x - 9x$

STOP • STOP • STOP • STOP • STOP • STOP • STOP • STOP • STOP

A2   **a.** x: $9x - 8x = (9 - 8)x = 1x = x$
      b. ^-x: $8x - 9x = (8 - 9)x = {}^-1x = {}^-x$

---

2   Recall that *equivalent* algebraic expressions are two expressions that have the same evaluation for all replacements of the variables. The fact that two expressions are equivalent can be verified by calling upon a property, definition, agreement, or number fact to justify each step of the simplification. That is, $8x - 9x$ is equivalent to ^-x, because

$8x - 9x$
$(8 - 9)x$ right distributive property of multiplication over subtraction
$\quad {}^-1x$ number fact
$\quad {}^-x$ multiplication property of $^-1$

At this point in your study of algebra, it should not be necessary to show each step, as in the above simplification. However, when in doubt as to whether two expressions are equivalent, you should be able to produce and justify each step in the simplification.

Q3   Simplify:

  a. $6y + 7y =$ _____   **b.** $7x + 9x =$ _____

  c. $x - 2x =$ _____   **d.** $10m^3 - 9m^3 =$ _____

  e. $12q^2 - 15q^2 =$ _____   **f.** $y + y =$ _____

  g. $5x - 5x =$ _____   **h.** $23z - 19z =$ _____

STOP • STOP • STOP • STOP • STOP • STOP • STOP • STOP • STOP

A3   **a.** $13y$ **b.** $16x$ **c.** ^-x
      d. m^3: $(10 - 9)m^3$ **e.** $^-3q^2$: $(12 - 15)q^2$ **f.** $2y$: $1y + 1y$
      g. 0: $(5 - 5)x = 0x = 0$ **h.** $4z$

3   When simplifying more complicated expressions, it is helpful (although not essential) to rearrange the terms so that like terms are together before combining like terms.

Example: Simplify $5x^3 - 4 + 4x^2 + 2x^3 + 6 - 7x^2$.

Solution

$5x^3 - 4 + 4x^2 + 2x^3 + 6 - 7x^2$
$5x^3 + 2x^3 + 4x^2 - 7x^2 - 4 + 6$
$7x^3 - 3x^2 + 2$

It is customary to arrange the terms of a polynomial in either ascending or descending order whenever possible. Although not essential, it is common to choose the order, if possible, that would make the first term of the polynomial positive.

Q4   Simplify:
   a. $4x - 7 + 2x + 3$ **b.** $5x^2 + 6 - 8x^2$

   c. $^-3t + 9s - 3s + 10$ **d.** $^-2x + 6x^2 + x^2 + 7x$

STOP • **STOP** • **STOP** • **STOP** • **STOP** • **STOP** • **STOP** • **STOP** • **STOP**

A4   **a.** $6x - 4$ **b.** $^-3x^2 + 6$ or $6 - 3x^2$
   c. $^-3t + 6s + 10$ or $6s - 3t + 10$ **d.** $7x^2 + 5x$ or $5x + 7x^2$
   (*Note:* Often many equivalent forms of the final result are possible. Recall that the terms of a polynomial can be rearranged in any order as long as the original sign of each term is left unchanged.)

4   The sum of two or more polynomials is found in a similar manner. For example,

$(x^2 + x - 7) + (2x^2 - 3x + 2)$
$x^2 + x - 7 + 2x^2 - 3x + 2$
$x^2 + 2x^2 + x - 3x - 7 + 2$
$3x^2 - 2x - 5$

The same calculations could be performed by arranging like terms in the same column.

Example: Simplify $(x^2 + x - 7) + (2x^2 - 3x + 2)$.

Solution

$\begin{array}{r} x^2 + x - 7 \\ 2x^2 - 3x + 2 \\ \hline 3x^2 - 2x - 5 \end{array}$

In either case, do as much work mentally as possible.

Q5   Find the sum by combining like terms (simplify):

a. $(4x^2 + 3x - 2) + (3x^2 + 4)$ b. $(7x^2 + 1 - 3x) + (x^2 - 4x)$

STOP • **STOP** • **STOP** • **STOP** • **STOP** • **STOP** • **STOP** • **STOP** • **STOP**

A5   a. $7x^2 + 3x + 2$ b. $8x^2 - 7x + 1$

Q6   Simplify:

a. $(4y^2 - 2y - 1) + (3y + 5)$ b. $(5x^2 - 6x + 9) + (^-4x^2 - 7x - 10)$

c. $(m^2 - 3) + (4 - 2m - 3m^2)$ d. $(t^2 + t) + (5t - 8) + (^-5t + 8)$

STOP • **STOP** • **STOP** • **STOP** • **STOP** • **STOP** • **STOP** • **STOP** • **STOP**

A6   a. $4y^2 + y + 4$ b. $x^2 - 13x - 1$

c. $^-2m^2 - 2m + 1$ or $1 - 2m - 2m^2$ d. $t^2 + t$

---

5   The addition property of opposites states that $x + {}^-x = 0$ for all rational-number replacements of x. It is also true that if $x + y = 0$, then $x = {}^-y$ and $y = {}^-x$. That is, if the sum of two values is zero, one value is the opposite of the other. For example,

$(2x - 5) + (^-2x + 5) = 0$

Hence,

$2x - 5 = {}^-(^-2x + 5)$ and $^-2x + 5 = {}^-(2x - 5)$

In general,

$(a + b) + (^-a - b) = 0$

Hence,

$^-(a + b) = {}^-a - b$

Therefore, to form the opposite of a polynomial, it is possible to simply change the sign of each term of the polynomial.

Q7   What is the opposite of:

a. $x + 5$?   _____   b. $x - 5$?   _____

c. $^-x + 5$?   _____   d. $^-x - 5$?   _____

STOP • **STOP** • **STOP** • **STOP** • **STOP** • **STOP** • **STOP** • **STOP** • **STOP**

A7   a. $^-x - 5$ b. $^-x + 5$ c. $x - 5$

d. $x + 5$

Q8   Find:

a. $^-(5x^2 - 2x + 4)$ b. $^-(^-3 + 7y)$ c. $^-(4m - 9)$

d. $^-(^-2 + 3t - t^2)$ e. $^-(^-6y^2 - 7y + 2)$ f. $^-(12 + 19z)$

STOP • **STOP** • **STOP** • **STOP** • **STOP** • **STOP** • **STOP** • **STOP** • **STOP**

A8   a. $^-5x^2 + 2x - 4$ b. $3 - 7y$
          c. $^-4m + 9$ or $9 - 4m$ d. $2 - 3t + t^2$
          e. $6y^2 + 7y - 2$ f. $^-12 - 19z$

---

6   The definition of subtraction states that $x - y = x + {}^-y$ where $x \in Q$ and $y \in Q$. That is, to subtract two values, add the opposite of the second (subtrahend) to the first (minuend). The difference between two polynomials can be found by applying this definition.

Example: Find $(2x + 5) - (3x - 7)$.

Solution

$(2x + 5) - (3x - 7)$
$2x + 5 + {}^-(3x - 7)$
$2x + 5 + {}^-3x + 7$
$^-x + 12$ or $12 - x$

Q9   Find each difference (simplify):
          a. $(7x^2 - 4) - (6x^2 + 2)$ b. $(y^3 - 3y) - (2y^2 - y)$

STOP • **STOP** • **STOP** • **STOP** • **STOP** • **STOP** • **STOP** • **STOP** • **STOP**

A9   a. $x^2 - 6$:   b. $y^3 - 2y^2 - 2y$:
             $(7x^2 - 4) - (6x^2 + 2)$ $(y^3 - 3y) - (2y^2 - y)$
             $7x^2 - 4 + {}^-(6x^2 + 2)$ $y^3 - 3y + {}^-(2y^2 - y)$
             $7x^2 - 4 + {}^-6x^2 - 2$ $y^3 - 3y + {}^-2y^2 + y$

Q10   Simplify:
          a. $(8x^2 - x + 3) - (^-2x^2 + 5x - 6)$ b. $(4 + x) - (3 - 2x + 5x^2)$

          c. $^-15q^3 + 11q^2 - 7q - 1 - (4q^2 + 8q - 5)$

STOP • **STOP** • **STOP** • **STOP** • **STOP** • **STOP** • **STOP** • **STOP** • **STOP**

A10   a. $10x^2 - 6x + 9$ b. $1 + 3x - 5x^2$
           c. $4 - 15q + 7q^2 - 15q^3$

---

7   Subtraction of polynomials can also be performed using columns.

Example: Find $5r^2 - 3r + 6 - (^-3r^2 - 3r + 4)$.

Solution

$$5r^2 - 3r + 6$$
$$\underline{3r^2 + 3r - 4} \qquad \text{(change signs of each term of}$$
$$8r^2 \qquad + 2 \qquad \text{the subtrahend)}$$

Hence, the difference is $8r^2 + 2$.

Q11   Simplify:

   a. $17x + 8x^2 - (12x^2 + 9x)$ **b.** $(10y^2 + 2y - 7) - (8y^2 - y - 2)$

   c. $6 + 4t - 4t^2 - (9t^2 + 4t - 6)$

STOP • STOP • STOP • STOP • STOP • STOP • STOP • STOP • STOP

A11   **a.** $8x - 4x^2$ **b.** $2y^2 + 3y - 5$ **c.** $12 - 13t^2$

This completes the instruction for this section.

11.2   Exercises

1. Simplify:
   a. $12x + 1 + 9x - 5$ **b.** $8y - 4 - 10y + 7$
   c. $16t - 8t - 4$ **d.** $12m^3 + 8m^2 - 6m^3 - 4m^2$
2. Simplify:
   a. $(^-5x^3 - 7x^2 + 8x - 7) + (^-5x^2 - 12x + 9)$
   b. $(6y - 13y^2 + 8) + (^-7y + 10 + 14y^2)$
   c. $^-(3x - 5)$ **d.** $^-(^-4x^2 - 3x + 9)$
   e. $(2x + 13) - (x + 15)$ **f.** $(2x^2 + 4x - 5) - (x - 4)$
   g. $(^-y^2 - 2) - (3y + 3)$ **h.** $15t^2 + 10t - (15t^2 + 12t)$
   i. $(2x + 7) - (x + 2) + (3x - 5)$
   j. $(5x^2 - 5) + (2x - 3) - (3x^2 - x)$

11.2   Exercise Answers

1. **a.** $21x - 4$ **b.** $3 - 2y$ **c.** $8t - 4$ **d.** $6m^3 + 4m^2$
2. **a.** $2 - 4x - 12x^2 - 5x^3$ **b.** $y^2 - y + 18$ **c.** $5 - 3x$
   d. $4x^2 + 3x - 9$ **e.** $x - 2$ **f.** $2x^2 + 3x - 1$
   g. $^-y^2 - 3y - 5$ **h.** ^-2t **i.** $4x$
   j. $2x^2 + 3x - 8$

11.3   **Multiplication**

---

1   Multiplication of monomial factors is the simplest product involving polynomials. In each of the following examples, the indicated product has been written in a simpler equivalent form by applying a property, agreement, or number fact.

1. $5 \cdot x = 5x$ agreement that the multiplication is understood between the numerical coefficient and the variable

2. $2(3x)$
    $(2 \cdot 3)x$ associative property of multiplication
    $6x$ number fact

3. $(^-2y)5$
    $5(^-2y)$ commutative property of multiplication
    $(5 \cdot {}^-2)y$ associative property of multiplication
    ^-10y number fact

Q1   Write each product in simpler form:

a. $3(5y)$ b. $^-7(2x)$

c. $(9r) \cdot {}^-3$ d. $(^-4m) \cdot {}^-2$

STOP • **STOP** • **STOP** • **STOP** • **STOP** • **STOP** • **STOP** • **STOP** • **STOP**

A1   a. $15y$ b. ^-14x c. ^-27r d. $8m$

---

2   Consider the product $x^2 \cdot x^3$. Since $x^2 = x \cdot x$ and $x^3 = x \cdot x \cdot x$, the product $x^2 \cdot x^3 = (x \cdot x)(x \cdot x \cdot x) = x^5$. Since the product, x^5, contains five factors of x, this result could have been obtained by adding the original exponents. In general,

$$x^m \cdot x^n = x^{m+n}$$

where x is a rational number and m and n are natural numbers. The above fact is called a *property of exponents*.

Q2   Find the product:

a. $x^3 \cdot x^4 = $ _____   b. $x \cdot x^5 = $ _____

c. $x \cdot x \cdot x^6 = $ _____   d. $y^4 \cdot y^7 = $ _____

STOP • **STOP** • **STOP** • **STOP** • **STOP** • **STOP** • **STOP** • **STOP** • **STOP**

A2   a. x^7: x^{3+4} b. x^6: x^{1+5} c. x^8: x^{1+1+6}
d. y^{11}: y^{4+7}

Q3   a. The product of two or more monomials with the same base can be written in simpler form as a power of that variable with an exponent that is the _____ of the exponents of the original monomials.

b. $x^p \cdot x^q = $ _____

STOP • **STOP** • **STOP** • **STOP** • **STOP** • **STOP** • **STOP** • **STOP** • **STOP**

A3   **a.** sum   **b.** x^{p+q}

---

3   When multiplying monomials, proceed as follows:

$(2x)(3x^2)$

$(2 \cdot 3)(x \cdot x^2)$ associative and commutative properties of multiplication

$6x^3$ number fact and property of exponents

$(-5x^4)(4x^5)$ $-x^2 \cdot 3x^3$

$(-5 \cdot 4)(x^4 \cdot x^5)$ $-1x^2 \cdot 3x^3$

$-20x^9$ $-1 \cdot 3 \cdot x^2 \cdot x^3$

$-3x^5$

Q4   Find the product:

a. $(5x^2)(-x^3)$ **b.** $(-3x^4)(-6x^3)$ **c.** $(-2y^4)(3y)$

d. $-7m^3 \cdot 2m \cdot 3m^2$ **e.** $x \cdot x^3 \cdot 4x^5$ **f.** $(2x^3)(-3x)(4x^3)$

STOP • **STOP** • **STOP** • **STOP** • **STOP** • **STOP** • **STOP** • **STOP** • **STOP**

A4   **a.** $-5x^5$ **b.** $18x^7$ **c.** $-6y^5$ **d.** $-42m^6$

e. $4x^9$ **f.** $-24x^7$

---

4   When "raising a monomial to a power," * two additional properties of exponents can be generalized. The first will be discussed here. The second will be discussed in Frame 5. Consider the product (power) $(2x)^3$.

$(2x)^3 = 2x \cdot 2x \cdot 2x$
$\qquad = 2 \cdot 2 \cdot 2 \cdot x \cdot x \cdot x$

Both 2 and x appear as a *factor* three times. Hence, $2 \cdot 2 \cdot 2 \cdot x \cdot x \cdot x$ could be rewritten 2^3x^3. Thus,

$(2x)^3 = 2^3x^3 = 8x^3$

In general,

$(xy)^m = x^m y^m$ where $x, y \in Q$ and $m \in N$

*"Raising to a power" is another way of speaking about powers. That is, x^5 is "x raised to the fifth power."

Q5   Write each power in simpler form:

a. $(3x)^2$ **b.** $(-4y)^3$ **c.** $(3y)^3$ **d.** $(-x)^5$

    e. $(^-5m)^2$ **f.** $(^-3t)^3$ **g.** $^-(^-2y)^3$ **h.** $^-(^-4x)^2$

    i. $^-(^-2x)^5$ **j.** $^-(3t)^2$

STOP • STOP • STOP • STOP • STOP • STOP • STOP • STOP • STOP

A5   **a.** $9x^2$: 3^2x^2 **b.** $^-64y^3$: $(^-4)^3y^3$
            c. $27y^3$: 3^3y^3 **d.** $^-x^5$: $(^-1x)^5 = (^-1)^5x^5 = {}^-1x^5$
            e. $25m^2$: $(^-5)^2m^2$ **f.** $^-27t^3$: $(^-3)^3t^3$
            g. $^-8y^3$: $^-(2^3y^3) = {}^-(8y^3)$ **h.** $^-16x^2$: $^-[(^-4)^2x^2] = {}^-(16x^2)$
            i. $32x^5$: $^-[(^-2)^5x^5]$ **j.** $^-9t^2$: $^-(3^2t^2)$

Q6   **a.** When raising a monomial to a power, each _____ of the monomial can be raised to that power.

         b. $(mn)^p = $ _____

STOP • STOP • STOP • STOP • STOP • STOP • STOP • STOP • STOP

A6   **a.** factor   **b.** m^pn^p

---

5   A second property of exponents that can be applied when raising a monomial to a power involves examples in which the monomial is itself a power. Such an example is $(x^2)^3$. $(x^2)^3$ means $x^2 \cdot x^2 \cdot x^2$, which can be further simplified to x^6. You should note that $x^2 \cdot x^2 \cdot x^2 = x^{2+2+2}$. Since $2 + 2 + 2 = 3 \cdot 2$, x^{2+2+2} can be written $x^{3 \cdot 2}$ or $x^{2 \cdot 3}$. Therefore, $(x^2)^3 = x^{2 \cdot 3} = x^6$. In general,

$$(x^m)^n = x^{mn} \qquad \text{where } x \in Q \text{ and } m,n \in N$$

Examples:

$$(x^3)^4 = x^{3 \cdot 4} = x^{12} \qquad (y^2)^5 = y^{2 \cdot 5} = y^{10}$$

Q7   Simplify:

         a. $(x^5)^3 = $ _____   **b.** $(t^4)^2 = $ _____

         c. $(y^4)^5 = $ _____   **d.** $(m^7)^4 = $ _____

STOP • STOP • STOP • STOP • STOP • STOP • STOP • STOP • STOP

A7   **a.** x^{15} **b.** t^8 **c.** y^{20} **d.** m^{28}

---

6   The properties of exponents presented thus far are summarized. For $x \in Q$, $m,n \in N$:

         1. $x^mx^n = x^{m+n}$
         2. $(xy)^m = x^my^m$
         3. $(x^m)^n = x^{mn}$

     The application of more than one property is often required when simplifying products involving exponents.

Examples:

1. $(2x^3)^2 = 2x^3 \cdot 2x^3$ or $(2x^3)^2 = 2^2(x^3)^2$
$$= 2 \cdot 2x^3 \cdot x^3 \qquad\qquad = 4x^6$$
$$= 4x^6$$

2. $(5x^2)^4(-2x)^3 = 5^4(x^2)^4(-2)^3x^3$
$$= 625x^8(-8)x^3$$
$$= 625(-8)x^8x^3$$
$$= {}^-5{,}000x^{11}$$

Regardless of the order of the steps performed, do as much mentally as possible. Example 2 could easily be shortened to

$(5x^2)^4(-2x)^3 = 625x^8 \cdot {}^-8x^3$
$$= {}^-5{,}000x^{11}$$

Q8   Simplify:

a. $(3x^2)^3$ **b.** $(^-5y^3)^3$ **c.** $(^-x^2)^5$ **d.** $(3m^3)^4$

e. $^-2(2y^2)^5$ **f.** $(7x^5)^2(2x^4)^2$ **g.** $^-(x^5)^4$ **h.** $(^-3r^3)^3(^-r^2)^2$

STOP • **STOP** • **STOP** • **STOP** • **STOP** • **STOP** • **STOP** • **STOP** • **STOP**

A8
a. $27x^6$ **b.** $^-125y^9$
c. $^-x^{10}$: $(^-x^2)^5 = (^-1x^2)^5$ **d.** $81m^{12}$
$$= (^-1)^5(x^2)^5$$
$$= ^-1x^{10}$$
$$= ^-x^{10}$$
e. $^-64y^{10}$: $^-2(32y^{10})$ **f.** $196x^{18}$: $49x^{10} \cdot 4x^8$
g. $^-x^{20}$ **h.** $^-27r^{13}$: $^-27r^9 \cdot r^4$

---

7   The simplification of the product involving a monomial and a polynomial requires the application of the distributive properties.

Example 1: Simplify $2(x + 5)$.

Solution

By the left distributive property of multiplication over addition,

$2(x + 5) = 2x + 2 \cdot 5$
$$= 2x + 10$$

Example 2: Simplify $(x - 3)5x$.

Solution

By the right distributive property of multiplication over subtraction and properties of exponents,

$$(x - 3)5x = x \cdot 5x - 3 \cdot 5x$$
$$= 5x^2 - 15x$$

Example 3: Simplify $^-x(2x + 5)$.

Solution

$$^-x(2x + 5) = {}^-1x(2x + 5)$$
$$= {}^-1x \cdot 2x + {}^-1x \cdot 5$$
$$= {}^-2x^2 + {}^-5x$$
$$= {}^-2x^2 - 5x$$

Example 4: Simplify $^-3x(x^2 - 2)$.

Solution

$$^-3x(x^2 - 2) = {}^-3x \cdot x^2 - {}^-3x \cdot 2$$
$$= {}^-3x^3 - {}^-6x$$
$$= {}^-3x^3 + 6x \text{ or } 6x - 3x^3$$

Q9   Simplify each product (do as much mentally as possible):
   a. $^-3(x - 4)$ **b.** $(x + 4)7x$ **c.** $(2x + 5)3x^2$ **d.** $3x(x - 2)$

   e. $^-x(x - 3)$ **f.** $(2x + 1) \cdot {}^-5$

STOP • **STOP** • **STOP** • **STOP** • **STOP** • **STOP** • **STOP** • **STOP** • **STOP**

A9   **a.** $^-3x + 12$ or $12 - 3x$: $^-3(x - 4) = {}^-3x - {}^-3 \cdot 4$
$$= {}^-3x - {}^-12$$
$$= {}^-3x + 12$$
   b. $7x^2 + 28x$: $(x + 4)7x = x \cdot 7x + 4 \cdot 7x$
   c. $6x^3 + 15x^2$: $(2x + 5)3x^2 = 2x \cdot 3x^2 + 5 \cdot 3x^2$
$$= 6x^3 + 15x^2$$
   d. $3x^2 - 6x$: $3x(x - 2) = 3x \cdot x - 3x \cdot 2$
$$= 3x^2 - 6x$$
   e. $^-x^2 + 3x$ or $3x - x^2$: $^-x(x - 3) = {}^-1x(x - 3)$
$$= {}^-1x \cdot x - {}^-1x \cdot 3$$
$$= {}^-1x^2 - {}^-3x$$
$$= {}^-x^2 + 3x$$
   f. $^-10x - 5$: $(2x + 1) \cdot {}^-5 = 2x \cdot {}^-5 + 1 \cdot {}^-5$
$$= {}^-10x + {}^-5$$
$$= {}^-10x - 5$$

8   The distributive properties can be extended to cover cases in which the polynomial has more than two terms. For example,

$$5(x^2 - 3x + 4) = 5x^2 - 5 \cdot 3x + 5 \cdot 4$$
$$= 5x^2 - 15x + 20$$

Q10   Simplify:
             a. $^-3(x^2 - 7x + 6)$ **b.** $2x(3x^2 + 4x - 5)$

STOP • STOP • STOP • STOP • STOP • STOP • STOP • STOP • STOP

A10   **a.** $^-3x^2 + 21x - 18$ **b.** $6x^3 + 8x^2 - 10x$

9   Now consider the product of two binomials such as $(2x + 3)(x - 4)$. Let $x - 4$ be replaced by n. $(2x + 3)(x - 4)$ would then become $(2x + 3)n$. By the right distributive property of multiplication over addition,

$$(2x + 3)n = 2x \cdot n + 3 \cdot n$$

Now, replace n by $x - 4$, producing

$$2x(x - 4) + 3(x - 4)$$

You should see that the quantity $x - 4$ has been multiplied times both terms of the quantity $2x + 3$. The simplification can now proceed:

$$2x(x - 4) + 3(x - 4)$$
$$2x^2 - 8x + 3x - 12$$
$$2x^2 - 5x - 12$$

It is not necessary to make the substitution suggested above. The quantity $x - 4$ can be multiplied times both terms of the other quantity immediately. For example,

$$(2x + 3)(\underline{x - 4})$$
$$2x(\underline{x - 4}) + 3(\underline{x - 4})$$
$$2x^2 - 8x + 3x - 12$$
$$2x^2 - 5x - 12$$

Q11   By the right distributive property of multiplication over addition, $(2x + 5)(x - 6)$ can be written

$$2x(\qquad) + 5(\qquad)$$

STOP • STOP • STOP • STOP • STOP • STOP • STOP • STOP • STOP

A11 $2x(x - 6) + 5(x - 6)$

Q12   Complete the simplification:
             $(2x + 5)(x - 6)$
             $2x(x - 6) + 5(x - 6)$

STOP • STOP • STOP • STOP • STOP • STOP • STOP • STOP • STOP

A12 $2x^2 - 12x + 5x - 30$
             $2x^2 - 7x - 30$

Q13   By the right distributive property of multiplication over subtraction, $(3x - 1)(2x + 5)$ can
be written
$3x($ $) - 1($ $).$

STOP • **STOP** • **STOP** • **STOP** • **STOP** • **STOP** • **STOP** • **STOP** • **STOP**

A13 $3x(2x + 5) - 1(2x + 5)$

Q14   Complete the simplification:
$(3x - 1)(2x + 5)$
$3x(2x + 5) - 1(2x + 5)$

STOP • **STOP** • **STOP** • **STOP** • **STOP** • **STOP** • **STOP** • **STOP** • **STOP**

A14 $6x^2 + 15x - 2x - 5$
$6x^2 + 13x - 5$

Q15   Use the procedure illustrated above to determine the following products:
a. $(x + 5)(x - 2)$ b. $(x - 3)(x + 7)$

c. $(x - 2)(x - 3)$ d. $(2x + 1)(x + 4)$

e. $(x - 5)(2x + 3)$ f. $(4x + 3)(x - 5)$

g. $(3x - 2)(2x - 5)$ h. $(x + 7)(x + 7)$

i. $(x + 4)(x - 4)$ j. $(5x + 1)(5x - 1)$

STOP • **STOP** • **STOP** • **STOP** • **STOP** • **STOP** • **STOP** • **STOP** • **STOP**

A15   a. $x^2 + 3x - 10$: $x(x - 2) + 5(x - 2)$ b. $x^2 + 4x - 21$: $x(x + 7) - 3(x + 7)$
                              $x^2 - 2x + 5x - 10$ $x^2 + 7x - 3x - 21$
       c. $x^2 - 5x + 6$: $x(x - 3) - 2(x - 3)$ d. $2x^2 + 9x + 4$: $2x(x + 4) + 1(x + 4)$
                              $x^2 - 3x - 2x + 6$ $2x^2 + 8x + 1x + 4$
       e. $2x^2 - 7x - 15$ f. $4x^2 - 17x - 15$ g. $6x^2 - 19x + 10$
       h. $x^2 + 14x + 49$ i. $x^2 - 16$: $x(x - 4) + 4(x - 4)$ j. $25x^2 - 1$
                                             $x^2 - 4x + 4x - 16$

---

10   Consider the following product between any two binomials.

$(a + b)(c + d)$
$a(c + d) + b(c + d)$
$ac + ad + bc + bd$

This simplification illustrates a method (called the *foil* method) by which you can multiply two binomials mentally:

1. ac is the product of the *f*irst terms of each binomial.
2. ad is the product of the *o*uter terms of each binomial.
3. bc is the product of the *i*nner terms of each binomial.
4. bd is the product of the *l*ast terms of each binomial.
5. The final result, $ac + ad + bc + bd$, is the *sum* of these individual products.

Q16   Given $(x + 2)(x - 7)$, indicate the product of:

    a. the first terms   _____   **b.** the outer terms   _____

    c. the inner terms   _____   **d.** the last terms   _____

    e. Write the indicated sum of the four individual terms. _____

    f. Write the polynomial of part e in simpler form.

    _____

STOP • **STOP** • **STOP** • **STOP** • **STOP** • **STOP** • **STOP** • **STOP** • **STOP**

A16   **a.** x^2 **b.** ^-7x **c.** $2x$ **d.** $^-14$
      e. $x^2 - 7x + 2x - 14$ **f.** $x^2 - 5x - 14$

---

11   When multiplying two binomials, the product of the outer terms and the product of the inner terms are often like terms. If so, their sum can easily be obtained mentally.

Example:

$(x + 3)(x + 7) = x^2 + 10x + 21$
               (1)   (2)   (3)
(1) product of the first terms
(2) sum of the products of the outer and inner terms $(7x + 3x)$
(3) product of the last terms

Q17   Using the technique of Frame 11, practice obtaining the following products mentally:

    a. $(x + 4)(x + 5) =$ _____   **b.** $(x - 2)(x - 3) =$ _____

    c. $(x + 7)(x - 4) =$ _____   **d.** $(x - 3)(x + 5) =$ _____

    e. $(2x + 5)(x - 2) =$ _____   **f.** $(3x - 2)(2x - 3) =$ _____

    g. $(x + 5)(x - 5) =$ _____   **h.** $(2x + 3)(2x + 3) =$ _____

STOP • **STOP** • **STOP** • **STOP** • **STOP** • **STOP** • **STOP** • **STOP** • **STOP**

A17   **a.** $x^2 + 9x + 20$ **b.** $x^2 - 5x + 6$ **c.** $x^2 + 3x - 28$ **d.** $x^2 + 2x - 15$
      e. $2x^2 + x - 10$ **f.** $6x^2 - 13x + 6$ **g.** $x^2 - 25$ **h.** $4x^2 + 12x + 9$

---

12   The product of a binomial and a trinomial or any two polynomials can be obtained using the distributive properties.

Examples:

$$(2x - 3)(x^2 + 2x - 4) = 2x(x^2 + 2x - 4) - 3(x^2 + 2x - 4)$$
$$= 2x^3 + 4x^2 - 8x - 3x^2 - 6x + 12$$
$$= 2x^3 + x^2 - 14x + 12$$

$(x^2 - 2x + 3)(2x^2 + 3x - 5)$
$x^2(2x^2 + 3x - 5) - 2x(2x^2 + 3x - 5) + 3(2x^2 + 3x - 5)$
$2x^4 + 3x^3 - 5x^2 - 4x^3 - 6x^2 + 10x + 6x^2 + 9x - 15$
$2x^4 - x^3 - 5x^2 + 19x - 15$

Q18   Simplify:

   a. $(x - 3)(x^2 - 2x + 3)$

   b. $(2x^2 - x - 3)(x^2 + 2x - 4)$

STOP • STOP • STOP • STOP • STOP • STOP • STOP • STOP • STOP

A18   **a.** $x^3 - 5x^2 + 9x - 9$: $(x - 3)(x^2 - 2x + 3)$
$$x(x^2 - 2x + 3) - 3(x^2 - 2x + 3)$$
$$x^3 - 2x^2 + 3x - 3x^2 + 6x - 9$$

   b. $2x^4 + 3x^3 - 13x^2 - 2x + 12$:
$(2x^2 - x - 3)(x^2 + 2x - 4)$
$2x^2(x^2 + 2x - 4) - x(x^2 + 2x - 4) - 3(x^2 + 2x - 4)$
$2x^4 + 4x^3 - 8x^2 - x^3 - 2x^2 + 4x - 3x^2 - 6x + 12$
$2x^4 + 4x^3 - x^3 - 8x^2 - 2x^2 - 3x^2 + 4x - 6x + 12$

13   When multiplying a monomial times two polynomials, multiply any two factors and then multiply this result times the third factor.

Example: Simplify $2x(x + 1)(2x - 3)$.

Solution

Use the "foil" method first on the product of the two binomials.

$$2x(x + 1)(2x - 3) = 2x(2x^2 - x - 3)$$
$$= 4x^3 - 2x^2 - 6x$$

It would be possible to multiply the monomial factor times either of the binomial factors and this result times the other binomial factor.

Q19   Simplify $^-3x(x - 1)(x - 2)$.

STOP • STOP • STOP • STOP • STOP • STOP • STOP • STOP • STOP

A19 $^-3x^3 + 9x^2 - 6x$: $^-3x(x^2 - 3x + 2)$

---

14   Study the following example carefully.
    Simplify $5x(x + 2)(x^2 + x + 1)$.

$5x(x + 2)(x^2 + x + 1)$
$5x[x(x^2 + x + 1) + 2(x^2 + x + 1)]$
$5x(x^3 + x^2 + x + 2x^2 + 2x + 2)$
$5x(x^3 + 3x^2 + 3x + 2)$
$5x^4 + 15x^3 + 15x^2 + 10x$

Q20   Simplify $^-3x(x + 2)(x^2 - 2x + 4)$.

STOP • STOP • STOP • STOP • STOP • STOP • STOP • STOP • STOP

A20 $^-3x^4 - 24x$: $^-3x[x(x^2 - 2x + 4) + 2(x^2 - 2x + 4)]$
    $^-3x(x^3 - 2x^2 + 4x + 2x^2 - 4x + 8)$
    $^-3x(x^3 + 8)$

---

15   The "foil" method can always be used when finding the product of two binomials. However, in special cases, other shortcuts will help. Consider the *square of a binomial*. By the "foil" method,

$$(a + b)^2 = (a + b)(a + b)$$
$$= a^2 + ab + ab + b^2$$
$$= a^2 + 2ab + b^2$$

Hence, $(a + b)^2 = a^2 + 2ab + b^2$, where a is the *first* term and b is the *second* term.
    $(a + b)^2 = a^2 + 2ab + b^2$ indicates that the square of any binomial is the square of the first term plus twice the product of the two terms plus the square of the second term.

Examples:

$$(x + 3)^2 = (x)^2 + 2(x)(3) + (3)^2$$
$$= x^2 + 6x + 9$$

$$(2x + 5)^2 = (2x)^2 + 2(2x)(5) + (5)^2$$
$$= 4x^2 + 20x + 25$$

$$(y - 6)^2 = (y)^2 + 2(y)(^-6) + (^-6)^2$$
$$= y^2 - 12y + 36$$

With practice the second step can be done mentally. Trinomials obtained from the square of a binomial are called *perfect square trinomials*.

Q21   Complete the following:

a. $(x + 1)^2 = (\quad)^2 + 2(\quad)(\quad) + (\quad)^2$

   $= \underline{\hspace{4cm}}$

b. $(x - 2)^2 = (\quad)^2 + 2(\quad)(\quad) + (\quad)^2$

   $= \underline{\hspace{4cm}}$

STOP • **STOP** • **STOP** • **STOP** • **STOP** • **STOP** • **STOP** • **STOP** • **STOP**

A21   a. $(x)^2 + 2(x)(1) + (1)^2$ b. $(x)^2 + 2(x)(^-2) + (^-2)^2$
        $x^2 + 2x + 1$ $x^2 - 4x + 4$

Q22   Use the technique of Frame 15 to simplify each product:

a. $(x + 7)^2$ b. $(x - 3)^2$

c. $(2x - 5)^2$ d. $(2x + 3)^2$

STOP • **STOP** • **STOP** • **STOP** • **STOP** • **STOP** • **STOP** • **STOP** • **STOP**

A22   a. $x^2 + 14x + 49$ b. $x^2 - 6x + 9$ c. $4x^2 - 20x + 25$ d. $4x^2 + 12x + 9$

Q23   Practice writing only the product:

a. $(x + 5)^2 =$ $\underline{\hspace{3cm}}$

b. $(y - 5)^2 =$ $\underline{\hspace{3cm}}$

c. $(3m - 2)^2 =$ $\underline{\hspace{3cm}}$

d. $(8 + t)^2 =$ $\underline{\hspace{3cm}}$

e. $(5r + 2)^2 =$ $\underline{\hspace{3cm}}$

f. $(x - 9)^2 =$ $\underline{\hspace{3cm}}$

STOP • **STOP** • **STOP** • **STOP** • **STOP** • **STOP** • **STOP** • **STOP** • **STOP**

A23   a. $x^2 + 10x + 25$ b. $y^2 - 10y + 25$ c. $9m^2 - 12m + 4$
        d. $64 + 16t + t^2$ e. $25r^2 + 20r + 4$ f. $x^2 - 18x + 81$

---

16   Another special case involving the product of two binomials is *the product of the sum and the difference of the same two values*. Such a product is $(a + b)(a - b)$. By the "foil" method,

$$(a + b)(a - b) = a^2 - ab + ab - b^2$$
$$= a^2 - b^2$$

The result, $a^2 - b^2$, is the *difference of two squares*. Observe that in this case the sum of the inner and outer terms will always be zero.

   $(a + b)(a - b) = a^2 - b^2$ indicates that the product of the sum and the difference of the same two values is the square of the first term minus the square of the second term.

Examples:

$$(x + 7)(x - 7) = (x)^2 - (7)^2 = x^2 - 49$$

$$(2x - 1)(2x + 1) = (2x)^2 - (1)^2 = 4x^2 - 1$$

Such products can easily be done mentally.

Q24 Simplify the following products mentally:

a. $(x + 10)(x - 10) =$ _____ **b.** $(x - 6)(x + 6) =$ _____

c. $(3y - 2)(3y + 2) =$ _____ **d.** $(5m + 7)(5m - 7) =$ _____

STOP • STOP • STOP • STOP • STOP • STOP • STOP • STOP • STOP

A24 **a.** $x^2 - 100$ **b.** $x^2 - 36$ **c.** $9y^2 - 4$ **d.** $25m^2 - 49$

This completes the instruction for this section.

11.3 Exercises

1. Find the following products:
   a. $(5x^2)(-3x^4)$ **b.** $(-2x^3)^4$ **c.** $(5y^3)^2(-2y^3)^3$
   d. $-(-4t)^2(3t^4)^2$ **e.** $-4(x^2 - 5x + 2)$ **f.** $5x(x - 3)$
   g. $-m(m^2 + m - 1)$ **h.** $-3y(y^2 - y + 2)$ **i.** $6x^2(2x^2 + 5x - 2)$
   j. $(x^2 - 9)5x$
2. Write the product:
   a. $(x - 7)(x + 5)$ **b.** $(y + 9)(y - 2)$ **c.** $(2x - 1)(3x - 2)$
   d. $(5x + 3)(2x - 7)$ **e.** $(m + 3)(4m + 5)$ **f.** $(2x + 3)(2x - 3)$
   g. $(x + 11)^2$ **h.** $(2x - 3)^2$ **i.** $(5 - 4t)^2$
   j. $(12 + y)(12 - y)$
3. Simplify:
   a. $(2x + 3)(x^2 - 3x - 1)$ **b.** $(y^2 - y + 2)(y - 4)$
   c. $(x^2 - 2x + 5)(2x - 1)$ **d.** $(m^2 + 2m - 1)(m^2 - m - 3)$
   e. $2(x + 2)(x - 4)$ **f.** $-3(y - 2)(2y + 1)$
   g. $x(2x + 1)(x - 1)$ **h.** $-2y(1 + 2y)^2$
   i. $-x(x - 3)(x + 3)$ **j.** $-(x - 1)(x^2 + x + 1)$

11.3 Exercise Answers

1. **a.** $-15x^6$ **b.** $16x^{12}$ **c.** $-200y^{15}$
   d. $-144t^{10}$ **e.** $-4x^2 + 20x - 8$ **f.** $5x^2 - 15x$
   g. $m - m^2 - m^3$ **h.** $-3y^3 + 3y^2 - 6y$
   i. $12x^4 + 30x^3 - 12x^2$ **j.** $5x^3 - 45x$
2. **a.** $x^2 - 2x - 35$ **b.** $y^2 + 7y - 18$ **c.** $6x^2 - 7x + 2$
   d. $10x^2 - 29x - 21$ **e.** $4m^2 + 17m + 15$ **f.** $4x^2 - 9$
   g. $x^2 + 22x + 121$ **h.** $4x^2 - 12x + 9$ **i.** $25 - 40t + 16t^2$
   j. $144 - y^2$
3. **a.** $2x^3 - 3x^2 - 11x - 3$ **b.** $y^3 - 5y^2 + 6y - 8$
   c. $2x^3 - 5x^2 + 12x - 5$ **d.** $m^4 + m^3 - 6m^2 - 5m + 3$
   e. $2x^2 - 4x - 16$ **f.** $6 + 9y - 6y^2$ **g.** $2x^3 - x^2 - x$
   h. $-2y - 8y^2 - 8y^3$ **i.** $9x - x^3$ **j.** $1 - x^3$

11.4   Factoring

---

1   Using one of the distributive properties to simplify a product is called "expanding" or "removing the parentheses."

Examples:

$2(x - 3) = 2x - 6$
$5x(x + 6) = 5x^2 + 30x$

$(2x - 1)x = 2x^2 - x$

$$(x + 3)(x - 2) = x(x - 2) + 3(x - 2)$$
$$= x^2 - 2x + 3x - 6$$
$$= x^2 + x - 6$$

In each example, the expression on the left of the equal sign is a *product*. The result on the right of the equal sign is a *sum*. Used in this manner, the distributive properties change products to sums.

Q1   **a.** $^-5(x - 7)$ is a _____ and $^-5x + 35$ is a _____ .
                              product/sum   product/sum

b. Expand $2x(x - 5)$._____

c. Simplify $x(x - 2) - 3(x - 2)$.

d. Remove the parentheses from $(x + 7)(x - 3)$.

STOP   •   **STOP**   •   **STOP**   •   **STOP**   •   **STOP**   •   **STOP**   •   **STOP**   •   **STOP**   •   **STOP**

A1   **a.** product, sum   **b.** $2x^2 - 10x$
c. $x^2 - 5x + 6$: $x(x - 2) - 3(x - 2)$
                        $x^2 - 2x - 3x + 6$
d. $x^2 + 4x - 21$: $(x + 7)(x - 3) = x(x - 3) + 7(x - 3)$
                                    $= x^2 - 3x + 7x - 21$
                                    $= x^2 + 4x - 21$

---

2   The distributive properties also change sums to products. For example,

1. $5x - 10 = 5(x - 2)$

The expression $5x - 10$ is a sum, because by the definition of subtraction $5x - 10 = 5x + {}^-10$.

2. $x^2 + x = x^2 + 1x$
          $= x(x + 1)$

3. $x^3 - 3x^2 + 5x = x(x^2 - 3x + 5)$

When the distributive properties are used to change sums to products, the process is called *factoring*. If each term of a polynomial contains the same monomial factor, the polynomial can be written as a product of the common factor times another polynomial. The common factor is said to be "factored from" the polynomial. For example,

$$3x + 12 = 3x + 3 \cdot 4$$
$$= 3(x + 4)$$

In this example, the common factor 3 is factored from $3x + 12$. It is wise to check the factoring by expanding: $3(x + 4) = 3x + 12$.

Q2 **a.** In $2x + 2 \cdot 5$, what is the common factor?_____
    b. Complete the factoring: $2x + 2 \cdot 5 = 2($ $)$

STOP • STOP • STOP • STOP • STOP • STOP • STOP • STOP • STOP

A2 **a.** 2   **b.** $2(x + 5)$

Q3 **a.** In $x^2 + 5x$, what is the common factor?_____
    b. Complete the factoring: $x^2 + 5x = x($ $)$

STOP • STOP • STOP • STOP • STOP • STOP • STOP • STOP • STOP

A3 **a.** x **b.** $x(x + 5)$

---

3 When factoring a common monomial from a polynomial, always consider the *greatest common factor*. The polynomial $4x^2 + 8x$ has common factors of 2, x, $2x$, 4, and $4x$. $4x$ would be the greatest common factor.* Hence,

$$4x^2 + 8x = 4x \cdot x + 4x \cdot 2$$
$$= 4x(x + 2)$$

The second step may be done mentally. The above result "checks," because $4x(x + 2) = 4x^2 + 8x$.

*Any rational number is a common factor of any polynomial. That is, $4x - 5 = \frac{1}{2}(8x - 10) = \frac{-2}{3}(^-6x + 15)$, etc. However, when the phrase "greatest common factor" is used, only positive integer values and variables are considered.

Q4 What is the greatest common factor present in each term of $6y^2 - 3y$? _____

STOP • STOP • STOP • STOP • STOP • STOP • STOP • STOP • STOP

A4 $3y$

Q5 Complete the factorization.
    $6y^2 - 3y = 3y($ $)$

STOP • STOP • STOP • STOP • STOP • STOP • STOP • STOP • STOP

A5 $3y(2y - 1)$: $6y^2 - 3y = 3y \cdot 2y - 3y \cdot 1$

Q6 **a.** What is the greatest common factor present in each term of $12x^2 + 18x$? _____
    b. Factor the greatest common factor from $12x^2 + 18x$.

STOP • STOP • STOP • STOP • STOP • STOP • STOP • STOP • STOP

A6   **a.** $6x$ **b.** $6x(2x + 3)$: $12x^2 + 18x = 6x \cdot 2x + 6x \cdot 3$

Q7   Factor the greatest common factor from each polynomial (check your results):

a. $3x + 6$ **b.** $9x^2 - 12x$

c. $12y - 4$ **d.** $16m + 4m^2$

e. $5y^2 - 10y + 20$ **f.** $3x^3 + 15x^2 - 6x$

STOP • STOP • STOP • STOP • STOP • STOP • STOP • STOP • STOP

A7   **a.** $3(x + 2)$ **b.** $3x(3x - 4)$ **c.** $4(3y - 1)$
           d. $4m(4 + m)$ **e.** $5(y^2 - 2y + 4)$ **f.** $3x(x^2 + 5x - 2)$

---

4   In many cases it is necessary to factor monomials from a polynomial in which the first term of the polynomial is negative.

Example 1: Factor $^-15x^2 + 10x$.

Solution

Choose the common factor to be ^-5x. Hence,

$^-15x^2 + 10x = {}^-5x(3x - 2)$

This choice was made so that the first term of the polynomial within the parentheses would be positive.

Example 2: Factor $^-5 - 15x$.

Solution

$^-5 - 15x = {}^-5(1 + 3x)$

Q8   Complete the factoring:

a. $^-3x + 6 = {}^-3(\qquad)$ **b.** $^-8 - 4x = {}^-4(\qquad)$

STOP • STOP • STOP • STOP • STOP • STOP • STOP • STOP • STOP

A8   **a.** $^-3(x - 2)$ **b.** $^-4(2 + x)$

Q9   Factor so that the first term of the polynomial within the parentheses will be positive:

a. $^-9 + 6x$ **b.** $^-5x + 10$

c. $^-15x - 3$ **d.** $^-10x - 15x^2$

STOP • STOP • STOP • STOP • STOP • STOP • STOP • STOP • STOP

A9   **a.** $^-3(3 - 2x)$ **b.** $^-5(x - 2)$

[*Note:* An alternative solution would be:

$$^-9 + 6x = 6x - 9 \qquad\qquad ^-5x + 10 = 10 - 5x$$
$$= 3(2x - 3) \qquad\qquad\qquad = 5(2 - x)]$$

c. $^-3(5x + 1)$ **d.** $^-5x(2 + 3x)$

---

5   The expression $^-5x + 3$ has a common factor of $^-1$. Therefore, it is possible to factor $^-1$ from both terms. That is,

$$^-5x + 3 = ^-1(5x - 3)$$

By the multiplication property of $^-1$, $^-1(5x - 3)$ could be written $^-(5x - 3)$. You should observe that $^-1$ is a factor of any polynomial. However, $^-1$ is factored from a polynomial only when it is necessary to make the first term of the polynomial (within the parentheses) positive.

Q10   Factor so that the first term of the polynomial within the parentheses will be positive:

a. $^-2x + 7$ **b.** $^-12 + 4x$ **c.** $7x - 21$ **d.** $^-12x^2 + 15x$

e. $^-8x^2 - 12x$ **f.** $5x^3 - 20x^2$ **g.** $^-35 + 7x$ **h.** $20 - 10x$

i. $4x^2 + 8x - 16$ **j.** $^-5x^2 - 10x + 30$

STOP • **STOP** • **STOP** • **STOP** • **STOP** • **STOP** • **STOP** • **STOP** • **STOP**

A10   **a.** $^-1(2x - 7)$ or $^-(2x - 7)$ **b.** $^-4(3 - x)$ **c.** $7(x - 3)$

d. $^-3x(4x - 5)$ **e.** $^-4x(2x + 3)$ **f.** $5x^2(x - 4)$

g. $^-7(5 - x)$ **h.** $10(2 - x)$ **i.** $4(x^2 + 2x - 4)$

j. $^-5(x^2 + 2x - 6)$

---

6   When factoring a common factor from a polynomial, the polynomial within the parentheses can be said to be the "sum of the remaining factors."

Examples:

$$2x + 10 = 2x + 2 \cdot 5 = 2(x + 5)$$

common factor   sum of the remaining factors

$$15x - 20x^2 = 5x \cdot 3 - 5x \cdot 4x = 5x(3 - 4x)$$

common factor   sum of the remaining factors

In the following expression, $(x + 3)$ is a common factor.

$$(x + 3)x + (x + 3)5$$

By the left distributive property of multiplication over addition, $(x + 3)x + (x + 3)5$ can be factored as

$$(x + 3)(x + 5) \qquad \text{where } x + 3 \text{ is the common factor}$$

and $x + 5$ is the sum of the remaining factors. That is,

$$(x + 3)x + (x + 3)5 = (x + 3)(x + 5)$$

common factor   sum of the remaining factors

Q11   **a.** What is the common factor in $(x - 2)x + (x - 2)4$?   _____

       b. What is the sum of the remaining factors?   _____

STOP • **STOP** • **STOP** • **STOP** • **STOP** • **STOP** • **STOP** • **STOP** • **STOP**

A11   **a.** $(x - 2)$ **b.** $(x + 4)$

Q12   Complete the factorization: $(x - 2)x + (x - 2)4$
                                           $(x - 2)(\qquad)$

STOP • **STOP** • **STOP** • **STOP** • **STOP** • **STOP** • **STOP** • **STOP** • **STOP**

A12 $(x - 2)(x + 4)$

---

7   In $(x + 3)x - (x + 3)5$, the common factor $x + 3$ appears on the left. Hence, by the left distributive property of multiplication over subtraction,

$$(x + 3)x - (x + 3)5 = (x + 3)(x - 5)$$

where the common factor, $x + 3$, has been placed on the left and the sum of the remaining factors, $x - 5$, has been placed on the right. You should notice that it makes no difference whether the common factor appears on the right or the left. For example, by the right distributive property of multiplication over subtraction,

$$x(x + 3) - 5(x + 3) = (x - 5)(x + 3)$$

Q13 $3x(x + 6) - 5(x + 6) = (\qquad)(x + 6)$

STOP • **STOP** • **STOP** • **STOP** • **STOP** • **STOP** • **STOP** • **STOP** • **STOP**

A13 $(3x - 5)(x + 6)$

Q14   Factor:

    a. $x(2x - 1) - 3(2x - 1) =$ _____

    b. $2x(x + 1) + 3(x + 1) =$ _____

    c. $x(2x + 3) - 1(2x + 3) =$ _____

STOP • **STOP** • **STOP** • **STOP** • **STOP** • **STOP** • **STOP** • **STOP** • **STOP**

A14   **a.** $(x - 3)(2x - 1)$ **b.** $(2x + 3)(x + 1)$ **c.** $(x - 1)(2x + 3)$

---

8   The factoring of second-degree polynomials of the form $ax^2 + bx + c$, $a = 1$, $b,c \in I$, will now be considered. Recall that

$(x + 2)(x + 3)$
$x(x + 3) + 2(x + 3)$
$x^2 + 3x + 2x + 6$
$x^2 + 5x + 6$

The process of factoring $x^2 + 5x + 6$ will reverse the above process to obtain the two

> binomials whose product is $x^2 + 5x + 6$. To make the process easier, the terms of $x^2 + 5x + 6$ will be examined to see how they were obtained. Notice that 6 (the constant term) is the *product* of 2 and 3, the last terms of the original binomials. Also notice that the first-degree term, $5x$, is the *sum* of $2x$ and $3x$.

Q15   Examine the following simplification and answer the questions:

$(x - 2)(x - 3)$
$x(x - 3) - 2(x - 3)$
$x^2 - 3x - 2x + 6$
$x^2 - 5x + 6$

a. The constant term, 6, is the _____ of $^-2$ and $^-3$, the last terms of the
   product/sum
original binomials.

b. The first-degree term, ^-5x, is the _____ of ^-3x and ^-2x.
   product/sum

STOP • **STOP** • **STOP** • **STOP** • **STOP** • **STOP** • **STOP** • **STOP** • **STOP**

A15   **a.** product   **b.** sum

Q16   Examine the following simplification and answer the questions:

$(x + 2)(x - 3)$
$x(x - 3) + 2(x - 3)$
$x^2 - 3x + 2x - 6$
$x^2 - x - 6$

a. The constant term, $^-6$, is the _____ of 2 and $^-3$.
   product/sum

b. The first-degree term, ^-x, is the _____ of ^-3x and $2x$.
   product/sum

STOP • **STOP** • **STOP** • **STOP** • **STOP** • **STOP** • **STOP** • **STOP** • **STOP**

A16   **a.** product   **b.** sum

---

9   To factor $x^2 + 5x + 6$ you must first find a pair of integers whose product is 6 and whose sum is 5. Possible combinations are:

$(1)(6)$ $(^-1)(^-6)$ $(2)(3)$ $(^-2)(^-3)$

The combination that gives the proper sum is $(2)(3)$ since $2 + 3 = 5$. These integers are used as coefficients of x terms whose sum is $5x$. That is, $2x + 3x = 5x$. The factoring now proceeds as follows:

$x^2 + 5x + 6$
$x^2 + 2x + 3x + 6$ ($5x$ is replaced by $2x + 3x$)
$(x^2 + 2x) + (3x + 6)$
$x(x + 2) + 3(x + 2)$
$\underbrace{(x + 3)}\underbrace{(x + 2)}$

sum of remaining factors   common factor

The order in which $2x$ and $3x$ are placed in the second step does not effect the factorization (except in the order of the factors of the final result). For example,

$$x^2 + 5x + 6$$
$$x^2 + 3x + 2x + 6$$
$$(x^2 + 3x) + (2x + 6)$$
$$x(x + 3) + 2(x + 3)$$
$$(x + 2)(x + 3)$$

Q17   To factor $x^2 - 5x + 6$, first find a pair of integers whose product is 6 (the constant term) and whose sum is _____ (the coefficient of the first-degree term).

STOP   •   STOP   •   STOP   •   STOP   •   STOP   •   STOP   •   STOP   •   STOP   •   STOP

A17 $^-5$

Q18   The possible combinations of integers whose product is 6 are (1)(   ), ($^-1$)(   ), (2)(   ), and ($^-2$)(   ).

STOP   •   STOP   •   STOP   •   STOP   •   STOP   •   STOP   •   STOP   •   STOP   •   STOP

A18   6, $^-6$, 3, $^-3$

Q19   The combination of factors (from Q18) that give a sum of $^-5$ is _____.

STOP   •   STOP   •   STOP   •   STOP   •   STOP   •   STOP   •   STOP   •   STOP   •   STOP

A19   ($^-2$)($^-3$):   because $^-2 + {}^-3 = {}^-5$

10   To factor $x^2 - 5x + 6$:

Step 1: Find a pair of integers whose product is 6 and whose sum is $^-5$. They are $^-2$ and $^-3$.

Step 2: Use these numbers as coefficients of x terms to form a sum equivalent to ^-5x: $^-2x - 3x$ or $^-3x - 2x$.

Step 3: Proceed as follows:

$$x^2 - 5x + 6 \qquad\qquad \text{or} \qquad x^2 - 5x + 6$$
$$x^2 - 2x - 3x + 6 \qquad\qquad\qquad x^2 - 3x - 2x + 6$$
$$x(x - 2) - 3(x - 2) \qquad\qquad\quad x(x - 3) - 2(x - 3)$$
$$(x - 3)(x - 2) \qquad\qquad\qquad (x - 2)(x - 3)$$

Q20   **a.** To factor $x^2 + 12x + 27$, find a pair of integers whose product is _____ and whose sum is _____.

   b. The possible combinations of factors are (1)(   ) and (3)(   ). (*Note:* The combinations involving negative factors have been ignored since the sum must be positive.)

   c. The combination (from part b) that gives the proper sum is _____.
   $$x^2 + 12x + 27$$

   d. $x^2 + 3x +$ _____ $+ 27$

   e. $x(x + 3) +$ _____

   f. (   )(   )

STOP   •   STOP   •   STOP   •   STOP   •   STOP   •   STOP   •   STOP   •   STOP   •   STOP

A20   **a.** 27, 12   **b.** (1)(27) and (3)(9)
       c. (3)(9): because $3 + 9 = 12$ **d.** $x^2 + 3x + 9x + 27$
       e. $x(x + 3) + 9(x + 3)$ **f.** $(x + 9)(x + 3)$

Q21   Complete the following factorization:
$x^2 + 12x + 27$
$x^2 + 9x + 3x + 27$

a. _____ + _____   b. (   )(   )

STOP • **STOP** • **STOP** • **STOP** • **STOP** • **STOP** • **STOP** • **STOP** • **STOP**

A21   a. $x(x + 9) + 3(x + 9)$ b. $(x + 3)(x + 9)$

Q22   a. To factor $x^2 - 8x + 15$, find a pair of integers whose product is _____ and whose sum

is _____.

b. The possible combinations of factors are $(^-1)($ $)$ and $(^-3)($ $)$. (*Note:* The combinations involving positive factors have been ignored because the sum must be negative.)

c. The combination (from part b) that gives the proper sum is _____.

$x^2 - 8x + 15$

d. $x^2 -$ _____ $- 5x + 15$

e. _____ $-$ _____

f. (   )(   )

STOP • **STOP** • **STOP** • **STOP** • **STOP** • **STOP** • **STOP** • **STOP** • **STOP**

A22   a. 15, $^-8$ b. $(^-1)(^-15)$ and $(^-3)(^-5)$
c. $(^-3)(^-5)$: because $^-3 + ^-5 = ^-8$ d. $x^2 - 3x - 5x + 15$
e. $x(x - 3) - 5(x - 3)$ f. $(x - 5)(x - 3)$

Q23   Factor:
a. $x^2 + 9x + 20$ b. $x^2 - 21x + 20$

STOP • **STOP** • **STOP** • **STOP** • **STOP** • **STOP** • **STOP** • **STOP** • **STOP**

A23   a. $(x + 5)(x + 4)$ or $(x + 4)(x + 5)$:
$x^2 + 9x + 20$ or $x^2 + 9x + 20$
$x^2 + 4x + 5x + 20$ $x^2 + 5x + 4x + 20$
$x(x + 4) + 5(x + 4)$ $x(x + 5) + 4(x + 5)$
$(x + 5)(x + 4)$ $(x + 4)(x + 5)$
b. $(x - 20)(x - 1)$ or $(x - 1)(x - 20)$:
$x^2 - 21x + 20$
$x^2 - 1x - 20x + 20$ Other possibility not shown.
$x(x - 1) - 20(x - 1)$
$(x - 20)(x - 1)$

Q24   Factor:
        a. $x^2 - 10x + 24$ **b.** $x^2 + 11x + 28$

STOP • **STOP** • **STOP** • **STOP** • **STOP** • **STOP** • **STOP** • **STOP** • **STOP**

A24   **a.** $(x - 4)(x - 6)$ **b.** $(x + 4)(x + 7)$
        (*Note:* The factors may be commuted.)

---

11   To factor $x^2 - x - 6$, first find a pair of integers whose product is $^-6$ and whose sum is $^-1$. Since the product is negative, the signs of the integers must be different, one positive and one negative. The possible choices that would produce a negative product and a negative sum are:

$(1)(^-6) = {}^-6$ and $1 + {}^-6 = {}^-5$
$(2)(^-3) = {}^-6$ and $2 + {}^-3 = {}^-1$

Hence, the proper combination is 2 and $^-3$. The factoring would be completed as follows:

$x^2 - x - 6$ or $x^2 - x - 6$
$x^2 + 2x - 3x - 6$ $x^2 - 3x + 2x - 6$
$x(x + 2) - 3(x + 2)$ $x(x - 3) + 2(x - 3)$
$(x - 3)(x + 2)$ $(x + 2)(x - 3)$

Q25   Factor $x^2 + 4x - 32$. (First find a pair of integers whose product is $^-32$ and whose sum is 4.)

STOP • **STOP** • **STOP** • **STOP** • **STOP** • **STOP** • **STOP** • **STOP** • **STOP**

A25 $(x + 8)(x - 4)$: $x^2 + 4x - 32$
                         $x^2 - 4x + 8x - 32$ Other possibility not shown.
                         $x(x - 4) + 8(x - 4)$
                         $(x + 8)(x - 4)$

Q26   Factor $x^2 - 9x - 22$.

STOP • **STOP** • **STOP** • **STOP** • **STOP** • **STOP** • **STOP** • **STOP** • **STOP**

A26 $(x - 11)(x + 2)$: $x^2 - 9x - 22$
                         $x^2 + 2x - 11x - 22$ Other possibility not shown.
                         $x(x + 2) - 11(x + 2)$
                         $(x - 11)(x + 2)$

Q27 Factor:

     a. $x^2 - 4x - 5$ **b.** $x^2 - 9x + 18$ **c.** $x^2 + 2x + 1$

     d. $x^2 - 2x - 3$ **e.** $x^2 - 2x - 15$ **f.** $x^2 - 18x + 32$

     g. $x^2 + 11x + 30$ **h.** $x^2 + 6x - 16$ **i.** $x^2 + 2x - 48$

     j. $x^2 - 10x + 25$

STOP • STOP • STOP • STOP • STOP • STOP • STOP • STOP • STOP

A27 Factors may be commuted.

     a. $(x - 5)(x + 1)$ **b.** $(x - 6)(x - 3)$
     c. $(x + 1)(x + 1)$ or $(x + 1)^2$ **d.** $(x - 3)(x + 1)$
     e. $(x - 5)(x + 3)$ **f.** $(x - 16)(x - 2)$
     g. $(x + 5)(x + 6)$ **h.** $(x + 8)(x - 2)$
     i. $(x + 8)(x - 6)$ **j.** $(x - 5)(x - 5)$ or $(x - 5)^2$

---

12   When factoring the previous polynomials, you may have noticed that the result could be written immediately once you found the correct factors which give the correct sum.

Examples:

$x^2 + 5x + 6 = (x + 2)(x + 3)$

$x^2 - x - 6 = (x - 3)(x + 2)$

$x^2 + 4x - 32 = (x + 8)(x - 4)$

$x^2 + 10x + 25 = (x + 5)(x + 5)$ or $(x + 5)^2$

Q28 Write the result of the factoring immediately:

     a. $x^2 - 9x + 8 =$ _____

     b. $x^2 - 7x - 18 =$ _____

     c. $x^2 + 2x - 24 =$ _____

     d. $x^2 + 5x - 36 =$ _____

     e. $x^2 + 20x + 36 =$ _____

     f. $x^2 - 2x + 1 =$ _____

     g. $x^2 + x - 6 =$ _____

     h. $x^2 + 6x + 9 =$ _____

     i. $x^2 - 11x + 10 =$ _____

     j. $x^2 - 11x - 12 =$ _____

STOP • STOP • STOP • STOP • STOP • STOP • STOP • STOP • STOP

A28   **a.** $(x - 8)(x - 1)$ **b.** $(x - 9)(x + 2)$
        c. $(x + 6)(x - 4)$ **d.** $(x + 9)(x - 4)$
        e. $(x + 18)(x + 2)$ **f.** $(x - 1)(x - 1)$ or $(x - 1)^2$
        g. $(x + 3)(x - 2)$ **h.** $(x + 3)(x + 3)$ or $(x + 3)^2$
        i. $(x - 10)(x - 1)$ **j.** $(x - 12)(x + 1)$

---

13   Although many trinomials can be factored mentally, an adaptation of the technique presented is important in order to factor more-difficult second-degree polynomials of the form $ax^2 + bx + c$, where $a \neq 1$, $a,b,c \in I$. Such a polynomial is $6x^2 + 7x - 3$. To factor this trinomial mentally would be difficult. However, consider the following procedure that may be used to factor it with relative ease.
        To factor $6x^2 + 7x - 3$:

   Step 1: Multiply 6 (the coefficient of the second-degree term) by $^-3$ (the constant term), which is $^-18$.

   Step 2: Find two integers whose product is $^-18$ and whose sum is 7 (the coefficient of the first degree term). They are 9 and $^-2$, because $(9)(^-2) = {}^-18$ and $9 + {}^-2 = 7$.

   Step 3: Use these integers as coefficients of x terms whose sum is equivalent to $7x$: $9x - 2x$ or $^-2x + 9x$.

   Step 4: Proceed as follows:

   $6x^2 + 7x - 3$ or $6x^2 + 7x - 3$
   $6x^2 + 9x - 2x - 3$ $6x^2 - 2x + 9x - 3$
   $3x(2x + 3) - 1(2x + 3)$ $2x(3x - 1) + 3(3x - 1)$
   $(3x - 1)(2x + 3)$ $(2x + 3)(3x - 1)$

Q29   To factor $2x^2 + 7x + 6$, first multiply 2 by 6, which gives 12.

   a. Now find two integers whose product is _____ and whose sum is _____.

   b. They are _____.

STOP • STOP • STOP • STOP • STOP • STOP • STOP • STOP • STOP

A29   **a.** 12, 7   **b.** 4 and 3 (either order)

Q30   Complete the factorization:
        $2x^2 + 7x + 6$
        $2x^2 + 4x + 3x + 6$

   a. _____ + _____   **b.** (_____)(_____)

STOP • STOP • STOP • STOP • STOP • STOP • STOP • STOP • STOP

A30   **a.** $2x(x + 2) + 3(x + 2)$ **b.** $(2x + 3)(x + 2)$

---

14   The justification of the procedure of Frame 13 is given next. Consider the product of the two binomials $ax + b$ and $cx + d$, where $a,b,c,d \in I$.

   $(ax + b)(cx + d)$
   $acx^2 + adx + bcx + bd$ "foil" method
   $acx^2 + (adx + bcx) + bd$
   $acx^2 + (ad + bc)x + bd$

Notice that the *sum ad + bc* (the coefficient of the first-degree term), is a combination of the factors *acbd* (the *product* of the coefficient of the second-degree term and the constant term).

In general, the second-degree polynomial $ax^2 + bx + c$, $a,b,c \in I$, $a \neq 0$, $c \neq 0$, is factorable into the product of two binomials with integral coefficients if two integers exist whose product is ac and whose sum is b.

Q31   To factor $5x^2 - 12x + 4$, multiply 5 by 4, which gives 20.

    a. Two integers whose product is 20 and whose sum is ⁻12 are _____.

    b. Complete the factorization.
$$5x^2 - 12x + 4$$
$$5x^2 - 10x - 2x + 4$$

STOP • STOP • STOP • STOP • STOP • STOP • STOP • STOP • STOP

A31   **a.** ⁻10 and ⁻2   **b.** $5x(x-2) - 2(x-2)$
        (either order) $(5x-2)(x-2)$

---

15   **Example:** Factor $4x^2 + 12x - 7$.

Solution

Step 1: $4(-7) = -28$.

Step 2: $(-2)(14) = -28$ and $-2 + 14 = 12$.

Now proceed as follows:

$4x^2 + 12x - 7$	or $4x^2 + 12x - 7$
$4x^2 - 2x + 14x - 7$	$4x^2 + 14x - 2x - 7$
$2x(2x-1) + 7(2x-1)$	$2x(2x+7) - 1(2x+7)$
$(2x+7)(2x-1)$	$(2x-1)(2x+7)$

Q32   Factor $3x^2 - 8x - 3$.

STOP • STOP • STOP • STOP • STOP • STOP • STOP • STOP • STOP

A32 $(3x+1)(x-3)$: $3x^2 - 8x - 3$
         $(3)(-3) = -9$ $3x^2 - 9x + 1x - 3$
         $(1)(-9) = -9$ and $3x(x-3) + 1(x-3)$
         $1 + -9 = -8$ $(3x+1)(x-3)$
         Other possibility not shown.

Q33   Factor:

    a. $6x^2 + 17x + 5$ **b.** $3x^2 - 14x + 8$

c. $6x^2 + 5x - 4$

d. $4x^2 - 9$
(*Hint:* $4x^2 - 9 = 4x^2 + 0x - 9$.)

e. $4x^2 - 20x + 25$

*f. $8x^2 + 14x - 15$

STOP • STOP • STOP • STOP • STOP • STOP • STOP • STOP • STOP

A33

a. $(2x + 5)(3x + 1)$

b. $(x - 4)(3x - 2)$

c. $(3x + 4)(2x - 1)$

d. $(2x + 3)(2x - 3)$

e. $(2x - 5)(2x - 5)$ or $(2x - 5)^2$

*f. $(4x - 3)(2x + 5)$:

$(8)(^-15) = ^-120$

$(^-1)(120) = ^-120$ but $^-1 + 12 \neq 14$

$(^-2)(60) = ^-120$ but $^-2 + 60 \neq 14$

$(^-3)(40) = ^-120$ but $^-3 + 40 \neq 14$

$(^-4)(30) = ^-120$ but $^-4 + 30 \neq 14$

$(^-5)(24) = ^-120$ but $^-5 + 24 \neq 14$

$(^-6)(20) = ^-120$ and $^-6 + 20 = 14$

$8x^2 + 14x - 15$

$8x^2 + 20x - 6x - 15$ Other possibility not shown.

$4x(2x + 5) - 3(2x + 5)$

$(4x - 3)(2x + 5)$

---

16   Not all polynomials of the form $ax^2 + bx + c$, $a,b,c \in I$, $a \neq 0$, $c \neq 0$, are factorable into the product of two binomials. Examine $3x^2 - 20x - 4$. The possible integers that produce a product of -12 and a negative sum are:

$(1)(^-12) = ^-12$ and $1 + ^-12 = ^-11$

$(2)(^-6) \ = ^-12$ and $2 + ^-6 = ^-4$

$(3)(^-4) \ = ^-12$ and $3 + ^-4 = ^-1$

In no case is the sum $^-20$. Hence, $3x^2 - 20x - 4$ is not factorable over the set of integers into the product of two binomials. Also, since there is no integer factor (other than 1 or $^-1$) or variable common to each term, $3x^2 - 20x - 4$ is said to be *prime* over the set of integers.* (In this text, when a polynomial is said to be prime, it will be understood to mean over the set of integers.)

In general, assuming there is no common monomial factor, the second-degree polynomial $ax^2 + bx + c$, $a,b,c \in I$, $a \neq 0$, $c \neq 0$, is prime if there are no integers whose product is ac and whose sum is b.

---

*A natural number other than 1 is said to be prime if it has only two different natural-number factors, itself and 1. Any polynomial with integer coefficients is said to be prime over the set of integers if: The polynomial does not contain an integer factor (other than 1 or $^-1$) common to each term, and the polynomial cannot be factored into two other polynomials with integer coefficients.

Q34   **a.** Is $3x^2 - 2x + 7$ prime? _____   **b.** Explain.

STOP • **STOP** • **STOP** • **STOP** • **STOP** • **STOP** • **STOP** • **STOP** • **STOP**

A34   **a.** yes
         b. There is no common monomial factor and there are no integers whose product is $^+21$ and whose sum is $^-2$.

Q35   **a.** Is $2x^2 + 7x + 5$ factorable? _____   **b.** Explain.

STOP • **STOP** • **STOP** • **STOP** • **STOP** • **STOP** • **STOP** • **STOP** • **STOP**

A35   **a.** yes
         b. Although there is no common monomial factor, there are two integers whose product is 10 and whose sum is 7, $(2)(5) = 10$ and $2 + 5 = 7$.

Q36   **a.** Is $3x^2 + 17x - 9$ factorable? _____   **b.** Explain.

STOP • **STOP** • **STOP** • **STOP** • **STOP** • **STOP** • **STOP** • **STOP** • **STOP**

A36   **a.** no
         b. There is no common monomial factor and there are no integers whose product is $^-27$ and whose sum is 17.

Q37   **a.** Is $4x^2 + 6x + 8$ factorable? _____   **b.** Explain.

STOP • **STOP** • **STOP** • **STOP** • **STOP** • **STOP** • **STOP** • **STOP** • **STOP**

A37   **a.** yes
         b. There is a common monomial factor 2. [Hence, $4x^2 + 6x + 8 = 2(2x^2 + 3x + 4)$. You should observe that $2x^2 + 3x + 4$ is prime.]

---

17   A trinomial such as $x^2 + 2x + 3$ is of the form $ax^2 + bx + c$, where $a = 1$. Therefore, the test to determine whether a trinomial of this form is factorable still applies.

Example 1: Is $x^2 + 2x + 3$ factorable? Explain.

Solution

No. $(1)(3) = 3$ and there are no integers whose product is 3 and whose sum is 2. Also, there is no common monomial factor.

Example 2: Is $x^2 - 5x - 6$ factorable? Explain.

Solution

Yes. Although there is no common monomial factor, there are two integers whose product is $^-6$ and whose sum is $^-5$: $(1)(^-6) = ^-6$ and $1 + ^-6 = ^-5$.

Q38   **a.** Is $x^2 - 3x - 5$ factorable? _____

      b. Is $x^2 - 3x - 5$ prime? _____

STOP • **STOP** • **STOP** • **STOP** • **STOP** • **STOP** • **STOP** • **STOP** • **STOP**

A38   **a.** no   **b.** yes

Q39   **a.** Is $3x^2 + 4x + 1$ prime? _____

      b. Is $2x^2 - 3x - 5$ prime? _____

STOP • **STOP** • **STOP** • **STOP** • **STOP** • **STOP** • **STOP** • **STOP** • **STOP**

A39   **a.** no   **b.** no

Q40   Factor where possible. Otherwise, write prime.
      a. $3x^2 + x - 2$ **b.** $4x^2 - 12x + 9$

      c. $x^2 + 9x - 20$ **d.** $4x^2 - 25$
                                 (*Hint:* $4x^2 - 25 = 4x^2 + 0x - 25$.)

STOP • **STOP** • **STOP** • **STOP** • **STOP** • **STOP** • **STOP** • **STOP** • **STOP**

A40   **a.** $(3x - 2)(x + 1)$ **b.** $(2x - 3)(2x - 3)$ or $(2x - 3)^2$
      c. prime   **d.** $(2x + 5)(2x - 5)$

18   If a polynomial contains a common factor it should be factored from the polynomial before attempting any other type of factoring.

Example 1: Factor $2x^2 + x$.

Solution

$2x^2 + x = x(2x + 1)$

Example 2: Factor $3x^2 - 6x - 9$.

Solution

$3x^2 - 6x - 9 = 3(x^2 - 2x - 3)$
$\qquad\qquad\quad = 3(x - 3)(x + 1)$

Example 3: Factor $2x^3 - 7x^2 + 5x$.

Solution

$2x^3 - 7x^2 + 5x = x(2x^2 - 7x + 5)$
$\qquad\qquad\qquad = x(2x^2 - 2x - 5x + 5)$
$\qquad\qquad\qquad = x[2x(x - 1) - 5(x - 1)]$
$\qquad\qquad\qquad = x(2x - 5)(x - 1)$

The above solution could be simplified by factoring $2x^2 - 7x + 5$ separately. That is,

$$2x^3 - 7x^2 + 5x = x(2x^2 - 7x + 5) \qquad 2x^2 - 7x + 5$$
$$= x(2x - 5)(x - 1) \qquad\quad 2x^2 - 2x - 5x + 5$$
$$2x(x - 1) - 5(x - 1)$$
$$(2x - 5)(x - 1)$$

Example 4: Factor $3x^3 - 27x$.

Solution

$$3x^3 - 27x = 3x(x^2 - 9)$$
$$= 3x(x + 3)(x - 3)$$

Example 5: Factor $^-x^2 - 10x - 25$.

Solution

$$^-x^2 - 10x - 25 = {}^-(x^2 + 10x + 25)$$
$$= {}^-(x + 5)(x + 5) \text{ or } {}^-(x + 5)^2$$

Example 6: Factor $2x^2 - 6x + 10$.

Solution

$$2x^2 - 6x + 10 = 2(x^2 - 3x + 5)$$
(*Note:* $x^2 - 3x + 5$ is prime.)

Example 7: Factor $5x^2 - 3x - 7$.

Solution

$5x^2 - 3x - 7$ is prime

There is no common monomial factor and there are no integers whose product is $^-35$ and whose sum is $^-3$. [*Remember:* Always remove a common monomial factor *first* (if it exists) before attempting any other type of factoring.]

Q41   Factor:

a. $^-5x^2 - 15x + 50$

b. $6x^2 - 3x - 45$

c. $2x^2 - 26x + 80$

d. $8x^3 - 17x^2 + 2x$

e. $2x^3 + 5x^2 - 2x$

f. $^-2x^2 - x + 15$

g. $8x^2 - 32$

h. $^-2x^2 + 24x - 72$

i. $2x^3 + 14x^2 + 12x$

j. $2x^2 + 7x - 3$

A41
a. $^-5(x + 5)(x - 2)$
d. $x(8x - 1)(x - 2)$
g. $8(x + 2)(x - 2)$
j. prime

b. $3(2x + 5)(x - 3)$
e. $x(2x^2 + 5x - 2)$
h. $^-2(x - 6)^2$

c. $2(x - 5)(x - 8)$
f. $^-(2x - 5)(x + 3)$
i. $2x(x + 6)(x + 1)$

This completes the instruction for this section.

11.4   Exercises

1. Factor:
a. $5x - 10$
c. $9x^2 - 3x + 15$
e. $x(x - 2) + 2(x - 2)$
g. $3x(3x - 1) - 1(3x - 1)$
i. $^-12x^2 + 18x^3$

b. $3x^2 - 6x$
d. $x^3 + 4x^2 - 3x$
f. $x(2x + 3) - 2(2x + 3)$
h. $^-x - 1$
j. $^-y + y^2$

2. Factor:
a. $x^2 + 10x + 21$
d. $x^2 + 5x - 36$
g. $24 - 14x + x^2$
j. $x^2 - 14x + 49$

b. $x^2 - 12x + 35$
e. $100 - x^2$
h. $12 + 9x + x^2$

c. $x^2 - 18x - 40$
f. $x^2 + x - 42$
i. $x^2 + 4x + 4$

3. Factor:
a. $3x^2 - 11x + 6$
d. $4x^2 - 3x + 9$
g. $2x^2 - 5x - 12$
j. $5x^3 - 20x^2 + 20x$

b. $y^2 + 19y + 84$
e. $5x^2 - 180$
h. $^-4x^2 + 6x + 40$

c. $x^3 + 5x^2 + 6x$
f. $3x^2 + 6x - 24$
i. $3x^3 + 5x^2 - 2x$

11.4   Exercise Answers (Binomial factors can be commuted)

1. a. $5(x - 2)$
d. $x(x^2 + 4x - 3)$
g. $(3x - 1)(3x - 1)$ or $(3x - 1)^2$
j. $^-y(1 - y)$

b. $3x(x - 2)$
e. $(x + 2)(x - 2)$
h. $^-(x + 1)$

c. $3(3x^2 - x + 5)$
f. $(x - 2)(2x + 3)$
i. $^-6x^2(2 - 3x)$

2. a. $(x + 3)(x + 7)$
d. $(x + 9)(x - 4)$
g. $(12 - x)(2 - x)$
j. $(x - 7)(x - 7)$ or $(x - 7)^2$

b. $(x - 5)(x - 7)$
e. $(10 + x)(10 - x)$
h. prime

c. $(x - 20)(x + 2)$
f. $(x + 7)(x - 6)$
i. $(x + 2)(x + 2)$ or $(x + 2)^2$

3. a. $(3x - 2)(x - 3)$
d. prime
g. $(2x + 3)(x - 4)$
j. $5x(x - 2)(x - 2)$ or $5x(x - 2)^2$

b. $(y + 7)(y + 12)$
e. $5(x + 6)(x - 6)$
h. $^-2(2x + 5)(x - 4)$

c. $x(x + 2)(x + 3)$
f. $3(x + 4)(x - 2)$
i. $x(x + 2)(3x - 1)$

11.5 Factoring the Difference of Two Squares and Perfect Square Trinomials

---

1 *Perfect square integers* are numbers obtained by squaring some integer. *Perfect square rational numbers* are numbers obtained by squaring some rational number.

Perfect square integers	Perfect square rational numbers
$(1)^2 = 1$	$\left(\dfrac{2}{3}\right)^2 = \dfrac{4}{9}$
$(^-2)^2 = 4$	$(0.1)^2 = 0.01$
$7^2 = 49$	$\left(\dfrac{^-4}{7}\right)^2 = \dfrac{16}{49}$
$(^-10)^2 = 100$	$6^2 = 36$
etc.	etc.

Since every integer is a rational number, every perfect square integer is likewise a perfect square rational number.

Q1 List the perfect square integers less than or equal to 400.

_____

_____

STOP • **STOP** • **STOP** • **STOP** • **STOP** • **STOP** • **STOP** • **STOP** • **STOP**

A1 1, 4, 9, 16, 25, 36, 49, 64, 81, 100, 121, 144, 169, 196, 225, 256, 289, 324, 361, 400

---

2 A *perfect square monomial* is the square of a monomial. Examples of perfect square monomials are:

$$(a)^2 = a^2 \qquad (^-2x)^2 = 4x^2$$
$$(3y)^2 = 9y^2 \qquad \left(\frac{4}{5}x\right)^2 = \frac{16}{25}x^2$$
$$(0.4x)^2 = 0.16x^2 \qquad (^-0.6m)^2 = 0.36m^2$$
$$(x^2)^2 = x^4 \qquad (2y^3)^2 = 4y^6$$

For a monomial to be a perfect square, the coefficient must be a perfect square rational number and all variables must be raised to an even power.

Q2 Write "yes" if the monomial is a perfect square. Otherwise, write "no."

a. x^2 _____ **b.** $9y^4$ _____ **c.** $^-16x^2$ _____

d. $0.64y^2$ _____ **e.** $225m^6$ _____ **f.** $\dfrac{25}{49}x^2$ _____

g. $36y^3$ _____ **h.** $0.1y^4$ _____

STOP • **STOP** • **STOP** • **STOP** • **STOP** • **STOP** • **STOP** • **STOP** • **STOP**

A2 **a.** yes: $(x)^2$ **b.** yes: $(3y^2)^2$
c. no: $^-16$ is not a perfect square rational number.

d. yes: $(0.8y)^2$ **e.** yes: $(15m^3)^2$ **f.** yes: $\left(\dfrac{5}{7}x\right)^2$

g. no: the exponent on y is odd.

h. no: 0.1 is not a perfect square rational number.

Q3   Complete each expression to form a true statement:

a. ($)^2 = 144x^2$ **b.** ($)^2 = \dfrac{1}{9} x^4$

c. ($)^2 = 81y^2$ **d.** ($)^2 = \dfrac{49}{100} x^2$

e. ($)^2 = 121x^2$ **f.** ($)^2 = 1.69y^6$

STOP   •   STOP   •   STOP   •   STOP   •   STOP   •   STOP   •   STOP   •   STOP   •   STOP

A3   **a.** $12x$ or ^-12x **b.** $\dfrac{1}{3} x^2$ or $\dfrac{^-1}{3} x^2$

c. $9y$ or ^-9y **d.** $\dfrac{7}{10} x$ or $\dfrac{^-7}{10} x$

e. $11x$ or ^-11x **f.** $1.3y^3$ or $^-1.3y^3$

---

3   A special product from Section 11.3 was

$(a + b)(a - b) = a^2 - b^2$

Expressions of the form $a^2 - b^2$ are called the *difference of two squares*.

Examples:

$(x + 7)(x - 7) = x^2 - 49$

$(3x + 4)(3x - 4) = 9x^2 - 16$

$(1 + 2y)(1 - 2y) = 1 - 4y^2$

The equation $a^2 - b^2 = (a + b)(a - b)$ is useful when factoring the difference of two squares.

Example 1: Factor $x^2 - 9$.

Solution

$x^2 - 9$
$(x)^2 - (3)^2$
$(x + 3)(x - 3)$

Example 2: Factor $16 - 25y^2$.

Solution

$16 - 25y^2$
$(4)^2 - (5y)^2$
$(4 + 5y)(4 - 5y)$

The second step is usually done mentally.

Q4   Factor:
          a. $x^2 - 1$ **b.** $x^2 - 4$

          c. $36 - x^2$ **d.** $4y^2 - 9$

          e. $64x^2 - 121$ **f.** $100 - 9x^2$

STOP • **STOP** • **STOP** • **STOP** • **STOP** • **STOP** • **STOP** • **STOP** • **STOP**

A4   **a.** $(x + 1)(x - 1)$ **b.** $(x + 2)(x - 2)$
          c. $(6 + x)(6 - x)$ **d.** $(2y + 3)(2y - 3)$
          e. $(8x + 11)(8x - 11)$ **f.** $(10 + 3x)(10 - 3x)$

---

4   In Section 11.4 it was emphasized that a common factor should be removed first (if it exists) before any other type of factoring is attempted.

          Example 1: Factor $^-18 + 2x^2$.   **Example 2:** Factor $x^3 - 81x$.

          Solution   *Solution*

          $^-18 + 2x^2$ $x^3 - 81x$
          $^-2(9 - x^2)$ $x(x^2 - 81)$
          $^-2(3 + x)(3 - x)$ $x(x + 9)(x - 9)$

Q5   Factor:
          a. $y^3 - 25y$ **b.** $9x^2 - 144$

          c. $^-4x + 100x^3$ **d.** $x^2 - x$

STOP • **STOP** • **STOP** • **STOP** • **STOP** • **STOP** • **STOP** • **STOP** • **STOP**

A5   **a.** $y(y + 5)(y - 5)$ **b.** $9(x + 4)(x - 4)$:
                                                      $9x^2 - 144 = 9(x^2 - 16)$
          c. $^-4x(1 + 5x)(1 - 5x)$:   **d.** $x(x - 1)$
               $^-4x + 100x^3 = ^-4x(1 - 25x^2)$
          [*Note:* An alternative solution to Q5c is as follows:
          $^-4x + 100x^3 = 100x^3 - 4x$
          $\qquad = 4x(25x^2 - 1)$
          $\qquad = 4x(5x + 1)(5x - 1)$]

5   A special product from Section 11.3 was

$$(a + b)^2 = a^2 + 2ab + b^2$$

It is likewise true that

$$(a - b)^2 = a^2 - 2ab + b^2$$

Examples:

$$(2x + 3)^2 = (2x)^2 + 2(2x)(3) + (3)^2 = 4x^2 + 12x + 9$$

$$(x - 7)^2 = (x)^2 - 2(x)(7) + (7)^2 = x^2 - 14x + 49$$

Trinomials obtained from the square of a binomial are called *perfect square trinomials*. Hence, $4x^2 + 12x + 9$ and $x^2 - 14x + 49$ are perfect square trinomials.

Q6   What perfect square trinomial is obtained from the square of each binomial:
a. $(x - 9)^2$ **b.** $(x + 1)^2$

c. $(2x - 3)^2$ **d.** $(3x + 5)^2$

STOP • **STOP** • **STOP** • **STOP** • **STOP** • **STOP** • **STOP** • **STOP** • **STOP**

A6   **a.** $x^2 - 18x + 81$ **b.** $x^2 + 2x + 1$
         c. $4x^2 - 12x + 9$ **d.** $9x^2 + 30x + 25$

6   If a trinomial can be factored as the square of a binomial, it is a perfect square trinomial.

Example 1: Is $x^2 - 10x + 25$ a perfect square trinomial?

Solution
Yes, $x^2 - 10x + 25 = (x - 5)^2$.

Example 2: Is $x^2 - 6x + 9$ a perfect square trinomial?

Solution
Yes, $x^2 - 6x + 9 = (x - 3)^2$.

Example 3: Is $x^2 + 10x + 16$ a perfect square trinomial?

Solution
No, $x^2 + 10x + 16 = (x + 8)(x + 2)$, and $x^2 + 10x + 16$ factors in no other way. Although $x^2 + 10x + 16$ can be factored, it is not factorable as the square of a binomial.

Example 4: Is $x^2 - 24x - 25$ a perfect square trinomial?

Solution
No, $x^2 - 24x - 25 = (x - 25)(x + 1)$, and $x^2 - 24x - 25$ factors in no other way. Although $x^2 - 24x - 25$ can be factored, it is not factorable as the square of a binomial.

> **Example 5:** Is $x^2 + 3x + 4$ a perfect square trinomial?
>
> *Solution*
>
> No, $x^2 + 3x + 4$ is prime.

Q7   **a.** Is $x^2 - 2x + 1$ a perfect square trinomial?   _____

b. Is $x^2 - 4x + 4$ a perfect square trinomial?   _____

c. Is $m^2 - 20m + 36$ a perfect square trinomial?   _____

d. Is $x^2 + 20x + 100$ a perfect square trinomial?   _____

e. Is $x^2 - 6x - 16$ a perfect square trinomial?   _____

f. Is $y^2 - 11y + 25$ a perfect square trinomial?   _____

STOP • **STOP** • **STOP** • **STOP** • **STOP** • **STOP** • **STOP** • **STOP** • **STOP**

A7   **a.** yes: $(x - 1)^2$ **b.** yes: $(x - 2)^2$
c. no: $m^2 - 20m + 36 = (m - 18)(m - 2)$
d. yes: $(x + 10)^2$ **e.** no: $x^2 - 6x - 16 = (x - 8)(x + 2)$
f. no: $y^2 + 11y + 25$ is prime

---

7   It is possible to identify a perfect square trinomial by means of a simple test. You know that $a^2 + 2ab + b^2$ is a perfect square trinomial, because $a^2 + 2ab + b^2 = (a + b)^2$.

a^2 is the square of a $(a)^2 = a^2$
b^2 is the square of b $(b)^2 = b^2$
$2ab$ is twice the product of a and b $2(a)(b) = 2ab$

To determine whether a trinomial is a perfect square trinomial, perform the following test.

Step 1: Two of the three terms must be perfect square monomials.

Step 2: If so, express these terms as squares of monomials.

Step 3: If twice the product of the two monomials thus formed equals the third term (disregarding its sign), the original trinomial is a perfect square.

Example 1: Is $x^2 + 14x + 49$ a perfect square trinomial?

Solution

Step 1: x^2 and 49 are perfect square monomials.

Step 2: $x^2 + 14x + 49 = (x)^2 + 14x + (7)^2$

Step 3: $2(x)(7) = 14x$. Hence, $x^2 + 14x + 49$ is a perfect square trinomial.

Example 2: Is $4x^2 - 20x + 25$ a perfect square trinomial?

Solution

Step 1: $4x^2$ and 25 are perfect square monomials.

Step 2: $4x^2 - 20x + 25 = (2x)^2 - 20x + (5)^2$

Step 3: $2(2x)(5) = 20x$. Hence, $4x^2 - 20x + 25$ is a perfect square trinomial.

Example 3: Is $x^2 + 10x - 25$ a perfect square trinomial?

Solution

Step 1: Only x^2 is a perfect square monomial. Hence, $x^2 + 10x - 25$ is not a perfect square trinomial.

Example 4: Is $y^2 + 10y + 16$ a perfect square trinomial?

Solution

Step 1: y^2 and 16 are perfect square monomials.

Step 2: $y^2 + 10y + 16 = (y)^2 + 10y + (4)^2$

Step 3: $2(y)(4) \neq 10y$. Hence, $y^2 + 10y + 16$ is not a perfect square trinomial.

Q8   Use the procedure of Frame 7 to determine whether each trinomial is a perfect square. If it is, write "yes." Otherwise, write "no."

a. $x^2 + 4x + 4$ 　　　　**b.** $x^2 - 2x + 1$

c. $x^2 - 5x + 16$ 　　　　**d.** $4s^2 + 12s + 9$

e. $n^2 - 12n + 36$ 　　　　**f.** $x^2 - 22x + 121$

g. $x^2 - 20x - 100$ 　　　　**h.** $y^2 - 8y + 16$

STOP　•　STOP　•　STOP　•　STOP　•　STOP　•　STOP　•　STOP　•　STOP　•　STOP

A8   **a.** yes: $2(x)(2) = 4x$ 　　**b.** yes: $2(x)(1) = 2x$
c. no: $2(x)(4) \neq 5x$ 　　**d.** yes: $2(2s)(3) = 12s$
e. yes: $2(n)(6) = 12n$ 　　**f.** yes: $2(x)(11) = 22x$
g. no: only x^2 is a perfect square monomial
h. yes: $2(y)(4) = 8y$

8   If a trinomial is a perfect square, it can be factored rapidly using either the equation $a^2 + 2ab + b^2 = (a + b)^2$ or $a^2 - 2ab + b^2 = (a - b)^2$. That is,

$a^2 + 2ab + b^2$ 　　or　　 $a^2 - 2ab + b^2$
$(a)^2 + 2ab + (b)^2$ 　　　　$(a)^2 - 2ab + (b)^2$
$(a + b)^2$ 　　　　　　$(a - b)^2$

Example 1: Factor $x^2 + 14x + 49$.

Solution

$x^2 + 14x + 49$
$(x)^2 + 14x + (7)^2$ $2(x)(7) = 14x$
$\quad (x + 7)^2$

Example 2: Factor $x^2 - 12x + 36$.

Solution

$x^2 - 12x + 36$
$(x)^2 - 12x + (6)^2$ $2(x)(6) = 12x$
$\quad (x - 6)^2$

Example 3: Factor $4x^2 - 20x + 25$.

Solution

$4x^2 - 20x + 25$
$(2x)^2 - 20x + (5)^2$ $2(2x)(5) = 20x$
$\quad (2x - 5)^2$

Q9   Factor:
   a. $x^2 - 20x + 100$ **b.** $x^2 + 16x + 64$

   c. $9x^2 + 42x + 49$ **d.** $9m^2 - 30m + 25$

   e. $y^2 - 24y + 144$ **f.** $49 - 14x + x^2$

   g. $36n^2 - 12n + 1$ **h.** $25x^2 + 60x + 36$

STOP • STOP • STOP • STOP • STOP • STOP • STOP • STOP • STOP

A9   **a.** $(x - 10)^2$ **b.** $(x + 8)^2$ **c.** $(3x + 7)^2$ **d.** $(3m - 5)^2$
        e. $(y - 12)^2$ **f.** $(7 - x)^2$ or $(x - 7)^2$ **g.** $(6n - 1)^2$ **h.** $(5x + 6)^2$

9   Naturally all trinomials are not perfect squares. However, it is wise to test each trinomial in which two of the three terms are perfect square monomials. If the trinomial is a perfect square, the factoring is then easy. If it is not, other means of factoring must be tried.

Regardless, remove common factors first. It is also possible that the trinomial might be prime.

Example 1: Factor $-8x^2 - 24x - 18$.

Solution

$$-8x^2 - 24x - 18 = -2(4x^2 + 12x + 9)$$
$$= -2(2x + 3)^2$$

Example 2: Factor $4x^2 - 5x + 1$.

Solution

$$2(2x)(1) \neq 5x$$

Hence, $4x^2 - 5x + 1$ is not a perfect square trinomial. Therefore, try the factoring process discussed in Section 11.4. Multiply 4 by 1, which is 4.

$$(-1)(-4) \quad \text{and} \quad -1 + -4 = -5$$
$$4x^2 - 5x + 1$$
$$4x^2 - 1x - 4x + 1$$
$$x(4x - 1) - 1(4x - 1)$$
$$(x - 1)(4x - 1)$$

Q10 Factor:

a. $2x^2 + 4x + 2$ b. $x^3 - 6x^2 + 9x$

c. $m^2 - 24m - 25$ d. $3y^2 - 12y - 6$

e. $x^2 + 20x + 64$ f. $6x^2 + x - 15$

g. $y^2 - 7y + 9$ h. $49 - m^2$

i. $^-x^2 + 10x - 25$ **j.** $16x^2 - 8x + 1$

STOP • STOP • STOP • STOP • STOP • STOP • STOP • STOP • STOP

A10   **a.** $2(x + 1)^2$ **b.** $x(x - 3)^2$
         c. $(m - 25)(m + 1)$ **d.** $3(y^2 - 4y - 2)$
         e. $(x + 16)(x + 4)$ **f.** $(3x + 5)(2x - 3)$
         g. prime   **h.** $(7 + m)(7 - m)$
         i. $^-(x - 5)^2$ **j.** $(4x - 1)^2$

---

10   As you have seen, the square of a binomial is a perfect square trinomial. An important relationship exists between the terms of a perfect square trinomial. For example, $(x + 5)^2 = x^2 + 10x + 25$. Notice that $\left[\dfrac{1}{2}(10)\right]^2 = (5)^2 = 25$. This indicates that the square of one half of 10 (the coefficient of the first-degree term) equals 25 (the constant term).

Q11   **a.** $x^2 + 18x + 81$ is a perfect square trinomial.

         Is $\left[\dfrac{1}{2}(18)\right]^2 = 81?$ _____

         b. $x^2 - 14x + 49$ is a perfect square trinomial.

         Is $\left[\dfrac{1}{2}(^-14)\right]^2 = 49?$ _____

STOP • STOP • STOP • STOP • STOP • STOP • STOP • STOP • STOP

A11   **a.** yes   **b.** yes

---

11   Consider the polynomial $x^2 - 6x$. By adding 9, $\left[\dfrac{1}{2}(^-6)\right]^2$, the perfect square trinomial $x^2 - 6x + 9$ is formed. Now, $x^2 - 6x + 9 = (x - 3)^2$. This process is called *completing the square.**

*In Chapter 14, the process of "completing the square" will be applied to the solution of quadratic (second-degree) equations.

Q12   Complete the square of $x^2 + 16x$ by performing the following steps:

         a. $\left[\dfrac{1}{2}(16)\right]^2 = $ _____

         b. $x^2 + 16x + $ _____ $= ($ $)^2$

STOP • STOP • STOP • STOP • STOP • STOP • STOP • STOP • STOP

A12   **a.** 64   **b.** $x^2 + 16x + 64 = (x + 8)^2$

Q13   Complete the square of $x^2 - 20x$ by performing the following steps:

         a. $\left[\dfrac{1}{2}(^-20)\right]^2 = $ _____

b. $x^2 - 20x +$ _____ $= ($ _____ $)^2$

STOP • STOP • STOP • STOP • STOP • STOP • STOP • STOP • STOP

A13   **a.** 100   **b.** $x^2 - 20x + 100 = (x - 10)^2$

---

12   Study the following examples of completing the square and the factoring of the resulting perfect square trinomial.

$x^2 - 5x +$ ___?___ $x^2 + \dfrac{1}{2}x +$ ___?___

$\left[\dfrac{1}{2}(^-5)\right]^2 = \left(\dfrac{-5}{2}\right)^2 = \dfrac{25}{4}$ $\left[\dfrac{1}{2}\left(\dfrac{1}{2}\right)\right]^2 = \left(\dfrac{1}{4}\right)^2 = \dfrac{1}{16}$

$x^2 - 5x + \dfrac{25}{4}$ $x^2 + \dfrac{1}{2}x + \dfrac{1}{16}$

$x^2 - 5x + \left(\dfrac{5}{2}\right)^2$ $x^2 + \dfrac{1}{2}x + \left(\dfrac{1}{4}\right)^2$

$\left(x - \dfrac{5}{2}\right)^2$ $\left(x + \dfrac{1}{4}\right)^2$

The above polynomials involve rational-number coefficients. The factoring has been completed over the set of rational numbers.

Q14   Complete the square and factor:

     a. $x^2 + 2x +$ _____   **b.** $x^2 - 8x +$ _____
         $($ $)^2$ $($ $)^2$

     c. $x^2 - 12x +$ _____   **d.** $x^2 + 14x +$ _____
         $($ $)^2$ $($ $)^2$

     e. $x^2 + 3x +$ _____   **f.** $x^2 - 7x +$ _____
         $($ $)^2$ $($ $)^2$

     g. $x^2 - \dfrac{2}{5}x +$ _____   **h.** $x^2 + \dfrac{1}{3}x +$ _____
         $($ $)^2$ $($ $)^2$

STOP • STOP • STOP • STOP • STOP • STOP • STOP • STOP • STOP

A14   **a.** $x^2 + 2x + 1$ **b.** $x^2 - 8x + 16$
         $(x + 1)^2$ $(x - 4)^2$
     c. $x^2 - 12x + 36$ **d.** $x^2 + 14x + 49$
         $(x - 6)^2$ $(x + 7)^2$

     e. $x^2 + 3x + \dfrac{9}{4}$ **f.** $x^2 - 7x + \dfrac{49}{4}$

         $\left(x + \dfrac{3}{2}\right)^2$ $\left(x - \dfrac{7}{2}\right)^2$

     g. $x^2 - \dfrac{2}{5}x + \dfrac{1}{25}$ **h.** $x^2 + \dfrac{1}{3}x + \dfrac{1}{36}$

         $\left(x - \dfrac{1}{5}\right)^2$ $\left(x + \dfrac{1}{6}\right)^2$

This completes the instruction for this section.

11.5   Exercises

1. Factor:
   a. $x^2 - 100$
   b. $x^2 - 20x + 100$
   c. $25x^2 + 10x + 1$
   d. $16 - x^2$
   e. $64m^2 - 25$
   f. $64 + 16y + y^2$
   g. $x^2 + 26x + 169$
   h. $x^2 - 50x + 49$
   i. $x^2 - 48x - 100$
   j. $y^2 - 121$

2. Factor:
   a. $^-3x^2 + 75$
   b. $^-x^2 + 6x - 9$
   c. $x^3 + 18x^2 + 81x$
   d. $y^2 + 6y - 14$
   e. $8y^2 - 18$
   f. $49 - 28x + 4x^2$
   g. $2x^2 - 3x - 2$
   h. $x^3 - x$

3. (1) Replace k by the value that will make the trinomial a perfect square and (2) factor the resulting trinomial over the set of rational numbers:
   a. $x^2 - 18x + k$
   b. $y^2 - 4y + k$
   c. $x^2 + 9x + k$
   d. $x^2 + \dfrac{2}{3}x + k$
   e. $x^2 - \dfrac{1}{4}x + k$
   f. $x^2 + \dfrac{3}{5}x + k$

11.5   Exercise Answers

1. a. $(x + 10)(x - 10)$
   b. $(x - 10)^2$
   c. $(5x + 1)^2$
   d. $(4 + x)(4 - x)$
   e. $(8m + 5)(8m - 5)$
   f. $(8 + y)^2$
   g. $(x + 13)^2$
   h. $(x - 49)(x - 1)$
   i. $(x - 50)(x + 2)$
   j. $(y + 11)(y - 11)$

2. a. $^-3(x + 5)(x - 5)$
   b. $^-(x - 3)^2$
   c. $x(x + 9)^2$
   d. prime
   e. $2(2y + 3)(2y - 3)$
   f. $(7 - 2x)^2$
   g. $(2x + 1)(x - 2)$
   h. $x(x + 1)(x - 1)$

3. a. (1) $x^2 - 18x + 81$
      (2) $(x - 9)^2$
   b. (1) $y^2 - 4y + 4$
      (2) $(y - 2)^2$

   c. (1) $x^2 + 9x + \dfrac{81}{4}$
      (2) $\left(x + \dfrac{9}{2}\right)^2$
   d. (1) $x^2 + \dfrac{2}{3}x + \dfrac{1}{9}$
      (2) $\left(x + \dfrac{1}{3}\right)^2$

   e. (1) $x^2 - \dfrac{1}{4}x + \dfrac{1}{64}$
      (2) $\left(x - \dfrac{1}{8}\right)^2$
   f. (1) $x^2 + \dfrac{3}{5}x + \dfrac{9}{100}$
      (2) $\left(x + \dfrac{3}{10}\right)^2$

Chapter 11 Sample Test

At the completion of Chapter 11 it is expected that you will be able to work the following problems.

11.1   **Introduction**

1. Identify each polynomial as a monomial, binomial, or trinomial. Otherwise, write polynomial.
   a. $x^2 - 25$
   b. $x^2 - x - 6$
   c. x^2y^3
   d. $5 - y + 2y^2 - 7y^3$

2. What are three different ways of reading each power?

    a. x^2 **b.** y^3 **c.** a^n

3. What is the degree of each polynomial?

    a. $x^4 - 16$ **b.** $1 + 2x + x^2$ **c.** $^-9$ **d.** 0

    e. $13y$ **f.** $m^3 - 6m^2 + 7m$

4. **a.** What is the coefficient of $19x^5$?

    b. What is the literal coefficient of $19x^5$?

    c. Arrange $14 + x^2 - 9x$ in descending order.

    d. Arrange $14 + x^2 - 9x$ in ascending order.

    e. Evaluate $x^2 - x + 5$ when $x = 3$.

    f. Evaluate x^3 when $x = ^-4$.

    g. Evaluate $^-x^2$ when $x = 7$.

    h. Evaluate $y^3 - 3y^2 + 4y - 5$ when $y = ^-1$.

11.2   Addition and Subtraction

5. Simplify:

    a. $^-3m + 8 + 5m - 12$ **b.** $9y^2 - 4y - 5y + 3y^2 - 4$

    c. $(4x - 5x^2 - 3) + (5 - 4x + 8x^2)$ **d.** $^-(2x - 1)$

    e. $(12y - 6) - (3y + 8)$ **f.** $4x^2 - 3x + 2 - (6x^2 - 2x - 7)$

11.3   Multiplication

6. Find the product:

    a. $(^-4x^3)(5x^2)$ **b.** $^-(^-3x^2)^4$

    c. $^-3(2x^2 - 5x + 2)$ **d.** $^-5y^2(y - 2)$

    e. $(x - 3)(x - 8)$ **f.** $(y + 5)(y - 5)$

    g. $(2x + 3)(3x - 5)$ **h.** $(m + 8)^2$

    i. $(3y - 1)^2$ **j.** $(x + 1)(x^2 - 2x + 3)$

    k. $(m^2 - 3m + 2)(m^2 + 4m - 1)$ **l.** $^-4(x + 1)(x - 5)$

    m. $x(x + 6)(3x - 1)$ **n.** $^-(x + 1)(x^2 - x + 1)$

11.4   Factoring

7. Factor:

    a. $3y - 24$ **b.** $36x - 72x^2$ **c.** $^-4x^2 + 12x - 4$

    d. $x(2x + 5) - 3(2x + 5)$ **e.** $24x^3 - 48x^2$ **f.** $x^2 + 2x - 35$

    g. $x^2 - 24x + 144$ **h.** $4x^2 - 4x - 15$ **i.** $^-3x^2 + 6x + 105$

    j. $^-2x^3 - 3x^2 + 2x$

11.5   Factoring the Difference of Two Squares and Perfect Square Trinomials

8. Factor:

    a. $9x^2 - 60x + 100$ **b.** $y^2 - 225$ **c.** $16 + 9x + x^2$

    d. $36 + 12m + m^2$ **e.** $4x^2 - 100$ **f.** $x^2 - 34x - 72$

    g. $10x^2 + 9x + 2$ **h.** $4x - x^3$ **i.** $25y^2 - 70y + 49$

    j. $3y^2 + 5y - 4$

9. Replace k by the value that will make the trinomial a perfect square and factor (over the set of rational numbers):

   a. $x^2 - 3x + k$ **b.** $x^2 - 22x + k$ **c.** $x^2 + \dfrac{3}{4}x + k$

   d. $x^2 - \dfrac{4}{9}x + k$

Chapter 11 Sample Test Answers

1. **a.** binomial   **b.** trinomial   **c.** monomial   **d.** polynomial

2. **a.** x squared, the square of x, x to the second power, x to the second, x raised to the second power, the second power of x (any three)

   b. y cubed, the cube of y, y to the third power, y to the third, y raised to the third power, the third power of y (any three)

   c. a to the nth power, a to the nth, a raised to the nth power, the nth power of a (any three)

3. **a.** 4   **b.** 2   **c.** 0   **d.** none   **e.** 1   **f.** 3

4. **a.** 19   **b.** x^5 **c.** $x^2 - 9x + 14$ **d.** $14 - 9x + x^2$

   e. 11   **f.** $^-64$ **g.** $^-49$ **h.** $^-13$

5. **a.** $2m - 4$ **b.** $12y^2 - 9y - 4$

   c. $3x^2 + 2$ **d.** $^-2x + 1$ or $1 - 2x$

   e. $9y - 14$ **f.** $^-2x^2 - x + 9$ or $9 - x - 2x^2$

6. **a.** $^-20x^5$ **b.** $^-81x^8$

   c. $^-6x^2 + 15x - 6$ **d.** $^-5y^3 + 10y^2$ or $10y^2 - 5y^3$

   e. $x^2 - 11x + 24$ **f.** $y^2 - 25$

   g. $6x^2 - x - 15$ **h.** $m^2 + 16m + 64$

   i. $9y^2 - 6y + 1$ **j.** $x^3 - x^2 + x + 3$

   k. $m^4 + m^3 - 11m^2 + 11m - 2$ **l.** $^-4x^2 + 16x + 20$

   m. $3x^3 + 17x^2 - 6x$ **n.** $^-x^3 - 1$

7. **a.** $3(y - 8)$ **b.** $36x(1 - 2x)$ **c.** $^-4(x^2 - 3x + 1)$

   d. $(x - 3)(2x + 5)$ **e.** $24x^2(x - 2)$ **f.** $(x + 7)(x - 5)$

   g. $(x - 12)^2$ **h.** $(2x - 5)(2x + 3)$ **i.** $^-3(x + 5)(x - 7)$

   j. $^-x(x + 2)(2x - 1)$

8. **a.** $(3x - 10)^2$ **b.** $(y + 15)(y - 15)$ **c.** prime

   d. $(6 + m)^2$ **e.** $4(x + 5)(x - 5)$ **f.** $(x - 36)(x + 2)$

   g. $(5x + 2)(2x + 1)$ **h.** $x(2 + x)(2 - x)$ **i.** $(5y - 7)^2$

   j. prime

9. **a.** $x^2 - 3x + \dfrac{9}{4} = \left(x - \dfrac{3}{2}\right)^2$ **b.** $x^2 - 22x + 121 = (x - 11)^2$

   c. $x^2 + \dfrac{3}{4}x + \dfrac{9}{64} = \left(x + \dfrac{3}{8}\right)^2$ **d.** $x^2 - \dfrac{4}{9}x + \dfrac{4}{81} = \left(x - \dfrac{2}{9}\right)^2$

Chapter 12

Rational Algebraic Expressions

12.1   Reduction to Lowest Terms

1   A rational number is a number which can be written in the form $\frac{p}{q}$, where p and q are integers and $q \neq 0$.* A *rational expression* is an expression that can be written as a fraction, where both numerator and denominator are polynomials. All possible values of the variables that would produce a zero denominator are excluded from the replacement set. Examples of rational expressions are:

1. 5 2. 0

3. $\dfrac{-4}{7}$ 4. $7x^2$

5. $\dfrac{1}{y^3}, \quad y \neq 0$ 6. $\dfrac{4}{x - 2}, \quad x \neq 2$

7. $\dfrac{x}{x - y}, \quad x \neq y$ 8. $\dfrac{4x}{(x - 1)(x + 7)}, \quad x \neq 1$ and $x \neq {}^-7$

9. $7x^2 + 3x - 4$ 10. $\dfrac{5}{8} x$

Notice the restriction on the variables in examples 5, 6, 7, and 8. By the definition of a rational expression, all possible values of the variables that would produce a zero denominator must be excluded from the replacement set. Thus, the value of y in example 5 cannot be zero, while in example 8 the value of x cannot be 1 or $^-7$. The value $y = 0$ in example 5 and $x = 1$ or $x = {}^-7$ in example 8 would cause the value of the denominator to be zero.

     A rational expression is said to be *undefined* for any value of the variable that makes the denominator zero.

*Division by zero is impossible.

Q1   **a.** The rational expression $\dfrac{1}{x - 4}$ is undefined for $x = $ _____.

          b. $\dfrac{5}{x(x + 3)}$ is undefined for $x = 0$ and $x = $ _____.

          c. The value of $\dfrac{2}{x - 8}$ is _____ for $x = 8$.

STOP • **STOP** • **STOP** • **STOP** • **STOP** • **STOP** • **STOP** • **STOP** • **STOP**

A1   **a.** 4   **b.** $^-3$ **c.** undefined

2   Rational numbers are reduced by dividing both the numerator and denominator by some common factor.

Example 1: Reduce $\dfrac{15}{27}$.

Solution

$$\frac{15}{27} = \frac{3 \cdot 5}{3 \cdot 9} = \frac{1 \cdot 5}{1 \cdot 9} = \frac{5}{9} \qquad \text{(common factor of 3)}$$

Often the above procedure is shortened as in the following:

Example 2: Reduce $\dfrac{21}{28}$.

Solution

$$\frac{\overset{3}{\cancel{21}}}{\underset{4}{\cancel{28}}} = \frac{3}{4} \qquad \text{(common factor of 7)}$$

A rational number is said to be in lowest terms when both numerator and denominator do not contain a common factor (other than 1).

Q2   Write each of the following in lowest terms:

a. $\dfrac{56}{72} =$ _____   **b.** $\dfrac{^{-}12}{30} =$ _____

c. $\dfrac{45}{35} =$ _____   **d.** $\dfrac{^{-}54}{^{-}81} =$ _____

STOP • **STOP** • **STOP** • **STOP** • **STOP** • **STOP** • **STOP** • **STOP** • **STOP**

A2   **a.** $\dfrac{7}{9}$ **b.** $\dfrac{^{-}2}{5}$ **c.** $\dfrac{9}{7}$ **d.** $\dfrac{2}{3}$

3   Simplifying rational expressions with monomial numerators and denominators also requires dividing both numerator and denominator by a common factor. For example,

1. $\dfrac{4x}{20} = \dfrac{\overset{1}{\cancel{4}}x}{\underset{5}{\cancel{20}}} = \dfrac{x}{5}$ 2. $\dfrac{32x^7}{20x^5} = \dfrac{\overset{8\ x^2}{\cancel{32x^7}}}{\underset{5\ \ 1}{\cancel{20x^5}}} = \dfrac{8 \cdot x^2}{5 \cdot 1} = \dfrac{8x^2}{5}$ *

   (common factor of 4)   (common factor of $4x^5$)

3. $\dfrac{36x^2y}{48x^3y} = \dfrac{\overset{3\ 1\ 1}{\cancel{36x^2y}}}{\underset{4\ \ x\ 1}{\cancel{48x^3y}}} = \dfrac{3 \cdot 1 \cdot 1}{4 \cdot x \cdot 1} = \dfrac{3}{4x} \qquad \text{(common factor of } 12x^2y)$

A rational expression is said to be in lowest terms when both numerator and denominator do not contain a common factor.

*The restriction $x \neq 0$ has been omitted. Henceforth, it will be assumed in all problems that a rational expression is defined for all permissible replacements of the variable.

Q3   Simplify (reduce to lowest terms):

a. $\dfrac{12x^4}{4x^2} = $ _____

b. $\dfrac{18y^5}{24y^7} = $ _____

c. $\dfrac{7x^2y}{2y} = $ _____

d. $\dfrac{4x^3y^2}{3x^3y^3} = $ _____

STOP • **STOP** • **STOP** • **STOP** • **STOP** • **STOP** • **STOP** • **STOP** • **STOP**

A3   a. $3x^2$ b. $\dfrac{3}{4y^2}$ c. $\dfrac{7x^2}{2}$ d. $\dfrac{4}{3y}$

---

4   Various properties of exponents can be used when simplifying rational expressions.

$$\frac{x^m}{x^n} = x^{m-n} \qquad x \neq 0,\ m,n \in N,\ m > n$$

Examples:

1. $\dfrac{x^5}{x^3} = x^{5-3} = x^2 \qquad \left(\dfrac{x^5}{x^3} = \dfrac{\overset{x^2}{\cancel{x^5}}}{\underset{1}{\cancel{x^3}}} = x^2 \right)$

2. $\dfrac{y^{10}}{y} = y^{10-1} = y^9 \qquad \left(\dfrac{y^{10}}{y} = \dfrac{\overset{y^9}{\cancel{y^{10}}}}{\underset{1}{\cancel{y}}} = y^9 \right)$

3. $\dfrac{(x+3)^4}{(x+3)^2} = (x+3)^{4-2} = (x+3)^2 \qquad \left[\dfrac{(x+3)^4}{(x+3)^2} = \dfrac{\overset{1}{\cancel{(x+3)}}\overset{1}{\cancel{(x+3)}}(x+3)(x+3)}{\underset{1}{\cancel{(x+3)}}\underset{1}{\cancel{(x+3)}}} \right.$

$$\left. = (x+3)(x+3) \right]$$

Notice that the statement $m > n$ means that this rule is applied when the exponent of the variable in the *numerator* is greater than the exponent of the same variable in the *denominator*.

Q4   Simplify (use the property of exponents from Frame 4):

a. $\dfrac{y^7}{y^3} = $ _____

b. $\dfrac{10x^3}{5x^2} = $ _____

c. $\dfrac{3t^{15}}{12t^{10}} = $ _____

d. $\dfrac{(y-1)^3}{(y-1)} = $ _____

STOP • **STOP** • **STOP** • **STOP** • **STOP** • **STOP** • **STOP** • **STOP** • **STOP**

A4   a. y^4 b. $2x$ c. $\dfrac{t^5}{4}$ d. $(y-1)^2$

5

A second property of exponents is used when the exponent of the variable in the denominator is greater than the exponent of the same variable in the numerator.

$$\frac{x^m}{x^n} = \frac{1}{x^{n-m}} \qquad x \neq 0,\ m,n \in N,\ n > m$$

Examples:

1. $\dfrac{x^2}{x^6} = \dfrac{1}{x^{6-2}} = \dfrac{1}{x^4}$ $\left[\dfrac{x^2}{x^6} = \dfrac{\overset{1}{\cancel{x^2}}}{\underset{x^4}{\cancel{x^6}}} = \dfrac{1}{x^4}\right]$

2. $\dfrac{x^5}{x^7} = \dfrac{1}{x^{7-5}} = \dfrac{1}{x^2}$ $\left[\dfrac{x^5}{x^7} = \dfrac{\overset{1}{\cancel{x^5}}}{\underset{x^2}{\cancel{x^7}}} = \dfrac{1}{x^2}\right]$

3. $\dfrac{y}{y^9} = \dfrac{1}{y^{9-1}} = \dfrac{1}{y^8}$ $\left[\dfrac{y}{y^9} = \dfrac{\overset{1}{\cancel{y}}}{\underset{y^8}{\cancel{y^9}}} = \dfrac{1}{y^8}\right]$

4. $\dfrac{(z+7)^5}{(z+7)^8} = \dfrac{1}{(z+7)^{8-5}} = \dfrac{1}{(z+7)^3}$ $\left[\dfrac{(z+7)^5}{(z+7)^8} = \dfrac{\overset{1}{\cancel{(z+7)^5}}}{\underset{(z+7)^3}{\cancel{(z+7)^8}}} = \dfrac{1}{(z+7)^3}\right]$

Q5

Simplify:

a. $\dfrac{x^4}{x^{15}} =$ _____

b. $\dfrac{(y-4)}{(y-4)^5} =$ _____

c. $\dfrac{z^7}{z^8} =$ _____

d. $\dfrac{32x^{10}}{16x^{13}} =$ _____

STOP • STOP • STOP • STOP • STOP • STOP • STOP • STOP • STOP

A5

a. $\dfrac{1}{x^{11}}$

b. $\dfrac{1}{(y-4)^4}$

c. $\dfrac{1}{z}$

d. $\dfrac{2}{x^3}$

6

The quotient of any expression and itself is equal to 1. That is, $\dfrac{x^n}{x^n} = 1$ for $x \neq 0$. Property one (Frame 4) was true for $m > n$. Since it is desirable to have property one true for $m = n$, as well, x^0 is defined as follows:

$$\frac{x^n}{x^n} = x^{n-n} = x^0 = 1 \qquad x \neq 0,\ n \in N$$

Examples:

1. $\dfrac{x^5}{x^5} = x^{5-5} = x^0 = 1$

2. $\dfrac{(t-5)^2}{(t-5)^2} = (t-5)^{2-2} = (t-5)^0 = 1$

3. $\dfrac{6z^4}{2z^4} = 3z^{4-4} = 3z^0 = 3 \cdot 1 = 3$

Q6   Simplify:

a. $\dfrac{x^9}{x^9} =$ _____

b. $\dfrac{72x^5}{12x^5} =$ _____

c. $\dfrac{(x-1)^3}{(x-1)^3} =$ _____

d. $\dfrac{28(y+10)^7}{32(y+10)^7} =$ _____

STOP • **STOP** • **STOP** • **STOP** • **STOP** • **STOP** • **STOP** • **STOP** • **STOP**

A6   a. 1   b. 6   c. 1   d. $\dfrac{7}{8}$

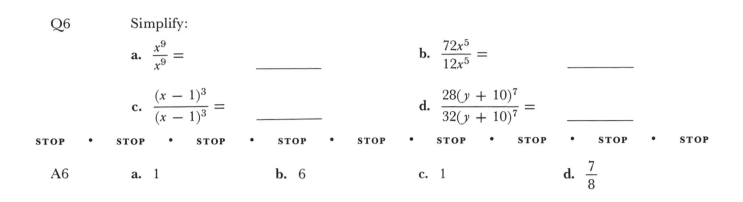

7   The properties of exponents to be used when simplifying rational expressions are now summarized. When $x \in Q$, $x \neq 0$, $m,n \in N$:

1. $\dfrac{x^m}{x^n} = x^{m-n}$ when $m > n$

2. $\dfrac{x^m}{x^n} = \dfrac{1}{x^{n-m}}$ when $n > m$

3. $\dfrac{x^m}{x^n} = x^0 = 1$ when $m = n$

Q7   Simplify (use the properties of Frame 7):

a. $\dfrac{x^{12}y^5}{x^7y^7} =$ _____

b. $\dfrac{x-1}{x-1} =$ _____

c. $\dfrac{y}{y^5} =$ _____

d. $\dfrac{(2y-3)^4}{(2y-3)^7} =$ _____

e. $\dfrac{12z^6}{8z^2} =$ _____

f. $\dfrac{56x^5y^7z^2}{16xy^7z^5} =$ _____

STOP • **STOP** • **STOP** • **STOP** • **STOP** • **STOP** • **STOP** • **STOP** • **STOP**

A7   a. $\dfrac{x^5}{y^2}$ b. 1   c. $\dfrac{1}{y^4}$ d. $\dfrac{1}{(2y-3)^3}$

e. $\dfrac{3z^4}{2}$ f. $\dfrac{7x^4}{2z^3}$

8   Simplification of rational expressions involving binomials or trinomials requires factoring unless numerator and denominator are already in factored form.

Example 1: Simplify $\dfrac{5x^2 + 10x}{x + 2}$.

Solution

$$\frac{5x^2 + 10x}{x + 2} = \frac{5x(\overset{1}{\cancel{x + 2}})}{\underset{1}{\cancel{x + 2}}} = 5x$$

Example 2: Simplify $\dfrac{x^2 - x - 12}{x^2 - 9}$.

Solution

$$\frac{x^2 - x - 12}{x^2 - 9} = \frac{(x - 4)(\overset{1}{\cancel{x + 3}})}{(x - 3)(\underset{1}{\cancel{x + 3}})} = \frac{x - 4}{x - 3}$$

Example 3:

$$\frac{(x - 5)^2}{(x - 5)(x + 3)} = \frac{(\overset{1}{\cancel{x - 5}})(x - 5)}{(\underset{1}{\cancel{x - 5}})(x + 3)} = \frac{x - 5}{x + 3}$$

(*Note:* In the second and third examples, the parentheses are no longer needed in the final answer.)

Q8   Simplify:

a. $\dfrac{2x - 8}{x^2 - 16}$

b. $\dfrac{x^2 + 5x + 4}{x^2 + x - 12}$

c. $\dfrac{5x - 15}{x^2 - 9x + 18}$

d. $\dfrac{x^2 - 25}{(x - 5)^2}$

e. $\dfrac{^-6x - 15}{4x^2 + 20x + 25}$

STOP • **STOP** • **STOP** • **STOP** • **STOP** • **STOP** • **STOP** • **STOP** • **STOP**

A8

a. $\dfrac{2}{x + 4}$: $\dfrac{2(\overset{1}{\cancel{x - 4}})}{(\underset{1}{\cancel{x - 4}})(x + 4)}$

b. $\dfrac{x + 1}{x - 3}$: $\dfrac{(\overset{1}{\cancel{x + 4}})(x + 1)}{(\underset{1}{\cancel{x + 4}})(x - 3)}$

c. $\dfrac{5}{x - 6}$: $\dfrac{5(\overset{1}{\cancel{x - 3}})}{(x - 6)(\underset{1}{\cancel{x - 3}})}$

d. $\dfrac{x + 5}{x - 5}$: $\dfrac{(\overset{1}{\cancel{x - 5}})(x + 5)}{(\underset{1}{\cancel{x - 5}})(x - 5)}$

e. $\dfrac{-3}{2x+5} : \dfrac{\overset{1}{-3(\cancel{2x+5})}}{\underset{1}{(\cancel{2x+5})}(2x+5)}$

---

9   Simplification of rational expressions by reducing to lowest terms is done by dividing common factors from the numerator and denominator. Thus, it must be possible to factor before reducing can occur.

A common error in reducing rational expressions is to attempt to divide out common terms rather than common factors. For example, the rational number $\dfrac{7}{8}$ is already reduced to lowest terms. If $\dfrac{7}{8}$ is rewritten as $\dfrac{4+3}{4+4}$, the 4's that appear in the numerator and denominator are common terms and not common factors and thus cannot be divided out. Similarly, in the rational expression $\dfrac{x+2}{x-5}$ the x's are common terms rather than common factors and thus cannot be divided out. Simplification of rational expressions by reducing to lowest terms is done by dividing common factors from both numerator and denominator. Thus, it must be possible to factor before reducing can occur.

Example 1: Simplify $\dfrac{x+3}{x}$.

Solution

The expression is already in lowest terms. There is no common factor in both the numerator and denominator.

Example 2: Simplify $\dfrac{x^2+3x}{x}$.

Solution

$$\frac{x^2+3x}{x} = \frac{\overset{1}{\cancel{x}}(x+3)}{\underset{1}{\cancel{x}}} = x+3$$

The numerator has factors x and $x+3$. Since the denominator also has a factor x,* numerator and denominator can be divided by x.

Example 3: Simplify $\dfrac{x^2+25}{x^2+10x+25}$.

Solution

$$\frac{x^2+25}{x^2+10x+25} = \frac{x^2+25}{(x+5)(x+5)}$$

The expression is already in lowest terms. The numerator is prime and there is no factor common to both numerator and denominator.

*Recall that $x = 1x$.

Example 4: Simplify $\dfrac{x^2 - 25}{x^2 + 10x + 25}$.

Solution

$$\frac{x^2 - 25}{x^2 + 10x + 25} = \frac{(x - 5)\overset{1}{\cancel{(x + 5)}}}{(x + 5)\underset{1}{\cancel{(x + 5)}}} = \frac{x - 5}{x + 5}$$

The expression is written in lowest terms by dividing both numerator and denominator by the common factor $x + 5$.

Q9   Simplify, if possible. If the expression is already in lowest terms, write "cannot be simplified."

a. $\dfrac{y^2 + 7}{7y}$

b. $\dfrac{y^2 + 7y}{7y}$

c. $\dfrac{x - 7}{x + 7}$

d. $\dfrac{x^2 - 49}{x + 7}$

STOP • STOP • STOP • STOP • STOP • STOP • STOP • STOP • STOP

A9   a. Cannot be simplified.

b. $\dfrac{y + 7}{7}$: $\dfrac{\overset{1}{\cancel{y}}(y + 7)}{7\underset{1}{\cancel{y}}}$

c. Cannot be simplified.

d. $x - 7$: $\dfrac{(x - 7)\overset{1}{\cancel{(x + 7)}}}{\underset{1}{\cancel{(x + 7)}}}$

10   The multiplication property of $^-1$ states that $^-1x = {}^-x$. In words, the property says that $^-1$ times a number is equal to the opposite of the number. This property can also be applied to rational expressions:

$$^-1\underbrace{(a - b)} = {}^-(a - b) = {}^-a - {}^-b = \underbrace{{}^-a + b} \text{ or } b - a$$
$$\qquad\qquad\text{opposites}$$

Recall that the opposite of an expression is formed by finding the opposite of each term of the expression.

Examples:

1. $^-1(x - 4) = {}^-(x - 4)$
    $\qquad\qquad = {}^-x - {}^-4$ (this step is usually omitted)
    $\qquad\qquad = {}^-x + 4 \text{ or } 4 - x$

2. $^-1(^-2x^2 - 3x + 7) = {}^-(^-2x^2 - 3x + 7)$
    $\qquad\qquad\qquad = 2x^2 + 3x - 7$

Q10   Complete each of the following:

  a. $^-1(y + 7) =$ _____   **b.** $^-1(4 - 2y) =$ _____

  c. $^-(8 + x) =$ _____   **d.** $^-(x^2 - 3x - 2) =$ _____

STOP • STOP • STOP • STOP • STOP • STOP • STOP • STOP • STOP

A10   **a.** $^-y - 7$ **b.** $^-4 + 2y$ or $2y - 4$
   c. $^-8 - x$ **d.** $^-x^2 + 3x + 2$

---

11   From Frame 10 it can be seen that opposites differ by a factor of $^-1$. Thus, it is possible to write any expression as the product of $^-1$ and the opposite of the expression.

Example 1: $x + 5 = {}^-1(?)$

Solution

The opposite of $x + 5$ is $^-x - 5$. Therefore, $x + 5 = {}^-1(^-x - 5)$.

Example 2: $2y - 7 = {}^-1(?)$

Solution

The opposite of $2y - 7$ is $^-2y + 7$. Therefore, $2y - 7 = {}^-1(^-2y + 7)$.

Example 3: Write $3 - 5x$ as the product of $^-1$ and its opposite.

Solution

The opposite of $3 - 5x$ is $^-3 + 5x$ or $5x - 3$. Therefore,

$3 - 5x = {}^-1(^-3 + 5x)$ or $3 - 5x = {}^-1(5x - 3)$.

Example 4: Factor $^-1$ from the expression $^-x - 3$.

Solution

The opposite of $^-x - 3$ is $x + 3$. Therefore, $^-x - 3 = {}^-1(x + 3)$.

Q11   **a.** $x + 5 = {}^-1($ $)$ **b.** $^-3 - y = {}^-1($ $)$

  c. Write $^-4y - 7$ as the product of $^-1$ and its opposite.   _____

  d. Factor $^-1$ from $6 - x$.   _____

  e. Factor $^-1$ from $y - 9$.   _____

STOP • STOP • STOP • STOP • STOP • STOP • STOP • STOP • STOP

A11   **a.** $^-x - 5$ **b.** $3 + y$ **c.** $^-1(4y + 7)$
   d. $^-1(^-6 + x)$ or $^-1(x - 6)$ **e.** $^-1(^-y + 9)$ or $^-1(9 - y)$

---

12   The multiplication property of $^-1$ can be used to reduce rational expressions to lowest terms when factors of the numerator and denominator are opposites of each other.

Examples:

1. $$\dfrac{x - 2}{2 - x} = \dfrac{x - 2}{^-1(^-2 + x)} = \dfrac{\overset{1}{\cancel{x - 2}}}{^-1(\underset{1}{\cancel{x - 2}})} = \dfrac{1}{^-1} = {}^-1$$

$^-1$ could also have been factored from the numerator.

$$\frac{x-2}{2-x} = \frac{^-1(^-x+2)}{2-x} = \frac{^-1(\overset{1}{2-x})}{\underset{1}{2-x}} = {}^-1$$

2. $$\frac{m-n}{n-m} = \frac{^-1(^-m+n)}{n-m} = \frac{^-1(\overset{1}{n-m})}{\underset{1}{n-m}} = {}^-1$$

3. $$\frac{2x-y}{y-2x} = \frac{^-1(^-2x+y)}{y-2x} = \frac{^-1(\overset{1}{y-2x})}{\underset{1}{y-2x}} = {}^-1$$

4. $$\frac{^-x-2}{x+2} = \frac{^-1(\overset{1}{x+2})}{\underset{1}{x+2}} = {}^-1$$

Q12 Simplify:

a. $$\frac{3-y}{y-3}$$

b. $$\frac{x+5}{^-x-5}$$

STOP • **STOP** • **STOP** • **STOP** • **STOP** • **STOP** • **STOP** • **STOP** • **STOP**

A12 a. $^-1$: $$\frac{3-y}{y-3} = \frac{^-1(^-3+y)}{y-3} = \frac{^-1(\overset{1}{y-3})}{\underset{1}{y-3}} = {}^-1$$

b. $^-1$: $$\frac{x+5}{^-x-5} = \frac{\overset{1}{x+5}}{\underset{1}{^-1(x+5)}} = \frac{1}{^-1} = {}^-1$$

13 The expressions $a-b$ and ^-a+b are *opposites* and their quotient can be shown to be $^-1$.

$$\frac{a-b}{^-a+b} = \frac{a-b}{^-1(a-b)} = \frac{\overset{1}{a-b}}{\underset{1}{^-1(a-b)}} = \frac{1}{^-1} = {}^-1$$

However, the expressions $a-b$ and $a+b$ are called *conjugates* * and their quotient cannot be simplified. That is, $\dfrac{a-b}{a+b}$ is already reduced to lowest terms.

Care should be taken when simplifying algebraic expressions so that opposites and conjugates are not confused.

Example 1: Simplify $\dfrac{x-2}{x+2}$.

*Conjugates are two binomials which differ only in the sign of the second term.

Solution

$x - 2$ and $x + 2$ are conjugates and thus the expression is already in lowest terms.

Example 2: Simplify $\dfrac{x - 2}{2 - x}$.

Solution

$$\frac{x - 2}{2 - x} = \frac{x - 2}{^-1(^-2 + x)} = \frac{\overset{1}{\cancel{x - 2}}}{^-1(\underset{1}{\cancel{x - 2}})} = {}^-1$$

Example 3: Simplify $\dfrac{^-2t + 3}{3 - 2t}$.

Solution

$$\frac{^-2t + 3}{3 - 2t} = \frac{3 - 2t}{3 - 2t} = 1$$

Q13   Simplify, if possible. Otherwise, write "cannot be simplified."

a. $\dfrac{x - 2y}{2y - x} = $ _____

b. $\dfrac{y - 3}{y + 3} = $ _____

c. $\dfrac{^-y + 5}{5 - y} = $ _____

d. $\dfrac{x^2 - 3x - 10}{^-x^2 + 3x + 10} = $ _____

STOP • **STOP** • **STOP** • **STOP** • **STOP** • **STOP** • **STOP** • **STOP** • **STOP**

A13   a. $^-1$ b. Cannot be simplified.
         c. 1   d. $^-1$

Q14   Simplify:

a. $\dfrac{x^2 + x}{x}$

b. $\dfrac{x^2 - 9}{x + 3}$

c. $\dfrac{2y^2 - y - 1}{y^2 - 1}$

d. $\dfrac{4x - 20}{12x + 36}$

e. $\dfrac{3 - 2y}{2y - 3}$

f. $\dfrac{^-5 + x}{x - 5}$

STOP • **STOP** • **STOP** • **STOP** • **STOP** • **STOP** • **STOP** • **STOP** • **STOP**

A14
    a. $x + 1$ **b.** $x - 3$

    c. $\dfrac{2y + 1}{y + 1} : \dfrac{(2y + 1)(y - 1)}{(y - 1)(y + 1)}$ **d.** $\dfrac{x - 5}{3(x + 3)} : \dfrac{4(x - 5)}{12(x + 3)}$

    e. $^-1$ **f.** 1

---

14   An algebraic expression is reduced to *lowest terms* if *no common factor* remains in both the numerator and denominator. In this text, answers involving rational expression may be left with parentheses as long as the result is in lowest terms.

This completes the instruction for this section.

12.1   Exercises

1. A rational expression is undefined for any value of a variable that produces a denominator of _____.
2. For which value(s) of the variable are each of the following undefined:

    a. $\dfrac{1}{x}$ _____   **b.** $\dfrac{3}{x - 4}$ _____

    c. $\dfrac{x}{x + 7}$ _____   **d.** $\dfrac{x - 2}{(x - 1)(x + 2)}$ _____

    e. $\dfrac{3}{x(x - 5)}$ _____   **f.** $\dfrac{x - 2}{5}$ _____

3. The process of reducing means to _____ both the numerator and the denominator by some common _____.
4. If both the numerator and denominator of a rational expression do not contain a common factor, the expression is said to be in _____ _____. (two words)
5. Simplify:

    a. $\dfrac{54x^2}{10x^5}$ **b.** $\dfrac{9x^7}{27x^3}$ **c.** $\dfrac{24x^3y}{36xy^5}$ **d.** $\dfrac{15mn}{4n}$

    e. $\dfrac{18x^7}{3x^7}$ **f.** $\dfrac{5y^3}{25y^3}$

6. Simplify:

    a. $\dfrac{(x - 2)^2}{(x - 2)^3}$ **b.** $\dfrac{(x - 2)^3}{(x - 2)^2}$ **c.** $\dfrac{(x - 2)^3}{(x - 2)^3}$ **d.** $\dfrac{x - 2}{(x - 2)^2}$

7. Simplify:

    a. $\dfrac{x^2 - 16}{3x - 12}$ **b.** $\dfrac{x^2 - 11x + 30}{x^2 - 2x - 15}$ **c.** $\dfrac{5x + 15}{x^2 - 9}$

    d. $\dfrac{(x - 1)^3}{x^2 - 1}$ **e.** $\dfrac{x^2 + 5x + 6}{x^2 + 7x + 12}$ **f.** $\dfrac{x^2 - 100}{x^2 - 7x - 30}$

    g. $\dfrac{^-x + 7}{x - 7}$ **h.** $\dfrac{2y - 6}{^-6 + 2y}$ **i.** $\dfrac{9x^2 - 30x + 25}{3x + 5}$

    j. $\dfrac{3x^2 - 12x}{x^2 + 7x - 44}$

8. Simplify, if possible. If the expression is already in lowest terms, write "cannot be simplified."

   a. $\dfrac{x+3}{3x}$ 　　　 b. $\dfrac{x^2+3x}{3x}$ 　　　 c. $\dfrac{y-7}{7}$ 　　　 d. $\dfrac{7y-7}{7}$

   e. $\dfrac{x-5}{x+5}$ 　　　 f. $\dfrac{^-x-5}{x+5}$ 　　　 g. $\dfrac{3-x}{x-3}$ 　　　 h. $\dfrac{x^2-4}{(x-4)^2}$

12.1 　　Exercise Answers

1. zero
2. a. $x=0$ 　　　　　　 b. $x=4$ 　　　　　　 c. $x=^-7$
   d. $x=1,\ x=^-2$ 　　 e. $x=0,\ x=5$
   f. The denominator is never zero and thus the expression is always defined.
3. divide, factor
4. lowest terms
5. a. $\dfrac{27}{5x^3}$ 　　　 b. $\dfrac{x^4}{3}$ 　　　 c. $\dfrac{2x^2}{3y^4}$ 　　　 d. $\dfrac{15m}{4}$

   c. 6 　　　　　　 f. $\dfrac{1}{5}$

6. a. $\dfrac{1}{x-2}$ 　　　 b. $x-2$ 　　　 c. 1 　　　 d. $\dfrac{1}{x-2}$

7. a. $\dfrac{x+4}{3}$ 　　　 b. $\dfrac{x-6}{x+3}$ 　　　 c. $\dfrac{5}{x-3}$ 　　　 d. $\dfrac{(x-1)^2}{x+1}$

   e. $\dfrac{x+2}{x+4}$ 　　　 f. $\dfrac{x+10}{x+3}$ 　　　 g. $^-1$ 　　　 h. 1

   i. $\dfrac{(3x-5)^2}{3x+5}$ 　　 j. $\dfrac{3x}{x+11}$

8. a. cannot be simplified 　　　 b. $\dfrac{x+3}{3}$

   c. cannot be simplified 　　　 d. $y-1$
   e. cannot be simplified 　　　 f. $^-1$
   g. $^-1$ 　　　　　　　　　　 h. cannot be simplified

12.2 　 **Multiplication and Division**

1 　 The product of any two rational numbers $\dfrac{a}{b}$ and $\dfrac{c}{d}$ is defined as the product of the numerators, ac, divided by the product of the denominators, bd. That is,

$$\frac{a}{b} \cdot \frac{c}{d} = \frac{ac}{bd} \qquad \text{where } \frac{a}{b} \text{ and } \frac{c}{d} \in Q$$

Example:

$$\frac{3}{5} \cdot \frac{4}{13} = \frac{3 \cdot 4}{5 \cdot 13} = \frac{12}{65}$$

(this step is usually done mentally

When finding a product of any two rational numbers, the final result should always be written in lowest terms. It is customary to divide out common factors from both numerator and denominator before the definition of the product is applied.

Example:

$$\frac{4}{15} \cdot \frac{25}{32} = \frac{\overset{1}{\cancel{4}}}{\underset{3}{\cancel{15}}} \cdot \frac{\overset{5}{\cancel{25}}}{\underset{8}{\cancel{32}}} = \frac{5}{24}$$

Q1   Find each of the following products:

a. $\dfrac{8}{18} \cdot \dfrac{6}{24}$ b. $\dfrac{13}{15} \cdot \dfrac{30}{26}$ c. $\dfrac{^{-}12}{56} \cdot \dfrac{8}{60}$ d. $\dfrac{^{-}9}{16} \cdot \dfrac{^{-}3}{5}$

STOP • **STOP** • **STOP** • **STOP** • **STOP** • **STOP** • **STOP** • **STOP** • **STOP**

A1   a. $\dfrac{1}{9}$ b. 1   c. $\dfrac{^{-}1}{35}$ d. $\dfrac{27}{80}$

2   The product of any two *rational expressions* is defined exactly as the product of two rational numbers. As with rational numbers, all answers are expressed in lowest terms by dividing common factors from both numerator and denominator.

Example 1: Multiply $\dfrac{5x}{12} \cdot \dfrac{4}{15x^2}$.

Solution

$$\frac{5x}{12} \cdot \frac{4}{15x^2} = \frac{\overset{1}{\cancel{5x}}}{\underset{3}{\cancel{12}}} \cdot \frac{\overset{1}{\cancel{4}}}{\underset{3x}{\cancel{15x^2}}} = \frac{1}{9x}$$

Example 2: Multiply $\dfrac{^{-}3y^3}{7x} \cdot \dfrac{21x^2}{27y}$.

Solution

$$\frac{^{-}3y^3}{7x} \cdot \frac{21x^2}{27y} = \frac{\overset{^{-}1y^2}{\cancel{^{-}3y^3}}}{\underset{1}{\cancel{7x}}} \cdot \frac{\overset{\overset{1}{\cancel{3}}x}{\cancel{21x^2}}}{\underset{\underset{3}{\cancel{9}}}{\cancel{27y}}} = \frac{^{-}xy^2}{3}$$

Example 3: Multiply $\dfrac{4x}{3y^2} \cdot \dfrac{6y^2}{8x}$.

Solution

$$\frac{4x}{3y^2} \cdot \frac{6y^2}{8x} = \frac{\overset{1}{\cancel{4x}}}{\underset{1}{\cancel{3y^2}}} \cdot \frac{\overset{\overset{1}{\cancel{2}}}{\cancel{6y^2}}}{\underset{\underset{1}{\cancel{2}}}{\cancel{8x}}} = \frac{1}{1} = 1$$

Q2   **a.** $\dfrac{5x^2}{4} \cdot \dfrac{3}{2x^3}$

b. $\dfrac{6x^3}{20y^2} \cdot \dfrac{5y}{18x}$

c. $\dfrac{-5x^4}{7y^3} \cdot \dfrac{49y^3}{15x^3}$

d. $\dfrac{-12x^7}{3x^3} \cdot \dfrac{12x^5}{48x^9}$

STOP • STOP • STOP • STOP • STOP • STOP • STOP • STOP • STOP

A2   **a.** $\dfrac{15}{8x}$ **b.** $\dfrac{x^2}{12y}$ **c.** $\dfrac{-7x}{3}$ **d.** $^{-}1$

3   When finding a product of rational expressions, it is often necessary to factor using the procedures of Chapter 11 before dividing by any common factors.

Example 1: Multiply $\dfrac{3x - 6}{8x} \cdot \dfrac{2}{x^2 - 4}$.

Solution

$$\frac{3x - 6}{8x} \cdot \frac{2}{x^2 - 4} = \frac{3\overset{1}{\cancel{(x - 2)}}}{\underset{1}{\cancel{2} \cdot 4x}} \cdot \frac{\overset{1}{\cancel{2}}}{\cancel{(x - 2)}(x + 2)} = \frac{3}{4x(x + 2)}$$

Example 2: Multiply $\dfrac{x^2 + 3x - 4}{5x + 15} \cdot \dfrac{5x}{2x + 8}$.

Solution

$$\frac{x^2 + 3x - 4}{5x + 15} \cdot \frac{5x}{2x + 8} = \frac{\overset{1}{\cancel{(x + 4)}}(x - 1)}{\underset{1}{\cancel{5}(x + 3)}} \cdot \frac{\overset{1}{\cancel{5}}x}{2\cancel{(x + 4)}} = \frac{x(x - 1)}{2(x + 3)}$$

Example 3: Multiply $\dfrac{x^2 - 81}{x^2 - x - 12} \cdot \dfrac{x^2 - 8x - 33}{2x + 18}$.

Solution

$$\frac{x^2 - 81}{x^2 - x - 12} \cdot \frac{x^2 - 8x - 33}{2x + 18} = \frac{(x - 9)\overset{1}{\cancel{(x + 9)}}}{(x - 4)\cancel{(x + 3)}} \cdot \frac{\cancel{(x + 3)}(x - 11)}{2\overset{1}{\cancel{(x + 9)}}} = \frac{(x - 9)(x - 11)}{2(x - 4)}$$

Q3   Find the product:

a. $\dfrac{y^2 - 49}{x} \cdot \dfrac{7x^2}{4y + 28}$

b. $\dfrac{5x - 19}{(x + 2)^2} \cdot \dfrac{x^2 + 4x + 4}{5x - 19}$

c. $\dfrac{15x + 25}{x + 5} \cdot \dfrac{x^2 - 25}{15x^2}$

d. $\dfrac{6x^3}{6 - x} \cdot \dfrac{x - 5}{5x}$

STOP • STOP • STOP • STOP • STOP • STOP • STOP • STOP • STOP

A3   a. $\dfrac{7x(y - 7)}{4}$: $\dfrac{(y - 7)(y + 7)}{x} \cdot \dfrac{7x^2}{4(y + 7)}$

b. 1: $\dfrac{5x - 19}{(x + 2)^2} \cdot \dfrac{(x + 2)^2}{5x - 19}$

c. $\dfrac{(3x + 5)(x - 5)}{3x^2}$: $\dfrac{5(3x + 5)}{x + 5} \cdot \dfrac{(x - 5)(x + 5)}{15x^2}$

d. $\dfrac{6x^2(x - 5)}{5(6 - x)}$

---

4   Recall from Section 12.1 that the quotient of opposite rational expressions is $^-1$.

Examples:

1. $\dfrac{^-5}{5} = {}^-1$

2. $\dfrac{^-x + 7}{x - 7} = {}^-1$ Verification: $\dfrac{^-x + 7}{x - 7} = \dfrac{^-1\overset{1}{\cancel{(x - 7)}}}{\cancel{x - 7}} = {}^-1$

3. $\dfrac{y - 5}{5 - y} = {}^-1$ Verification: $\dfrac{y - 5}{5 - y} = \dfrac{^-1(^-y + 5)}{5 - y} = \dfrac{^-1\overset{1}{\cancel{(5 - y)}}}{\cancel{5 - y}} = {}^-1$

4. $\dfrac{x^2 - 3x + 9}{^-x^2 + 3x - 9} = {}^-1$ Verification: $\dfrac{x^2 - 3x + 9}{^-x^2 + 3x - 9} = \dfrac{\overset{1}{\cancel{x^2 - 3x + 9}}}{^-1\cancel{(x^2 - 3x + 9)}} = {}^-1$

Q4 Verify that $\dfrac{y-6}{6-y} = {}^-1$.

STOP • STOP • STOP • STOP • STOP • STOP • STOP • STOP • STOP

A4 $\dfrac{y-6}{6-y} = \dfrac{{}^-(y+6)}{(6-y)} = \dfrac{{}^-1(\overset{1}{\cancel{6-y}})}{\underset{1}{\cancel{6-y}}} = {}^-1$

or

$\dfrac{y-6}{6-y} = \dfrac{y-6}{{}^-1({}^-6+y)} = \dfrac{\overset{1}{\cancel{y-6}}}{\underset{1}{{}^-1(\cancel{y-6})}} = {}^-1$

Q5 Verify that $\dfrac{{}^-x+3}{x-3} = {}^-1$.

STOP • STOP • STOP • STOP • STOP • STOP • STOP • STOP • STOP

A5 $\dfrac{{}^-x+3}{x-3} = \dfrac{{}^-1(\overset{1}{\cancel{x-3}})}{\underset{1}{\cancel{x-3}}} = {}^-1$ or $\dfrac{{}^-x+3}{x-3} = \dfrac{\overset{1}{\cancel{{}^-x+3}}}{\underset{1}{{}^-1(\cancel{{}^-x+3})}} = {}^-1$

5 When finding a product it is often possible to divide out factors that are opposites of each other.

Example 1: Multiply $\dfrac{x-3}{4} \cdot \dfrac{8}{3-x}$ ($x-3$ and $3-x$ are opposites).

Solution

$\dfrac{\overset{-1}{\cancel{x-3}}}{\underset{1}{4}} \cdot \dfrac{\overset{2}{\cancel{8}}}{\underset{1}{\cancel{3-x}}} = \dfrac{{}^-2}{1} = {}^-2$

Example 2: Multiply $\dfrac{x^2-1}{x+7} \cdot \dfrac{x+7}{1-x}$.

Solution

$\dfrac{x^2-1}{x+7} \cdot \dfrac{x+7}{1-x} = \dfrac{\overset{-1}{(\cancel{x-1})}(x+1)}{\underset{1}{\cancel{x+7}}} \cdot \dfrac{\overset{1}{\cancel{x+7}}}{\underset{1}{\cancel{1-x}}} = {}^-1(x+1) = {}^-x-1$

($x-1$ and $1-x$ are opposites.)

Example 3: Multiply $\dfrac{3x^2}{^-x + 5} \cdot \dfrac{25 - x^2}{x^3}$.

Solution

$$\frac{3x^2}{^-x + 5} \cdot \frac{25 - x^2}{x^3} = \frac{\overset{1}{3\cancel{x^2}}}{\cancel{^-x + 5}} \cdot \frac{\overset{1}{(\cancel{5 - x})}(5 + x)}{\cancel{x^2} \cdot x} = \frac{3(5 + x)}{x}$$

(Notice that $^-x + 5$ and $5 - x$ are equivalent expressions and not opposites.)

Q6 Find the product:

a. $\dfrac{7}{y - 11} \cdot \dfrac{11 - y}{3}$

b. $\dfrac{x^2 - 16}{8x} \cdot \dfrac{4x^2}{^-x - 4}$

c. $\dfrac{8 - x}{5x} \cdot \dfrac{x - 5}{x^2 + 16x + 64}$

d. $\dfrac{^-x + 3}{x^2 - 9} \cdot \dfrac{x^2 + x - 6}{x - 2}$

STOP • STOP • STOP • STOP • STOP • STOP • STOP • STOP • STOP

A6 a. $\dfrac{^-7}{3}$: $\dfrac{7}{\underset{1}{\cancel{y - 11}}} \cdot \dfrac{\overset{^-1}{\cancel{11 - y}}}{3} = \dfrac{^-7}{3}$

b. $\dfrac{^-x(x - 4)}{2}$: $\dfrac{(x - 4)\overset{^-1}{\cancel{(x + 4)}}}{\underset{2}{\cancel{8x}}} \cdot \dfrac{\overset{1x}{\cancel{4x^2}}}{\underset{1}{\cancel{^-x - 4}}}$

c. $\dfrac{(8 - x)(x - 5)}{5x(x + 8)^2}$: $\dfrac{8 - x}{5x} \cdot \dfrac{x - 5}{(x + 8)^2}$ (already in lowest terms)

d. $^-1$: $\dfrac{\overset{^-1}{\cancel{^-x + 3}}}{\underset{1}{\cancel{(x - 3)}}\underset{1}{\cancel{(x + 3)}}} \cdot \dfrac{\overset{1}{\cancel{(x + 3)}}\overset{1}{\cancel{(x - 2)}}}{\underset{1}{\cancel{(x - 2)}}}$

6 Recall from Chapter 3 that the reciprocal of a rational number $\dfrac{a}{b}$ for $a \neq 0$ is $\dfrac{b}{a}$.

Examples:

1. The reciprocal of $\dfrac{5}{3}$ is $\dfrac{3}{5}$.

2. The reciprocal of $^-2$ $\left(\text{or } \dfrac{^-2}{1}\right)$ is $\dfrac{1}{^-2}$ or $\dfrac{^-1}{2}$.

(The single negative sign is usually placed in the numerator.) The reciprocal of a rational expression is defined in the same way.

Examples:

3. The reciprocal of $\dfrac{x-3}{x^2+7x-6}$ is $\dfrac{x^2+7x-6}{x-3}$.

4. The reciprocal of x^2-9 $\left(\text{or }\dfrac{x^2-9}{1}\right)$ is $\dfrac{1}{x^2-9}$.

Q7   Give the reciprocal:

a. $\dfrac{-5}{9}$ _____

b. $\dfrac{x-7}{5y^2}$ _____

c. $\dfrac{x^2-25}{x^2-9x+8}$ _____

d. $y-2$ _____

e. $\dfrac{1}{3-x}$ _____

f. $(y-3)^2$ _____

STOP • **STOP** • **STOP** • **STOP** • **STOP** • **STOP** • **STOP** • **STOP** • **STOP**

A7   **a.** $\dfrac{-9}{5}$

b. $\dfrac{5y^2}{x-7}$

c. $\dfrac{x^2-9x+8}{x^2-25}$

d. $\dfrac{1}{y-2}$

e. $3-x$

f. $\dfrac{1}{(y-3)^2}$

---

7   The quotient of two rational numbers is defined as follows: For any two rational numbers $\dfrac{a}{b}$ and $\dfrac{c}{d}$ where $\dfrac{c}{d} \neq 0$,

$$\frac{a}{b} \div \frac{c}{d} = \frac{a}{b} \cdot \frac{d}{c} = \frac{ad}{bc}$$

Examples:

$$\frac{4}{9} \div \frac{5}{18} = \frac{4}{\cancel{9}_1} \cdot \frac{\cancel{18}^{2}}{5} = \frac{8}{5}$$

$$\frac{5}{7} \div 3 = \frac{5}{7} \div \frac{3}{1} = \frac{5}{7} \cdot \frac{1}{3} = \frac{5}{21}$$

To find the quotient of two rational expressions, use the definition of this frame. That is, to divide two rational expressions, multiply the dividend (first expression) by the reciprocal of the divisor (second expression).

Example 1: Divide $\dfrac{x-3}{x^2+2x+1} \div \dfrac{3x}{x^2-1}$.

Solution

$$\frac{x - 3}{x^2 + 2x + 1} \div \frac{3x}{x^2 - 1} = \frac{x - 3}{(x + 1)\cancel{(x + 1)}} \cdot \frac{\overset{1}{\cancel{(x - 1)}}\cancel{(x + 1)}}{3x} = \frac{(x - 3)(x - 1)}{3x(x + 1)}$$

(*Note:* It is common to form the reciprocal and factor in the same step.)

Example 2: Divide $\dfrac{15x^2y}{4 - x} \div \dfrac{5xy}{x - 4}$.

Solution

$$\frac{15x^2y}{4 - x} \div \frac{5xy}{x - 4} = \frac{\overset{3x}{\cancel{15x^2y}}}{\underset{1}{\cancel{4 - x}}} \cdot \frac{\overset{-1}{\cancel{x - 4}}}{\underset{1}{\cancel{5xy}}} = {}^-3x$$

The final product is again always expressed in lowest terms by dividing out common factors.

Q8   Find the quotient:

a. $\dfrac{3x}{42y} \div \dfrac{12x}{7y^2}$

b. $\dfrac{x^2 - 16}{x^2 - x - 20} \div \dfrac{4x}{x - 5}$

c. $\dfrac{x^2 - 10x + 21}{(x - 7)^2} \div \dfrac{9x - 27}{x^2 - 14x + 49}$

d. $\dfrac{x^2 + 3x - 10}{x^2 - 2x - 3} \div \dfrac{x^2 + 4x - 5}{x^2 - 4x + 3}$

e. $\dfrac{3x - 15}{27x} \div (x - 5)$

STOP • **STOP** • **STOP** • **STOP** • **STOP** • **STOP** • **STOP** • **STOP** • **STOP**

A8   a. $\dfrac{y}{24}$ b. $\dfrac{x - 4}{4x}$ c. $\dfrac{x - 7}{9}$ d. $\dfrac{x - 2}{x + 1}$

e. $\dfrac{1}{9x}$

8   A common error in finding the quotient of two algebraic expressions is to try to divide out common factors before finding the reciprocal of the divisor. Common factors should only be divided out after the quotient has been changed to a product by inverting the divisor.

Example 1: Find $\dfrac{x + 6}{4x} \div \dfrac{4x}{x + 6}$. It is incorrect to divide out the common factors $x + 6$ and $4x$. The quotient must first be changed to a product.

Solution

$$\frac{x + 6}{4x} \div \frac{4x}{x + 6} = \frac{x + 6}{4x} \cdot \frac{x + 6}{4x} = \frac{(x + 6)^2}{16x^2}$$

Example 2: Find $\dfrac{4(x + 5)}{x^2 - 25} \div \dfrac{x + 5}{8x^2}$. To solve, first change the quotient to a product and then divide out any common factors.

Solution

$$\frac{4(x + 5)}{x^2 - 25} \div \frac{x + 5}{8x^2} = \frac{4\overset{1}{\cancel{(x + 5)}}}{(x - 5)(x + 5)} \cdot \frac{8x^2}{\underset{1}{\cancel{x + 5}}} = \frac{32x^2}{(x - 5)(x + 5)}$$

Q9   Find the quotients:

a. $\dfrac{35x^2}{7y} \div \dfrac{7y}{21x}$

b. $\dfrac{3x - 27}{x^2 - 81} \div \dfrac{x + 9}{3x - 27}$

STOP • **STOP** • **STOP** • **STOP** • **STOP** • **STOP** • **STOP** • **STOP** • **STOP**

A9   a. $\dfrac{15x^3}{y^2}$ b. $\dfrac{9(x - 9)}{(x + 9)^2}$

Q10   Find the product or quotient as indicated:

a. $\dfrac{32x^2}{44} \cdot \dfrac{10x - 4}{24x}$

b. $\dfrac{x - 5}{3x} \div \dfrac{3x}{15x}$

c. $\dfrac{x^2 - 4x + 4}{x^2 - 4} \div \dfrac{x^2 - 4}{4}$

d. $\dfrac{2x^2 - 7x - 15}{x - 2} \cdot \dfrac{18 - 9x}{4x - 20}$

STOP • **STOP** • **STOP** • **STOP** • **STOP** • **STOP** • **STOP** • **STOP** • **STOP**

A10   a. $\dfrac{x(10x - 4)}{33}$ b. $\dfrac{5(x - 5)}{3x}$ c. $\dfrac{4}{(x + 2)^2}$ d. $\dfrac{{}^-9(2x + 3)}{4}$

This completes the instruction for this section.

12.2 Exercises

1. Find the product:

a. $\dfrac{4}{18} \cdot \dfrac{^-9}{12}$

b. $\dfrac{^-15}{27} \cdot \dfrac{^-9}{5}$

c. $\dfrac{24}{56} \cdot \dfrac{16}{28}$

d. $\dfrac{^-17}{51} \cdot 3$

e. $\dfrac{18x}{42y^4} \cdot \dfrac{7x^2y^2}{54x}$

f. $\dfrac{5x}{^-x+3} \cdot \dfrac{x-3}{^-x}$

g. $\dfrac{3x^2-48}{9x^2} \cdot \dfrac{x^3}{x+4}$

h. $\dfrac{x^2-12x+36}{x-7} \cdot \dfrac{7-x}{(x+6)^2}$

i. $\dfrac{x^2-36}{x^2-6x+5} \cdot \dfrac{3x-15}{4x-24}$

j. $\dfrac{x^2-1}{x^2+6x-7} \cdot \dfrac{2x^2-x-3}{3x^2+3x}$

2. Give the reciprocal:

a. $\dfrac{4}{17}$

b. $\dfrac{^-6}{5}$

c. 8

d. $\dfrac{x}{y+3}$

e. y^2-3y-4

f. $\dfrac{1}{4x^2+32}$

3. Find the quotient:

a. $\dfrac{4}{46} \div \dfrac{26}{23}$

b. $\dfrac{24}{72} \div {}^-18$

c. $\dfrac{15x^2}{12y} \div \dfrac{40x^3}{36y^3}$

d. $\dfrac{2x}{x+4} \div \dfrac{8x}{x^2-16}$

e. $\dfrac{x^2-2x+1}{5x} \div \dfrac{5x-5}{2x^3}$

f. $\dfrac{(x+7)^2}{x^2-49} \div \dfrac{7x}{x-7}$

g. $\dfrac{x^2-13x+40}{2x-14} \div \dfrac{x^2-64}{3x-21}$

h. $\dfrac{y^2-7x}{^-x+3} \div \dfrac{^-y^2+7x}{x-3}$

i. $\dfrac{x^2-100}{x^2-4} \div \dfrac{x^2-17x+70}{x^2-9x+14}$

j. $\dfrac{4x+32}{32} \div \dfrac{8x^2}{32x}$

4. Find the product or quotient as indicated:

a. $\dfrac{^-x+7}{2x} \cdot \dfrac{4x^3}{7-x}$

b. $\dfrac{x^2-1}{x^2+2x+1} \div \dfrac{x-1}{3x+3}$

c. $\dfrac{65x^2}{35y} \cdot \dfrac{14y^2}{13x^2y}$

d. $\dfrac{4x^2-36}{x-9} \cdot \dfrac{9-x}{x+3}$

e. $\dfrac{x^2-x-6}{x^2-3x-10} \div \dfrac{x^2+4x-21}{x^2+2x-35}$

f. $\dfrac{5x^2-80}{25x^3} \div \dfrac{10x+40}{50x}$

g. $\dfrac{6x^2-24}{24x^2} \div (x+2)$

h. $\dfrac{x^2-3x-54}{6x-54} \cdot (x-9)$

i. $\dfrac{x^2+36}{x^2+10x-24} \div \dfrac{1}{x+12}$

j. $\dfrac{x+4}{x-3} \cdot \dfrac{3}{4}$

12.2 Exercise Answers

1. a. $\dfrac{^-1}{6}$

b. 1

c. $\dfrac{12}{49}$

d. $^-1$

e. $\dfrac{x^2}{18y^2}$ **f.** 5 **g.** $\dfrac{x(x-4)}{3}$ **h.** $\dfrac{^-(x-6)^2}{(x+6)^2}$

i. $\dfrac{3(x+6)}{4(x-1)}$ **j.** $\dfrac{(2x-3)(x+1)}{3x(x+7)}$

2. a. $\dfrac{17}{4}$ **b.** $\dfrac{^-5}{6}$ **c.** $\dfrac{1}{8}$ **d.** $\dfrac{y+3}{x}$

e. $\dfrac{1}{y^2-3y-4}$ **f.** $4x^2+32$

3. a. $\dfrac{1}{13}$ **b.** $\dfrac{^-1}{54}$ **c.** $\dfrac{9y^2}{8x}$ **d.** $\dfrac{x-4}{4}$

e. $\dfrac{2x^2(x-1)}{25}$ **f.** $\dfrac{x+7}{7x}$ **g.** $\dfrac{3(x-5)}{2(x+8)}$ **h.** 1

i. $\dfrac{x+10}{x+2}$ **j.** $\dfrac{x+8}{2x}$

4. a. $2x^2$ **b.** 3 **c.** 2 **d.** $4(x-3)$

e. 1 **f.** $\dfrac{x-4}{x^2}$ **g.** $\dfrac{x-2}{4x^2}$ **h.** $\dfrac{(x-9)(x+6)}{6}$

i. $\dfrac{x^2+36}{x-2}$ **j.** $\dfrac{3(x+4)}{4(x-3)}$

12.3   Addition and Subtraction

1

The sum or difference of two rational numbers with *like denominators* can be found using either of the following generalizations. For integers x, y, and z, $z \neq 0$,

(1) $\dfrac{x}{z} + \dfrac{y}{z} = \dfrac{x+y}{z}$

(2) $\dfrac{x}{z} - \dfrac{y}{z} = \dfrac{x-y}{z}$

Examples:

1. $\dfrac{5}{14} + \dfrac{3}{14} = \dfrac{5+3}{14} = \dfrac{\overset{4}{\cancel{8}}}{\underset{7}{\cancel{14}}} = \dfrac{4}{7}$ **2.** $\dfrac{^-3}{11} - \dfrac{6}{11} = \dfrac{^-3-6}{11} = \dfrac{^-9}{11}$

The same principles apply to the sum or difference of rational expressions with like denominators.

Examples:

$\dfrac{4x}{9} + \dfrac{2x}{9} = \dfrac{4x+2x}{9} = \dfrac{6x}{9} = \dfrac{2x}{3}$

$\dfrac{8}{y} - \dfrac{5}{y} = \dfrac{8-5}{y} = \dfrac{3}{y}$

$$\frac{2x - 2}{x - 2} + \frac{x - 4}{x - 2} = \frac{2x - 2 + x - 4}{x - 2} = \frac{3x - 6}{x - 2} = \frac{3(x - 2)}{x - 2} = 3$$

$$\frac{4x - 7}{8y} - \frac{2x - 1}{8y} = \frac{4x - 7 - (2x - 1)}{8y} = \frac{4x - 7 - 2x + 1}{8y} = \frac{2x - 6}{8y}$$

$$= \frac{\overset{1}{\cancel{2}}(x - 3)}{\underset{4}{\cancel{8}y}} = \frac{x - 3}{4y}$$

Q1   Find:

a. $\dfrac{6x}{11} + \dfrac{3}{11}$

b. $\dfrac{8}{x} - \dfrac{2}{x}$

c. $\dfrac{2y}{25} + \dfrac{13y}{25}$

d. $\dfrac{x + 7}{4y} + \dfrac{3 - 5x}{4y}$

e. $\dfrac{3m - 2}{2n} - \dfrac{m + 3}{2n}$

STOP • STOP • STOP • STOP • STOP • STOP • STOP • STOP • STOP

A1   a. $\dfrac{6x + 3}{11}$ b. $\dfrac{6}{x}$ c. $\dfrac{3y}{5}$: $\dfrac{2y}{25} + \dfrac{13y}{25} = \dfrac{\overset{3}{\cancel{15}}y}{\underset{5}{\cancel{25}}} = \dfrac{3y}{5}$

d. $\dfrac{5 - 2x}{2y}$: $\dfrac{x + 7}{4y} + \dfrac{3 - 5x}{4y} = \dfrac{x + 7 + 3 - 5x}{4y} = \dfrac{10 - 4x}{4y} = \dfrac{2(5 - 2x)}{4y} = \dfrac{5 - 2x}{2y}$

e. $\dfrac{2m - 5}{2n}$: $\dfrac{3m - 2}{2n} - \dfrac{m + 3}{2n} = \dfrac{3m - 2 - (m + 3)}{2n} = \dfrac{3m - 2 - m - 3}{2n} = \dfrac{2m - 5}{2n}$

2   Notice that when finding the difference of two rational expressions it is necessary to take the opposite of the numerator of the subtrahend (second expression). It is especially important that the opposite of *each term* in the numerator (if there are more than one) is taken and not merely the first term of the numerator.

Examples:

$$\frac{x + 5}{3x} - \frac{x + 3}{3x} = \frac{x + 5 - (x + 3)}{3x} = \frac{x + 5 - x - 3}{3x} = \frac{2}{3x}$$

$$\frac{^-x}{x^2 - 8x + 7} - \frac{^-2x + 7}{x^2 - 8x + 7} = \frac{^-x - (^-2x + 7)}{x^2 - 8x + 7} = \frac{^-x + 2x - 7}{x^2 - 8x + 7} = \frac{x - 7}{x^2 - 8x + 7}$$

$$= \frac{\overset{1}{\cancel{x - 7}}}{(\cancel{x - 7})(x - 1)} = \frac{1}{x - 1}$$

$$\frac{x^2 + 6x - 1}{x^2 - 16} - \frac{x^2 + 2x - 2}{x^2 - 16} = \frac{x^2 + 6x - 1 - (x^2 + 2x - 2)}{x^2 - 16}$$

$$= \frac{x^2 + 6x - 1 - x^2 - 2x + 2}{x^2 - 16} = \frac{4x + 1}{x^2 - 16}$$

Q2   Find:

a. $\dfrac{8x - 5}{x + 9} - \dfrac{3x + 1}{x + 9}$

b. $\dfrac{x^2 - 5x + 7}{x - 11} - \dfrac{x^2 - 9x - 3}{x - 11}$

c. $\dfrac{7x - 5}{3x} - \dfrac{4x - 4}{3x}$

d. $\dfrac{x^2 - 19x + 3}{x + 2} - \dfrac{x^2 + 14x - 1}{x + 2}$

STOP • **STOP** • **STOP** • **STOP** • **STOP** • **STOP** • **STOP** • **STOP** • **STOP**

A2   a. $\dfrac{5x - 6}{x + 9}$: $\dfrac{8x - 5}{x + 9} - \dfrac{3x + 1}{x + 9} = \dfrac{8x - 5 - (3x + 1)}{x + 9} = \dfrac{8x - 5 - 3x - 1}{x + 9} = \dfrac{5x - 6}{x + 9}$

b. $\dfrac{4x + 10}{x - 11}$: $\dfrac{x^2 - 5x + 7}{x - 11} - \dfrac{x^2 - 9x - 3}{x - 11} = \dfrac{x^2 - 5x + 7 - (x^2 - 9x - 3)}{x - 11}$

$$= \frac{x^2 - 5x + 7 - x^2 + 9x + 3}{x - 11} = \frac{4x + 10}{x - 11}$$

c. $\dfrac{3x - 1}{3x}$: $\dfrac{7x - 5}{3x} - \dfrac{4x - 4}{3x} = \dfrac{7x - 5 - (4x - 4)}{3x} = \dfrac{7x - 5 - 4x + 4}{3x} = \dfrac{3x - 1}{3x}$

d. $\dfrac{4 - 33x}{x + 2}$: $\dfrac{x^2 - 19x + 3}{x + 2} - \dfrac{x^2 + 14x - 1}{x + 2} = \dfrac{x^2 - 19x + 3 - (x^2 + 14x - 1)}{x + 2}$

$$= \frac{x^2 - 19x + 3 - x^2 - 14x + 1}{x + 2}$$

$$= \frac{^-33x + 4}{x + 2} \text{ or } \frac{4 - 33x}{x + 2}$$

3

Just as with products and quotients, the answer for a sum or difference of rational algebraic expressions is always expressed in lowest terms. This is done by factoring the numerator and denominator of each answer and dividing out any common factors.

Examples:

$$\frac{2x}{3x+15} + \frac{10}{3x+15} = \frac{2x+10}{3x+15} = \frac{2\overset{1}{\cancel{(x+5)}}}{3\underset{1}{\cancel{(x+5)}}} = \frac{2}{3}$$

$$\frac{x^2}{x^2-10x+9} - \frac{81}{x^2-10x+9} = \frac{x^2-81}{x^2-10x+9} = \frac{\overset{1}{\cancel{(x-9)}}(x+9)}{\underset{1}{\cancel{(x-9)}}(x-1)} = \frac{x+9}{x-1}$$

$$\frac{^-x}{x^2-1} + \frac{1}{x^2-1} = \frac{^-x+1}{x^2-1} = \frac{\overset{^-1}{\cancel{^-x+1}}}{\underset{1}{\cancel{(x-1)}}(x+1)} = \frac{^-1}{x+1}$$

(*Note:* $^-x+1$ and $x-1$ are opposites.)

$$\frac{4(x-3)}{x-8} - \frac{2(x+3)}{x-8} = \frac{4(x-3)-2(x+3)}{x-8} = \frac{4x-12-2x-6}{x-8} = \frac{2x-18}{x-8}$$
$$= \frac{2(x-9)}{x-8}$$

Q3

Find:

a. $\dfrac{5x}{x^2-25} + \dfrac{25}{x^2-25}$

b. $\dfrac{x^2}{x-3} - \dfrac{7x-12}{x-3}$

c. $\dfrac{3x+7}{2(x+2)} - \dfrac{x+3}{2(x+2)}$

d. $\dfrac{3(x-5)}{x^2-15x-34} - \dfrac{2(x+1)}{x^2-15x-34}$

STOP • **STOP** • **STOP** • **STOP** • **STOP** • **STOP** • **STOP** • **STOP** • **STOP**

A3

a. $\dfrac{5}{x-5}$: $\dfrac{5x+25}{x^2-25} = \dfrac{5\overset{1}{\cancel{(x+5)}}}{(x-5)\underset{1}{\cancel{(x+5)}}} = \dfrac{5}{x-5}$

b. $x-4$: $\dfrac{x^2-7x+12}{x-3} = \dfrac{\overset{1}{\cancel{(x-3)}}(x-4)}{\underset{1}{\cancel{x-3}}} = x-4$

c. 1: $\dfrac{3x + 7 - x - 3}{2(x + 2)} = \dfrac{2x + 4}{2(x + 2)} = \dfrac{2(x + 2)}{2(x + 2)} = 1$

d. $\dfrac{1}{x + 2}$: $\dfrac{3(x - 5) - 2(x + 1)}{x^2 - 15x - 34} = \dfrac{3x - 15 - 2x - 2}{x^2 - 15x - 34} = \dfrac{x - 17}{(x - 17)(x + 2)} = \dfrac{1}{x + 2}$

4

To find the sum or difference of two rational numbers with *unlike denominators,* it is first necessary to find the equivalent of each with a common denominator, preferably, the least common denominator (LCD). Recall that the LCD for two or more rational numbers is the least common multiple (LCM) of the denominators. That is, it is the smallest number that all the denominators will divide into evenly. For small-number denominators, it is often possible to guess the LCD.

Example 1: Find the LCD for $\dfrac{1}{4} - \dfrac{1}{3}$.

Solution

The LCM for 4 and 3 is $4 \cdot 3$ or 12. Therefore, the LCD is 12.

(*Note:* The product of the denominators will *always* yield *a* common denominator, but it will not always be the *least* common denominator.)

Example 2: Find the LCD for $\dfrac{5}{8} + \dfrac{7}{16}$.

Solution

The LCM for 8 and 16 is 16. Therefore, the LCD is 16.

(*Note:* Often one of the denominators is the LCD.)

Example 3: Find the LCD for $\dfrac{3}{5} + \dfrac{2}{6} - \dfrac{1}{2}$.

Solution

The LCM for 5, 6, and 2 is 30. Therefore, the LCD is 30.

(*Note:* Although $5 \cdot 6 \cdot 2 = 60$, 60 is not the LCM. The reason is that 6 is a multiple of 2 and thus we need only find $5 \cdot 6 = 30$ for the LCM.)

Q4

Complete:

a. The LCD of two or more rational numbers is the _____ of the denominators.

b. The LCD for $\dfrac{3}{15} - \dfrac{7}{5}$ is _____.

c. The LCD for $\dfrac{1}{6} + \dfrac{3}{4} - \dfrac{7}{8}$ is _____.

STOP • STOP • STOP • STOP • STOP • STOP • STOP • STOP • STOP

A4   **a.** LCM   **b.** 15   **c.** 24

5

To find the sum or difference of two or more rational numbers with *unlike denominators:*

1. Find the equivalent of each rewritten over their LCD.

2. Proceed using the rules for the sum or difference of rational numbers with like denominators.

Example 1: Find $\dfrac{4}{5} - \dfrac{2}{9}$.

Solution

The LCD is 45.

$$\dfrac{4}{5} - \dfrac{2}{9}$$

$$\dfrac{4}{5} \cdot \dfrac{9}{9} - \dfrac{2}{9} \cdot \dfrac{5}{5} \qquad \text{(this step may be done mentally)}$$

$$\dfrac{36}{45} - \dfrac{10}{45}$$

$$\dfrac{26}{45}$$

Example 2: Find $\dfrac{^{-}1}{6} + \dfrac{^{-}3}{10}$.

Solution

The LCD is 30.

$$\dfrac{^{-}1}{6} + \dfrac{^{-}3}{10}$$

$$\dfrac{^{-}1}{6} \cdot \dfrac{5}{5} + \dfrac{^{-}3}{10} \cdot \dfrac{3}{3} \qquad \text{(do this step mentally, if possible)}$$

$$\dfrac{^{-}5}{30} + \dfrac{^{-}9}{30}$$

$$\dfrac{^{-}14}{30}$$

$$\dfrac{^{-}7}{15}$$

Q5   Complete:

a. $\dfrac{3}{4} + \dfrac{2}{5}$ b. $\dfrac{7}{9} - \dfrac{5}{12}$

STOP   •   STOP   •   STOP   •   STOP   •   STOP   •   STOP   •   STOP   •   STOP   •   STOP

A5   a. $\dfrac{23}{20}$ b. $\dfrac{13}{36}$: $\dfrac{28}{36} - \dfrac{15}{36} = \dfrac{13}{36}$

6   Procedures discussed thus far are adequate if the LCD is obvious. For more difficult problems, the prime factorization method for finding the LCD is helpful. To find the LCD using the prime factorization method, use the following steps.

1. Determine the prime factorization of each denominator.
2. The LCM is the *product* of the *different prime factors* that appear the *greatest number of times* in any *one* factorization.

Example 1: Find the LCD for $\dfrac{1}{18} + \dfrac{1}{15}$.

Solution

Step 1: List the prime factorizations of 18 and 15.

$18 = 2 \cdot 3 \cdot 3$
$15 = 3 \cdot 5$

Step 2: The different prime factors that appear are 2, 3, and 5.

```
                          ┌─ 2 appears once as a factor of 18
                          ├─ 3 appears (the greatest number of times)
18 = ②  · ③·③◄──            twice as a factor of 18
15 = 3 · ⑤◄──────────── 5 appears once as a factor of 15
LCM = 2 · 3 · 3 · 5 = 90
```

Therefore, the LCD for $\dfrac{1}{18}$ and $\dfrac{1}{15}$ is 90.

Example 2: Find the LCD for $\dfrac{5}{12} - \dfrac{2}{45} + \dfrac{9}{10}$.

Solution

Step 1: List the prime factorizations of 12, 45, and 10.

$12 = 2 \cdot 2 \cdot 3$
$45 = 3 \cdot 3 \cdot 5$
$10 = 2 \cdot 5$

Step 2: The different prime factors that occur are 2, 3, and 5.

```
                          ┌─ 2 appears (the greatest number of times) twice
                          │    as a factor of 12
                          ├─ 3 appears (the greatest number of times) twice
12 = ②·②  · 3◄──           as a factor of 45
45 = ③·③ · ⑤ ◄──────── 5 appears once as a factor of 45 or 10 (circle only
10 = 2 · 5                 one factor of 5)
LCM = 2 · 2 · 3 · 3 · 5 = 180
```

Therefore, the LCD for $\dfrac{5}{12}, \dfrac{2}{45},$ and $\dfrac{9}{10}$ is 180.

Q6   Complete: the LCM of two or more numbers is the _____ of the different _____ factors that appear the _____ number of times in any _____ factorization.

STOP • **STOP** • **STOP** • **STOP** • **STOP** • **STOP** • **STOP** • **STOP** • **STOP**

A6   product, prime, greatest, one

Q7   Find the LCD for $\dfrac{7}{20} + \dfrac{-9}{35}$ using the method of Frame 6.

    a. Determine the prime factorizations for 20 and 35.

        $20 =$ _____

        $35 =$ _____

    b. Write the factorizations from part a and circle the different prime factors that appear the greatest number of times in any one factorization.

        $20 =$ _____

        $35 =$ _____

    c. LCM = _____

    d. The LCD for $\dfrac{7}{20} + \dfrac{-9}{35}$ is _____.

STOP   •   STOP   •   STOP   •   STOP   •   STOP   •   STOP   •   STOP   •   STOP   •   STOP

A7   **a.** $20 = 2 \cdot 2 \cdot 5, \ 35 = 5 \cdot 7$ **b.** $20 = \boxed{2 \cdot 2} \cdot \boxed{5}, \ 35 = 5 \cdot \boxed{7}$
       c. 140   **d.** 140

Q8   Find the LCD for:

    a. $\dfrac{11}{24} - \dfrac{7}{36}$ **b.** $\dfrac{1}{18} + \dfrac{3}{40}$

STOP   •   STOP   •   STOP   •   STOP   •   STOP   •   STOP   •   STOP   •   STOP   •   STOP

A8   **a.** 72: $24 = \boxed{2 \cdot 2 \cdot 2} \cdot 3$ **b.** 360: $18 = 2 \cdot \boxed{3 \cdot 3}$
            $36 = 2 \cdot 2 \cdot \boxed{3 \cdot 3}$ $40 = \boxed{2 \cdot 2 \cdot 2} \cdot \boxed{5}$
      LCD $= 2 \cdot 2 \cdot 2 \cdot 3 \cdot 3$ LCD $= 2 \cdot 2 \cdot 2 \cdot 3 \cdot 3 \cdot 5$

Q9   Use the LCDs from Q8 to solve:

    a. $\dfrac{11}{24} - \dfrac{7}{36}$ **b.** $\dfrac{1}{18} + \dfrac{3}{40}$

STOP   •   STOP   •   STOP   •   STOP   •   STOP   •   STOP   •   STOP   •   STOP   •   STOP

A9   **a.** $\dfrac{19}{72}$: $\dfrac{11}{24} - \dfrac{7}{36}$ **b.** $\dfrac{47}{360}$: $\dfrac{1}{18} + \dfrac{3}{40}$

               $\dfrac{33}{72} - \dfrac{14}{72}$ $\dfrac{20}{360} + \dfrac{27}{360}$

                   $\dfrac{19}{72}$ $\dfrac{47}{360}$

7   When variables appear in the denominator, the same method can be used to find the LCD.

Example 1: Find the LCD for $\dfrac{7}{10x} - \dfrac{5}{6x}$.

Solution

Step 1: List the prime factorizations for

$10x$ and $6x$.
$10x = 2 \cdot 5 \cdot x$
$6x = 2 \cdot 3 \cdot x$

Step 2: The different prime factors that appear are 2, 3, 5, and x.

$10x = ②\cdot ⑤ \cdot x$

$6x = 2 \cdot ③ \cdot ⓧ$

- 2 appears *once* as a factor of both $10x$ and $6x$
- 3 appears *once* as a factor of $6x$
- 5 appears *once* as a factor of $10x$
- x appears *once* as a factor of both $10x$ and $6x$

LCM $= 2 \cdot 3 \cdot 5 \cdot x = 30x$

Therefore, the LCD for $\dfrac{7}{10x}$ and $\dfrac{5}{6x}$ is $30x$.

Example 2: Find the LCD for $\dfrac{5}{2y^2} + \dfrac{3}{4y}$.

Solution

Step 1: List the prime factorizations for $2y^2$ and $4y$.

$2y^2 = 2 \cdot y \cdot y$
$4y = 2 \cdot 2 \cdot y$

Step 2: The different prime factors that appear are 2 and y.

$2y^2 = 2 \cdot \overline{(y \cdot y)}$

$4y = \overline{(2 \cdot 2)} \cdot y$

- 2 appears *twice* as a factor of $4y$
- y appears *twice* as a factor of $2y^2$

LCM $= 2 \cdot 2 \cdot y \cdot y = 4y^2$

Therefore, the LCD for $\dfrac{5}{2y^2}$ and $\dfrac{3}{4y}$ is $4y^2$.

Q10   Find the LCD for:

a. $\dfrac{7}{15x} + \dfrac{23}{21x}$

b. $\dfrac{3y + 1}{3y} - \dfrac{y + 2}{9y^2}$

c. $\dfrac{1}{6x} + \dfrac{x-2}{10x^2}$

d. $\dfrac{x+3}{4x} + \dfrac{1}{3y}$

STOP • STOP • STOP • STOP • STOP • STOP • STOP • STOP • STOP

A10 **a.** $105x$: $15x = ③ \cdot ⑤ \cdot ⓧ$
$ 21x = 3 \cdot ⑦ \cdot x$
$\text{LCD} = 3 \cdot 5 \cdot 7 \cdot x$

b. $9y^2$: $3y = 3 \cdot y$
$ 9y^2 = \overset{\frown}{3 \cdot 3} \overset{\frown}{y \cdot y}$
$\text{LCD} = 3 \cdot 3 \cdot y \cdot y$
(*Note:* $9y^2$ is a multiple of $3y$.)

c. $30x^2$: $6x = ② \cdot ③ \cdot x$
$ 10x^2 = 2 \cdot ⑤ \cdot \overset{\frown}{x \cdot x}$
$\text{LCD} = 2 \cdot 3 \cdot 5 \cdot x \cdot x$

d. $12xy$: $4x = \overset{\frown}{2 \cdot 2} \cdot ⓧ$
$ 3y = ③ \cdot ⓨ$
$\text{LCD} = 2 \cdot 2 \cdot 3 \cdot x \cdot y$
(*Note:* $12xy$ is the product of the original denominators.)

---

8 Consider the LCD for $\dfrac{3}{4x^2y} - \dfrac{4}{7xy^3}$.

$4x^2y = \overset{\frown}{2 \cdot 2} \cdot \overset{\frown}{x \cdot x} \cdot y$
$7xy^3 = ⑦ \cdot x \cdot \overset{\frown}{y \cdot y \cdot y}$
$\text{LCD} = 2 \cdot 2 \cdot 7 \cdot x \cdot x \cdot y \cdot y \cdot y$
$\phantom{\text{LCD}} = 28x^2y^3$

Notice the following things:

1. 28 is the LCM for 4 and 7, the number parts of the denominators.
2. The largest exponent on each variable determines the number of factors needed in the variable part of the LCD.

Example:

The variable parts above are x^2y and xy^3. The largest exponent on an x is 2 (x^2); the largest exponent on a y is 3 (y^3). Thus, the variable part of the LCD is x^2y^3.

3. The LCD for two or more denominators is the product of the *LCM of the number parts* of the denominators *and* the *LCM of the variable parts* of the denominators.

Example: Find the LCD for $\dfrac{2}{8x^2y^2} - \dfrac{3}{24xy^3}$.

Solution

The LCM for 8 and 24 is 24 $\left(\begin{array}{l} 8 = 2 \cdot 2 \cdot 2 \\ 24 = 2 \cdot 2 \cdot 2 \cdot 3 \end{array}\right)$. The LCM for x^2y^2 and xy^3 is x^2y^3 (since the largest exponent for x is 2 and for y is 3). Therefore, the LCD $= 24x^2y^3$.

Q11   Find the LCD for:

   a. $\dfrac{2}{9x^2y^3} - \dfrac{5}{27xy^4}$ **b.** $\dfrac{m+9}{12m^2} + \dfrac{2m+7}{14m^3}$

STOP • **STOP** • **STOP** • **STOP** • **STOP** • **STOP** • **STOP** • **STOP** • **STOP**

A11   **a.** $27x^2y^4$: $\begin{aligned} 9 &= 3 \cdot 3 \\ 27 &= \boxed{3 \cdot 3 \cdot 3} \end{aligned}$ The largest exponent on x is 2 (x^2). The largest exponent on y is 4 (y^4). Thus, the variable part of the LCD is x^2y^4.

   b. $84m^3$: $\begin{aligned} 12 &= \boxed{2 \cdot 2} \cdot \boxed{3} \\ 14 &= 2 \cdot \boxed{7} \end{aligned}$ The largest exponent on m is 3 (m^3).

---

9   To find the LCD for denominators involving two or more terms:

Step 1: Factor each denominator completely.
Step 2: Form the LCM by indicating the product of the different factors that appear the greatest number of times in any one factorization.

Example: Find the LCD for $\dfrac{x+2}{3x-6} + \dfrac{x+3}{2x-4}$.

Solution

Step 1: Factor each denominator completely.

$3x - 6 = 3(x - 2)$
$2x - 4 = 2(x - 2)$

Step 2: The different prime factors that appear are 2, 3, and $x - 2$.

$3x - 6 = \boxed{3}\,(x - 2)$ 2 appears *once* as a factor of $2x - 4$
$2x - 4 = \boxed{2}\,\boxed{(x - 2)}$ 3 appears *once* as a factor of $3x - 6$
                   $x - 2$ appears *once* as a factor of both $3x - 6$ and $2x - 4$

LCM $= 2 \cdot 3 \cdot (x - 2) = 6(x - 2)$

Therefore, the LCD for $\dfrac{x+2}{3x-6}$ and $\dfrac{x+3}{2x-4}$ is $6(x - 2)$.

Q12   Find the LCD for $\dfrac{5}{2x+2} - \dfrac{7}{5x+5}$.

STOP • **STOP** • **STOP** • **STOP** • **STOP** • **STOP** • **STOP** • **STOP** • **STOP**

A12 $10(x + 1)$: $\begin{aligned} 2x + 2 &= \boxed{2}\,(x + 1) \\ 5x + 5 &= \boxed{5}\,\boxed{(x + 1)} \end{aligned}$

---

10   **Example:** Find the LCD for $\dfrac{5x}{x^2 - 16} + \dfrac{7x}{2x + 8}$.

Solution

Step 1: Factor each denominator completely:

$x^2 - 16 = (x - 4)(x + 4)$
$2x + 8 = 2(x + 4)$

Step 2: The different prime factors that appear are $x - 4$, $x + 4$, and 2.

$x^2 - 16 = (x - 4)(x + 4)$ $x - 4$ appears *once* as a factor of $x^2 - 16$

$2x + 8 = 2(x + 4)$ $x + 4$ appears *once* as a factor of $x^2 - 16$ or $2x + 8$ (circle one)

LCM $= 2(x - 4)(x + 4)$ 2 appears *once* as a factor of $2x + 8$

Therefore, the LCD for $\dfrac{5x}{x^2 - 16}$ and $\dfrac{7x}{2x + 8}$ is $2(x - 4)(x + 4)$.

Q13   Find the LCD for $\dfrac{1}{x^2 - 25} + \dfrac{7x}{5x - 25}$.

STOP • **STOP** • **STOP** • **STOP** • **STOP** • **STOP** • **STOP** • **STOP** • **STOP**

A13 $5(x - 5)(x + 5)$: $x^2 - 25 = (x - 5)(x + 5)$
                   $5x - 25 = 5(x - 5)$

11   **Example:** Find the LCD for $\dfrac{x + 5}{x^2 + 5x + 6} - \dfrac{x + 3}{x^2 - x - 6}$.

Solution

Step 1: Factor each denominator completely.

$x^2 + 5x + 6 = (x + 3)(x + 2)$
$x^2 - x - 6 = (x - 3)(x + 2)$

Step 2: The different prime factors that appear are $x + 3$, $x + 2$, and $x - 3$.

$x^2 + 5x + 6$ $x + 3$ appears *once* as a factor of $x^2 + 5x + 6$

$= (x + 3)(x + 2)$ $x + 2$ appears once as a factor of $x^2 + 5x + 6$, or

$x^2 - x - 6$ $x^2 - x - 6$ (circle one)

$= (x - 3)(x + 2)$ $x - 3$ appears *once* as a factor of $x^2 - x - 6$

LCM $= (x + 3)(x + 2)(x - 3)$

Therefore, the LCD for $\dfrac{x + 5}{x^2 + 5x + 6} - \dfrac{x + 3}{x^2 - x - 6}$ is $(x + 3)(x + 2)(x - 3)$.

Q14   Find the LCD for:

a. $\dfrac{x + 1}{x - 1} + \dfrac{x - 1}{x + 1}$ b. $\dfrac{1}{x^2 - 4} + \dfrac{5}{2x + 4}$

c. $\dfrac{4}{2x} - \dfrac{6}{3y}$

d. $\dfrac{7}{3y} - \dfrac{2x}{3y^2 - 27}$

STOP • STOP • STOP • STOP • STOP • STOP • STOP • STOP • STOP

A14   **a.** $(x - 1)(x + 1)$ **b.** $2(x - 2)(x + 2)$

   c. $6xy$ **d.** $3y(y^2 - 9)$ or $3y(y + 3)(y - 3)$

---

12   The LCD for $\dfrac{7}{15x} + \dfrac{23}{21x}$ is $105x$ (see Q10a). To find the sum, proceed as follows:

Step 1: Rewrite each rational expression over the LCD, $105x$.

$$\dfrac{7}{15x} + \dfrac{23}{21x}$$

$$\dfrac{7}{15x} \cdot \dfrac{7}{7} + \dfrac{23}{21x} \cdot \dfrac{5}{5} \qquad \text{(do this step mentally, if possible)}$$

$$\dfrac{49}{105x} + \dfrac{115}{105x}$$

Step 2: Write the sum or difference of the numerators over the LCD.

$$\dfrac{49 + 115}{105x}$$

Step 3: Simplify the numerator.

$$\dfrac{164}{105x}$$

Step 4: Determine whether the result is in simplest form which requires factoring (if possible) and reducing. $\dfrac{164}{105x}$ is in simplest form. Thus,

$$\dfrac{7}{15x} + \dfrac{23}{21x} = \dfrac{164}{105x}$$

Q15   Find $\dfrac{2}{15x} - \dfrac{3}{10x}$.

STOP • STOP • STOP • STOP • STOP • STOP • STOP • STOP • STOP

A15 $\dfrac{-1}{6x}$:   the LCD is $30x$.

$$\frac{2}{15x} - \frac{3}{10x}$$

$$\frac{2}{15x} \cdot \frac{2}{2} - \frac{3}{10x} \cdot \frac{3}{3} \qquad \text{(do this step mentally if possible)}$$

$$\frac{4}{30x} - \frac{9}{30x}$$

$$\frac{4-9}{30x} = \frac{-5}{30x} = \frac{-1}{6x}$$

---

13   The LCD for $\dfrac{1}{6x} + \dfrac{x-2}{10x^2}$ is $30x^2$ (see Q10c). To find the sum, proceed as follows:

Step 1: Rewrite each rational expression over the LCD, $30x^2$.

$$\frac{1}{6x} + \frac{x-2}{10x^2}$$

$$\frac{1}{6x} \cdot \frac{5x}{5x} + \frac{x-2}{10x^2} \cdot \frac{3}{3} \qquad \text{(do this step mentally, if possible)}$$

$$\frac{5x}{30x^2} + \frac{3(x-2)}{30x^2}$$

Step 2: Write the sum or difference of the numerators over the LCD.

$$\frac{5x + 3(x-2)}{30x^2}$$

Step 3: Simplify the numerator.

$$\frac{5x + 3x - 6}{30x^2} = \frac{8x - 6}{30x^2}$$

Step 4: Determine whether the result is in simplest form which requires factoring (if possible) and reducing.

$$\frac{8x-6}{30x^2} = \frac{2(4x-3)}{30x^2} = \frac{4x-3}{15x^2}$$

Thus

$$\frac{1}{6x} + \frac{x-2}{10x^2} = \frac{4x-3}{15x^2}.$$

Q16   Find $\dfrac{x+4}{9x^2} + \dfrac{2}{5x}$.

A16 $\dfrac{23x + 20}{45x^2}$: the LCD is $45x^2$.

$$\frac{x + 4}{9x^2} \cdot \frac{5}{5} + \frac{2}{5x} \cdot \frac{9x}{9x}$$

$$\frac{5(x + 4)}{45x^2} + \frac{18x}{45x^2}$$

$$\frac{5(x + 4) + 18x}{45x^2}$$

$$\frac{5x + 20 + 18x}{45x^2}$$

$$\frac{23x + 20}{45x^2}$$

Q17   Find:

    a. $\dfrac{2}{a} + \dfrac{3}{b}$ **b.** $\dfrac{5}{2x} - \dfrac{3}{5}$

    c. $\dfrac{3}{4x} + \dfrac{x - 3}{16x^2}$ **d.** $\dfrac{x + 5}{10x} + \dfrac{x - 2}{2}$

STOP • **STOP** • **STOP** • **STOP** • **STOP** • **STOP** • **STOP** • **STOP** • **STOP**

A17   **a.** $\dfrac{2b + 3a}{ab}$: $\dfrac{2}{a} \cdot \dfrac{b}{b} + \dfrac{3}{b} \cdot \dfrac{a}{a}$ **b.** $\dfrac{25 - 6x}{10x}$: $\dfrac{5}{2x} \cdot \dfrac{5}{5} - \dfrac{3}{5} \cdot \dfrac{2x}{2x}$

$$\frac{2b}{ab} + \frac{3a}{ab} \qquad\qquad\qquad \frac{25}{10x} - \frac{6x}{10x}$$

$$\frac{2b + 3a}{ab} \qquad\qquad\qquad\qquad \frac{25 - 6x}{10x}$$

c. $\dfrac{13x - 3}{16x^2}$: the LCD is $16x^2$.

$$\frac{3}{4x} \cdot \frac{4x}{4x} + \frac{x - 3}{16x^2}$$

$$\frac{12x}{16x^2} + \frac{x - 3}{16x^2}$$

$$\frac{12x + x - 3}{16x^2}$$

$$\frac{13x - 3}{16x^2}$$

d. $\dfrac{5x^2 - 9x + 5}{10x}$: the LCD is $10x$.

$$\frac{x + 5}{10x} + \frac{x - 2}{2} \cdot \frac{5x}{5x}$$

$$\frac{x + 5}{10x} + \frac{5x(x - 2)}{10x}$$

$$\frac{(x + 5) + 5x(x - 2)}{10x}$$

$$\frac{x + 5 + 5x^2 - 10x}{10x}$$

$$\frac{5x^2 - 9x + 5}{10x}$$

14   Consider the following examples of a sum and difference of two rational algebraic expressions.

Example 1: Find $\dfrac{3}{x^2y} - \dfrac{4}{xy^3}$.

Solution

The LCD is x^2y^3.

$$\frac{3}{x^2y} \cdot \frac{y^2}{y^2} - \frac{4}{xy^3} \cdot \frac{x}{x}$$

$$\frac{3y^2}{x^2y^3} - \frac{4x}{x^2y^3}$$

$$\frac{3y^2 - 4x}{x^2y^3}$$

Example 2: Find $\dfrac{x + 1}{18x} + \dfrac{x - 1}{2x^2}$.

Solution

The LCD is $18x^2$.

$$\frac{x + 1}{18x} \cdot \frac{x}{x} + \frac{x - 1}{2x^2} \cdot \frac{9}{9}$$

$$\frac{x(x + 1)}{18x^2} + \frac{9(x - 1)}{18x^2}$$

$$\frac{x(x + 1) + 9(x - 1)}{18x^2}$$

$$\frac{x^2 + x + 9x - 9}{18x^2}$$

$$\frac{x^2 + 10x - 9}{18x^2}$$

Example 3: Find $\dfrac{m + 9}{12m^2} + \dfrac{2m + 7}{14m^3}$.

Solution

The LCD is $84m^3$ (see A11b).

$$\frac{m + 9}{12m^2} + \frac{2m + 7}{14m^3}$$

$$\frac{m + 9}{12m^2} \cdot \frac{7m}{7m} + \frac{2m + 7}{14m^3} \cdot \frac{6}{6}$$

$$\frac{7m(m + 9)}{84m^3} + \frac{6(2m + 7)}{84m^3}$$

$$\frac{7m(m + 9) + 6(2m + 7)}{84m^3}$$

$$\frac{7m^2 + 63m + 12m + 42}{84m^3}$$

$$\frac{7m^2 + 75m + 42}{84m^3}$$

Q18   Find:

a. $\dfrac{2}{xy^2} + \dfrac{9}{x^2y}$

b. $\dfrac{x + 2}{8x^2} + \dfrac{x - 1}{3x}$

STOP • STOP • STOP • STOP • STOP • STOP • STOP • STOP • STOP

A18   **a.** $\dfrac{2x + 9y}{x^2y^2}$:   the LCD is x^2y^2.

b. $\dfrac{8x^2 - 5x + 6}{24x^2}$:   the LCD is $24x^2$.

$$\frac{2}{xy^2} \cdot \frac{x}{x} + \frac{9}{x^2y} \cdot \frac{y}{y}$$

$$\frac{2x}{x^2y^2} + \frac{9y}{x^2y^2}$$

$$\frac{2x + 9y}{x^2y^2}$$

$$\frac{x + 2}{8x^2} \cdot \frac{3}{3} + \frac{x - 1}{3x} \cdot \frac{8x}{8x}$$

$$\frac{3(x + 2)}{24x^2} + \frac{8x(x - 1)}{24x^2}$$

$$\frac{3(x + 2) + 8x(x - 1)}{24x^2}$$

$$\frac{3x + 6 + 8x^2 - 8x}{24x^2}$$

$$\frac{8x^2 - 5x + 6}{24x^2}$$

15   When subtracting, special care must be taken to avoid errors with the signs.

Example: Find $\dfrac{x - 6}{12} - \dfrac{x + 7}{8}$.

Solution

The LCD is 24.

$$\frac{x-6}{12} - \frac{x+7}{8}$$

$$\frac{x-6}{12}\cdot\frac{2}{2} - \frac{x+7}{8}\cdot\frac{3}{3}$$

$$\frac{2(x-6)}{24} - \frac{3(x+7)}{24}$$

$$\frac{2(x-6) - 3(x+7)}{24}$$

$$\frac{2x - 12 - 3x - 21}{24}$$ — mistakes in signs are most common; recall that $^-3(x+7) = {}^-3x - 21$

$$\frac{^-x - 33}{24}$$

Thus,

$$\frac{x-6}{12} - \frac{x+7}{8} = \frac{^-x - 33}{24}$$

Q19　　Find $\dfrac{x+2}{8} - \dfrac{x-3}{10}$.

STOP • STOP • STOP • STOP • STOP • STOP • STOP • STOP • STOP

A19　　$\dfrac{x+22}{40}$: the LCD is 40.

$$\frac{x+2}{8}\cdot\frac{5}{5} - \frac{x-3}{10}\cdot\frac{4}{4}$$

$$\frac{5(x+2)}{40} - \frac{4(x-3)}{40}$$

$$\frac{5(x+2) - 4(x-3)}{40}$$

$$\frac{5x + 10 - 4x + 12}{40}$$

$$\frac{x+22}{40}$$

16 Some additional examples involving the difference of two rational expressions are as follows:

Example 1: Find $\dfrac{7}{xy^3} - \dfrac{2y}{x^2}$.

Solution

The LCD is x^2y^3.

$$\dfrac{7}{xy^3} - \dfrac{2y}{x^2}$$

$$\dfrac{7}{xy^3}\cdot\dfrac{x}{x} - \dfrac{2y}{x^2}\cdot\dfrac{y^3}{y^3}$$

$$\dfrac{7x}{x^2y^3} - \dfrac{2y^4}{x^2y^3}$$

$$\dfrac{7x - 2y^4}{x^2y^3}$$

Hence,

$$\dfrac{7}{xy^3} - \dfrac{2y}{x^2} = \dfrac{7x - 2y^4}{x^2y^3}$$

Example 2: Find $\dfrac{5}{2x} - \dfrac{6x - 9}{3x^2}$.

Solution

The LCD is $6x^2$.

$$\dfrac{5}{2x} - \dfrac{6x - 9}{3x^2}$$

$$\dfrac{5}{2x}\cdot\dfrac{3x}{3x} - \dfrac{6x - 9}{3x^2}\cdot\dfrac{2}{2}$$

$$\dfrac{15x}{6x^2} - \dfrac{2(6x - 9)}{6x^2}$$

$$\dfrac{15x - 2(6x - 9)}{6x^2}$$

$$\dfrac{15x - 12x + 18}{6x^2}$$

$$\dfrac{3x + 18}{6x^2}$$

$$\dfrac{\overset{1}{3}(x + 6)}{\underset{2}{6}x^2}$$

$$\dfrac{x + 6}{2x^2}$$

Hence,

$$\dfrac{5}{2x} - \dfrac{6x - 9}{3x^2} = \dfrac{x + 6}{2x^2}$$

Q20 Find:

a. $\dfrac{4y}{18x^3} - \dfrac{2x}{3y^2}$

b. $\dfrac{3y + 1}{3y} - \dfrac{y + 2}{9y^2}$

STOP • STOP • STOP • STOP • STOP • STOP • STOP • STOP • STOP

A20

a. $\dfrac{2(y^3 - 3x^4)}{9x^3y^2}$: the LCD is $18x^3y^2$.

$$\frac{4y}{18x^3} - \frac{2x}{3y^2}$$

$$\frac{4y}{18x^3} \cdot \frac{y^2}{y^2} - \frac{2x}{3y^2} \cdot \frac{6x^3}{6x^3}$$

$$\frac{4y^3}{18x^3y^2} - \frac{12x^4}{18x^3y^2}$$

$$\frac{4y^3 - 12x^4}{18x^3y^2}$$

$$\frac{\overset{2}{\cancel{4}}(y^3 - 3x^4)}{\underset{9}{\cancel{18}}x^3y^2}$$

$$\frac{2(y^3 - 3x^4)}{9x^3y^2}$$

b. $\dfrac{9y^2 + 2y - 2}{9y^2}$: the LCD is $9y^2$.

$$\frac{3y + 1}{3y} - \frac{y + 2}{9y^2}$$

$$\frac{3y + 1}{3y} \cdot \frac{3y}{3y} - \frac{y + 2}{9y^2}$$

$$\frac{3y(3y + 1)}{9y^2} - \frac{y + 2}{9y^2}$$

$$\frac{3y(3y + 1) - (y + 2)}{9y^2}$$

$$\frac{9y^2 + 3y - y - 2}{9y^2}$$

$$\frac{9y^2 + 2y - 2}{9y^2}$$

17

The LCD for $\dfrac{4}{x + 2} + \dfrac{3}{x - 5}$ is $(x + 2)(x - 5)$. To find the sum, proceed as before.

Step 1: Rewrite each rational expression over the LCD, $(x + 2)(x - 5)$.

$$\frac{4}{x + 2} + \frac{3}{x - 5}$$

$$\frac{4}{x + 2} \cdot \frac{x - 5}{x - 5} + \frac{3}{x - 5} \cdot \frac{x + 2}{x + 2}$$

Step 2: Write the sum of the numerators over the LCD.

$$\frac{4(x - 5) + 3(x + 2)}{(x + 2)(x - 5)}$$

Step 3: Simplify the numerator.

$$\frac{4x - 20 + 3x + 6}{(x + 2)(x - 5)}$$

$$\frac{7x - 14}{(x + 2)(x - 5)}$$

Step 4: Determine whether the result is in simplest form which requires factoring (if possible) and reducing.

$$\frac{7x - 14}{(x + 2)(x - 5)}$$

$$\frac{7(x - 2)}{(x + 2)(x - 5)}$$

Since $\dfrac{7(x - 2)}{(x + 2)(x - 5)}$ is in simplest form,

$$\frac{4}{x + 2} + \frac{3}{x - 5} = \frac{7(x - 2)}{(x + 2)(x - 5)}$$

Q21   Find $\dfrac{1}{y+7}+\dfrac{6}{y-3}$

* * *

STOP • **STOP** • **STOP** • **STOP** • **STOP** • **STOP** • **STOP** • **STOP** • **STOP**

A21 $\dfrac{7y+39}{(y+7)(y-3)}$: $\dfrac{1}{y+7}+\dfrac{6}{y-3}$

$$\frac{1}{y+7}\cdot\frac{y-3}{y-3}+\frac{6}{y-3}\cdot\frac{y+7}{y+7}$$

$$\frac{1(y-3)}{(y+7)(y-3)}+\frac{6(y+7)}{(y+7)(y-3)}$$

$$\frac{1(y-3)+6(y+7)}{(y+7)(y-3)}$$

$$\frac{y-3+6y+42}{(y+7)(y-3)}$$

$$\frac{7y+39}{(y+7)(y-3)}$$

---

18   To find $\dfrac{5}{2x+2}-\dfrac{7}{3x+3}$ it is necessary to first find the LCD. This is done using the method of Frame 9.

$2x+2=2(x+1)$
$3x+3=3(x+1)$
$\text{LCD}=2\cdot3\cdot(x+1)=6(x+1)$

The solution is then completed using steps similar to those of the previous frame.

Step 1: Rewrite each rational expression over the LCD, $6(x+1)$.

$$\frac{5}{2x+2}-\frac{7}{3x+3}$$

$$\frac{5}{2(x+1)}-\frac{7}{3(x+1)}$$

$$\frac{5}{2(x+1)}\cdot\frac{3}{3}-\frac{7}{3(x+1)}\cdot\frac{2}{2}$$

$$\frac{15}{6(x+1)}-\frac{14}{6(x+1)}$$

Step 2: Write the difference of the numerators over the LCD.

$$\frac{15-14}{6(x+1)}$$

Step 3: Simplify the numerator.

$$\frac{1}{6(x + 1)}$$

Step 4: Determine whether the result is in simplest form which requires factoring (if possible) and reducing. $\dfrac{1}{6(x + 1)}$ is in simplest form. Hence,

$$\frac{5}{2x + 2} - \frac{7}{3x + 3} = \frac{1}{6(x + 1)}$$

Q22   Find:

a. $\dfrac{1}{4x - 8} - \dfrac{3}{2x - 4}$

b. $\dfrac{x + 2}{15x - 5} - \dfrac{x + 3}{3x - 1}$

STOP • STOP • STOP • STOP • STOP • STOP • STOP • STOP • STOP

A22   a. $\dfrac{^-5}{4(x - 2)}$: $4x - 8 = 4(x - 2)$

$2x - 4 = 2(x - 2)$

$\text{LCD} = 4(x - 2)$

$\dfrac{1}{4(x - 2)} - \dfrac{3}{2(x - 2)}$

$\dfrac{1}{4(x - 2)} - \dfrac{3}{2(x - 2)} \cdot \dfrac{2}{2}$

$\dfrac{1}{4(x - 2)} - \dfrac{6}{4(x - 2)}$

$\dfrac{1 - 6}{4(x - 2)}$

$\dfrac{^-5}{4(x - 2)}$

b. $\dfrac{^-4x - 13}{5(3x - 1)}$: $15x - 5 = 5(3x - 1)$

$3x - 1 = 1(3x - 1)$

$\text{LCD} = 5(3x - 1)$

$\dfrac{x + 2}{5(3x - 1)} - \dfrac{x + 3}{3x - 1}$

$\dfrac{x + 2}{5(3x - 1)} - \dfrac{x + 3}{3x - 1} \cdot \dfrac{5}{5}$

$\dfrac{x + 2}{5(3x - 1)} - \dfrac{5(x + 3)}{5(3x - 1)}$

$\dfrac{x + 2 - 5(x + 3)}{5(3x - 1)}$

$\dfrac{x + 2 - 5x - 15}{5(3x - 1)}$

$\dfrac{^-4x - 13}{5(3x - 1)}$

19   **Example:** Find $\dfrac{1}{x^2 - 25} - \dfrac{3x}{4x - 20}$.

Solution

$x^2 - 25 = (x - 5)(x + 5)$
$4x - 20 = 4(x - 5)$
$\text{LCD} = 4(x - 5)(x + 5)$

$$\frac{1}{(x - 5)(x + 5)} - \frac{3x}{4(x - 5)}$$

$$\frac{1}{(x - 5)(x + 5)} \cdot \frac{4}{4} - \frac{3x}{4(x - 5)} \cdot \frac{x + 5}{x + 5}$$

$$\frac{4}{4(x - 5)(x + 5)} - \frac{3x(x + 5)}{4(x - 5)(x + 5)}$$

$$\frac{4 - 3x(x + 5)}{4(x - 5)(x + 5)}$$

$$\frac{4 - 3x^2 - 15x}{4(x - 5)(x + 5)}$$

$$\frac{^-3x^2 - 15x + 4}{4(x - 5)(x + 5)}$$

$$\frac{^-(3x^2 + 15x - 4)}{4(x - 5)(x + 5)}$$

Since $3x^2 + 15x - 4$ is prime, the above result is in simplest form. Thus,

$$\frac{1}{x^2 - 25} - \frac{3x}{4x - 20} = \frac{^-3x^2 - 15x + 4}{4(x - 5)(x + 5)}$$

Q23   Find:

a. $\dfrac{1}{3x + 6} - \dfrac{5}{3x - 6}$ **b.** $\dfrac{3x}{x^2 + 7x + 12} - \dfrac{2}{4x + 12}$

A23 **a.** $\dfrac{^{-}4(x+3)}{3(x-2)(x+2)}$:

$3x + 6 = 3(x + 2)$

$3x - 6 = 3(x - 2)$

$\text{LCD} = 3(x - 2)(x + 2)$

$$\dfrac{1}{3(x+2)} - \dfrac{5}{3(x-2)}$$

$$\dfrac{1}{3(x+2)} \cdot \dfrac{x-2}{x-2} - \dfrac{5}{3(x-2)} \cdot \dfrac{x+2}{x+2}$$

$$\dfrac{1(x-2)}{3(x-2)(x+2)} - \dfrac{5(x+2)}{3(x-2)(x+2)}$$

$$\dfrac{1(x-2) - 5(x+2)}{3(x-2)(x+2)}$$

$$\dfrac{x - 2 - 5x - 10}{3(x-2)(x+2)}$$

$$\dfrac{^{-}4x - 12}{3(x-2)(x+2)}$$

$$\dfrac{^{-}4(x+3)}{3(x-2)(x+2)}$$

b. $\dfrac{5x-4}{2(x+3)(x+4)}$:

$x^2 + 7x + 12 = (x + 3)(x + 4)$

$4x + 12 = 4(x + 3)$

$\text{LCD} = 4(x + 3)(x + 4)$

$$\dfrac{3x}{(x+3)(x+4)} - \dfrac{2}{4(x+3)}$$

$$\dfrac{3x}{(x+3)(x+4)} \cdot \dfrac{4}{4} - \dfrac{2}{4(x+3)} \cdot \dfrac{x+4}{x+4}$$

$$\dfrac{12x}{4(x+3)(x+4)} - \dfrac{2(x+4)}{4(x+3)(x+4)}$$

$$\dfrac{12x - 2(x+4)}{4(x+3)(x+4)}$$

$$\dfrac{12x - 2x - 8}{4(x+3)(x+4)}$$

$$\dfrac{10x - 8}{4(x+3)(x+4)}$$

$$\dfrac{\overset{1}{\cancel{2}}(5x-4)}{\underset{2}{\cancel{4}}(x+3)(x+4)}$$

$$\dfrac{5x - 4}{2(x+3)(x+4)}$$

---

20   **Example:** Find $\dfrac{x+5}{x^2+5x+6} - \dfrac{x+3}{x^2-x-6}$.

Solution

The LCD is $(x + 3)(x + 2)(x - 3)$. (See Frame 11 for finding the LCD.)

$$\dfrac{x+5}{(x+3)(x+2)} - \dfrac{x+3}{(x-3)(x+2)}$$

$$\dfrac{x+5}{(x+3)(x+2)} \cdot \dfrac{x-3}{x-3} - \dfrac{x+3}{(x-3)(x+2)} \cdot \dfrac{x+3}{x+3}$$

$$\dfrac{(x+5)(x-3)}{(x+3)(x+2)(x-3)} - \dfrac{(x+3)(x+3)}{(x+3)(x+2)(x-3)}$$

$$\dfrac{x^2+2x-15}{(x+3)(x+2)(x-3)} - \dfrac{x^2+6x+9}{(x+3)(x+2)(x-3)}$$

$$\dfrac{x^2+2x-15-(x^2+6x+9)}{(x+3)(x+2)(x-3)}$$

$$\dfrac{x^2+2x-15-x^2-6x-9}{(x+3)(x+2)(x-3)}$$

$$\dfrac{^{-}4x-24}{(x+3)(x+2)(x-3)}$$

$$\dfrac{^{-}4(x+6)}{(x+3)(x+2)(x-3)}$$

Since $\dfrac{^-4(x + 6)}{(x + 3)(x + 2)(x - 3)}$ is in simplest form,

$$\frac{x + 5}{x^2 + 5x + 6} - \frac{x + 3}{x^2 - x - 6} = \frac{^-4(x + 6)}{(x + 3)(x + 2)(x - 3)}$$

Q24   Find $\dfrac{x - 1}{x^2 - 25} - \dfrac{x + 2}{x^2 + 3x - 10}$.

STOP • **STOP** • **STOP** • **STOP** • **STOP** • **STOP** • **STOP** • **STOP** • **STOP**

A24 $\dfrac{12}{(x - 5)(x + 5)(x - 2)}$: the LCD is $(x - 5)(x + 5)(x - 2)$.

$$\frac{x - 1}{(x - 5)(x + 5)} - \frac{x + 2}{(x + 5)(x - 2)}$$

$$\frac{x - 1}{(x - 5)(x + 5)} \cdot \frac{x - 2}{x - 2} - \frac{x + 2}{(x + 5)(x - 2)} \cdot \frac{x - 5}{x - 5}$$

$$\frac{(x - 1)(x - 2)}{(x - 5)(x + 5)(x - 2)} - \frac{(x + 2)(x - 5)}{(x - 5)(x + 5)(x - 2)}$$

$$\frac{x^2 - 3x + 2}{(x - 5)(x + 5)(x - 2)} - \frac{x^2 - 3x - 10}{(x - 5)(x + 5)(x - 2)}$$

$$\frac{x^2 - 3x + 2 - (x^2 - 3x - 10)}{(x - 5)(x + 5)(x - 2)}$$

$$\frac{x^2 - 3x + 2 - x^2 + 3x + 10}{(x - 5)(x + 5)(x - 2)}$$

$$\frac{12}{(x - 5)(x + 5)(x - 2)}$$

21   When the LCD for a sum or difference is the same as the product of the denominators, a convenient method to use is the *cross-product rule for addition or subtraction of rational numbers,* which is stated as follows: For any two rational numbers $\dfrac{x}{u}$ and $\dfrac{y}{v}$:

$$\frac{x}{u} + \frac{y}{v} = \frac{xv + uy}{uv}$$

$$\frac{x}{u} - \frac{y}{v} = \frac{xv - uy}{uv}$$

Notice that the name "cross-product" results from the fact that the products taken follow a cross $\diagdown\!\!\!\!\diagup$ pattern.

Example 1: Find $\dfrac{3}{5} - \dfrac{4}{9}$.

Solution

$$\dfrac{3}{5} - \dfrac{4}{9} = \dfrac{3 \cdot 9 - 4 \cdot 5}{5 \cdot 9}$$

$$= \dfrac{27 - 20}{45}$$

$$= \dfrac{7}{45}$$

Example 2: Find $\dfrac{2}{3} + \dfrac{4}{7}$.

Solution

$$\dfrac{2}{3} + \dfrac{4}{7} = \dfrac{2 \cdot 7 + 3 \cdot 4}{3 \cdot 7}$$

$$= \dfrac{14 + 12}{21}$$

$$= \dfrac{26}{21}$$

Q25   Use the cross-product rule for the addition or subtraction of rational numbers to find:

a. $\dfrac{1}{2} + \dfrac{1}{3}$

b. $\dfrac{5}{8} - \dfrac{7}{9}$

STOP • STOP • STOP • STOP • STOP • STOP • STOP • STOP • STOP

A25   **a.** $\dfrac{5}{6}$: $\dfrac{1}{2} + \dfrac{1}{3} = \dfrac{1 \cdot 3 + 2 \cdot 1}{2 \cdot 3}$

$$= \dfrac{3 + 2}{6}$$

$$= \dfrac{5}{6}$$

b. $\dfrac{^{-}11}{72}$: $\dfrac{5}{8} - \dfrac{7}{9} = \dfrac{5 \cdot 9 - 8 \cdot 7}{8 \cdot 9}$

$$= \dfrac{45 - 56}{72}$$

$$= \dfrac{^{-}11}{72}$$

22   The cross-product rule is also useful when finding the sum or difference of *rational algebraic expressions.*

Example 1: Find $\dfrac{x}{x - 3} - \dfrac{5}{x + 1}$.

Solution

The LCD is $(x - 3)(x + 1)$, the product of the denominators.

$$\dfrac{x}{x - 3} - \dfrac{5}{x + 1}$$

$$\dfrac{x(x + 1) - 5(x - 3)}{(x - 3)(x + 1)}$$

$$\dfrac{x^2 + x - 5x + 15}{(x - 3)(x + 1)}$$

$$\dfrac{x^2 - 4x + 15}{(x - 3)(x + 1)}$$

Example 2: Find $\dfrac{x-3}{x+1} + \dfrac{6}{2x-5}$.

Solution

The LCD is $(x+1)(2x-5)$, the product of the denominators.

$$\dfrac{x-3}{x+1} + \dfrac{6}{2x-5}$$

$$\dfrac{(x-3)(2x-5) + 6(x+1)}{(x+1)(2x-5)}$$

$$\dfrac{2x^2 - 11x + 15 + 6x + 6}{(x+1)(2x-5)}$$

$$\dfrac{2x^2 - 5x + 21}{(x+1)(2x-5)}$$

Q26   Use the cross-product rule to find:

a. $\dfrac{2}{x-7} + \dfrac{3}{x-5}$ b. $\dfrac{4x}{2x+3} - \dfrac{x}{x+5}$

c. $\dfrac{x-2}{x+3} + \dfrac{x+2}{x-6}$ d. $\dfrac{3y}{y} - \dfrac{2}{2y-7}$

STOP • STOP • STOP • STOP • STOP • STOP • STOP • STOP • STOP

A26   a. $\dfrac{5x-31}{(x-7)(x-5)}$: $\dfrac{2(x-5) + 3(x-7)}{(x-7)(x-5)}$ b. $\dfrac{x(2x+17)}{(2x+3)(x+5)}$

$$\dfrac{2x - 10 + 3x - 21}{(x-7)(x-5)}$$

$$\dfrac{5x-31}{(x-7)(x-5)}$$

c. $\dfrac{2x^2 - 3x + 18}{(x+3)(x-6)}$

d. $\dfrac{5y^2 - 20y + 14}{y(2y-7)}$: $\dfrac{(3y-2)(2y-7) - y(y-5)}{y(2y-7)}$

$$\dfrac{6y^2 - 25y + 14 - y^2 + 5y}{y(2y-7)}$$

$$\dfrac{5y^2 - 20y + 14}{y(2y-7)}$$

23   It is important to notice that the cross-product rule is not always the most convenient method. This is usually true when the product of the denominators is not the least common denominator. For example, consider the difference $\dfrac{x-1}{x^2-25} - \dfrac{x+2}{x^2+3x-10}$ worked using the LCD and the cross-product rule.

LCD method

$$\frac{x-1}{x^2-25} - \frac{x+2}{x^2+3x-10}$$

$$\frac{x-1}{(x-5)(x+5)}\cdot\frac{x-2}{x-2} - \frac{x+2}{(x+5)(x-2)}\cdot\frac{x-5}{x-5}$$

$$\frac{(x-1)(x-2)}{(x-5)(x+5)(x-2)} - \frac{(x+2)(x-5)}{(x-5)(x+5)(x-2)}$$

$$\frac{x^2-3x+2}{(x-5)(x+5)(x-2)} - \frac{x^2-3x-10}{(x-5)(x+5)(x-2)}$$

$$\frac{(x-1)(x^2+3x-10)-(x+2)(x^2-25)}{(x^2-25)(x^2-3x-10)}$$

$$\frac{12}{(x-5)(x+5)(x-2)}$$

Cross-product rule

$$\frac{x-1}{x^2-25} - \frac{x+2}{x^2+3x-10}$$

$$\frac{(x-1)(x^2+3x-10)-(x+2)(x^2-25)}{(x^2-25)(x^2-3x-10)}$$

$$\frac{x^3+2x^2-13x+10-(x^3+2x^2-25x-50)}{(x^2-25)(x^2+3x-10)}$$

$$\frac{x^3+2x^2-13x+10-x^3-2x^2+25x+50}{(x^2-25)(x^2+3x-10)}$$

$$\frac{12x+60}{(x^2-25)(x^2+3x-10)}$$

$$\frac{12\overset{1}{\cancel{(x+5)}}}{(x-5)(x+5)\underset{1}{\cancel{(x+5)}}(x-2)}$$

$$\frac{12}{(x-5)(x+5)(x-2)}$$

Although the cross-product rule appears here to be only somewhat longer, it usually involves working with higher-degree expressions (here third degree as opposed to second) and an extra step to reduce the fraction.

In general, if the product of the denominators is the LCD, the cross-product method is usually the easiest one to use. If not, the LCD method is best.

Q27   Find:

a. $\dfrac{3}{x-1} - \dfrac{5x}{x+7}$ b. $\dfrac{x+3}{x^2-36} - \dfrac{x}{x^2+8x+12}$

STOP • **STOP** • **STOP** • **STOP** • **STOP** • **STOP** • **STOP** • **STOP** • **STOP**

A27   a. $\dfrac{^-(5x+7)(x-3)}{(x-1)(x+7)}$: the

LCD is the product of the denominators. Hence, the cross-product method can be used conveniently.

$\dfrac{3}{x-1} - \dfrac{5x}{x+7}$

$\dfrac{3(x+7) - 5x(x-1)}{(x-1)(x+7)}$

$\dfrac{3x+21-5x^2+5x}{(x-1)(x+7)}$

$\dfrac{^-5x^2+8x+21}{(x-1)(x+7)}$

$\dfrac{^-(5x^2-8x-21)}{(x-1)(x+7)}$

$\dfrac{^-(5x+7)(x-3)}{(x-1)(x+7)}$

b. $\dfrac{11x+6}{(x+6)(x-6)(x+2)}$: the LCD is $(x-6)(x+6)(x+2)$.

$\dfrac{x+3}{(x+6)(x-6)} - \dfrac{x}{(x+6)(x+2)}$

$\dfrac{x+3}{(x+6)(x-6)} \cdot \dfrac{x+2}{x+2} - \dfrac{x}{(x+6)(x+2)} \cdot \dfrac{x-6}{x-6}$

$\dfrac{(x+3)(x+2)}{(x+6)(x-6)(x+2)} - \dfrac{x(x-6)}{(x+6)(x-6)(x+2)}$

$\dfrac{(x+3)(x+2) - x(x-6)}{(x+6)(x-6)(x+2)}$

$\dfrac{x^2+5x+6-x^2+6x}{(x+6)(x-6)(x+2)}$

$\dfrac{11x+6}{(x+6)(x-6)(x+2)}$

---

24   If opposites are present in the denominators, a simple change can be made to simplify the work.

Example 1: Find $\dfrac{1}{x-1} + \dfrac{3}{1-x}$. Notice that $x-1$ and $1-x$ are opposites.

Solution

$$\frac{1}{x-1} + \frac{3}{1-x} \cdot \frac{^-1}{^-1}$$

$$\frac{1}{x-1} + \frac{^-3}{x-1} \qquad \left(\textit{Note:}\ \frac{3}{1-x} \cdot \frac{^-1}{^-1} = \frac{^-3}{^-1+x} = \frac{^-3}{x-1}\right)$$

$$\frac{1+^-3}{x-1}$$

$$\frac{^-2}{x-1}$$

Example 2: Find $\dfrac{x-3}{5-x} - \dfrac{x+7}{x-5}$ ($5-x$ and $x-5$ are opposites).

Solution

$$\frac{x-3}{5-x} \cdot \frac{^-1}{^-1} - \frac{x+7}{x-5}$$

$$\frac{^-x+3}{x-5} - \frac{x+7}{x-5}$$

$$\frac{^-x+3-(x+7)}{x-5}$$

$$\frac{^-x+3-x-7}{x-5}$$

$$\frac{^-2x-4}{x-5}$$

$$\frac{^-2(x+2)}{x-5}$$

Q28   Find:

a. $\dfrac{3}{y-4} + \dfrac{2}{4-y}$

b. $\dfrac{x+2}{x-2} - \dfrac{x-3}{2-x}$

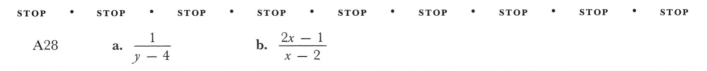

STOP   •   STOP   •   STOP   •   STOP   •   STOP   •   STOP   •   STOP   •   STOP   •   STOP

A28   a. $\dfrac{1}{y-4}$

b. $\dfrac{2x-1}{x-2}$

This completes the instruction for this section.

12.3   Exercises

1. Find:

a. $\dfrac{4}{x} + \dfrac{3}{x}$

b. $\dfrac{7}{5y} - \dfrac{2}{5y}$

c. $\dfrac{x-2}{x-3} - \dfrac{x-1}{x-3}$

d. $\dfrac{3x+4}{2x} + \dfrac{x-3}{2x}$

e. $\dfrac{5x}{x^2+2x+3} + \dfrac{x-2}{x^2+2x+3}$

f. $\dfrac{x^2+3x+1}{x-3} - \dfrac{x^2+2x+4}{x-3}$

g. $\dfrac{2x}{x^2+2x+1} - \dfrac{x-1}{x^2+2x+1}$

h. $\dfrac{x}{x^2-y^2} - \dfrac{y}{x^2-y^2}$

2. a. To find the sum or difference of two rational algebraic expressions with unlike denominators it is necessary to find a _____ _____, preferably the _____ _____ _____ (LCD).

b. The least common denominator (LCD) is the _____ _____ _____ (LCM) of the denominators.

3. Find the LCM for:

a. 12 and 18

b. 15 and 50

c. a and b

d. $3x$ and $4y$

e. $5x^2$ and $7x$

f. $5x^2y$ and $15xy^2$

g. $x^2 - 9$ and $x + 3$

h. $x^2 - 25$ and $4x - 20$

i. $x^2 + 7x + 12$ and $x^2 + x - 6$

j. $3x^2 - 5x - 12$ and $9x^2 - 16$

4. Find:

a. $\dfrac{3}{x^2} - \dfrac{4}{x}$

b. $\dfrac{5}{a} + \dfrac{7}{b}$

c. $\dfrac{8}{xy^2} - \dfrac{3}{x^2y}$

d. $\dfrac{x+5}{5} + \dfrac{2x-7}{6}$

e. $\dfrac{x+4}{3} - \dfrac{x-3}{4}$

f. $\dfrac{5}{3x} + \dfrac{4}{7y}$

g. $\dfrac{2x}{y} + 5 \quad \left(\textit{Note: } 5 = \dfrac{5}{1}\right)$

h. $9 - \dfrac{9y}{5}$

i. $\dfrac{6}{x+1} + \dfrac{3}{x-1}$

j. $\dfrac{6}{x+1} - \dfrac{3}{x-1}$

k. $\dfrac{7}{x-2} - \dfrac{8}{x^2-4}$

l. $7 - \dfrac{2a-b}{2a+b}$

m. $\dfrac{3}{x} + \dfrac{5x}{3x-9}$

n. $\dfrac{y-1}{2y-3} + \dfrac{y+2}{y+4}$

o. $\dfrac{-3x-9}{x^2+7x+12} + \dfrac{x}{x+4}$

p. $\dfrac{-3}{x^2+2x+1} + \dfrac{4}{x^2-1}$

q. $\dfrac{x+3}{x^2+7x+10} + \dfrac{x-4}{x^2-x-6}$

r. $\dfrac{x-2}{x+3} - \dfrac{x+3}{x-2}$

s. $\dfrac{3}{5-x} - \dfrac{8}{x-5}$

t. $\dfrac{x+2}{x-3} + \dfrac{5}{3-x}$

*5. a. $\dfrac{3x}{x} - \dfrac{4}{xy^2} + \dfrac{5}{xy}$

b. $\dfrac{-x}{3-x} - \dfrac{18}{x^2-9}$

c. $\dfrac{2}{y^2+y} + \dfrac{3}{5y-15}$

d. $\dfrac{5}{3x^2} + \dfrac{2}{x} + 9$

12.3   Exercise Answers

1. a. $\dfrac{7}{x}$

    b. $\dfrac{1}{y}$

    c. $\dfrac{-1}{x-3}$

    d. $\dfrac{4x+1}{2x}$

    e. $\dfrac{2(3x-1)}{x^2+2x+3}$

    f. 1

    g. $\dfrac{1}{x+1}$

    h. $\dfrac{1}{x+y}$

2. a. common denominator, least common denominator
    b. least common multiple

3. a. 36

    b. 150

    c. ab

    d. $12xy$

    e. $35x^2$

    f. $15x^2y^2$

    g. $(x-3)(x+3)$

    h. $4(x-5)(x+5)$

    i. $(x+3)(x+4)(x-2)$

    j. $(3x+4)(3x-4)(x-3)$

    (*Note:* The binomial factors can be written in any order.)

4. a. $\dfrac{3-4x}{x^2}$

    b. $\dfrac{5b+7a}{ab}$

    c. $\dfrac{8x-3y}{x^2y^2}$

    d. $\dfrac{16x-5}{30}$

    e. $\dfrac{x+25}{12}$

    f. $\dfrac{35y+12x}{21xy}$

    g. $\dfrac{2x+5y}{y}$

    h. $\dfrac{9(5-y)}{5}$

    i. $\dfrac{3(3x-1)}{(x+1)(x-1)}$

    j. $\dfrac{3(x-3)}{(x+1)(x-1)}$

    k. $\dfrac{7x+6}{(x-2)(x+2)}$

    l. $\dfrac{4(3a+2b)}{2a+b}$

    m. $\dfrac{5x^2+9x-27}{3x(x-3)}$

    n. $\dfrac{3y^2+4y-10}{(2y-3)(y+4)}$

    o. $\dfrac{x-3}{x+4}$

    p. $\dfrac{x+7}{(x-1)(x+1)^2}$

    q. $\dfrac{2x^2+x-29}{(x+2)(x+5)(x-3)}$

    r. $\dfrac{-5(2x+1)}{(x+3)(x-2)}$

    s. $\dfrac{-11}{x-5}$

    t. 1

*5. a. $\dfrac{3xy^2-4+5y}{xy^2}$

    b. $\dfrac{x+6}{x+3}$

    c. $\dfrac{3y^2+13y-30}{5y(y-3)(y+1)}$ or $\dfrac{(3y-5)(y+6)}{5y(y-3)(x+1)}$

    d. $\dfrac{5+6x+27x^2}{3x^2}$

12.4   Complex Fractions and Fractional Equations

> 1
>
> A *complex fraction* is a rational expression that has a fraction in its numerator or its denominator or both. It is sometimes described as a fraction within a fraction. Examples of complex fractions are:
>
> $$\dfrac{\frac{1}{2}}{5}, \quad \dfrac{\frac{2x}{3}}{\frac{4}{5}}, \quad \dfrac{1}{x+\frac{1}{3}}, \quad \dfrac{x}{\frac{2}{3}-x}, \quad \dfrac{5-\frac{x}{y}}{5+\frac{x}{y}}, \quad \dfrac{\frac{1}{x}-\frac{1}{y}}{\frac{1}{x}+\frac{1}{y}}$$

Q1   Write "yes" if the expression is a complex fraction and "no" otherwise.

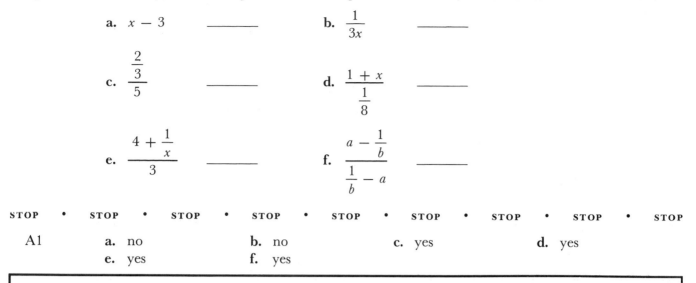

a. $x - 3$ _____

b. $\dfrac{1}{3x}$ _____

c. $\dfrac{\frac{2}{3}}{5}$ _____

d. $\dfrac{1 + x}{\frac{1}{8}}$ _____

e. $\dfrac{4 + \frac{1}{x}}{3}$ _____

f. $\dfrac{a - \frac{1}{b}}{\frac{1}{b} - a}$ _____

STOP • **STOP** • **STOP** • **STOP** • **STOP** • **STOP** • **STOP** • **STOP** • **STOP**

A1   **a.** no   **b.** no   **c.** yes   **d.** yes
e. yes   **f.** yes

2   Any complex fraction may be simplified. One method for simplifying complex fractions involves the following steps:

Step 1: Write the *numerator* as a single rational expression.
Step 2: Write the *denominator* as a single rational expression.
Step 3: Use the definition of division to invert the divisor.
Step 4: Find the product.
Step 5: Reduce the product to lowest terms.

Example 1: Simplify $\dfrac{\frac{1}{2}}{5}$.

Solution

Steps 1 and 2 are already satisfied.

Step 3: $\dfrac{\frac{1}{2}}{5} = \dfrac{1}{2} \div \dfrac{5}{1} = \dfrac{1}{2} \cdot \dfrac{1}{5}$

Step 4: $\qquad = \dfrac{1}{10}$

Step 5: $\dfrac{1}{10}$ is in lowest terms.

Thus,

$$\dfrac{\frac{1}{2}}{5} = \dfrac{1}{10}$$

Example 2: Simplify $\dfrac{\frac{2x}{3}}{\frac{-4x}{5}}$.

Solution

Steps 1 and 2 are already satisfied.

Step 3: $\dfrac{\dfrac{2x}{3}}{\dfrac{-4x}{5}} = \dfrac{2x}{3} \div \dfrac{-4x}{5}$

$$= \dfrac{\overset{1}{\cancel{2x}}}{3} \cdot \dfrac{-5}{\underset{2}{\cancel{4x}}}$$

Step 4: $\qquad = \dfrac{-5}{6}$

Step 5: $\dfrac{-5}{6}$ is in lowest terms.

Thus,

$$\dfrac{\dfrac{2x}{3}}{\dfrac{-4x}{5}} = \dfrac{-5}{6}$$

Q2   Simplify:

a. $\dfrac{\dfrac{2}{3}}{6}$ b. $\dfrac{\dfrac{-5x^2}{6}}{\dfrac{2x}{3}}$

STOP • STOP • STOP • STOP • STOP • STOP • STOP • STOP • STOP

A2   a. $\dfrac{1}{9}$: $\quad \dfrac{\dfrac{2}{3}}{6} = \dfrac{2}{3} \div \dfrac{6}{1}$ b. $\dfrac{-5x}{4}$: $\quad \dfrac{\dfrac{-5x^2}{6}}{\dfrac{2x}{3}} = \dfrac{-5x^2}{6} \div \dfrac{2x}{3}$

$$= \dfrac{2}{3} \cdot \dfrac{1}{6} \qquad\qquad\qquad = \dfrac{-5x\overset{x}{\cancel{{}^2}}}{\underset{2}{\cancel{6}}} \cdot \dfrac{\overset{1}{\cancel{3}}}{\underset{1}{\cancel{2x}}}$$

$$= \dfrac{1}{9} \qquad\qquad\qquad\qquad = \dfrac{-5x}{4}$$

3   When writing a numerator or denominator as a single rational expression, use the procedures of the previous sections to find a sum or difference.

Example 1: Simplify $\dfrac{x}{\dfrac{2}{3} - x}$.

Solution

Step 1 is already satisfied.

Step 2: $\dfrac{x}{\dfrac{2}{3} - x} = \dfrac{x}{\dfrac{2}{3} - \dfrac{x}{1}}$

$$= \dfrac{x}{\dfrac{2 - 3x}{3}}$$

Step 3: $\qquad = x \div \dfrac{2 - 3x}{3}$

Step 4: $\qquad = \dfrac{x}{1} \cdot \dfrac{3}{2 - 3x}$

$$= \dfrac{3x}{2 - 3x}$$

Step 5: $\dfrac{3x}{2 - 3x}$ is in lowest terms.

Thus,

$$\dfrac{x}{\dfrac{2}{3} - x} = \dfrac{3x}{2 - 3x}$$

Example 2: Simplify $\dfrac{5 - \dfrac{x}{y}}{5 + \dfrac{x}{y}}$.

Solution

$\left(\text{\textit{Note:} Write 5 as } \dfrac{5}{1}.\right)$

Step 1: $\dfrac{\dfrac{5}{1} - \dfrac{x}{y}}{\dfrac{5}{1} + \dfrac{x}{y}} = \dfrac{\dfrac{5y - x}{y}}{\dfrac{5}{1} + \dfrac{x}{y}}$

Step 2: $\qquad = \dfrac{\dfrac{5y - x}{y}}{\dfrac{5y + x}{y}}$

Steps 1 and 2 are usually combined into a single step.

Step 3: $\qquad = \dfrac{5y - x}{y} \div \dfrac{5y + x}{y}$

$$= \dfrac{5y - x}{\cancel{y}} \cdot \dfrac{\overset{1}{\cancel{y}}}{5y + x}$$

Step 4: $\qquad = \dfrac{5y - x}{5y + x}$

Step 5: $\dfrac{5y - x}{5y + x}$ is in lowest terms.

Thus,

$$\frac{5 - \dfrac{x}{y}}{5 + \dfrac{x}{y}} = \frac{5y - x}{5y + x}$$

(*Note:* Although most of the reduction will be done in step 4 before finding the product, it is advisable to check the final answer to make sure it is reduced to lowest terms.)

Q3   Simplify:

a. $\dfrac{y^2}{y - \dfrac{3}{5}}$

b. $\dfrac{1 - \dfrac{m}{n}}{\dfrac{m^2}{n^2}}$

c. $\dfrac{a - \dfrac{1}{b}}{a + \dfrac{1}{b}}$

d. $\dfrac{\dfrac{1}{x} - \dfrac{1}{y}}{\dfrac{1}{x} + \dfrac{1}{y}}$

A3

a. $\dfrac{5y^2}{5y-3} : \dfrac{y^2}{\dfrac{y}{1}-\dfrac{3}{5}} = \dfrac{y^2}{\dfrac{5y-3}{5}}$

$$= y^2 \div \dfrac{5y-3}{5}$$

$$= \dfrac{y^2}{1} \cdot \dfrac{5}{5y-3}$$

$$= \dfrac{5y^2}{5y-3}$$

b. $\dfrac{n(n-m)}{m^2} : \dfrac{\dfrac{1}{1}-\dfrac{m}{n}}{\dfrac{m^2}{n^2}} = \dfrac{\dfrac{n-m}{n}}{\dfrac{m^2}{n^2}}$

$$= \dfrac{n-m}{n} \div \dfrac{m^2}{n^2}$$

$$= \dfrac{n-m}{\cancel{n}} \cdot \dfrac{\overset{n}{\cancel{n^2}}}{m^2}$$

$$= \dfrac{n(n-m)}{m^2}$$

c. $\dfrac{ab-1}{ab+1} : \dfrac{\dfrac{a}{1}-\dfrac{1}{b}}{\dfrac{a}{1}+\dfrac{1}{b}} = \dfrac{\dfrac{ab-1}{b}}{\dfrac{ab+1}{b}}$

$$= \dfrac{ab-1}{b} \div \dfrac{ab+1}{b}$$

$$= \dfrac{ab-1}{b} \cdot \dfrac{b}{ab+1}$$

$$= \dfrac{ab-1}{ab+1}$$

d. $\dfrac{y-x}{y+x} : \dfrac{\dfrac{1}{x}-\dfrac{1}{y}}{\dfrac{1}{x}+\dfrac{1}{y}} = \dfrac{\dfrac{y-x}{xy}}{\dfrac{y+x}{xy}}$

$$= \dfrac{y-x}{xy} \div \dfrac{y+x}{xy}$$

$$= \dfrac{y-x}{\cancel{xy}} \cdot \dfrac{\overset{1}{\cancel{xy}}}{y+x}$$

$$= \dfrac{y-x}{y+x}$$

---

4   An alternative method that may be used to simplify complex fractions is: Multiply both the numerator and denominator of the complex fraction by the LCD (LCM) of all denominators of individual terms of the complex fraction.

Example 1: Simplify $\dfrac{\dfrac{1}{2}}{5}$.

Solution

$$\dfrac{\dfrac{1}{2}}{5} = \dfrac{\dfrac{1}{2}\cdot 2}{5\cdot 2} \qquad \text{(the LCD is 2)}$$

$$= \dfrac{1}{10}$$

Example 2: Simplify $\dfrac{x}{\dfrac{2}{3}-x}$.

Solution

$$\frac{x}{\frac{2}{3} - x} = \frac{x \cdot 3}{\left(\frac{2}{3} - x\right)3} \qquad \text{(the LCD is 3)}$$

$$= \frac{3x}{\frac{2}{3} \cdot 3 - x \cdot 3}$$

$$= \frac{3x}{2 - 3x}$$

Example 3: Simplify $\dfrac{3 + \frac{a}{b}}{3 - \frac{b}{a}}$.

Solution

$$\frac{3 + \frac{a}{b}}{3 - \frac{b}{a}} = \frac{\left(3 + \frac{a}{b}\right)ab}{\left(3 - \frac{b}{a}\right)ab} \qquad \text{(the LCD is } ab\text{)}$$

$$= \frac{3ab + \frac{a}{b}(ab)}{3ab - \frac{b}{a}(ab)}$$

$$= \frac{3ab + a^2}{3ab - b^2}$$

$$= \frac{a(3b + a)}{b(3a - b)}$$

When simplifying complex fractions, either of the two methods may be used.

Q4 Simplify (use the method of your choice):

a. $\dfrac{5x}{\frac{3}{4}}$

b. $\dfrac{x - \frac{2}{3}}{x + \frac{1}{9}}$

c. $\dfrac{2 - \dfrac{x}{5}}{\dfrac{7}{8}}$

d. $\dfrac{x + \dfrac{1}{y}}{x - \dfrac{1}{y}}$

e. $\dfrac{\dfrac{4x}{9} - 3}{3 - \dfrac{4x}{9}}$

STOP • STOP • STOP • STOP • STOP • STOP • STOP • STOP • STOP

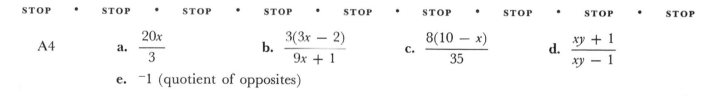

A4   **a.** $\dfrac{20x}{3}$ **b.** $\dfrac{3(3x - 2)}{9x + 1}$ **c.** $\dfrac{8(10 - x)}{35}$ **d.** $\dfrac{xy + 1}{xy - 1}$

e. $^-1$ (quotient of opposites)

5   Section 5.5 was a study of the procedures to use when solving a fractional equation. This section will again develop procedures for solving fractional equations of a different sort. The equations of Section 5.5 always contained the variable in the *numerator* of the fractions present in the equation. The equations of this section will contain a variable in the denominator of one or more of the fractions present in the equation.

The procedures for solving fractional equations with variable denominators will be the same as those used in Section 5.5. The steps to follow are:

Step 1: Find the LCD for the fractions present in the equation.
Step 2: Multiply every term of the equation by the LCD.
Step 3: Solve the resulting equation.
Step 4: Check the solution by substitution into the original equation.

Example: Solve $\dfrac{6}{x} - \dfrac{1}{2} = \dfrac{1}{4}$.

Solution

Step 1: The LCD is $4x$.

Step 2: $4x\left(\dfrac{6}{x}\right) - 4x\left(\dfrac{1}{2}\right) = 4x\left(\dfrac{1}{4}\right)$

$$4(6) - 2x(1) = x(1)$$

$$24 - 2x = x$$

Step 3: $24 - 2x + 2x = x + 2x$

$$24 = 3x$$

$$8 = x$$

Step 4: Check: $\dfrac{6}{x} - \dfrac{1}{2} = \dfrac{1}{4}$

$$\dfrac{6}{8} - \dfrac{1}{2} \overset{?}{=} \dfrac{1}{4}$$

$$\dfrac{3}{4} - \dfrac{1}{2} \overset{?}{=} \dfrac{1}{4}$$

$$\dfrac{1}{4} = \dfrac{1}{4}$$

Q5   Solve $\dfrac{3}{x} - \dfrac{1}{2} = \dfrac{5}{6}$.

STOP   •   STOP   •   STOP   •   STOP   •   STOP   •   STOP   •   STOP   •   STOP   •   STOP

A5 $x = \dfrac{9}{4}$:   The LCD is $6x$.

$$6x\left(\dfrac{3}{x}\right) - 6x\left(\dfrac{1}{2}\right) = 6x\left(\dfrac{5}{6}\right)$$

$$6(3) - 3x(1) = x(5)$$

$$18 - 3x = 5x$$

$$18 = 8x$$

$$\dfrac{9}{4} = x$$

6   **Example:** Solve $\dfrac{4}{5x} - \dfrac{2}{3} = \dfrac{1}{3x}$.

Solution

The LCD is $15x$.

$$15x\left(\dfrac{4}{5x}\right) - 15x\left(\dfrac{2}{3}\right) = 15x\left(\dfrac{1}{3x}\right)$$

$$3(4) - 5x(2) = 5(1)$$

$$12 - 10x = 5$$

$$12 - 10x - 12 = 5 - 12$$

$$^{-}10x = {}^{-}7$$

$$x = \dfrac{7}{10}$$

Check: $\dfrac{4}{5x} - \dfrac{2}{3} = \dfrac{1}{3x}$

$$\dfrac{4}{5\left(\dfrac{7}{10}\right)} - \dfrac{2}{3} \overset{?}{=} \dfrac{1}{3\left(\dfrac{7}{10}\right)}$$

$$\dfrac{4}{\dfrac{7}{2}} - \dfrac{2}{3} \overset{?}{=} \dfrac{1}{\dfrac{21}{10}}$$

$$\dfrac{8}{7} - \dfrac{2}{3} \overset{?}{=} \dfrac{10}{21}$$

$$\dfrac{24}{21} - \dfrac{14}{21} \overset{?}{=} \dfrac{10}{21}$$

$$\dfrac{10}{21} = \dfrac{10}{21}$$

Q6   Solve:

a. $\dfrac{4}{3x} - \dfrac{5}{4x} = \dfrac{1}{6}$ **b.** $\dfrac{2}{7} - \dfrac{3}{x} = \dfrac{-5}{7}$

c. $\dfrac{4}{3x} + \dfrac{1}{2} = \dfrac{2}{x}$ **d.** $\dfrac{1}{x} - \dfrac{3}{5} = \dfrac{3}{x} - \dfrac{2}{5}$

STOP • **STOP** • **STOP** • **STOP** • **STOP** • **STOP** • **STOP** • **STOP** • **STOP**

A6   **a.** $x = \dfrac{1}{2}$:   The LCD is $12x$.   **b.** $x = 3$

$$12x\left(\dfrac{4}{3x}\right) - 12x\left(\dfrac{5}{4x}\right) = 12x\left(\dfrac{1}{6}\right)$$

$$4(4) - 3(5) = 2x(1)$$
$$16 - 15 = 2x$$
$$1 = 2x$$
$$\dfrac{1}{2} = x$$

c. $x = \dfrac{4}{3}$ **d.** $x = {}^-10$

---

7   The procedures of Frame 5 can also be used to solve equations with more complicated denominators.

Example: Solve $\dfrac{3}{x-2} + 2 = \dfrac{1}{x-2}$.

Solution

Step 1: The LCD is $x - 2$.

Step 2: $(x-2)\left(\dfrac{3}{x-2}\right) + (x-2)2 = \left(\dfrac{1}{x-2}\right)(x-2)$

$$3 + 2(x-2) = 1$$

[*Note:* $(x-2)2 = 2(x-2)$ by the commutative property of multiplication.]

$$3 + 2x - 4 = 1 \qquad \text{Check: } \dfrac{3}{x-2} + 2 = \dfrac{1}{x-2}$$

$$2x - 1 = 1 \qquad\qquad \dfrac{3}{1-2} + 2 \overset{?}{=} \dfrac{1}{1-2}$$

$$2x = 2 \qquad\qquad\qquad \dfrac{3}{^-1} + 2 \overset{?}{=} \dfrac{1}{^-1}$$

$$x = 1 \qquad\qquad\qquad\qquad {}^-3 + 2 \overset{?}{=} {}^-1$$

$$\qquad\qquad\qquad\qquad\qquad {}^-1 = {}^-1$$

Q7   Solve $\dfrac{3}{x-4} - 1 = \dfrac{1}{x-4}$

STOP • **STOP** • **STOP** • **STOP** • **STOP** • **STOP** • **STOP** • **STOP** • **STOP**

A7 $x = 6$:   The LCD is $x - 4$.

$$(x-4)\left(\dfrac{3}{x-4}\right) - (x-4)1 = (x-4)\left(\dfrac{1}{x-4}\right)$$

$$3 - 1(x-4) = 1$$

$$3 - x + 4 = 1$$

$${}^-x + 7 = 1$$

$${}^-x = {}^-6$$

$$x = 6$$

---

8   Recall that a rational number with denominator zero is said to be undefined, because division by zero is impossible. For this reason, it is important that the replacement set for the variable in a fractional equation *not include* any numbers for which the value of any denominator is zero.

Examples:

$$\frac{3}{x} + 5 = 2$$

The replacement set for this equation cannot include zero, because $\frac{3}{0}$ is undefined. That is, $x \neq 0$ is a restriction on the replacement set.

$$\frac{8}{x - 3} = \frac{12}{x + 3}$$

The replacement set for this equation cannot include 3 or $^-3$, because for each value *one* of the fractions is undefined. Thus, the restrictions on the replacement set are $x \neq 3$ and $x \neq {}^-3$ or simply $x \neq \pm 3$.

$$\frac{x - 5}{3} - 1 = 7x$$

The replacement set for this equation has no restrictions, because none of the denominators contain variables.

Q8   Write the restrictions on the replacement set for each of the following:

a. $\dfrac{7}{x} = 4$ _____

b. $\dfrac{1}{x - 1} = \dfrac{3}{x - 5}$ _____

c. $\dfrac{x - 5}{4} = 7$ _____

d. $\dfrac{3}{x(x + 1)} = \dfrac{1}{x - 3}$ _____

STOP • STOP • STOP • STOP • STOP • STOP • STOP • STOP • STOP

A8   a. $x \neq 0$ b. $x \neq 1, x \neq 5$
      c. no restrictions   d. $x \neq 0, x \neq {}^-1, x \neq 3$

9   Any value for the variable in an equation that produces a zero denominator cannot be used as a solution to the equation. Consider the solution of the equation

$$\frac{2}{x + 1} - 5 = \frac{2}{x + 1}$$

The LCD is $x + 1$.

$$(x + 1)\left(\frac{2}{x + 1}\right) - (x + 1)5 = (x + 1)\left(\frac{2}{x + 1}\right)$$

$$2 - 5(x + 1) = 2$$
$$2 - 5x - 5 = 2$$
$$^-5x - 3 = 2$$
$$^-5x = 5$$
$$x = {}^-1$$

Check: $\dfrac{2}{x + 1} - 5 = \dfrac{2}{x + 1}$

$\dfrac{2}{^-1 + 1} - 5 \overset{?}{=} \dfrac{2}{^-1 + 1}$

$\dfrac{2}{0} - 5 \overset{?}{=} \dfrac{2}{0}$

Since the replacement $x = {}^-1$ produces a zero denominator in the equation, it cannot be used as a solution to the equation. Thus, there is *no solution* to the equation $\dfrac{2}{x + 1} - 5 = \dfrac{2}{x + 1}$.

Q9   Solve:

a. $\dfrac{3}{x-4} + 2 = \dfrac{5}{x-4}$ **b.** $\dfrac{3}{x+5} - 4 = \dfrac{3}{x+5}$

STOP • **STOP** • **STOP** • **STOP** • **STOP** • **STOP** • **STOP** • **STOP** • **STOP**

A9   **a.** $x = 5$

         b. no solution: The value of $x = {}^{-}5$ (found by solving the equation) produces a zero denominator in the equation.

---

10   To solve $\dfrac{9}{x+3} - \dfrac{2}{x^2-x-12} = \dfrac{4}{x-4}$, factor the denominators to find the LCD and proceed as in Frame 5. First factor $x^2 - x - 12$.

$$\frac{9}{x+3} - \frac{2}{(x+3)(x-4)} = \frac{4}{x-4}$$

The LCD is now seen to be $(x+3)(x-4)$, and the solution can be completed.

$$(x+3)(x-4)\left(\frac{9}{x+3}\right) - (x+3)(x-4)\left[\frac{2}{(x+3)(x-4)}\right] = (x+3)(x-4)\left(\frac{4}{x-4}\right)$$

$$9(x-4) - 2 = 4(x+3)$$
$$9x - 36 - 2 = 4x + 12$$
$$9x - 38 = 4x + 12$$
$$5x = 50$$
$$x = 10$$

Q10   Solve $\dfrac{5}{x+1} + \dfrac{1}{x^2-1} = \dfrac{2}{x-1}$.

STOP • **STOP** • **STOP** • **STOP** • **STOP** • **STOP** • **STOP** • **STOP** • **STOP**

A10 $x = 2$:   The LCD is $(x+1)(x-1)$ or $x^2 - 1$.

$$(x^2-1)\left(\frac{5}{x+1}\right) + (x^2-1)\left(\frac{1}{x^2-1}\right) = (x^2-1)\left(\frac{2}{x-1}\right)$$

$$5(x - 1) + 1 = 2(x + 1)$$
$$5x - 5 + 1 = 2x + 2$$
$$5x - 4 = 2x + 2$$
$$3x = 6$$
$$x = 2$$

Check: $\dfrac{5}{x + 1} + \dfrac{1}{x^2 - 1} = \dfrac{2}{x - 1}$

$\dfrac{5}{2 + 1} + \dfrac{1}{2^2 - 1} \overset{?}{=} \dfrac{2}{2 - 1}$

$\dfrac{5}{3} + \dfrac{1}{3} \overset{?}{=} \dfrac{2}{1}$

$2 = 2$

Q11   Solve $\dfrac{2}{x + 2} + \dfrac{x}{x - 2} = \dfrac{x^2 + 4}{x^2 - 4}$.

STOP • **STOP** • **STOP** • **STOP** • **STOP** • **STOP** • **STOP** • **STOP** • **STOP**

A11   no solution: The LCD is $(x - 2)(x + 2)$ or $x^2 - 4$.

$$(x^2 - 4)\left(\frac{2}{x + 2}\right) + (x^2 - 4)\left(\frac{x}{x - 2}\right) = (x^2 - 4)\left(\frac{x^2 + 4}{x^2 - 4}\right)$$
$$2(x - 2) + x(x + 2) = x^2 + 4$$
$$2x - 4 + x^2 + 2x = x^2 + 4$$
$$x^2 + 4x - 4 = x^2 + 4 \quad \text{(subtract } x^2 \text{ from both}$$
$$4x - 4 = 4 \quad\quad\quad\quad\quad\quad \text{sides)}$$
$$4x = 8$$
$$x = 2$$

Check: $\dfrac{2}{x + 2} + \dfrac{x}{x - 2} = \dfrac{x^2 + 4}{x^2 - 4}$

$\dfrac{2}{2 + 2} + \dfrac{2}{2 - 2} = \dfrac{2^2 + 4}{2^2 - 4}$

$\dfrac{2}{4} + \dfrac{2}{0} = \dfrac{8}{0}$

The value $x = 2$ produces a denominator of zero. Hence, there is no solution.

Q12 Solve:

a. $\dfrac{3}{x-7} = \dfrac{2}{x+5}$

b. $\dfrac{x}{x-1} - \dfrac{3}{x+1} = 1$

STOP • **STOP** • **STOP** • **STOP** • **STOP** • **STOP** • **STOP** • **STOP** • **STOP**

A12 a. $x = {}^-29$: The LCD is $(x-7)(x+5)$.

$$(x-7)(x+5)\left(\dfrac{3}{x-7}\right) = (x-7)(x+5)\left(\dfrac{2}{x+5}\right)$$
$$3(x+5) = 2(x-7)$$
$$3x + 15 = 2x - 14$$
$$x = {}^-29$$

b. $x = 2$: The LCD is $(x-1)(x+1)$ or $x^2 - 1$.

$$(x^2 - 1)\left(\dfrac{x}{x-1}\right) - (x^2 - 1)\left(\dfrac{3}{x+1}\right) = (x^2 - 1)1$$
$$x(x+1) - 3(x-1) = 1(x^2 - 1)$$
$$x^2 + x - 3x + 3 = x^2 - 1$$
$$x^2 - 2x + 3 = x^2 - 1$$
$${}^-2x + 3 = {}^-1$$
$${}^-2x = {}^-4$$
$$x = 2$$

This completes the instruction for this section.

12.4 Exercises

1. Simplify:

a. $\dfrac{\dfrac{2}{3}}{\dfrac{4}{5}}$

b. $\dfrac{\dfrac{-4}{7x^3}}{\dfrac{20}{21x}}$

c. $\dfrac{12}{\dfrac{4}{9}}$

d. $\dfrac{\dfrac{a}{5} - \dfrac{2}{5}}{\dfrac{a}{3} - \dfrac{2}{3}}$

e. $\dfrac{\dfrac{1}{x}}{1 - \dfrac{1}{x}}$

f. $\dfrac{\dfrac{3}{x} + \dfrac{4}{y}}{\dfrac{2}{x} - \dfrac{5}{y}}$

g. $\dfrac{\dfrac{2x}{3} - 4}{3 - \dfrac{4x}{9}}$

h. $\dfrac{\dfrac{x - y}{5}}{\dfrac{x^2 - y^2}{10}}$

i. $\dfrac{\dfrac{1}{x^2} - \dfrac{1}{y^2}}{\dfrac{1}{x} - \dfrac{1}{y}}$

j. $\dfrac{\dfrac{x^2 + 14x + 49}{x + 2}}{\dfrac{x^2 + 5x - 14}{x^2 - 4}}$

2. Write the restrictions on the replacement sets for:

a. $\dfrac{4}{x} - 3 = 0$

b. $\dfrac{2}{x - 5} = 6 - \dfrac{3}{x + 1}$

c. $\dfrac{5}{x^2 - 9} = \dfrac{1}{9}$

d. $\dfrac{1}{x(x - 7)} = \dfrac{4}{x + 3}$

e. $\dfrac{5}{3} + \dfrac{x}{2} = \dfrac{x + 7}{10}$

f. $\dfrac{4}{x^2 - x - 12} = \dfrac{3}{4}$

3. Solve:

a. $\dfrac{2}{y} + 5 = 0$

b. $\dfrac{3}{x} - \dfrac{2}{5} = \dfrac{3}{10}$

c. $\dfrac{6}{x + 1} + 2 = \dfrac{12}{x + 1}$

d. $\dfrac{2}{y} - \dfrac{1}{y} + 6 = 0$

e. $\dfrac{3}{x + 2} - 7 = \dfrac{10}{x + 2}$

f. $\dfrac{1}{2x} - \dfrac{2}{3x} = 7 - \dfrac{5}{x}$

g. $\dfrac{3}{x + 6} = \dfrac{5}{x + 10}$

h. $\dfrac{2}{x - 3} - 1 = \dfrac{2}{x - 3}$

i. $\dfrac{1}{4y} + \dfrac{1}{12} = \dfrac{2}{3y}$

j. $\dfrac{12}{y} = \dfrac{12}{y + 1} + \dfrac{1}{y}$

k. $\dfrac{5}{x - 1} + \dfrac{3}{x^2 + 3x - 4} = \dfrac{3}{x + 4}$

l. $\dfrac{2}{x + 2} + \dfrac{x}{x - 2} = \dfrac{x^2 + 4}{x^2 - 4}$

m. $\dfrac{5}{x - 2} - \dfrac{8}{(x + 4)(x - 2)} = \dfrac{3}{x + 4}$

n. $5 - \dfrac{3}{x + 7} = \dfrac{-3}{x + 7}$

o. $\dfrac{6}{5 - x} - \dfrac{7}{5 + x} = \dfrac{21}{25 - x^2}$

p. $\dfrac{5}{y + 3} - \dfrac{7}{2(y + 3)} = 2$

q. $\dfrac{3}{x - 4} - \dfrac{4}{x + 1} = \dfrac{1}{x^2 - 3x - 4}$

r. $\dfrac{2}{5x - 10} - 3 = \dfrac{4}{x - 2}$

12.4 Exercise Answers

1. a. $\dfrac{5}{6}$

b. $\dfrac{-3}{5x^2}$

c. 27

d. $\dfrac{3}{5}$

e. $\dfrac{1}{x - 1}$

f. $\dfrac{3y + 4x}{2y - 5x}$

g. $\dfrac{6(x - 6)}{27 - 4x}$

h. $\dfrac{2}{x + y}$

i. $\dfrac{y + x}{xy}$

j. $x + 7$

2. **a.** $x \neq 0$ **b.** $x \neq 5, x \neq {}^-1$ **c.** $x \neq \pm 3$

    d. $x \neq 0, x \neq 7, x \neq {}^-3$ **e.** no restrictions   **f.** $x \neq 4, x \neq {}^-3$

3. **a.** $y = \dfrac{{}^-2}{5}$ **b.** $x = \dfrac{30}{7}$ **c.** $x = 2$

    d. $y = \dfrac{{}^-1}{6}$ **e.** $x = {}^-3$ **f.** $x = \dfrac{29}{42}$

    g. $x = 0$ **h.** no solution   **i.** $y = 5$

    j. $y = 11$ **k.** $x = {}^-13$ **l.** no solution

    m. $x = {}^-9$ **n.** no solution   **o.** $x = 2$

    p. $y = \dfrac{{}^-9}{4}$ **q.** $x = 18$ **r.** $x = \dfrac{4}{5}$

Chapter 12 Sample Test

At the completion of Chapter 12 it is expected that you will be able to work the following problems.

12.1   Reduction to Lowest Terms

1. A rational expression is undefined for any value of a variable that produces a denominator of _____.

2. For what value(s) of the variable are each of the following undefined:

    a. $\dfrac{4}{x - 3}$ **b.** $\dfrac{x - 5}{x + 7}$ **c.** $\dfrac{8}{x}$

    d. $\dfrac{1}{x(x - 2)}$

3. Simplify:

    a. $\dfrac{24x^5}{36x}$ **b.** $\dfrac{64x^3y^7}{96xy^{15}}$ **c.** $\dfrac{4x - 20}{x^2 - 25}$

    d. $\dfrac{x + 5}{(x + 5)^3}$ **e.** $\dfrac{(x + 2)^3}{x^2 + 4x + 4}$ **f.** $\dfrac{3x^2 + 20x - 7}{x^2 + 5x - 14}$

    g. $\dfrac{x - 3}{3 - x}$ **h.** $\dfrac{{}^-x + 7}{7 - x}$ **i.** $\dfrac{5x + 2}{10x}$

    j. $\dfrac{5x^2 - 20}{5x - 10}$

12.2   Multiplication and Division

4. Find the product:

    a. $\dfrac{15x^2}{x + 3} \cdot \dfrac{x^2 - 9}{3x}$ **b.** $\dfrac{x^2 - 1}{x^2 + 6x - 7} \cdot \dfrac{2x^2 - x - 3}{3x^2 + 3x}$

5. Give the reciprocal:

    a. $\dfrac{{}^-5}{9}$ **b.** x **c.** $\dfrac{x}{2 - y}$

   d. $x^2 - 36$ **e.** $\dfrac{1}{x^2 - 2x + 1}$

6. Find the quotient:

   a. $\dfrac{20x^5y}{3xy^3} \div \dfrac{10x^4}{9xy}$ **b.** $\dfrac{x^2 - 36}{5x - 40} \div \dfrac{3x - 18}{x^2 - 16x + 64}$

7. Find the product or quotient as indicated:

   a. $\dfrac{3x - 15}{x^2 - 25} \cdot \dfrac{x + 5}{3}$ **b.** $\dfrac{3 - x}{x + 7} \div \dfrac{x - 3}{7 + x}$

   c. $\dfrac{81x^2 - 1}{5x^2 + 5x} \cdot \dfrac{x + 1}{9x + 1}$ **d.** $\dfrac{x^2 - 5x + 4}{5x - 15} \div \dfrac{3x - 12}{15x^2}$

   e. $\dfrac{x^2 + x - 6}{x^2 - 4x - 21} \div (x^2 - x - 2)$ **f.** $\dfrac{x^2 - 49}{5x^2 + 37x + 14} \cdot (5x^2 - 33x - 14)$

12.3   Addition and Subtraction

8. Find the sum or difference as indicated:

   a. $\dfrac{5}{y} + \dfrac{3}{y}$ **b.** $\dfrac{x^2}{x + 3} - \dfrac{9}{x + 3}$

   c. $\dfrac{3x - 6}{2y - 5} + \dfrac{x - 7}{2y - 5}$ **d.** $\dfrac{5x + 7}{x^2 - 16} - \dfrac{4x + 3}{x^2 - 16}$

   e. $\dfrac{x^2 + 5x - 6}{x^2 - 3x - 7} - \dfrac{x^2 - 2x + 3}{x^2 - 3x - 7}$ **f.** $\dfrac{x^2}{x - 7} - \dfrac{4x + 21}{x - 7}$

9. Find the LCM for:
   a. 36 and 45   **b.** $4z$ and $30x$
   c. $15x^2y$ and $40xy^3$ **d.** $x^2 - 81$ and $x + 2$
   e. $x^2 - 4x + 4$ and $x^2 - 4$ **f.** $x^2 - 1$ and $x + 1$

10. Find:

   a. $\dfrac{5}{x} + \dfrac{7}{y}$ **b.** $\dfrac{7}{y^2} - \dfrac{2}{y}$

   c. $\dfrac{3}{x^2y} - \dfrac{1}{xy^3}$ **d.** $\dfrac{x + 5}{2} + \dfrac{x - 3}{3}$

   e. $\dfrac{x + 5}{2} - \dfrac{x - 3}{3}$ **f.** $7 - \dfrac{4}{y}$

   g. $\dfrac{5}{x + 2} + \dfrac{3}{x - 1}$ **h.** $\dfrac{7}{x + 4} - \dfrac{2}{x + 3}$

   i. $\dfrac{4}{x} - \dfrac{4}{2x - 10}$ **j.** $\dfrac{x + 1}{x - 3} + \dfrac{x + 2}{x - 7}$

12.4   Complex Fractions and Fractional Equations

11. Simplify:

   a. $\dfrac{\dfrac{3}{7}}{\dfrac{-27}{21}}$ **b.** $\dfrac{\dfrac{x}{12y}}{15xy}$ **c.** $\dfrac{\dfrac{1}{y}}{1 - \dfrac{1}{y}}$

d. $\dfrac{\dfrac{2}{x} - \dfrac{3}{y}}{\dfrac{5}{x} - \dfrac{7}{y}}$
  e. $\dfrac{\dfrac{1}{x^2} - \dfrac{1}{y^2}}{\dfrac{y + x}{xy}}$

12. Write the restrictions on the replacement sets for:

a. $3 - \dfrac{4}{x} = 0$
  b. $\dfrac{1}{x - 2} = \dfrac{4}{x + 3}$

c. $\dfrac{3}{x^2 - 1} = 6$
  d. $\dfrac{x + 3}{7} - 4 = \dfrac{2x}{3}$

13. Solve:

a. $\dfrac{3}{x} + 7 = 0$
  b. $\dfrac{2}{x} - \dfrac{3}{5} = \dfrac{1}{10}$

c. $\dfrac{1}{4x} - \dfrac{3}{5x} = \dfrac{1}{10}$
  d. $\dfrac{1}{x - 3} + 3 = \dfrac{1}{x - 3}$

e. $\dfrac{5}{x - 2} - \dfrac{8}{(x + 4)(x - 2)} = \dfrac{3}{x + 4}$

f. $\dfrac{12}{y} = \dfrac{12}{y + 1} + \dfrac{1}{y}$

Chapter 12 Sample Test Answers

1. zero

2. a. $x = 3$
  b. $x = {}^-7$
  c. $x = 0$
  d. $x = 0, x = 2$

3. a. $\dfrac{2x^4}{3}$
  b. $\dfrac{2x^2}{3y^8}$
  c. $\dfrac{4}{x + 5}$

  d. $\dfrac{1}{(x + 5)^2}$
  e. $x + 2$
  f. $\dfrac{3x - 1}{x - 2}$

  g. ${}^-1$
  h. 1

  i. cannot be simplified
  j. $x + 2$

4. a. $5x(x - 3)$
  b. $\dfrac{(2x - 3)(x + 1)}{3x(x + 7)}$

5. a. $\dfrac{{}^-9}{5}$
  b. $\dfrac{1}{x}$
  c. $\dfrac{2 - y}{x}$

  d. $\dfrac{1}{x^2 - 36}$
  e. $x^2 - 2x + 1$

6. a. $\dfrac{6x}{y}$
  b. $\dfrac{(x + 6)(x - 8)}{15}$

7. a. 1
  b. ${}^-1$
  c. $\dfrac{9x - 1}{5x}$

  d. $\dfrac{x^2(x - 1)}{x - 3}$
  e. $\dfrac{1}{(x - 7)(x + 1)}$
  f. $(x - 7)^2$

8. a. $\dfrac{8}{y}$ **b.** $x - 3$ **c.** $\dfrac{4x - 13}{2y - 5}$

    d. $\dfrac{1}{x - 4}$ **e.** $\dfrac{7x - 9}{x^2 - 3x - 7}$ **f.** $x + 3$

9. a. 180   **b.** $60xz$ **c.** $120x^2y^3$
    d. $(x - 9)(x + 9)(x + 2)$ **e.** $(x - 2)^2(x + 2)$
    f. $x^2 - 1$ or $(x + 1)(x - 1)$

10. a. $\dfrac{5y + 7x}{xy}$ **b.** $\dfrac{7 - 2y}{y^2}$ **c.** $\dfrac{3y^2 - x}{x^2y^3}$

    d. $\dfrac{5x + 9}{6}$ **e.** $\dfrac{x + 21}{6}$ **f.** $\dfrac{7y - 4}{y}$

    g. $\dfrac{8x + 1}{(x + 2)(x - 1)}$ **h.** $\dfrac{5x + 13}{(x + 4)(x + 3)}$ **i.** $\dfrac{2(x - 10)}{x(x - 5)}$

    j. $\dfrac{2x^2 - 7x - 13}{(x - 3)(x - 7)}$

11. a. $\dfrac{^-1}{3}$ **b.** $\dfrac{1}{180y^2}$ **c.** $\dfrac{1}{y - 1}$

    d. $\dfrac{2y - 3x}{5y - 7x}$ **e.** $\dfrac{y - x}{xy}$

12. a. $x \neq 0$ **b.** $x \neq 2, x \neq {}^-3$
    c. $x \neq 1, x \neq {}^-1$ $(x \neq \pm 1)$
    d. no restrictions

13. a. $x = \dfrac{^-3}{7}$ **b.** $x = \dfrac{20}{7}$ **c.** $x = \dfrac{^-7}{2}$

    d. no solution   **e.** $x = {}^-9$ **f.** $y = 11$

Chapter 13

Roots and Radicals

13.1   Introduction to Roots, Radicals, and Irrational Numbers

1	If each of the integers is squared, a set of numbers called the *perfect square integers* is formed.

$$\text{integers} = \{\ldots, {}^-3, {}^-2, {}^-1, 0, 1, 2, 3, \ldots\}$$
$$\text{perfect square integers} = \{0, 1, 4, 9, 16, 25, 36, \ldots\}$$

Each of the perfect square integers, except zero, equals the square of two different numbers. These two numbers are each *square roots*.

Examples:

1. The square roots of 16 are 4 and $^-4$, because $4^2 = 16$ and $(^-4)^2 = 16$.
2. The square roots of 225 are $^-15$ and 15, because $15^2 = 225$ and $(^-15)^2 = 225$.
3. The square root of 0 is 0, because $0^2 = 0$. There is only one square root of 0.

Q1   Let P represent the set of perfect square integers. Insert $\in$ or $\notin$ to form a true statement:

    a. 36 _____P **b.** 18 _____P

    c. $^-25$ _____P **d.** 49 _____P

    e. 1 _____P **f.** 200 _____P

STOP • STOP • STOP • STOP • STOP • STOP • STOP • STOP • STOP

A1   **a.** $\in$ **b.** $\notin$ **c.** $\notin$: there are no negative integers in P.

      d. $\in$ **e.** $\in$ **f.** $\notin$

Q2   Write the square roots of the following integers:

    a. 25 _____,_____   **b.** 100 _____,_____

    c. 4 _____,_____   **d.** 9 _____,_____

    e. 0 _____,_____   **f.** 1 _____,_____

STOP • STOP • STOP • STOP • STOP • STOP • STOP • STOP • STOP

A2   **a.** 5, $^-5$ **b.** 10, $^-10$ **c.** 2, $^-2$ **d.** 3, $^-3$

      e. 0: should only be written once   **f.** 1, $^-1$

Q3   Every positive perfect square integer has how many square roots? _____

STOP • STOP • STOP • STOP • STOP • STOP • STOP • STOP • STOP

A3   two: zero has only one square root but zero is not positive.

---

2   Every positive perfect square integer has a positive and a negative square root. The symbol $\sqrt{25}$ is used to represent the positive square root of 25 ($\sqrt{25} = 5$). The negative square root of 25 is written $^{-}\sqrt{25}$ ($^{-}\sqrt{25} = {}^{-}5$). The symbol $\sqrt{}$ is called a *radical sign*.

Examples:

1. $\sqrt{36} = 6$.
2. $^{-}\sqrt{49} = {}^{-}7$.
3. The square roots of 81 are 9 and $^{-}9$.

Q4   Express each of the following as integers:

     a. $\sqrt{64} = $ _____   **b.** $^{-}\sqrt{1} = $ _____

     c. $\sqrt{0} = $ _____   **d.** $^{-}\sqrt{144} = $ _____

     e. The square roots of 16 are _____.

STOP • STOP • STOP • STOP • STOP • STOP • STOP • STOP • STOP

A4   **a.** 8   **b.** $^{-}1$ **c.** 0   **d.** $^{-}12$

     e. 4 and $^{-}4$

Q5   The symbol $\sqrt{}$ is called _____.

STOP • STOP • STOP • STOP • STOP • STOP • STOP • STOP • STOP

A5   a radical sign.

---

3   The symbol $\sqrt{16}$ is read "the principal square root of 16." It is sometimes shortened to "the square root of 16." (Notice the singular word "root.") The principal square root of a positive number is the positive square root and the principal square root of zero is zero. Since the principal square root of a positive number a is written $\sqrt{a}$, the other square root can be written $^{-}\sqrt{a}$. Therefore, the two square roots of 16 are written $\sqrt{16}$ and $^{-}\sqrt{16}$. This is sometimes shortened to $\pm\sqrt{16}$. For example,

$$\pm\sqrt{49} = {}^{+}\sqrt{49} \text{ and } {}^{-}\sqrt{49}$$
$$= {}^{+}7 \text{ and } {}^{-}7 \text{ (or } \pm 7)$$

Q6   Express each of the following as integers:

     a. The principal square root of 81 = _____

     b. The square roots of 81 = _____

     c. The square root of 64 = _____

     d. $\pm\sqrt{121} = $ _____

     e. $^{-}\sqrt{121} = $ _____

STOP • STOP • STOP • STOP • STOP • STOP • STOP • STOP • STOP

A6   **a.** 9   **b.** 9 and $^{-}9$ **c.** 8

     d. 11 and $^{-}11$ or ± 11 **e.** $^{-}11$

---

4   One can write the square root of numbers other than perfect square integers. Think of the principal square root as being the positive number which when squared is equal to the number under the radical sign. $\dfrac{1}{9}$ is not a perfect square integer, but since $\dfrac{1}{3} \cdot \dfrac{1}{3} =$

$\frac{1}{9}$, $\sqrt{\frac{1}{9}} = \frac{1}{3}$. Notice that $^-\sqrt{\frac{1}{9}} = \frac{^-1}{3}$, because $\left(\frac{^-1}{3}\right)\left(\frac{^-1}{3}\right) = \frac{1}{9}$. $\frac{^-1}{3}$ is a square root of $\frac{1}{9}$, but is it not the principal square root of $\frac{1}{9}$.

Examples:

1. $\sqrt{0.04} = 0.2$, because $(0.2)(0.2) = 0.04$.

2. $^-\sqrt{\frac{1}{16}} = \frac{^-1}{4}$, because $\left(\frac{^-1}{4}\right)\left(\frac{^-1}{4}\right) = \frac{1}{16}$.

Numbers such as $\frac{1}{16}$ whose square root is a rational number can be called *perfect square rational numbers.*

Q7   Express each of the following as rational numbers:

a. $\sqrt{\frac{1}{36}} =$ _____

b. $^-\sqrt{\frac{1}{4}} =$ _____

c. $\sqrt{0.09} =$ _____

d. $\sqrt{0.0049} =$ _____

e. $\sqrt{0.0001} =$ _____

f. $^-\sqrt{\frac{1}{121}} =$ _____

STOP • **STOP** • **STOP** • **STOP** • **STOP** • **STOP** • **STOP** • **STOP** • **STOP**

A7   a. $\frac{1}{6}$ b. $\frac{^-1}{2}$ c. 0.3 d. 0.07

e. 0.01 f. $\frac{^-1}{11}$

5   The number under the radical sign is called the *radicand.* So far the radicand has been very carefully selected. Each time a rational number could be found which when squared was equal to the radicand. This is not always the case.

Consider $\sqrt{2}$. The value of the square root of 2 must be larger than 1 and smaller than 2, because $1^2 = 1$ and $2^2 = 4$. The true value of $\sqrt{2}$ cannot be written exactly as a decimal or a fraction. Various approximations of the square root of 2 can be written, but each is seen to be slightly inaccurate when it is squared. ($\doteq$ is read "is approximately equal to.")

$\sqrt{2} \doteq 1.4$ because $(1.4)^2 = 1.96$
$\sqrt{2} \doteq 1.41$ because $(1.41)^2 = 1.9881$
$\sqrt{2} \doteq 1.414$ because $(1.414)^2 = 1.999396$

Square roots of non-perfect-square rational numbers are not rational numbers. They belong to a set of numbers called the *irrational numbers.* They can be located on the number line. The location of $\sqrt{2}$ is as follows:

Q8   Indicate whether the number is rational or irrational:

a. $\sqrt{7}$ _____

b. $\sqrt{36}$ _____

c. $\sqrt{1}$ _____

d. $\sqrt{3}$ _____

e. $\sqrt{8}$ _____

f. $\sqrt{9}$ _____

STOP • **STOP** • **STOP** • **STOP** • **STOP** • **STOP** • **STOP** • **STOP** • **STOP**

A8   **a.** irrational   **b.** rational   **c.** rational
        d. irrational   **e.** irrational   **f.** rational

---

6   Recall that rational numbers could be written as the quotient of two integers. Examples are: $\frac{1}{2}, \frac{-3}{14}, \frac{4}{206}$, 5, 0.027, and 0. All rational numbers can be expressed in decimal form by dividing the numerator by the denominator. Some fractions have a decimal representation that terminates, such as $\frac{3}{4} = 0.75$. Others have a decimal representation that repeats, such as $\frac{1}{7} = 0.\overline{142857}$. The bar over a block of numbers indicates that that block of numbers is repeated endlessly.

Numbers represented by decimals can be divided into the two sets rational numbers, Q, and irrational numbers, L.

1. Terminating and repeating decimals represent rational numbers.
2. Decimals that do not terminate but also do not repeat in blocks are irrational numbers (nonterminating, nonrepeating decimals).

Examples:

1. The following terminating decimals represent rational numbers: 43.7, 0.25, 0.0076, 10.007, 5, 830.
2. The following repeating decimals represent rational numbers: $0.3\overline{3}$, $0.4\overline{7}$, $0.\overline{45}$, $0.\overline{67}$, $0.06\overline{6}$.
3. The following decimal represents an irrational number: 0.10110111011110. . . . Notice that the pattern guarantees that there will not be a block of digits which when repeated would form the complete number.

Q9   Place the set Q or the set L on the line to make a true statement:

a. $43.6 \in$ _____   **b.** $52.\overline{7} \in$ _____

c. $0.212212221\cdots \in$ _____   **d.** $34.\overline{34} \in$ _____

STOP • **STOP** • **STOP** • **STOP** • **STOP** • **STOP** • **STOP** • **STOP** • **STOP**

A9   **a.** Q **b.** Q **c.** L **d.** Q

---

7   The number π which is used to find the circumference and area of a circle is another example of an irrational number. The number π is the ratio of the circumference to the diameter of any circle. Various rational approximations are used for π. However, none can be said to equal π, for an irrational number cannot equal a rational number. Some common rational approximations follow:

$\pi \doteq 3.14$ precise to 2 decimal places

$\pi \doteq \frac{22}{7}$ precise to 2 decimal places $\left(\frac{22}{7} = 3.\overline{142857}\right)$

$\pi \doteq \frac{355}{113}$ precise to 6 decimal places; this is an easy approximation to remember (think 1, 1, 3, 3, 5, 5) for use with an electronic calculator

$\pi \doteq 3.14159265$ precise to 8 decimal places

Q10 **a.** Give two approximations to π which are precise to 2 decimal places. _____ and

_____

b. Give two approximations to π which are precise to 6 decimal places. _____

and _____

STOP • STOP • STOP • STOP • STOP • STOP • STOP • STOP • STOP

A10 **a.** 3.14 and $\dfrac{22}{7}$ **b.** 3.141593 and $\dfrac{355}{113}$

Q11 Is π a rational number? _____ .

STOP • STOP • STOP • STOP • STOP • STOP • STOP • STOP • STOP

A11 no: the true decimal value (not approximation) of π is a nonterminating, nonrepeating decimal.

---

8 Other roots can be defined besides the square root. These are the cube root, fourth root, fifth root, . . . , nth root. . . . Their definition follows:

For every natural number n greater than 1, the number b is called an nth *root of a* if $b^n = a$. The principal nth root of a is written $\sqrt[n]{a}$. For even numbers, n, the principal nth root is the positive nth root. The number n is called the *index* of the radical sign. $\sqrt[3]{8}$ is called the cube root of 8 and has index 3. $\sqrt[3]{8} = 2$, because $2^3 = 8$.

Q12 Indicate the index of each radical:

a. $\sqrt[4]{16}$ _____   **b.** $\sqrt[3]{-27}$ _____

c. $\sqrt{4}$ _____   **d.** $\sqrt[6]{242}$ _____

STOP • STOP • STOP • STOP • STOP • STOP • STOP • STOP • STOP

A12 **a.** 4   **b.** 3
c. 2: the index is understood to be 2 when the radical sign is used as a square-root symbol.
d. 6

Q13 Write the following in mathematical symbols:

a. 3rd root of 14   _____   **b.** 7th root of 23   _____

c. cube root of $^-27$ _____   **d.** square root of 7   _____

STOP • STOP • STOP • STOP • STOP • STOP • STOP • STOP • STOP

A13 **a.** $\sqrt[3]{14}$ **b.** $\sqrt[7]{23}$
c. $\sqrt[3]{-27}$ **d.** $\sqrt{7}$

---

9 Roots can be simplified more easily if the first few powers of 2, 3, and 5 are memorized. The following are used so often that it would be wise to spend a few minutes learning them rather than always having to take the time to look them up.

$2^2 = 4$	$3^2 = 9$	$5^2 = 25$	$10^2 = 100$
$2^3 = 8$	$3^3 = 27$	$5^3 = 125$	$10^3 = 1,000$
$2^4 = 16$	$3^4 = 81$	$5^4 = 625$	$10^4 = 10,000$
$2^5 = 32$	$3^5 = 243$		$10^5 = 100,000$
$2^6 = 64$			$10^6 = 1,000,000$
$2^7 = 128$			

> *Examples:*
>
> $\sqrt[4]{16} = 2$
>
> $\sqrt[6]{1,000,000} = 10$

Q14   Evaluate the following roots:

    a. $\sqrt[5]{32} =$ _____   **b.** $\sqrt[4]{81} =$ _____

    c. $\sqrt[3]{1000} =$ _____   **d.** $\sqrt[7]{128} =$ _____

    e. $\sqrt[4]{16} =$ _____   **f.** $\sqrt[3]{125} =$ _____

STOP • STOP • STOP • STOP • STOP • STOP • STOP • STOP • STOP

A14   **a.** 2   **b.** 3   **c.** 10
        d. 2   **e.** 2   **f.** 5

---

10   The number $\sqrt[3]{25}$ is irrational because there is no rational number that when cubed is equal to 25. Irrational numbers represented with radical signs can be approximated to many decimal places, but the approximation is not exactly equal to the irrational number. Tables of square roots and cube roots are used to approximate these numbers. Slide rules and some calculators also can be used.

Examples:

1. $\sqrt[3]{8}$ is rational, because $(2)^3 = 8$.
2. $\sqrt[3]{9}$ is irrational, because $(\ \underset{\uparrow}{\ }\ )^3 = 9$.
                  no rational number
3. $\sqrt[4]{16}$ is rational, because $(2)^4 = 16$.
4. $\sqrt[4]{17}$ is irrational, because $(\ \underset{\uparrow}{\ }\ )^4 = 17$.
                  no rational number
5. $\sqrt[3]{0.001}$ is rational, because $(0.1)^3 = 0.001$.
6. $\sqrt[3]{0.0001}$ is irrational, because $(\ \underset{\uparrow}{\ }\ )^3 = 0.0001$.
                  no rational number

Q15   Label the irrational numbers and simplify the rational numbers:

    a. $\sqrt{0.01}$ _____   **b.** $\sqrt[3]{100}$ _____

    c. $\sqrt[3]{\dfrac{1}{8}}$ _____   **d.** $\sqrt[5]{40}$ _____

    e. $\sqrt[4]{0.0625}$ _____   **f.** $\sqrt{1,000}$ _____

STOP • STOP • STOP • STOP • STOP • STOP • STOP • STOP • STOP

A15   **a.** 0.1   **b.** irrational   **c.** $\dfrac{1}{2}$

        d. irrational   **e.** 0.5   **f.** irrational

---

11   It is possible to find cube roots of negative numbers, because a negative number cubed is negative. The same is true for other roots of odd index.

Examples:

1. $\sqrt[3]{-125} = {}^-5$, because $({}^-5)^3 = {}^-125$.
2. $\sqrt[5]{-32} = {}^-2$, because $({}^-2)^5 = {}^-32$.
3. $\sqrt[7]{-0.0000001} = {}^-0.1$, because $({}^-0.1)^7 = {}^-0.0000001$.

Q16   Simplify:

   a. $\sqrt[3]{-8} =$ _____   **b.** $\sqrt[5]{-1} =$ _____

   c. $\sqrt[3]{-27} =$ _____   **d.** $\sqrt[5]{-100{,}000} =$ _____

STOP • STOP • STOP • STOP • STOP • STOP • STOP • STOP • STOP

A16   **a.** $^-2$ **b.** $^-1$ **c.** $^-3$ **d.** $^-10$

---

12   Irrational numbers are located in all sections of the number line. The set of irrational numbers is infinite. Some examples of the location of some irrational numbers are as follows:

If all rational numbers and all irrational numbers are located on the number line, each point on the line is associated with a number and each number is associated with a point. This is called a *one-to-one correspondence* between the numbers and the points on a line.

    The set of all rational and irrational numbers is called the set of *real numbers* and is denoted by R. In set notation $R = Q \cup L$. That is, the set of real numbers equals the union of the set of rational numbers and the set of irrational numbers. Also, $Q \cap L = \varnothing$, because no number is both rational and irrational. The set of rational numbers and the set of irrational numbers are disjoint.

Q17   Label the following statements true or false:

   a. There is a one-to-one correspondence between the rational numbers and the number line. _____

   b. There is a one-to-one correspondence between the irrational numbers and the number line. _____

   c. There is a one-to-one correspondence between the real numbers and the number line.

_____

STOP • STOP • STOP • STOP • STOP • STOP • STOP • STOP • STOP

A17   **a.** false   **b.** false   **c.** true

---

13   So far numbers such as $\sqrt{-25}$ have not been discussed. The number represented by $\sqrt{-25}$ would be a number which when squared would equal $^-25$. The two most likely guesses are 5 or $^-5$. However, $5^2 \neq {}^-25$ and $({}^-5)^2 \neq {}^-25$. Therefore, neither 5 or $^-5$ are equal to $\sqrt{-25}$.

    To discover if there is a number on the number line which when squared is $^-25$, separate the numbers into three sets. Consider positive numbers, negative numbers, and zero.

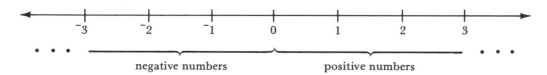

The square root of $^-25$ could not be a negative number, because a negative times itself is positive. The square root of $^-25$ could not be positive, because a positive number times itself is positive. The square root of $^-25$ is not zero, because $0^2 \neq {}^-25$. It must be concluded that there is no number on the number line equal to $\sqrt{^-25}$. Another way to say the same conclusion is that $\sqrt{^-25}$ is not a real number.

Q18   Insert $\in$ or $\notin$ to form a true statement:

a. $\sqrt{^-16}$ _____R **b.** $\dfrac{2}{3}$ _____R

c. $\sqrt{25}$ _____R **d.** $\sqrt{^-10}$ _____R

e. $\sqrt{^-3}$ _____R **f.** $^-\sqrt{7}$ _____R

STOP • **STOP** • **STOP** • **STOP** • **STOP** • **STOP** • **STOP** • **STOP** • **STOP**

A18   **a.** $\notin$ **b.** $\in$ **c.** $\in$ **d.** $\notin$
        e. $\notin$ **f.** $\in$

---

14   Some of the properties of square roots are also true of fourth roots, sixth roots, and other even-number roots.

1. If n is *even* and b is *positive,* then there are two real numbers that are nth roots of b.
    a. 4th roots of 625 are $^+5$ and $^-5$, because $(^+5)^4 = 625$ and $(^-5)^4 = 625$; therefore, $\sqrt[4]{625} = 5, {}^-\sqrt[4]{625} = {}^-5$.
    b. The square roots of 12 are $\sqrt{12}$ and $^-\sqrt{12}$.
2. If n is *even* and b is *negative,* then $\sqrt[n]{b}$ is not a real number. This is because even powers of nonzero numbers are positive:
    a. $\sqrt[6]{^-64}$ is not a real number, because $(\underset{\underset{\text{no real number}}{\uparrow}}{\quad})^6 = {}^-64$.
    b. $\sqrt[4]{^-10}$ is not a real number, because $(\underset{\underset{\text{no real number}}{\uparrow}}{\quad})^4 = {}^-10$.

Q19   Insert $\in$ or $\notin$ on each blank to form a true statement:

a. $\sqrt[4]{^-16}$ _____R **b.** $\sqrt[3]{^-8}$ _____R

c. $\sqrt[6]{^-15}$ _____R **d.** $\sqrt{^-9}$ _____R

e. $\sqrt[5]{^-32}$ _____R **f.** $\sqrt[9]{^-1}$ _____R

STOP • **STOP** • **STOP** • **STOP** • **STOP** • **STOP** • **STOP** • **STOP** • **STOP**

A19   **a.** $\notin$ **b.** $\in$: $^-2$ **c.** $\notin$ **d.** $\notin$
        e. $\in$: $^-2$ **f.** $\in$: $^-1$

This completes the instruction for this section.

13.1 Exercises

1. Consider the symbol $\sqrt[3]{27}$.
   a. What is the name of the symbol $\sqrt[3]{\ }$?
   b. What is the 3 called?
   c. What is the number 27 called?
   d. How is $\sqrt[3]{27}$ read?

2. List the first 12 perfect square integers.

3. In the problems below simplify each radical expression if possible. If the expression does not represent a real number, write "is not a real number."
   a. square roots of 9   b. square roots of 64
   c. square root of 25   d. square root of 100
   e. $\sqrt{49}$ f. $\sqrt{36}$
   g. $\sqrt{121}$ h. $\pm\sqrt{144}$
   i. $-\sqrt{16}$ j. $-\sqrt{64}$
   k. $\sqrt{0.04}$ l. $\sqrt{0.0016}$
   m. $\sqrt{\dfrac{4}{25}}$ n. $\sqrt{\dfrac{1}{81}}$
   o. $\sqrt{-9}$ p. $-\sqrt{-25}$
   q. $\sqrt[3]{125}$ r. $\sqrt[3]{\dfrac{1}{8}}$
   s. $\sqrt[3]{-8}$ t. $\sqrt[5]{-32}$
   u. $\sqrt[4]{81}$ v. $\sqrt[4]{16}$
   w. $\sqrt[4]{-1}$ x. $\sqrt[6]{-4}$

4. Indicate whether each of the following numbers is rational, irrational, or not real:
   a. $\sqrt{9}$ b. $\sqrt{5}$
   c. $\sqrt{-4}$ d. $\sqrt{12}$
   e. $\sqrt{4}$ f. π
   g. $0.27\overline{7}$ h. 3.14
   i. $0.010010001\cdots$ j. $0.011011\overline{011}$
   k. $\sqrt[3]{7}$ l. $\sqrt[3]{125}$
   m. $\sqrt[4]{-16}$ n. $-\sqrt[4]{16}$

5. $Q \cup L = $ _____

6. $Q \cap L = $ _____

7. Answer rational or irrational to the following statements:
   a. A terminating decimal is _____.
   b. A nonterminating, nonrepeating decimal is _____.
   c. A repeating decimal is _____.
   d. The number π is _____.

8. There is a one-to-one correspondence between the number line and the _____ numbers.

9. Where is the number $\sqrt{-100}$ on the number line?

13.1 Exercise Answers

1. a. the radical sign
   b. the index
   c. the radicand
   d. the cube root of 27

2. 0, 1, 4, 9, 16, 25, 36, 49, 64, 81, 100, 121

3. **a.** 3 and ⁻3   **b.** 8 and ⁻8   **c.** 5

     d. 10   **e.** 7   **f.** 6

     g. 11   **h.** ±12   **i.** ⁻4

     j. ⁻8   **k.** 0.2   **l.** 0.04

     m. $\dfrac{2}{5}$ **n.** $\dfrac{1}{9}$ **o.** not a real number

     p. not a real number   **q.** 5   **r.** $\dfrac{1}{2}$

     s. ⁻2   **t.** ⁻2   **u.** 3

     v. 2   **w.** not a real number   **x.** not a real number

4. **a.** rational   **b.** irrational   **c.** not real

     d. irrational   **e.** rational   **f.** irrational

     g. rational   **h.** rational   **i.** irrational

     j. rational   **k.** irrational   **l.** rational

     m. not real   **n.** rational

5. *R*

6. ∅:   the empty set, because no numbers are in both sets

7. **a.** rational   **b.** irrational   **c.** rational   **d.** irrational

8. real

9. It is not on the number line.

13.2   Use of Tables to Approximate Irrational Numbers

1   Irrational numbers are sometimes found in radical form. For example, $\sqrt{7}$, $-\sqrt{3}$, and $\sqrt[3]{10}$ are irrational numbers. A rational approximation of these numbers can be found by using tables of powers and roots. Such a table is found on page 600. The first column contains the numbers from 1 to 100. In the following columns the square of the number, the cube of the number, the square root of the number, and the cube root of the number is located. Each of these will be discussed so that the most effective use of the table is apparent.

     The columns headed "Square" and "Cube" can be used to raise a number to the second power and third power.

Example 1: $52^2 = ?$.

Solution

Find 52 in the column headed "No." and read the value of the number beside it on the "square" column. $52^2 = 2{,}704$.

No.	Square	Cube	Square Root	Cube Root	No.	Square	Cube	Square Root	Cube Root
1	1	1	1.000	1.000	51	2,601	132,651	7.141	3.708
2	4	8	1.414	1.260	52	2,704	140,608	7.211	3.732
3	9	27	1.732	1.442	53	2,809	148,877	7.280	3.756
4	16	64	2.000	1.587	54	2,916	157,464	7.348	3.780

Example 2: $97^3 = ?$

Solution

Find 97 in the "No." column and read the value in the "Cube" column opposite it. $97^3 = 912,673.$

Q1   Use the table to find the following numbers:

   a. $15^2 =$ _____   **b.** $31^2 =$ _____

   c. $63^2 =$ _____   **d.** $19^3 =$ _____

   e. $88^3 =$ _____   **f.** $3^3 =$ _____

STOP • STOP • STOP • STOP • STOP • STOP • STOP • STOP • STOP

A1   **a.** 225   **b.** 961   **c.** 3,969
          d. 6,859   **e.** 681,472   **f.** 27

---

2   The numbers in the "Square" column are called perfect square integers and the numbers in the "Cube" column are called perfect cube integers. These columns can be used to find the square root or cube root of perfect squares and perfect cubes.

To find the square root of 961, find 961 in the "Square" column. The square root of 961 is in the "No." column. $\sqrt{961} = 31.$

No.	Square	Cube	Square Root	Cube Root	No.	Square	Cube	Square Root	Cube Root
28	784	21,952	5.292	3.037	78	6,084	474,552	8.832	4.273
29	841	24,389	5.385	3.072	79	6,241	493,039	8.888	4.291
30	900	27,000	5.477	3.107	80	6,400	512,000	8.944	4.309
31	961	29,791	5.568	3.141	81	6,561	531,441	9.000	4.327
32	1,024	32,768	5.657	3.175	82	6,724	551,368	9.055	4.344

Examples:

$\sqrt{6,241} = 79$ (located above).

$\sqrt[3]{24,389} = 29$ (located above).

Q2   Find the square root or cube root of the following numbers:

   a. $\sqrt{1,156} =$ _____   **b.** $\sqrt[3]{132,651} =$ _____

   c. $\sqrt[3]{343} =$ _____   **d.** $\sqrt{8,649} =$ _____

STOP • STOP • STOP • STOP • STOP • STOP • STOP • STOP • STOP

A2   **a.** 34   **b.** 51
          c. 7   **d.** 93

---

3   The "Square Root" and "Cube Root" columns may be used to find rational approximations to the square roots or cube roots of the integers from 1 to 100. The number is located in the "No." column and the square root of the number is in the "Square Root" column opposite the number. Notice that the symbol "$\doteq$" is used to say "is approximately equal to."

No.	Square	Cube	Square Root	Cube Root	No.	Square	Cube	Square Root	Cube Root
11	121	1,331	3.317	2.221	61	3,721	226,981	7.810	3.936
12	144	1,728	3.464	2.239	62	3,844	238,328	7.874	3.958
13	169	2,197	3.606	2.351	63	3,969	250,047	7.937	3.979
14	196	2,744	3.742	2.410	64	4,096	262,144	8.000	4.000
15	225	3,375	3.873	2.466	65	4,225	274,625	8.062	4.021

Examples:

$$\sqrt{12} \doteq 3.464$$

$$\sqrt{62} \doteq 7.874$$

$$\sqrt[3]{14} \doteq 2.410$$

$$\sqrt[3]{64} = 4.000 \text{ (this is not an approximation because } \sqrt{64} \text{ is rational)}$$

Q3 Find the square root or cube root of the following numbers:

a. $\sqrt{14} \doteq$ _____   b. $\sqrt{93} \doteq$ _____

c. $\sqrt{81} =$ _____   d. $\sqrt[3]{42} \doteq$ _____

e. $\sqrt[3]{96} \doteq$ _____   f. $\sqrt[3]{27} =$ _____

STOP   •   STOP   •   STOP   •   STOP   •   STOP   •   STOP   •   STOP   •   STOP   •   STOP

A3   a. 3.742   b. 9.644   c. 9.000
     d. 3.476   e. 4.579   f. 3.000

4   If square roots of other numbers are required, a different table may have to be consulted. However, there is a way to extend the present table which is sometimes helpful to know.

   Consider the following square roots, some of which can be found in the table. It is customary to use "=" even when "$\doteq$" is more accurate. This will be done in this text.

$$\sqrt{0.04} = 0.2000 \qquad \sqrt{0.40} = 0.6325$$
$$\sqrt{4.00} = 2.000 \qquad \sqrt{40.00} = 6.325$$
$$\sqrt{400.00} = 20.00 \qquad \sqrt{4,000.00} = 63.25$$

Notice that the numbers whose square roots involve the digits 6, 3, 2, and 5 were 0.40, 40.00, and 4,000.00. The numbers whose square roots involve the digit 2 were 0.04, 4.00, and 400.00.

   It can be concluded that if the square root of a number is known, the square root of a number that is 100 times as great or as small has the same digits and is 10 times as great or as small.

Example: If $\sqrt{35.} = 5.916$, find $\sqrt{3,500.}$ and $\sqrt{0.35.}$

Solution

$$\sqrt{3,500.} = 59.16 \qquad \text{if the radicand is 100 times as large, the root is 10 times as large}$$
$$\sqrt{0.35} = 0.5916 \qquad \text{if the radicand is 0.01 times as large, the root is 0.1 as large}$$

Q4   Use the table to find $\sqrt{300.}$

STOP   •   STOP   •   STOP   •   STOP   •   STOP   •   STOP   •   STOP   •   STOP   •   STOP

A4   17.32: $300 = 3 \times 100$. Look up $\sqrt{3} = 1.732$, then multiply by 10.

Q5   Use the table to find $\sqrt{3,000}$.

STOP • STOP • STOP • STOP • STOP • STOP • STOP • STOP • STOP

A5   54.77: $3,000 = 30 \times 100$. Look up $\sqrt{30} = 5.477$, then multiply by 10.

---

5   To decide what number to look up in the table, subdivide the digits of the radicand *starting at the decimal* into groups of two digits. For example,

$$\sqrt{53_\wedge 00.00_\wedge 00_\wedge}$$

Look up the number in the first nonzero block of two digits starting at the left end of the radicand. In the example, $\sqrt{53} = 7.280$ would be found in the table. The digits from $\sqrt{53}$ are then placed over the original square root with one digit above each block of two digits. The decimal point is then placed in the answer above the original decimal point.

$$\frac{7\quad 2.\ 8\quad 0}{\sqrt{53_\wedge 00.00_\wedge 00}}$$

In this case $\sqrt{5,300.} = 72.80$

Example 1: Find $\sqrt{700,000.00}$.

Solution

$$\frac{8\quad 3\quad 6.\ 7}{\sqrt{70_\wedge 00_\wedge 00.00_\wedge}}\qquad \text{(looking up } \sqrt{70})$$

Therefore, $\sqrt{700,000.} = 836.7$.

Example 2: Find $\sqrt{80,000}$.

Solution

$$\frac{2\quad 8\quad 2.\ 8}{\sqrt{8_\wedge 00_\wedge 00.00_\wedge}}\qquad \text{(looking up } \sqrt{8})$$

Therefore, $\sqrt{80,000} = 282.8$.

Q6   Find the square roots:

a. $\sqrt{4,700.} = $ _____   **b.** $\sqrt{800.} = $ _____

c. $\sqrt{700,000.} = $ _____   **d.** $\sqrt{70,000.} = $ _____

STOP • STOP • STOP • STOP • STOP • STOP • STOP • STOP • STOP

A6   **a.** 68.56   **b.** 28.28   **c.** 836.7   **d.** 264.6

---

6   Square roots of numbers less than 1 can be found with the same procedure. Mark off the radicand into groups of two from the decimal and look up the square root of the leftmost nonzero group. For example,

$$\sqrt{0.00400} = \sqrt{0.00_\wedge 40_\wedge 00_\wedge}$$

The digits of $\sqrt{40}$ are placed above the radical sign with one digit above each group of two. A zero must be placed over the first group of two zeros.

$$\begin{array}{c} 0. \quad 0 \quad 6 \quad 3 \quad 2 \quad 5 \\ \hline \sqrt{0.00_\wedge 40_\wedge 00} \end{array}$$

Therefore, $\sqrt{0.00400} = 0.06325$.

Q7   Find the square roots of the following numbers:

     a. $\sqrt{0.7500} =$ _____   **b.** $\sqrt{0.0900} =$ _____

     c. $\sqrt{0.0070} =$ _____   **d.** $\sqrt{0.000007} =$ _____

STOP • **STOP** • **STOP** • **STOP** • **STOP** • **STOP** • **STOP** • **STOP** • **STOP**

A7   **a.** 0.8660   **b.** 0.30:   this is not an approximation.

            c. 0.08367   **d.** 0.002646

---

7   The cube-root table can be extended by a similar procedure. This time mark the radicand into groups of three from the decimal point

$$\sqrt[3]{25_\wedge 000.000}$$

The cube root of the leftmost group is located in the table. This group sometimes has three digits, sometimes two digits, and sometimes one digit, depending upon the location of the decimal. The digits of the cube root are placed above the radical with one digit assigned to each group.

$$\begin{array}{c} 2 \quad\quad 9. \quad 2 \quad 4 \\ \hline \sqrt[3]{25_\wedge 000.000} \end{array}$$

Therefore, $\sqrt[3]{25,000.} = 29.24$.

Q8   Find the following cube roots from the table:

     a. $\sqrt[3]{16,000.} =$ _____   **b.** $\sqrt[3]{9,000.} =$ _____

     c. $\sqrt[3]{10,000.} =$ _____   **d.** $\sqrt[3]{1,000.} =$ _____

STOP • **STOP** • **STOP** • **STOP** • **STOP** • **STOP** • **STOP** • **STOP** • **STOP**

A8   **a.** 25.20   **b.** 20.80

            c. 21.54   **d.** 10.00

---

8   The same procedure is extended to find cube roots of numbers less than 1. Mark off the number into groups of three digits from the decimal and look up the cube root of the leftmost nonzero group.

Example: Find $\sqrt[3]{0.0000220}$.

Solution

$$\begin{array}{c} 0. \quad 0 \quad 2 \quad 8 \quad 0 \quad 2 \\ \hline \sqrt[3]{0.000_\wedge 022_\wedge 000_\wedge 000_\wedge 000} \end{array} \quad \text{(looking up } \sqrt[3]{22}\text{)}$$

Therefore, $\sqrt[3]{0.0000220} = 0.02802$.

Q9   Find the cube roots of the numbers:

a. $\sqrt[3]{0.0560} =$ _____   b. $\sqrt[3]{0.08100} =$ _____

c. $\sqrt[3]{0.000070} =$ _____   d. $\sqrt[3]{0.000100} =$ _____

STOP • **STOP** • **STOP** • **STOP** • **STOP** • **STOP** • **STOP** • **STOP** • **STOP**

A9   a. 0.3826   b. 0.4327

             c. 0.04121   d. 0.04642

---

9   The last technique for extending the table that will be discussed involves the column of perfect squares and perfect cubes.

One can find $\sqrt{961}$ in the table by looking in the "Square" column. Notice that $\sqrt{961} = 31$. From this value other square roots can be obtained, for example:

$$\sqrt{9{_\wedge}61.} = 31. \qquad\qquad \sqrt{9.61} = 3.1$$
$$\sqrt{9{_\wedge}61{_\wedge}00.} = 310 \qquad\qquad \sqrt{0.09{_\wedge}61} = 0.31$$
$$\sqrt{9{_\wedge}61{_\wedge}00{_\wedge}00.} = 3{,}100 \qquad \sqrt{0.00{_\wedge}09{_\wedge}61} = 0.031$$

When looking for the square root of a perfect square rational number, look for a perfect square integer whose square root would have the same digits.

Example 1: Find $\sqrt{37.21}$.

Solution

3,721 is a perfect square and $\sqrt{3{,}721} = 61$. Hence, $\sqrt{37.21} = 6.1$

Example 2: Find $\sqrt{0.0484}$.

Solution

484 is a perfect square and $\sqrt{484} = 22$. Therefore, $\sqrt{0.04{_\wedge}84} = 0.22$.

Q10   Find the square roots:

a. $\sqrt{1.96}$ b. $\sqrt{0.3481}$ c. $\sqrt{0.009216}$

STOP • **STOP** • **STOP** • **STOP** • **STOP** • **STOP** • **STOP** • **STOP** • **STOP**

A10   a. 1.4: use $\sqrt{196}$.   b. 0.59: use $\sqrt{3{,}481}$.   c. 0.096: use $\sqrt{9{,}216}$.

---

10   Cube roots may be found using a procedure similar to that of Frame 9.

Example 1: Find $\sqrt[3]{5.832}$.

Solution

5,832 is a perfect cube. $\sqrt[3]{5{_\wedge}832.} = 18$; hence, $\sqrt[3]{5.832} = 1.8$.

Example 2: Find $\sqrt[3]{0.000064}$.

Solution

64 is a perfect cube. $\sqrt[3]{64.000} = 4.0$; hence, $\sqrt[3]{0.000{_\wedge}064} = 0.04$.

Q11   Find the cube roots:

a. $\sqrt[3]{74.088} =$ _____

b. $\sqrt[3]{0.830584} =$ _____

c. $\sqrt[3]{0.000729} = $ _____

STOP • STOP • STOP • STOP • STOP • STOP • STOP • STOP • STOP

A11   **a.** 4.2: use $\sqrt[3]{74,088}$.   **b.** 0.94: use $\sqrt[3]{830,584}$.   **c.** 0.09: use $\sqrt[3]{729}$.

This completes the instruction for this section.

13.2   Exercises

By using a table of powers and roots, find the following square roots and cube roots:

1. a. $\sqrt{121}$ b. $\sqrt{961}$ c. $\sqrt{6,084}$ d. $\sqrt{1,849}$
2. a. $\sqrt[3]{512}$ b. $\sqrt[3]{3,375}$ c. $\sqrt[3]{314,432}$ d. $\sqrt[3]{97,336}$
3. a. $\sqrt{21}$ b. $\sqrt{47}$ c. $\sqrt{86}$ d. $\sqrt{59}$
4. a. $\sqrt[3]{78}$ b. $\sqrt[3]{21}$ c. $\sqrt[3]{7}$ d. $\sqrt[3]{38}$
5. a. $\sqrt{800.}$ b. $\sqrt{3,200.}$ c. $\sqrt{20,000.}$ d. $\sqrt{7,100.}$
6. a. $\sqrt{0.05}$ b. $\sqrt{0.50}$ c. $\sqrt{0.0038}$ d. $\sqrt{0.0008}$
7. a. $\sqrt[3]{6,000.}$ b. $\sqrt[3]{48,000.}$ c. $\sqrt[3]{4,000.}$ d. $\sqrt[3]{99,000.}$
8. a. $\sqrt[3]{0.060}$ b. $\sqrt[3]{0.006}$ c. $\sqrt[3]{0.077}$ d. $\sqrt[3]{0.000062}$
9. a. $\sqrt{3.24}$ b. $\sqrt{32.49}$ c. $\sqrt{0.3249}$ d. $\sqrt{0.0064}$
10. a. $\sqrt[3]{2.744}$ b. $\sqrt[3]{59.319}$ c. $\sqrt[3]{0.001331}$ d. $\sqrt[3]{0.804357}$

13.2   Exercise Answers

1. a. 11   b. 31   c. 78   d. 43
2. a. 8   b. 15   c. 68   d. 46
3. a. 4.583   b. 6.856   c. 9.274   d. 7.681
4. a. 4.273   b. 2.759   c. 1.913   d. 3.362
5. a. 28.28   b. 56.57   c. 141.4   d. 84.26
6. a. 0.2236   b. 0.7071   c. 0.06164   d. 0.02828
7. a. 18.17   b. 36.34   c. 15.87   d. 46.26
8. a. 0.3915   b. 0.1817   c. 0.4254   d. 0.03958
9. a. 1.8   b. 5.7   c. 0.57   d. 0.08
10. a. 1.4   b. 3.9   c. 0.11   d. 0.93

13.3   Simplest Radical Form

1   In most cases answers to problems that happen to be irrational numbers will be left in radical form. If a rational approximation is wanted in this text, you will be asked for it. In order for answers to be comparable, a standard procedure has been developed for simplifying radicals. The resulting irrational number is said to be in *simplest radical form.* The procedure will be developed in this section.

A basic property of radical expressions for real numbers a and b is the following:

$$\sqrt{ab} = \sqrt{a} \cdot \sqrt{b} \qquad \text{for all } a \geqslant 0, b \geqslant 0$$

Notice that the radicands are assumed to be nonnegative. In this chapter it will be assumed that all variables in the radicand represent nonnegative numbers. With this assumption all expressions will represent real numbers. An example of this property follows:

$$\sqrt{100} = \sqrt{4} \cdot \sqrt{25}$$

This property can be used to simplify radical expressions which have factors that are perfect squares. The goal is to find a factor of the radicand which is a perfect square. The square root of the perfect square factor can be written without the radical sign.

Example 1: Simplify $\sqrt{8}$.

Solution

Look for a factor of 8 that is a perfect square. Since $8 = 4 \cdot 2$ and 4 is a perfect square,

$$\sqrt{8} = \sqrt{4 \cdot 2} = \sqrt{4} \cdot \sqrt{2} = 2\sqrt{2}$$

(*Note:* $2\sqrt{2}$ means 2 *times* the square root of 2.)

Example 2: Simplify $\sqrt{18}$.

Solution

Factors of 18 are 6 and 3 or 9 and 2. The factors of 6 and 3 are not useful, because neither are perfect squares. However, 9 is a perfect square.

$$\sqrt{18} = \sqrt{9 \cdot 2} = \sqrt{9} \cdot \sqrt{2} = 3\sqrt{2}$$

Q1   Simplify:
    a. $\sqrt{50}$
    b. $\sqrt{45}$

STOP • STOP • STOP • STOP • STOP • STOP • STOP • STOP • STOP

A1   **a.** $5\sqrt{2}$: $\sqrt{50} = \sqrt{25 \cdot 2} = \sqrt{25} \cdot \sqrt{2} = 5\sqrt{2}$
    b. $3\sqrt{5}$: $\sqrt{45} = \sqrt{9 \cdot 5} = \sqrt{9} \cdot \sqrt{5} = 3\sqrt{5}$

Q2   Simplify:
    a. $\sqrt{44}$
    b. $\sqrt{200}$

STOP • STOP • STOP • STOP • STOP • STOP • STOP • STOP • STOP

A2   **a.** $2\sqrt{11}$: $\sqrt{44} = \sqrt{4 \cdot 11} = \sqrt{4} \cdot \sqrt{11} = 2\sqrt{11}$
    b. $10\sqrt{2}$: $\sqrt{200} = \sqrt{100 \cdot 2} = \sqrt{100} \cdot \sqrt{2} = 10\sqrt{2}$

2   It is sometimes possible to factor a radicand into more than one factor which are perfect squares.

Example 1: Simplify $\sqrt{48}$.

Solution

$$\sqrt{48} = \sqrt{4 \cdot 12} = \sqrt{4 \cdot 4 \cdot 3} = \sqrt{4} \cdot \sqrt{4} \cdot \sqrt{3} = 2 \cdot 2\sqrt{3} = 4\sqrt{3}$$

Another solution would have been to factor 48 into $16 \cdot 3$ so that $\sqrt{48} = \sqrt{16 \cdot 3} = \sqrt{16} \cdot \sqrt{3} = 4\sqrt{3}$. This solution is quicker but requires that you recognize a larger perfect square. If you see a factor of the radicand that is a perfect square, do not hesitate to factor just because it might not be the largest perfect square. If the first perfect square is not the largest, the remaining factor will contain another perfect square.

Example 2: Simplify $\sqrt{300}$.

Solution

$$\sqrt{300} = \sqrt{100 \cdot 3} = \sqrt{100} \cdot \sqrt{3} = 10\sqrt{3} \quad\text{or}\quad \sqrt{300} = \sqrt{4 \cdot 75} = \sqrt{4 \cdot 25 \cdot 3}$$
$$= \sqrt{4} \cdot \sqrt{25} \cdot \sqrt{3}$$
$$= 2 \cdot 5\sqrt{3}$$
$$= 10\sqrt{3}$$

Q3   Simplify:

a. $\sqrt{192}$

b. $\sqrt{450}$

STOP • STOP • STOP • STOP • STOP • STOP • STOP • STOP • STOP

A3   a. $8\sqrt{3}$: solution 1: $\sqrt{192} = \sqrt{64 \cdot 3} = \sqrt{64} \cdot \sqrt{3} = 8\sqrt{3}$
solution 2: $\sqrt{192} = \sqrt{4 \cdot 48} = \sqrt{4 \cdot 16 \cdot 3} = \sqrt{4} \cdot \sqrt{16} \cdot \sqrt{3}$
$$= 2 \cdot 4\sqrt{3} = 8\sqrt{3}$$
b. $15\sqrt{2}$: solution 1: $\sqrt{450} = \sqrt{225 \cdot 2} = \sqrt{225} \cdot \sqrt{2} = 15\sqrt{2}$
solution 2: $\sqrt{450} = \sqrt{25 \cdot 18} = \sqrt{25 \cdot 9 \cdot 2} = \sqrt{25} \cdot \sqrt{9} \cdot \sqrt{2}$
$$= 5 \cdot 3 \cdot \sqrt{2} = 15\sqrt{2}$$

3   In a way similar to square roots, nth roots may be simplified by finding factors of the radicand that are the nth power of a number and factoring by use of the following property:

$$\sqrt[n]{ab} = \sqrt[n]{a} \cdot \sqrt[n]{b} \quad\text{where } a \geqslant 0,\, b \geqslant 0 \quad\text{if } n \text{ is even}$$

Example 1: Simplify $\sqrt[3]{24}$.

Solution
$$\sqrt[3]{24} = \sqrt[3]{8 \cdot 3} = \sqrt[3]{8} \cdot \sqrt[3]{3} = 2\sqrt[3]{3}$$

Example 2: Simplify $\sqrt[3]{-54}$.

Solution
$$\sqrt[3]{-54} = \sqrt[3]{-27 \cdot 2} = \sqrt[3]{-27} \cdot \sqrt[3]{2} = -3\sqrt[3]{2}$$

Q4   Simplify:
a. $\sqrt[3]{40}$
b. $\sqrt[4]{80}$
c. $\sqrt[3]{-250}$
d. $\sqrt[3]{-5}$

STOP • STOP • STOP • STOP • STOP • STOP • STOP • STOP • STOP

A4   a. $2\sqrt[3]{5}$: $\sqrt[3]{40} = \sqrt[3]{8 \cdot 5} = \sqrt[3]{8} \cdot \sqrt[3]{5} = 2\sqrt[3]{5}$
b. $2\sqrt[4]{5}$: $\sqrt[4]{80} = \sqrt[4]{16 \cdot 5} = \sqrt[4]{16} \cdot \sqrt[4]{5} = 2\sqrt[4]{5}$

c. $^-5\sqrt[3]{2}$: $\sqrt[3]{-250} = \sqrt[3]{-125 \cdot 2} = \sqrt[3]{-125} \cdot \sqrt[3]{2} = {}^-5\sqrt[3]{2}$

d. $-\sqrt[3]{5}$: $\sqrt[3]{-5} = \sqrt[3]{-1 \cdot 5} = \sqrt[3]{-1} \cdot \sqrt[3]{5} = {}^-1\sqrt[3]{5} = -\sqrt[3]{5}$

Q5 Simplify:

a. $\sqrt[3]{16}$

b. $\sqrt[4]{32}$

c. $\sqrt{27}$

d. $\sqrt[3]{-1}$

e. $\sqrt[3]{-16}$

f. $\sqrt[4]{162}$

g. $\sqrt[5]{64}$

h. $\sqrt[5]{-2}$

i. $\sqrt[3]{2,000}$

j. $\sqrt{200}$

STOP • **STOP** • **STOP** • **STOP** • **STOP** • **STOP** • **STOP** • **STOP** • **STOP**

A5 **a.** $2\sqrt[3]{2}$: $\sqrt[3]{16} = \sqrt[3]{8} \cdot \sqrt[3]{2} = 2\sqrt[3]{2}$

b. $2\sqrt[4]{2}$: $\sqrt[4]{32} = \sqrt[4]{16} \cdot \sqrt[4]{2} = 2\sqrt[4]{2}$

c. $3\sqrt{3}$: $\sqrt{27} = \sqrt{9} \cdot \sqrt{3} = 3\sqrt{3}$

d. $^-1$: $\sqrt[3]{-1} = {}^-1$

e. $^-2\sqrt[3]{2}$: $\sqrt[3]{-16} = \sqrt[3]{-8} \cdot \sqrt[3]{2} = {}^-2\sqrt[3]{2}$

f. $3\sqrt[4]{2}$: $\sqrt[4]{162} = \sqrt[4]{81} \cdot \sqrt[4]{2} = 3\sqrt[4]{2}$

g. $2\sqrt[5]{2}$: $\sqrt[5]{64} = \sqrt[5]{32} \cdot \sqrt[5]{2} = 2\sqrt[5]{2}$

h. $^-\sqrt[5]{2}$: $\sqrt[5]{-2} = \sqrt[5]{-1} \cdot \sqrt[5]{2} = {}^-1\sqrt[5]{2} = {}^-\sqrt[5]{2}$

i. $10\sqrt[3]{2}$: $\sqrt[3]{2,000} = \sqrt[3]{1,000} \cdot \sqrt[3]{2} = 10\sqrt[3]{2}$

j. $10\sqrt{2}$: $\sqrt{200} = \sqrt{100} \cdot \sqrt{2} = 10\sqrt{2}$

---

4   Roots may be found of expressions that contain variables. In this text all variables in the radicand represent positive numbers. Some fundamental examples are:

$$\sqrt[3]{x^3} = x$$

$$\sqrt{x^4} = x^2$$

$$\sqrt{4x^2} = 2x$$

$$\sqrt[3]{8x^6} = 2x^2$$

$$\sqrt[3]{27x^3y^6} = 3xy^2$$

Q6 Simplify:

a. $\sqrt{x^2}$ **b.** $\sqrt{16x^4}$

c. $\sqrt[3]{x^6}$ **d.** $\sqrt[3]{-8x^6}$

e. $\sqrt{x^2y^4}$ **f.** $\sqrt[3]{8x^3y^3}$

g. $\sqrt{16x^4y^{10}}$ **h.** $\sqrt[3]{125x^3y^6z^9}$

STOP • **STOP** • **STOP** • **STOP** • **STOP** • **STOP** • **STOP** • **STOP** • **STOP**

A6 **a.** x **b.** $4x^2$ **c.** x^2 **d.** $^-2x^2$

e. xy^2 **f.** $2xy$ **g.** $4x^2y^5$ **h.** $5xy^2z^3$

---

5   Roots of variable expressions may be found by finding the largest nth power under the nth-root radical sign which is a factor of the radicand.

Example 1: Simplify $\sqrt{8x^2y^3}$.

Solution

$$\sqrt{8x^2y^3} = \sqrt{4\cdot2\cdot x^2\cdot y^2\cdot y} = \sqrt{4\cdot x^2\cdot y^2\cdot 2y}$$
$$= \sqrt{4x^2y^2}\cdot\sqrt{2y} = 2xy\sqrt{2y}$$

Example 2: Simplify $2\sqrt{3a^3b^6}$.

Solution

$$2\sqrt{3a^3b^6} = 2\sqrt{a^2b^6\cdot3a} = 2\sqrt{a^2b^6}\cdot\sqrt{3a} = 2ab^3\sqrt{3a}$$

Example 3: Simplify $\sqrt[3]{8p^3q^4}$.

Solution

$$\sqrt[3]{8p^3q^4} = \sqrt[3]{8p^3q^3\cdot q} = \sqrt[3]{8p^3q^3}\cdot\sqrt[3]{q} = 2pq\sqrt[3]{q}$$

Q7   Simplify:

a. $\sqrt{x^7y^8}$

b. $\sqrt[3]{x^8}$

c. $\sqrt{2x^4y^5}$

d. $2x\sqrt{4x^3y^2}$

e. $6\sqrt{x^6y^3z}$

f. $\sqrt[3]{9a^3}$

g. $\sqrt[3]{54a^3b^5}$

STOP • **STOP** • **STOP** • **STOP** • **STOP** • **STOP** • **STOP** • **STOP** • **STOP**

A7
a. $x^3y^4\sqrt{x}$: $\sqrt{x^7y^8} = \sqrt{x^6y^8}\cdot\sqrt{x} = x^3y^4\sqrt{x}$
b. $x^2\sqrt[3]{x^2}$: $\sqrt[3]{x^8} = \sqrt[3]{x^6\cdot x^2} = \sqrt[3]{x^6}\cdot\sqrt[3]{x^2} = x^2\sqrt[3]{x^2}$
c. $x^2y^2\sqrt{2y}$: $\sqrt{2x^4y^5} = \sqrt{x^4y^4\cdot2y} = \sqrt{x^4y^4}\cdot\sqrt{2y} = x^2y^2\sqrt{2y}$
d. $4x^2y\sqrt{x}$: $2x\sqrt{4x^3y^2} = 2x\sqrt{4x^2y^2\cdot x} = 2x\sqrt{4x^2y^2}\cdot\sqrt{x} = 2x\cdot2xy\sqrt{x} = 4x^2y\sqrt{x}$
e. $6x^3y\sqrt{yz}$: $6\sqrt{x^6y^3z} = 6\sqrt{x^6y^2\cdot yz} = 6\sqrt{x^6y^2}\cdot\sqrt{yz} = 6x^3y\sqrt{yz}$
f. $a\sqrt[3]{9}$: $\sqrt[3]{9a^3} = \sqrt[3]{a^3\cdot9} = \sqrt[3]{a^3}\cdot\sqrt[3]{9} = a\sqrt[3]{9}$
g. $3ab\sqrt[3]{2b^2}$: $\sqrt[3]{54a^3b^5} = \sqrt[3]{27a^3b^3\cdot2b^2} = \sqrt[3]{27a^3b^3}\cdot\sqrt[3]{2b^2} = 3ab\sqrt[3]{2b^2}$

---

6   A radical expression of index n is not in simplest radical form as long as the radicand contains a factor of some number or variable to the nth power. For example, $\sqrt{36} = 2\sqrt{9}$, but $2\sqrt{9}$ is not the simplest radical form, because $2\sqrt{9} = 2\cdot3 = 6$. Another example is $\sqrt{x^7} = \sqrt{x^4x^3} = x^2\sqrt{x^3}$, which is not in simplest radical form because $x^2\sqrt{x^3} = x^2\sqrt{x^2x} = x^2\cdot x\sqrt{x} = x^3\sqrt{x}$.

Q8   Write equivalent expressions in simplest radical form:

a. $2\sqrt{32}$

b. $2y\sqrt{4y^3}$

c. $\sqrt{50x^3}$

d. $5\sqrt{4y}$

e. $\sqrt[3]{81x^{12}y^3}$

f. $2x^2\sqrt{x^7y^5}$

g. $\sqrt{72xy^2z^4}$

STOP   •   STOP   •   STOP   •   STOP   •   STOP   •   STOP   •   STOP   •   STOP   •   STOP

A8   a. $8\sqrt{2}$: $2\sqrt{32} = 2\sqrt{16\cdot2} = 2\sqrt{16}\cdot\sqrt{2} = 2\cdot4\sqrt{2} = 8\sqrt{2}$

b. $4y^2\sqrt{y}$: $2y\sqrt{4y^3} = 2y\sqrt{4y^2\cdot y} = 2y\sqrt{4y^2}\cdot\sqrt{y} = 2y\cdot2y\sqrt{y} = 4y^2\sqrt{y}$

c. $5x\sqrt{2x}$: $\sqrt{50x^3} = \sqrt{25x^2\cdot2x} = \sqrt{25x^2}\cdot\sqrt{2x} = 5x\sqrt{2x}$

d. $10\sqrt{y}$: $5\sqrt{4y} = 5\sqrt{4}\cdot\sqrt{y} = 5\cdot2\sqrt{y} = 10\sqrt{y}$

e. $3x^4y\sqrt[3]{3}$: $\sqrt[3]{81x^{12}y^3} = \sqrt[3]{27x^{12}y^3\cdot3} = 3x^4y\sqrt[3]{3}$

f. $2x^5y^2\sqrt{xy}$: $2x^2\sqrt{x^7y^5} = 2x^2\sqrt{x^6\cdot y^4\cdot xy} = 2x^2\cdot x^3y^2\sqrt{xy} = 2x^5y^2\sqrt{xy}$

g. $6yz^2\sqrt{2x}$: $\sqrt{72xy^2z^4} = \sqrt{36y^2z^4\cdot2x} = 6yz^2\sqrt{2x}$

---

7   Another property of radicals involves quotients.

$$\sqrt{\frac{a}{b}} = \frac{\sqrt{a}}{\sqrt{b}} \qquad \text{where } a \geqslant 0, \quad b > 0$$

An example that illustrates the property is

$$\sqrt{\frac{16}{25}} = \frac{\sqrt{16}}{\sqrt{25}} = \frac{4}{5}$$

The accuracy of the result can be checked by squaring $\frac{4}{5}$ to obtain $\frac{16}{25}$.
Other examples are:

1. $\sqrt{\dfrac{10}{16}} = \dfrac{\sqrt{10}}{\sqrt{16}} = \dfrac{\sqrt{10}}{4}$ 2. $\sqrt{\dfrac{45}{49}} = \dfrac{\sqrt{45}}{\sqrt{49}} = \dfrac{\sqrt{9\cdot5}}{7} = \dfrac{3\sqrt{5}}{7}$

Q9   Simplify:

a. $\sqrt{\dfrac{25}{81}}$

b. $\sqrt{\dfrac{14}{64}}$

c. $\sqrt{\dfrac{40}{49}}$

d. $\sqrt{\dfrac{9x^2}{16y^4}}$

e. $\sqrt{\dfrac{12x}{49z^4}}$

STOP   •   STOP   •   STOP   •   STOP   •   STOP   •   STOP   •   STOP   •   STOP   •   STOP

A9

a. $\dfrac{5}{9}$: $\sqrt{\dfrac{25}{81}} = \dfrac{\sqrt{25}}{\sqrt{81}} = \dfrac{5}{9}$

b. $\dfrac{\sqrt{14}}{8}$: $\sqrt{\dfrac{14}{64}} = \dfrac{\sqrt{14}}{\sqrt{64}} = \dfrac{\sqrt{14}}{8}$

c. $\dfrac{2\sqrt{10}}{7}$: $\sqrt{\dfrac{40}{49}} = \dfrac{\sqrt{40}}{\sqrt{49}} = \dfrac{\sqrt{4\cdot 10}}{7} = \dfrac{2\sqrt{10}}{7}$

d. $\dfrac{3x}{4y^2}$: $\sqrt{\dfrac{9x^2}{16y^4}} = \dfrac{\sqrt{9x^2}}{\sqrt{16y^4}} = \dfrac{3x}{4y^2}$

e. $\dfrac{2\sqrt{3x}}{7z^2}$: $\sqrt{\dfrac{12x}{49z^4}} = \dfrac{\sqrt{12x}}{\sqrt{49z^4}} = \dfrac{\sqrt{4\cdot 3x}}{7z^2} = \dfrac{2\sqrt{3x}}{7z^2}$

8

If a radical expression is to be left in simplest radical form, there should be no radical expressions in the denominator of a fraction. Notice that $\sqrt{7}\cdot\sqrt{7} = 7$. Therefore, one technique for removing a square-root sign is to multiply by the same radical. To ensure that the value of the expression is not changed, it is necessary to multiply both the numerator and the denominator by the same radical expression.

Example 1: Simplify $\sqrt{\dfrac{9}{2}}$.

Solution

$$\sqrt{\dfrac{9}{2}} = \dfrac{\sqrt{9}}{\sqrt{2}} = \dfrac{3}{\sqrt{2}} = \left(\dfrac{3}{\sqrt{2}}\cdot\dfrac{\sqrt{2}}{\sqrt{2}}\right) = \dfrac{3\sqrt{2}}{2}$$

The important step for you to notice is the one in parentheses. In order to remove the radical in the denominator, the fraction was multiplied by $\dfrac{\sqrt{2}}{\sqrt{2}}$. This did not change the value of the fraction, because $\dfrac{\sqrt{2}}{\sqrt{2}} = 1$. The process of removing the radical from the denominator is called *rationalizing the denominator*. Try not to write $\sqrt{2}\cdot\sqrt{2} = \sqrt{4} = 2$; rather, write $\sqrt{2}\cdot\sqrt{2} = 2$ and save yourself some work. For example, you know that $\sqrt{648}\cdot\sqrt{648} = 648$ without writing $\sqrt{648}\cdot\sqrt{648} = \sqrt{419{,}904} = 648$.

Example 2: Simplify $\sqrt{\dfrac{32}{45}}$.

Solution

$$\sqrt{\dfrac{32}{45}} = \dfrac{\sqrt{32}}{\sqrt{45}} = \dfrac{\sqrt{16\cdot 2}}{\sqrt{9\cdot 5}} = \dfrac{4\sqrt{2}}{3\sqrt{5}} = \dfrac{4\sqrt{2}}{3\sqrt{5}}\cdot\dfrac{\sqrt{5}}{\sqrt{5}} = \dfrac{4\sqrt{10}}{3\cdot 5} = \dfrac{4\sqrt{10}}{15}$$

Q10   Simplify (rationalize the denominator):

a. $\sqrt{\dfrac{1}{2}}$

b. $\sqrt{\dfrac{4}{3}}$

c. $\sqrt{\dfrac{3}{2}}$

d. $\sqrt{\dfrac{5}{8}}$

STOP • STOP • STOP • STOP • STOP • STOP • STOP • STOP • STOP

A10   a. $\dfrac{\sqrt{2}}{2}$: $\sqrt{\dfrac{1}{2}} = \dfrac{\sqrt{1}}{\sqrt{2}} = \dfrac{1}{\sqrt{2}} \cdot \dfrac{\sqrt{2}}{\sqrt{2}} = \dfrac{\sqrt{2}}{2}$

b. $\dfrac{2\sqrt{3}}{3}$: $\sqrt{\dfrac{4}{3}} = \dfrac{\sqrt{4}}{\sqrt{3}} = \dfrac{2}{\sqrt{3}} = \dfrac{2}{\sqrt{3}} \cdot \dfrac{\sqrt{3}}{\sqrt{3}} = \dfrac{2\sqrt{3}}{3}$

c. $\dfrac{\sqrt{6}}{2}$: $\sqrt{\dfrac{3}{2}} = \dfrac{\sqrt{3}}{\sqrt{2}} = \dfrac{\sqrt{3}}{\sqrt{2}} \cdot \dfrac{\sqrt{2}}{\sqrt{2}} = \dfrac{\sqrt{6}}{2}$

d. $\dfrac{\sqrt{10}}{4}$: $\sqrt{\dfrac{5}{8}} = \dfrac{\sqrt{5}}{\sqrt{8}} = \dfrac{\sqrt{5}}{\sqrt{4 \cdot 2}} = \dfrac{\sqrt{5}}{2\sqrt{2}} = \dfrac{\sqrt{5}}{2\sqrt{2}} \cdot \dfrac{\sqrt{2}}{\sqrt{2}} = \dfrac{\sqrt{10}}{2 \cdot 2} = \dfrac{\sqrt{10}}{4}$

9        $\dfrac{\sqrt{2}}{2}$ does not look much simpler than $\dfrac{1}{\sqrt{2}}$ to many people, and yet $\dfrac{\sqrt{2}}{2}$ is considered to be in simplest radical form. One reason is that a rational approximation could be found by two methods: by the first method,

$$\frac{\sqrt{2}}{2} = \frac{1.414}{2} = 0.707$$

By the second method,

$$\frac{1}{\sqrt{2}} = \frac{1}{1.414} = 1.414\overline{)1.000{,}000}$$

```
              0.707
    1.414)1.000,000
          9898
         ──────
          10200
           9898
         ──────
            302
```

Most people agree that the first method is simpler.

Q11   Simplify:

a. $\sqrt{\dfrac{25}{32}}$

b. $\sqrt{\dfrac{72}{50}}$

c. $\sqrt{\dfrac{72}{75}}$

STOP • STOP • STOP • STOP • STOP • STOP • STOP • STOP • STOP

A11 a. $\dfrac{5\sqrt{2}}{8}$: $\sqrt{\dfrac{25}{32}} = \dfrac{\sqrt{25}}{\sqrt{32}} = \dfrac{5}{\sqrt{16\cdot2}} = \dfrac{5}{4\sqrt{2}}\cdot\dfrac{\sqrt{2}}{\sqrt{2}} = \dfrac{5\sqrt{2}}{4\cdot2} = \dfrac{5\sqrt{2}}{8}$

b. $\dfrac{6}{5}$: $\sqrt{\dfrac{72}{50}} = \dfrac{\sqrt{72}}{\sqrt{50}} = \dfrac{\sqrt{36\cdot2}}{\sqrt{25\cdot2}} = \dfrac{6\sqrt{2}}{5\sqrt{2}} = \dfrac{6}{5}$, or reducing first gives an alternative so-

lution: $\sqrt{\dfrac{72}{50}} = \sqrt{\dfrac{36}{25}} = \dfrac{\sqrt{36}}{\sqrt{25}} = \dfrac{6}{5}$

c. $\dfrac{2\sqrt{6}}{5}$: $\sqrt{\dfrac{72}{75}} = \dfrac{\sqrt{72}}{\sqrt{75}} = \dfrac{\sqrt{36\cdot2}}{\sqrt{25\cdot3}} = \dfrac{6\sqrt{2}}{5\sqrt{3}}\cdot\dfrac{\sqrt{3}}{\sqrt{3}} = \dfrac{6\sqrt{6}}{5\cdot3} = \dfrac{2\sqrt{6}}{5}$, or reduce first:

$\sqrt{\dfrac{72}{75}} = \sqrt{\dfrac{24}{25}} = \dfrac{\sqrt{24}}{\sqrt{25}} = \dfrac{\sqrt{4\cdot6}}{5} = \dfrac{2\sqrt{6}}{5}$

10 A similar property holds for radicals of higher indexes and is as follows:

$\sqrt[n]{\dfrac{a}{b}} = \dfrac{\sqrt[n]{a}}{\sqrt[n]{b}}$ where $a \geqslant 0$, $b > 0$ if n is even

For example,

1. $\sqrt[3]{\dfrac{8}{125}} = \dfrac{\sqrt[3]{8}}{\sqrt[3]{125}} = \dfrac{2}{5}$

2. $\sqrt[3]{\dfrac{-4}{27}} = \dfrac{\sqrt[3]{-4}}{\sqrt[3]{27}} = \dfrac{-\sqrt[3]{4}}{3}$

Q12 Simplify:

a. $\sqrt[3]{\dfrac{27}{64}}$

b. $\sqrt[4]{\dfrac{1}{16}}$

c. $\sqrt[3]{\dfrac{5}{64}}$

d. $\sqrt[4]{\dfrac{3}{16}}$

STOP • STOP • STOP • STOP • STOP • STOP • STOP • STOP • STOP

A12 a. $\dfrac{3}{4}$ b. $\dfrac{1}{2}$ c. $\dfrac{\sqrt[3]{5}}{4}$ d. $\dfrac{\sqrt[4]{3}}{2}$

11 One must be careful in converting an expression of radicals with index higher than 2 to simplest radical form. If $\sqrt[3]{2}$ is in the denominator, multiplying by $\sqrt[3]{2}$ gives $\sqrt[3]{2}\cdot\sqrt[3]{2} = \sqrt[3]{4}$, which is still irrational. One must multiply by $\sqrt[3]{4}$ to obtain a perfect cube as a radicand. That is, $\sqrt[3]{2}\cdot\sqrt[3]{4} = \sqrt[3]{8} = 2$.

Example 1: Simplify $\sqrt[3]{\dfrac{5}{2}}$.

Solution

$$\sqrt[3]{\frac{5}{2}} = \frac{\sqrt[3]{5}}{\sqrt[3]{2}} \cdot \frac{\sqrt[3]{4}}{\sqrt[3]{4}} = \frac{\sqrt[3]{20}}{\sqrt[3]{8}} = \frac{\sqrt[3]{20}}{2}$$

Example 2: Simplify $\sqrt[3]{\dfrac{8}{5}}$.

Solution

$$\sqrt[3]{\frac{8}{5}} = \frac{\sqrt[3]{8}}{\sqrt[3]{5}} = \frac{2}{\sqrt[3]{5}} \cdot \frac{\sqrt[3]{25}}{\sqrt[3]{25}} = \frac{2\sqrt[3]{25}}{\sqrt[3]{125}} = \frac{2\sqrt[3]{25}}{5}$$

Notice that the numerator and the denominator are multiplied by the radical expression that will convert the radicand of the denominator into a perfect power of degree equal to the index of the radical.

Q13   Simplify:

a. $\sqrt[3]{\dfrac{1}{2}}$

b. $\sqrt[3]{\dfrac{8}{9}}$

c. $\sqrt[3]{\dfrac{7}{4}}$

STOP • STOP • STOP • STOP • STOP • STOP • STOP • STOP • STOP

A13   a. $\dfrac{\sqrt[3]{4}}{2}$: $\sqrt[3]{\dfrac{1}{2}} = \dfrac{\sqrt[3]{1}}{\sqrt[3]{2}} = \dfrac{1}{\sqrt[3]{2}} \cdot \dfrac{\sqrt[3]{4}}{\sqrt[3]{4}} = \dfrac{\sqrt[3]{4}}{\sqrt[3]{8}} = \dfrac{\sqrt[3]{4}}{2}$

b. $\dfrac{2\sqrt[3]{3}}{3}$: $\sqrt[3]{\dfrac{8}{9}} = \dfrac{\sqrt[3]{8}}{\sqrt[3]{9}} = \dfrac{2}{\sqrt[3]{9}} \cdot \dfrac{\sqrt[3]{3}}{\sqrt[3]{3}} = \dfrac{2\sqrt[3]{3}}{\sqrt[3]{27}} = \dfrac{2\sqrt[3]{3}}{3}$

c. $\dfrac{\sqrt[3]{14}}{2}$: $\sqrt[3]{\dfrac{7}{4}} = \dfrac{\sqrt[3]{7}}{\sqrt[3]{4}} = \dfrac{\sqrt[3]{7}}{\sqrt[3]{4}} \cdot \dfrac{\sqrt[3]{2}}{\sqrt[3]{2}} = \dfrac{\sqrt[3]{14}}{\sqrt[3]{8}} = \dfrac{\sqrt[3]{14}}{2}$

12   Radicals that involve variables can also be converted to simplest radical form by using the same techniques.

Example 1: Simplify $\sqrt{\dfrac{4x^2}{3y}}$.

Solution

$$\sqrt{\frac{4x^2}{3y}} = \frac{\sqrt{4x^2}}{\sqrt{3y}} = \frac{2x}{\sqrt{3y}} \cdot \frac{\sqrt{3y}}{\sqrt{3y}} = \frac{2x\sqrt{3y}}{3y}$$

Example 2: Simplify $\sqrt{\dfrac{2x}{3y^5}}$.

Solution

$$\sqrt{\frac{2x}{3y^5}} = \frac{\sqrt{2x}}{\sqrt{3y^5}} \cdot \frac{\sqrt{3y}}{\sqrt{3y}} = \frac{\sqrt{6xy}}{\sqrt{9y^6}} = \frac{\sqrt{6xy}}{3y^3}$$

Q14   Simplify:

a. $\sqrt{\dfrac{3x^2}{4y^6}}$

b. $\sqrt{\dfrac{4x^2}{25y^2z^4}}$

c. $\sqrt[3]{\dfrac{3x}{8y}}$

d. $\sqrt{\dfrac{5x}{12xy^5}}$

STOP • STOP • STOP • STOP • STOP • STOP • STOP • STOP • STOP

A14   a. $\dfrac{x\sqrt{3}}{2y^3}$: $\sqrt{\dfrac{3x^2}{4y^6}} = \dfrac{\sqrt{3x^2}}{\sqrt{4y^6}} = \dfrac{x\sqrt{3}}{2y^3}$

b. $\dfrac{2x}{5yz^2}$: $\sqrt{\dfrac{4x^2}{25y^2z^4}} = \dfrac{\sqrt{4x^2}}{\sqrt{25y^2z^4}} = \dfrac{2x}{5yz^2}$

c. $\dfrac{\sqrt[3]{3xy^2}}{2y}$: $\sqrt[3]{\dfrac{3x}{8y}} = \dfrac{\sqrt[3]{3x}}{\sqrt[3]{8y}} = \dfrac{\sqrt[3]{3x}}{2\sqrt[3]{y}} \cdot \dfrac{\sqrt[3]{y^2}}{\sqrt[3]{y^2}} = \dfrac{\sqrt[3]{3xy^2}}{2\sqrt[3]{y^3}} = \dfrac{\sqrt[3]{3xy^2}}{2y}$

d. $\dfrac{\sqrt{15y}}{6y^3}$: $\sqrt{\dfrac{5x}{12xy^5}} = \sqrt{\dfrac{5}{12y^5}} = \dfrac{\sqrt{5}}{\sqrt{4y^4 \cdot 3y}} = \dfrac{\sqrt{5} \cdot \sqrt{3y}}{2y^2\sqrt{3y} \cdot \sqrt{3y}} = \dfrac{\sqrt{15y}}{2y^2 \cdot 3y} = \dfrac{\sqrt{15y}}{6y^3}$

13   The answers to the previous exercises may be obtained by alternative solutions. Do not feel you must use the method shown in the sample solution. If you are consistent in your reasoning, you may find shorter methods. The procedures for finding the simplest radical form presented in this section are summarized below.

A radical expression of index n is in *simplest radical form* if the following are satisfied:

1. No radicand contains a factor of a number or variable to the nth power.
2. No radical appears in a denominator.
3. No fraction appears within a radical.

Q15   Convert the following expressions to simplest radical form:

a. $\sqrt{72}$

b. $3\sqrt{4ab^8}$

c. $\sqrt{\dfrac{a^2c^2}{9b^4}}$

d. $\sqrt{45x^2y^3}$

e. $5\sqrt[3]{16x^4y^6}$

f. $\sqrt{\dfrac{4x^2}{9y^{10}}}$

g. $\sqrt{\dfrac{2x}{18x^4}}$

h. $\sqrt{\dfrac{6x}{7y}}$

i. $\sqrt[3]{\dfrac{27m}{4n}}$

STOP • **STOP** • **STOP** • **STOP** • **STOP** • **STOP** • **STOP** • **STOP** • **STOP**

A15

a. $6\sqrt{2}$: $\sqrt{72}=\sqrt{36\cdot2}=6\sqrt{2}$

b. $6b^4\sqrt{a}$: $3\sqrt{4ab^8}=3\sqrt{4b^8\cdot a}=3\cdot2b^4\sqrt{a}=6b^4\sqrt{a}$

c. $\dfrac{ac}{3b^2}$: $\sqrt{\dfrac{a^2c^2}{9b^4}}=\dfrac{\sqrt{a^2c^2}}{\sqrt{9b^4}}=\dfrac{ac}{3b^2}$

d. $3xy\sqrt{5y}$: $\sqrt{45x^2y^3}=\sqrt{9x^2y^2\cdot5y}=3xy\sqrt{5y}$

e. $10xy^2\sqrt[3]{2x}$: $5\sqrt[3]{16x^4y^6}=5\sqrt[3]{8x^3y^6\cdot2x}=5\cdot2xy^2\sqrt[3]{2x}=10xy^2\sqrt[3]{2x}$

f. $\dfrac{2x}{3y^5}$: $\sqrt{\dfrac{4x^2}{9y^{10}}}=\dfrac{\sqrt{4x^2}}{\sqrt{9y^{10}}}=\dfrac{2x}{3y^5}$

g. $\dfrac{\sqrt{x}}{3x^2}$: $\sqrt{\dfrac{2x}{18x^4}}=\sqrt{\dfrac{x}{9x^4}}=\dfrac{\sqrt{x}}{\sqrt{9x^4}}=\dfrac{\sqrt{x}}{3x^2}$

h. $\dfrac{\sqrt{42xy}}{7y}$: $\sqrt{\dfrac{6x}{7y}}=\dfrac{\sqrt{6x}}{\sqrt{7y}}=\dfrac{\sqrt{6x}}{\sqrt{7y}}\cdot\dfrac{\sqrt{7y}}{\sqrt{7y}}=\dfrac{\sqrt{42xy}}{7y}$

i. $\dfrac{3\sqrt[3]{2mn^2}}{2n}$: $\sqrt[3]{\dfrac{27m}{4n}}=\dfrac{\sqrt[3]{27m}}{\sqrt[3]{4n}}=\dfrac{3\sqrt[3]{m}}{\sqrt[3]{4n}}\cdot\dfrac{\sqrt[3]{2n^2}}{\sqrt[3]{2n^2}}=\dfrac{3\sqrt[3]{2mn^2}}{2n}$

This completes the instruction for this section.

13.3 Exercises

Simplify and write in simplest radical form:

1. $\sqrt{81}$
2. $\sqrt{108}$
3. $\sqrt{175}$
4. $\sqrt[3]{-125}$
5. $\sqrt[4]{81x^8}$
6. $\sqrt[3]{54}$
7. $3\sqrt{16xy^2z^3}$
8. $5\sqrt{4a^2b^3c^5}$
9. $4\sqrt{162a^5}$
10. $\sqrt[3]{125x^3y^6}$
11. $\sqrt[3]{24x^5}$
12. $\sqrt[3]{-8x^3y^7}$
13. $\sqrt{\dfrac{25}{81}}$
14. $\sqrt{\dfrac{49}{50}}$
15. $\sqrt{\dfrac{24}{81}}$
16. $\sqrt[3]{\dfrac{-1}{27}}$
17. $\sqrt[3]{\dfrac{16}{125}}$
18. $\sqrt[3]{\dfrac{-8}{27}}$
19. $\sqrt{\dfrac{2}{7}}$
20. $\sqrt{\dfrac{12}{5}}$
21. $\sqrt{\dfrac{9}{12}}$
22. $\sqrt{\dfrac{8x^2}{25y}}$
23. $\sqrt{\dfrac{9x}{8y^2}}$
24. $\sqrt{\dfrac{10a}{25b}}$
25. $\sqrt{\dfrac{32x^2}{48x^6}}$
26. $\sqrt[3]{\dfrac{2x}{4y^3}}$
27. $\sqrt[3]{\dfrac{8x^3}{3z^2}}$

13.3 Exercise Answers

1. 9
2. $6\sqrt{3}$
3. $5\sqrt{7}$
4. -5
5. $3x^2$
6. $3\sqrt[3]{2}$
7. $12yz\sqrt{xz}$
8. $10abc^2\sqrt{bc}$
9. $36a^2\sqrt{2a}$
10. $5xy^2$
11. $2x\sqrt[3]{3x^2}$
12. $-2xy^2\sqrt[3]{y}$
13. $\dfrac{5}{9}$
14. $\dfrac{7\sqrt{2}}{10}$
15. $\dfrac{2\sqrt{6}}{9}$
16. $\dfrac{-1}{3}$
17. $\dfrac{2\sqrt[3]{2}}{5}$
18. $\dfrac{-2}{3}$
19. $\dfrac{\sqrt{14}}{7}$
20. $\dfrac{2\sqrt{15}}{5}$
21. $\dfrac{\sqrt{3}}{2}$
22. $\dfrac{2x\sqrt{2y}}{5y}$
23. $\dfrac{3\sqrt{2x}}{4y}$
24. $\dfrac{\sqrt{10ab}}{5b}$
25. $\dfrac{\sqrt{6}}{3x^2}$
26. $\dfrac{\sqrt[3]{4x}}{2y}$
27. $\dfrac{2x\sqrt[3]{9z}}{3z}$

13.4 Operations on Radical Expressions

1   Radical expressions represent real numbers. All arithmetic operations apply to these numbers. In this section addition, subtraction, multiplication, and division of radical expressions will be discussed.

To simplify $3\sqrt{5} + 4\sqrt{5}$ requires the application of the right distributive property of multiplication over addition:

$$3\sqrt{5} + 4\sqrt{5} = (3 + 4)\sqrt{5}$$
$$= 7\sqrt{5}$$

Radical expressions with the same radicand are said to be *like radicals*. Like radicals can be combined by addition or subtraction. However, radical expressions that do not have the same radicand cannot be combined unless they can be changed into like radicals.

Example 1: Simplify $3\sqrt{3} + 5\sqrt{3} + 2\sqrt{2}$.

Solution

$$3\sqrt{3} + 5\sqrt{3} + 2\sqrt{2} = 8\sqrt{3} + 2\sqrt{2}$$

This solution cannot be combined further.

Example 2: Simplify $5\sqrt{3} - \sqrt{a} + \sqrt{3} - 4\sqrt{a}$.

Solution

Recall that $\sqrt{a} = 1\sqrt{a}$ and $\sqrt{3} = 1\sqrt{3}$.

$$5\sqrt{3} - \sqrt{a} + \sqrt{3} - 4\sqrt{a} = 6\sqrt{3} - 5\sqrt{a}$$

Example 3: Simplify $2 + 5\sqrt{2} + \sqrt{2}$.

Solution

$$2 + 5\sqrt{2} + \sqrt{2} = 2 + 6\sqrt{2}$$

The solution cannot be simplified further.

Q1   Simplify:

    a. $3\sqrt{2} + 6\sqrt{2} =$ _____   **b.** $3\sqrt{5} + 2\sqrt{5} - 4\sqrt{5} =$ _____

    c. $6\sqrt{7} - 2\sqrt{7} - 8\sqrt{7} =$ _____   **d.** $5\sqrt{3} - \sqrt{3} =$ _____

    e. $\sqrt{5} + \sqrt{5} + \sqrt{5} =$ _____   **f.** $\sqrt{3} + \sqrt{3} - 2\sqrt{3} =$ _____

STOP · STOP · STOP · STOP · STOP · STOP · STOP · STOP · STOP

A1   **a.** $9\sqrt{2}$ **b.** $\sqrt{5}$ **c.** $^-4\sqrt{7}$

    d. $4\sqrt{3}$ **e.** $3\sqrt{5}$ **f.** 0

Q2   Simplify:

    a. $5\sqrt{2} + 7\sqrt{2} - 3\sqrt{3} =$ _____

    b. $5 + 6\sqrt{2} - 2\sqrt{2} =$ _____

    c. $3\sqrt{5} - 2\sqrt{3} + \sqrt{5} - \sqrt{3} =$ _____

    d. $^-3\sqrt{5} + 3 - \sqrt{5} =$ _____

STOP · STOP · STOP · STOP · STOP · STOP · STOP · STOP · STOP

A2   **a.** $12\sqrt{2} - 3\sqrt{3}$ **b.** $5 + 4\sqrt{2}$

    c. $4\sqrt{5} - 3\sqrt{3}$ **d.** $3 - 4\sqrt{5}$

Q3   Combine the like radical expressions:

    a. $2\sqrt{x} + 5\sqrt{x} =$ _____

    b. $3\sqrt{x} + \sqrt{y} - \sqrt{x} =$ _____

  c. $5\sqrt{x} - \sqrt{xy} + 3\sqrt{xy} + 2\sqrt{x} =$ _____

  d. $3\sqrt{x} - 3 + \sqrt{x} + 2 =$ _____

STOP • **STOP** • **STOP** • **STOP** • **STOP** • **STOP** • **STOP** • **STOP** • **STOP**

A3   **a.** $7\sqrt{x}$ **b.** $2\sqrt{x} + \sqrt{y}$

      c. $2\sqrt{xy} + 7\sqrt{x}$ **d.** $^-1 + 4\sqrt{x}$ or $4\sqrt{x} - 1$

---

2   Some radical expressions are not like radicals, but when converted to simplest radical form they become like radicals. When combining radical expressions, always convert to simplest radical form first.

Example 1: Simplify $\sqrt{5} + \sqrt{20}$.

Solution

$$\sqrt{5} + \sqrt{20} = \sqrt{5} + \sqrt{4 \cdot 5} = \sqrt{5} + 2\sqrt{5} = 3\sqrt{5}$$

Example 2: Simplify $\sqrt{16b^3} - b\sqrt{25b} + 6b\sqrt{b}$.

Solution

$\sqrt{16b^3} - b\sqrt{25b} + 6b\sqrt{b}$

$\sqrt{16b^2 \cdot b} - b\sqrt{25 \cdot b} + 6b\sqrt{b}$

$4b\sqrt{b} - 5b\sqrt{b} + 6b\sqrt{b}$

$5b\sqrt{b}$

Q4   Combine the irrational numbers below:

  a. $2\sqrt{3} + \sqrt{27}$

  b. $2\sqrt{6} - 2\sqrt{24} + \sqrt{54}$

  c. $\sqrt{125} - \sqrt{80}$

  d. $3\sqrt{8} - 5\sqrt{50} - 6\sqrt{2} + \sqrt{98}$

STOP • **STOP** • **STOP** • **STOP** • **STOP** • **STOP** • **STOP** • **STOP** • **STOP**

A4   **a.** $5\sqrt{3}$: $2\sqrt{3} + \sqrt{27} = 2\sqrt{3} + \sqrt{9 \cdot 3} = 2\sqrt{3} + 3\sqrt{3} = 5\sqrt{3}$

      b. $\sqrt{6}$: $2\sqrt{6} - 2\sqrt{24} + \sqrt{54} = 2\sqrt{6} - 2\sqrt{4 \cdot 6} + \sqrt{9 \cdot 6} = 2\sqrt{6} - 4\sqrt{6} + 3\sqrt{6}$

      c. $\sqrt{5}$: $\sqrt{125} - \sqrt{80} = \sqrt{25 \cdot 5} - \sqrt{16 \cdot 5}$

      d. $^-18\sqrt{2}$: $3\sqrt{8} - 5\sqrt{50} - 6\sqrt{2} + \sqrt{98} = 3\sqrt{4 \cdot 2} - 5\sqrt{25 \cdot 2} - 6\sqrt{2} + \sqrt{49 \cdot 2}$

Q5   Simplify:

  a. $\sqrt{36} - 2\sqrt{32} + \sqrt{49} + 3\sqrt{18}$

b. $\sqrt{4a} + \sqrt{9a}$

c. $5\sqrt{2x} - \sqrt{8x}$

d. $2\sqrt{4a} - 7\sqrt{9b} - 3\sqrt{16a} - 10\sqrt{25b}$

STOP • STOP • STOP • STOP • STOP • STOP • STOP • STOP • STOP

A5 **a.** $13 + \sqrt{2}$: $\sqrt{36} - 2\sqrt{32} + \sqrt{49} + 3\sqrt{18} = 6 - 2\sqrt{16 \cdot 2} + 7 + 3\sqrt{9 \cdot 2}$
b. $5\sqrt{a}$: $\sqrt{4a} + \sqrt{9a} = 2\sqrt{a} + 3\sqrt{a}$
c. $3\sqrt{2x}$: $5\sqrt{2x} - \sqrt{4 \cdot 2x} = 5\sqrt{2x} - 2\sqrt{2x} = 3\sqrt{2x}$
d. $^-8\sqrt{a} - 71\sqrt{b}$: $2\sqrt{4a} - 7\sqrt{9b} - 3\sqrt{16a} - 10\sqrt{25b} = 4\sqrt{a} - 21\sqrt{b} - 12\sqrt{a}$
$\qquad - 50\sqrt{b} = {}^-8\sqrt{a} - 71\sqrt{b}$

3

Remember to remove all fractions from under the radical and do not leave radicals in the denominator when changing to simplest radical form.

Example: Simplify $\sqrt{2a} + \sqrt{\dfrac{1}{2}a}$.

Solution

$$\sqrt{2a} + \sqrt{\dfrac{1}{2}a} = \sqrt{2a} + \sqrt{\dfrac{a}{2}}$$

$$= \sqrt{2a} + \dfrac{\sqrt{a}}{\sqrt{2}} \cdot \dfrac{\sqrt{2}}{\sqrt{2}}$$

$$= \sqrt{2a} + \dfrac{\sqrt{2a}}{2}$$

$$= \dfrac{2\sqrt{2a}}{2} + \dfrac{\sqrt{2a}}{2}$$

$$= \dfrac{3\sqrt{2a}}{2}$$

The solution could also be written $\dfrac{3}{2}\sqrt{2a}$.

Q6 Simplify:

a. $\sqrt{\dfrac{1}{3}} + \dfrac{1}{3}\sqrt{12}$

b. $3\sqrt{8a^3} + 2a\sqrt{\dfrac{a}{2}}$

c. $y\sqrt{28x^3y} + x\sqrt{63xy^3} - \sqrt{175x^3y^3}$

STOP • STOP • STOP • STOP • STOP • STOP • STOP • STOP • STOP

A6 **a.** $\sqrt{3}$: $\sqrt{\dfrac{1}{3}} + \dfrac{1}{3}\sqrt{12} = \dfrac{\sqrt{1}}{\sqrt{3}} \cdot \dfrac{\sqrt{3}}{\sqrt{3}} + \dfrac{1}{3}\sqrt{4 \cdot 3} = \dfrac{\sqrt{3}}{3} + \dfrac{2}{3}\sqrt{3} = \dfrac{3\sqrt{3}}{3}$

b. $7a\sqrt{2a}$: $3\sqrt{8a^3} + 2a\sqrt{\dfrac{a}{2}} = 3\sqrt{4a^2 \cdot 2a} + 2a\dfrac{\sqrt{a}}{\sqrt{2}} \cdot \dfrac{\sqrt{2}}{\sqrt{2}} = 6a\sqrt{2a} + a\sqrt{2a}$

c. 0: $y\sqrt{4x^2 \cdot 7xy} + x\sqrt{9y^2 \cdot 7xy} - \sqrt{25x^2y^2 \cdot 7xy}$

$2xy\sqrt{7xy} + 3xy\sqrt{7xy} - 5xy\sqrt{7xy} = 0$

---

4   Irrational numbers written with radicals can be multiplied by using the same property that was used to simplify radical expressions.

$$\sqrt{a} \cdot \sqrt{b} = \sqrt{ab} \qquad \text{where } a \geqslant 0, b \geqslant 0 \quad \text{if } n \text{ is even}$$

Example 1: Simplify $\sqrt{7} \cdot \sqrt{28}$.

Solution

$$\sqrt{7} \cdot \sqrt{28} = \sqrt{7 \cdot 28} = \sqrt{7 \cdot 7 \cdot 4} = \sqrt{49 \cdot 4} = 7 \cdot 2 = 14$$

Example 2: Simplify $2\sqrt{5} \cdot 8\sqrt{2}$.

Solution

$$2\sqrt{5} \cdot 8\sqrt{2} = 2 \cdot 8 \cdot \sqrt{5} \cdot \sqrt{2} = 16\sqrt{10}$$

Notice that a whole number outside the radical sign *cannot* be multiplied times a number inside the radical sign. For instance,

$$16\sqrt{10} \neq \sqrt{160}$$

Q7   Multiply the irrational numbers and simplify:

a. $\sqrt{3} \cdot \sqrt{12}$

b. $\sqrt{5} \cdot \sqrt{45}$

c. $3\sqrt{2} \cdot 4$

d. $\sqrt{4} \cdot \sqrt{18}$

e. $\sqrt{5a} \cdot \sqrt{3a}$

f. $5\sqrt{2b} \cdot \sqrt{8}$

g. $\sqrt{2x} \cdot \sqrt{6xy}$

STOP • STOP • STOP • STOP • STOP • STOP • STOP • STOP • STOP

A7 **a.** 6

b. 15: $\sqrt{5} \cdot \sqrt{45} = \sqrt{5 \cdot 45} = \sqrt{5 \cdot 5 \cdot 9} = \sqrt{25 \cdot 9} = 5 \cdot 3$, or alternatively:

$\sqrt{5} \cdot \sqrt{45} = \sqrt{5} \cdot \sqrt{5 \cdot 9} = \sqrt{5} \cdot \sqrt{5} \cdot \sqrt{9} = 5 \cdot 3 = 15$

c. $12\sqrt{2}$ **d.** $6\sqrt{2}$ **e.** $a\sqrt{15}$ **f.** $20\sqrt{b}$

g. $2x\sqrt{3y}$: $\sqrt{2x} \cdot \sqrt{6xy} = \sqrt{12x^2y} = \sqrt{4x^2 \cdot 3y} = 2x\sqrt{3y}$

Q8   Simplify:

a. $\sqrt{8a} \cdot \sqrt{18a}$

b. $2\sqrt{6}\cdot\sqrt{8}$

c. $(5\sqrt{5})(3\sqrt{10})(\sqrt{4})$

d. $(x^2\sqrt{x})(2x\sqrt{x})(x\sqrt{x^2})$

STOP • STOP • STOP • STOP • STOP • STOP • STOP • STOP • STOP

A8   **a.** $12a$: $\sqrt{8a}\cdot\sqrt{18a}=\sqrt{8\cdot18a^2}=\sqrt{4\cdot2\cdot2\cdot9a^2}=2\cdot2\cdot3a$
       b. $8\sqrt{3}$: $2\sqrt{6}\cdot\sqrt{8}=2\sqrt{48}=2\sqrt{16\cdot3}=2\cdot4\sqrt{3}=8\sqrt{3}$
       c. $150\sqrt{2}$: $(5\sqrt{5})(3\sqrt{10})(\sqrt{4})=15\sqrt{50}(\sqrt{4})=15\sqrt{25\cdot2}(2)=15\cdot5\cdot2\sqrt{2}$ or
              $(5\sqrt{5})(3\sqrt{10})(\sqrt{4})=15\sqrt{5}\cdot\sqrt{10}\cdot\sqrt{4}=15\cdot\sqrt{5}\cdot\sqrt{5}\cdot\sqrt{2}\cdot\sqrt{4}=$
              $15\cdot5\sqrt{2}\cdot2$
       d. $2x^6$: $(x^2\sqrt{x})(2x\sqrt{x})(x\sqrt{x^2})=2x^4\sqrt{x^4}=2x^4\cdot x^2=2x^6$

---

5   To multiply a monomial times a binomial, use the distributive law as shown.

Example: Simplify $\sqrt{5}(4+\sqrt{10})$.

Solution

$$\sqrt{5}(4+\sqrt{10})=\sqrt{5}\cdot4+\sqrt{5}\cdot\sqrt{10}$$
$$=4\sqrt{5}+\sqrt{50}$$
$$=4\sqrt{5}+\sqrt{25\cdot2}$$
$$=4\sqrt{5}+5\sqrt{2}$$

Q9   Multiply:
       a. $\sqrt{3}(\sqrt{3}+\sqrt{2})$

       b. $\sqrt{3}(\sqrt{2}+\sqrt{6})$

       c. $\sqrt{3}(\sqrt{6}+\sqrt{15})$

STOP • STOP • STOP • STOP • STOP • STOP • STOP • STOP • STOP

A9   **a.** $3+\sqrt{6}$: $\sqrt{3}(\sqrt{3}+\sqrt{2})=\sqrt{3}\cdot\sqrt{3}+\sqrt{3}\cdot\sqrt{2}=3+\sqrt{6}$
       b. $\sqrt{6}+3\sqrt{2}$: $\sqrt{3}(\sqrt{2}+\sqrt{6})=\sqrt{3}\cdot\sqrt{2}+\sqrt{3}\cdot\sqrt{6}=\sqrt{6}+\sqrt{3}\cdot\sqrt{3}\cdot\sqrt{2}$
       c. $3\sqrt{2}+3\sqrt{5}$: $\sqrt{3}(\sqrt{6}+\sqrt{15})=\sqrt{3}\cdot\sqrt{6}+\sqrt{3}\cdot\sqrt{15}=\sqrt{3}\cdot\sqrt{3}\cdot\sqrt{2}+$
              $\sqrt{3}\cdot\sqrt{3}\cdot\sqrt{5}$

---

6   Although several steps are shown in the solutions to the problems, many of these steps can be done mentally. Try to examine a product and write the result.
$\sqrt{5}(4+\sqrt{10})=4\sqrt{5}+5\sqrt{2}$. Think: $\sqrt{5}$ times 4 cannot be simplified; therefore, the product is written $4\sqrt{5}$. $\sqrt{10}$ is $\sqrt{5}\cdot\sqrt{2}$; therefore, $\sqrt{5}\cdot\sqrt{5}\cdot\sqrt{2}=5\sqrt{2}$.

Q10   Multiply mentally:
        a. $\sqrt{2}\cdot\sqrt{10}=$ _____

        b. $\sqrt{3}(\sqrt{6}+2)=$ _____

        c. $\sqrt{2}(\sqrt{10}-\sqrt{2})=$ _____

    d. $\sqrt{3}(\sqrt{3} + \sqrt{15}) =$ _____

STOP • **STOP** • **STOP** • **STOP** • **STOP** • **STOP** • **STOP** • **STOP** • **STOP**

A10   **a.** $2\sqrt{5}$ **b.** $3\sqrt{2} + 2\sqrt{3}$
            c. $2\sqrt{5} - 2$ **d.** $3 + 3\sqrt{5}$

---

7   It is sometimes necessary to multiply binomial expressions that involve radicals. Recall the method of multiplying binomials by using the distributive laws.

$$(a + b)(c + d) = a(c + d) + b(c + d)$$
$$= ac + ad + bc + bd$$

Notice that there are four products to find. When multiplying radical binomials, the same four products occur.

$$(4 + \sqrt{3})(5 + \sqrt{2}) = 4(5 + \sqrt{2}) + \sqrt{3}(5 + \sqrt{2})$$
$$= 4 \cdot 5 + 4 \cdot \sqrt{2} + \sqrt{3} \cdot 5 + \sqrt{3} \cdot \sqrt{2}$$
$$= 20 + 4\sqrt{2} + 5\sqrt{3} + \sqrt{6}$$

In this example the product contains four unlike terms and cannot be simplified. In other instances it will be possible to simplify further.

Example: Multiply $(\sqrt{5} + 1)(\sqrt{5} + 3)$.

Solution
$$(\sqrt{5} + 1)(\sqrt{5} + 3) = \sqrt{5}(\sqrt{5} + 3) + 1(\sqrt{5} + 3)$$
$$= 5 + 3\sqrt{5} + \sqrt{5} + 3$$
$$= 8 + 4\sqrt{5}$$

Q11   Multiply:
        a. $(2 - \sqrt{3})(5 + \sqrt{3})$ **b.** $(1 + 2\sqrt{5})(3 + 2\sqrt{5})$

STOP • **STOP** • **STOP** • **STOP** • **STOP** • **STOP** • **STOP** • **STOP** • **STOP**

A11   **a.** $7 - 3\sqrt{3}$: $(2 - \sqrt{3})(5 + \sqrt{3})$
                     $2(5 + \sqrt{3}) - \sqrt{3}(5 + \sqrt{3})$
                     $10 + 2\sqrt{3} - 5\sqrt{3} - \sqrt{3} \cdot \sqrt{3}$
                     $7 - 3\sqrt{3}$
    b. $23 + 8\sqrt{5}$: $(1 + 2\sqrt{5})(3 + 2\sqrt{5})$
                     $1(3 + 2\sqrt{5}) + 2\sqrt{5}(3 + 2\sqrt{5})$
                     $3 + 2\sqrt{5} + 6\sqrt{5} + 4\sqrt{5} \cdot \sqrt{5}$
                     $23 + 8\sqrt{5}$

Q12   Find the products:
        a. $(\sqrt{7} + 3)(\sqrt{7} - 5)$ **b.** $(4 - 3\sqrt{6})(4 + 3\sqrt{6})$

STOP • **STOP** • **STOP** • **STOP** • **STOP** • **STOP** • **STOP** • **STOP** • **STOP**

A12 **a.** $^-8 - 2\sqrt{7}$: $(\sqrt{7} + 3)(\sqrt{7} - 5)$
$$7 - 5\sqrt{7} + 3\sqrt{7} - 15$$
$$^-8 - 2\sqrt{7}$$

b. $^-38$: $(4 - 3\sqrt{6})(4 + 3\sqrt{6})$
$$16 + 12\sqrt{6} - 12\sqrt{6} - 54$$
$$^-38$$

Q13 Find the products:

a. $(2 + \sqrt{5})^2$ **b.** $(\sqrt{3} + \sqrt{5})^2$

STOP • STOP • STOP • STOP • STOP • STOP • STOP • STOP • STOP

A13 **a.** $9 + 4\sqrt{5}$: $(2 + \sqrt{5})^2$
$$(2 + \sqrt{5})(2 + \sqrt{5})$$
$$4 + 2\sqrt{5} + 2\sqrt{5} + 5$$
$$9 + 4\sqrt{5}$$

b. $8 + 2\sqrt{15}$: $(\sqrt{3} + \sqrt{5})^2$
$$(\sqrt{3} + \sqrt{5})(\sqrt{3} + \sqrt{5})$$
$$3 + \sqrt{15} + \sqrt{15} + 5$$
$$8 + 2\sqrt{15}$$

---

8   The variables in a polynomial can be given values and the polynomial can then be evaluated.

Example 1: Evaluate $x^2 + 3x + 2$ if $x = 7$.

Solution

$$\begin{aligned} x^2 + 3x + 2 &= (7)^2 + 3(7) + 2 \\ &= 49 + 21 + 2 \\ &= 72 \end{aligned}$$

A variable can represent an irrational value as well.

Example 2: Evaluate $x^2 + 3x + 2$ if $x = 2\sqrt{3}$.

Solution

$$\begin{aligned} x^2 + 3x + 2 &= (2\sqrt{3})^2 + 3(2\sqrt{3}) + 2 \\ &= 4 \cdot 3 + 6\sqrt{3} + 2 \\ &= 14 + 6\sqrt{3} \end{aligned}$$

Similarly, the variable can represent a binomial expression containing radicals.

Example 3: Evaluate $x^2 + 3x + 2$ if $x = 2 - \sqrt{3}$.

Solution

$$\begin{aligned} x^2 + 3x + 2 &= (2 - \sqrt{3})^2 + 3(2 - \sqrt{3}) + 2 \\ &= (2 - \sqrt{3})(2 - \sqrt{3}) + 6 - 3\sqrt{3} + 2 \\ &= 4 - 2\sqrt{3} - 2\sqrt{3} + 3 + 6 - 3\sqrt{3} + 2 \\ &= 15 - 7\sqrt{3} \end{aligned}$$

Q14   Evaluate $x^2 - x + 3$:
  a. when $x = 2\sqrt{5}$ **b.** when $x = 2 - \sqrt{2}$

STOP • **STOP** • **STOP** • **STOP** • **STOP** • **STOP** • **STOP** • **STOP** • **STOP**

A14   **a.** $23 - 2\sqrt{5}$: $x^2 - x + 3$
                      $(2\sqrt{5})^2 - 2\sqrt{5} + 3$
                      $20 - 2\sqrt{5} + 3$
                      $23 - 2\sqrt{5}$

  b. $7 - 3\sqrt{2}$: $x^2 - x + 3$
                      $(2 - \sqrt{2})^2 - (2 - \sqrt{2}) + 3$
                      $4 - 2\sqrt{2} - 2\sqrt{2} + 2 - 2 + \sqrt{2} + 3$
                      $7 - 3\sqrt{2}$

Q15   Evaluate $x^2 + 2x - 2$ when $x = {}^{-}1 + \sqrt{3}$.

STOP • **STOP** • **STOP** • **STOP** • **STOP** • **STOP** • **STOP** • **STOP** • **STOP**

A15 0: $x^2 + 2x - 2 = ({}^{-}1 + \sqrt{3})^2 + 2({}^{-}1 + \sqrt{3}) - 2$
                      $= 1 - 2\sqrt{3} + 3 - 2 + 2\sqrt{3} - 2$
                      $= 0$

---

9   A number is a solution of an equation if a true statement results from substituting the number for the variable in the equation. The value of $2\sqrt{3}$ can be shown to be a solution to $x^2 - 12 = 0$ by substituting.

$x^2 - 12 = (2\sqrt{3})^2 - 12$
$\quad\quad\quad = (2\sqrt{3})(2\sqrt{3}) - 12$
$\quad\quad\quad = 12 - 12$
$\quad\quad\quad = 0$

Q16   Show that ${}^{-}1 - \sqrt{3}$ is a solution to $x^2 + 2x - 2 = 0$.

STOP • **STOP** • **STOP** • **STOP** • **STOP** • **STOP** • **STOP** • **STOP** • **STOP**

A16 $x^2 + 2x - 2 = ({}^{-}1 - \sqrt{3})^2 + 2({}^{-}1 - \sqrt{3}) - 2$
                      $= 1 + 2\sqrt{3} + 3 - 2 - 2\sqrt{3} - 2$
                      $= 0$

Q17   By substituting and evaluating the result, find the two solutions to the equation $x^2 - 4x - 7 = 0$ from the replacement set $\{2 + \sqrt{11},\ 1 - 3\sqrt{2},\ 2 - \sqrt{11},\ 3 - \sqrt{2}\}$.

STOP • **STOP** • **STOP** • **STOP** • **STOP** • **STOP** • **STOP** • **STOP** • **STOP**

A17 $2 + \sqrt{11}$ and $2 - \sqrt{11}$

---

10 To divide radical expressions, use the property

$$\frac{\sqrt{a}}{\sqrt{b}} = \sqrt{\frac{a}{b}} \qquad \text{where } a \geqslant 0, \, b > 0 \quad \text{if } n \text{ is even}$$

Example 1: Simplify $\dfrac{\sqrt{12}}{\sqrt{3}}$.

Solution

$$\frac{\sqrt{12}}{\sqrt{3}} = \sqrt{\frac{12}{3}} = \sqrt{4} = 2$$

Example 2: Simplify $\dfrac{\sqrt{7abc}}{\sqrt{bc}}$.

Solution

$$\frac{\sqrt{7abc}}{\sqrt{bc}} = \sqrt{\frac{7abc}{bc}} = \sqrt{7a}$$

An alternative method is to multiply by a number equivalent to one that will remove the radical in the denominator. This process is called *rationalizing the denominator*.

Example 3: Simplify $\dfrac{\sqrt{20}}{\sqrt{3}}$ (rationalize the denominator).

Solution

$$\frac{\sqrt{20}}{\sqrt{3}} = \frac{\sqrt{20}}{\sqrt{3}} \cdot \frac{\sqrt{3}}{\sqrt{3}} = \frac{\sqrt{4} \cdot \sqrt{5} \cdot \sqrt{3}}{3} = \frac{2\sqrt{15}}{3}$$

Q18 Divide and simplify:

a. $\dfrac{\sqrt{80}}{\sqrt{5}}$

b. $\dfrac{\sqrt{21}}{\sqrt{7}}$

c. $\dfrac{6\sqrt{12}}{\sqrt{2}}$

d. $\dfrac{3\sqrt{6}}{3\sqrt{2}}$

STOP • **STOP** • **STOP** • **STOP** • **STOP** • **STOP** • **STOP** • **STOP** • **STOP**

A18 a. 4: $\dfrac{\sqrt{80}}{\sqrt{5}} = \sqrt{\dfrac{80}{5}} = \sqrt{16} = 4$

b. $\sqrt{3}$: $\dfrac{\sqrt{21}}{\sqrt{7}} = \sqrt{\dfrac{21}{7}} = \sqrt{3}$

c. $6\sqrt{6}$: $\dfrac{6\sqrt{12}}{\sqrt{2}} = 6\sqrt{\dfrac{12}{2}} = 6\sqrt{6}$

d. $\sqrt{3}$: $\dfrac{3\sqrt{6}}{3\sqrt{2}} = \dfrac{\sqrt{6}}{\sqrt{2}} = \sqrt{\dfrac{6}{2}} = \sqrt{3}$

---

11   There are several ways of simplifying some radical expressions. You may use whichever method you prefer, although one is usually shorter than the others. A radical expression will be simplified by three methods below. Practice these methods by looking for alternative ways of simplifying the radical expressions in the rest of this section.

Example: Simplify $\dfrac{\sqrt{2}}{\sqrt{8}}$.

Solution

Method 1: $\dfrac{\sqrt{2}}{\sqrt{8}} = \sqrt{\dfrac{2}{8}} = \sqrt{\dfrac{1}{4}} = \dfrac{1}{2}$

Method 2: $\dfrac{\sqrt{2}}{\sqrt{8}} = \dfrac{\sqrt{2}}{\sqrt{2}\cdot\sqrt{4}} = \dfrac{\overset{1}{\cancel{\sqrt{2}}}}{\cancel{\sqrt{2}}\cdot\sqrt{4}} = \dfrac{1}{2}$

Method 3: $\dfrac{\sqrt{2}}{\sqrt{8}} = \dfrac{\sqrt{2}}{\sqrt{8}}\cdot\dfrac{\sqrt{2}}{\sqrt{2}} = \dfrac{\sqrt{4}}{\sqrt{16}} = \dfrac{2}{4} = \dfrac{1}{2}$

Q19   Simplify:

a. $\dfrac{\sqrt{3}}{\sqrt{75}}$

b. $\dfrac{2\sqrt{3}}{\sqrt{12}}$

c. $\dfrac{4\sqrt{24}}{2\sqrt{6}}$

d. $\dfrac{10}{\sqrt[3]{9}}$

STOP • STOP • STOP • STOP • STOP • STOP • STOP • STOP • STOP

A19   **a.** $\dfrac{1}{5}$: $\dfrac{\sqrt{3}}{\sqrt{75}} = \dfrac{\sqrt{3}}{\sqrt{3}\cdot\sqrt{25}} = \dfrac{1}{\sqrt{25}} = \dfrac{1}{5}$ or $\dfrac{\sqrt{3}}{\sqrt{75}} = \sqrt{\dfrac{3}{75}} = \sqrt{\dfrac{1}{25}} = \dfrac{1}{5}$

b. 1: $\dfrac{2\sqrt{3}}{\sqrt{12}} = \dfrac{2\sqrt{3}}{\sqrt{4}\cdot\sqrt{3}} = \dfrac{2\sqrt{3}}{2\sqrt{3}} = 1$ or $\dfrac{2\sqrt{3}}{\sqrt{12}} = 2\sqrt{\dfrac{3}{12}} = 2\sqrt{\dfrac{1}{4}} = 2\cdot\dfrac{1}{2} = 1$

c. 4: $\dfrac{4\sqrt{24}}{2\sqrt{6}} = 2\sqrt{\dfrac{24}{6}} = 2\sqrt{4} = 2\cdot 2 = 4$

d. $\dfrac{10\sqrt[3]{3}}{3}$: $\dfrac{10}{\sqrt[3]{9}} = \dfrac{10}{\sqrt[3]{9}}\cdot\dfrac{\sqrt[3]{3}}{\sqrt[3]{3}} = \dfrac{10\sqrt[3]{3}}{\sqrt[3]{27}} = \dfrac{10\sqrt[3]{3}}{3}$

Q20   Simplify:

a. $\dfrac{\sqrt{21b}}{\sqrt{7b}}$

b. $\dfrac{a\sqrt{2}}{\sqrt{a}}$

c. $\dfrac{3\sqrt{6}}{\sqrt{8}}$

d. $\dfrac{\sqrt{4abc}}{\sqrt{24ab}}$

STOP • **STOP** • **STOP** • **STOP** • **STOP** • **STOP** • **STOP** • **STOP** • **STOP**

A20

a. $\sqrt{3}$: $\dfrac{\sqrt{21b}}{\sqrt{7b}} = \sqrt{\dfrac{21b}{7b}} = \sqrt{3}$

b. $\sqrt{2a}$: $\dfrac{a\sqrt{2}}{\sqrt{a}} = \dfrac{a\sqrt{2}}{\sqrt{a}} \cdot \dfrac{\sqrt{a}}{\sqrt{a}} = \dfrac{a\sqrt{2a}}{a} = \sqrt{2a}$

c. $\dfrac{3\sqrt{3}}{2}$: $\dfrac{3\sqrt{6}}{\sqrt{8}} = \dfrac{3\sqrt{6}}{\sqrt{8}} \cdot \dfrac{\sqrt{2}}{\sqrt{2}} = \dfrac{3\sqrt{12}}{\sqrt{16}} = \dfrac{3\sqrt{4 \cdot 3}}{4} = \dfrac{3 \cdot 2\sqrt{3}}{4} = \dfrac{3\sqrt{3}}{2}$

Alternatively: $\dfrac{3\sqrt{6}}{\sqrt{8}} = 3\sqrt{\dfrac{6}{8}} = 3\sqrt{\dfrac{3}{4}} = \dfrac{3\sqrt{3}}{\sqrt{4}} = \dfrac{3\sqrt{3}}{2}$

d. $\dfrac{\sqrt{6c}}{6}$: $\dfrac{\sqrt{4abc}}{\sqrt{24ab}} = \sqrt{\dfrac{4abc}{24ab}} = \sqrt{\dfrac{c}{6}} = \dfrac{\sqrt{c}}{\sqrt{6}} \cdot \dfrac{\sqrt{6}}{\sqrt{6}} = \dfrac{\sqrt{6c}}{6}$

---

12 When reducing, it is important to identify common factors in the numerator and the denominator. Recall that $\dfrac{x+2}{4}$ does not reduce, whereas $\dfrac{2(x+2)}{4}$ does.

Example 1: Simplify $\dfrac{4 + \sqrt{12}}{2}$.

Solution

$\dfrac{4 + \sqrt{12}}{2} = \dfrac{4 + \sqrt{4 \cdot 3}}{2} = \dfrac{4 + 2\sqrt{3}}{2} = \dfrac{2(2 + \sqrt{3})}{2} = 2 + \sqrt{3}$

Example 2: Simplify $\dfrac{\sqrt{2} + \sqrt{10}}{\sqrt{2}}$.

Solution

$\dfrac{\sqrt{2} + \sqrt{10}}{\sqrt{2}} = \dfrac{\sqrt{2} + \sqrt{2} \cdot \sqrt{5}}{\sqrt{2}} = \dfrac{\sqrt{2}(1 + \sqrt{5})}{\sqrt{2}} = 1 + \sqrt{5}$

Example 3: Simplify $\dfrac{5 + \sqrt{7}}{10}$.

Solution

Since the numerator cannot be factored, the fraction is already in simplest form.

Q21 Reduce:

a. $\dfrac{6 + 3\sqrt{2}}{3}$

b. $\dfrac{8 + \sqrt{24}}{2}$

c. $\dfrac{5 - \sqrt{50}}{10}$

d. $\dfrac{\sqrt{15} + \sqrt{6}}{\sqrt{3}}$

STOP • STOP • STOP • STOP • STOP • STOP • STOP • STOP • STOP

A21   a. $2 + \sqrt{2}$: $\dfrac{6 + 3\sqrt{2}}{3} = \dfrac{3(2 + \sqrt{2})}{3} = 2 + \sqrt{2}$

b. $4 + \sqrt{6}$: $\dfrac{8 + \sqrt{24}}{2} = \dfrac{8 + \sqrt{4 \cdot 6}}{2} = \dfrac{8 + 2\sqrt{6}}{2} = \dfrac{2(4 + \sqrt{6})}{2}$

c. $\dfrac{1 - \sqrt{2}}{2}$: $\dfrac{5 - \sqrt{50}}{10} = \dfrac{5 - 5\sqrt{2}}{10} = \dfrac{5(1 - \sqrt{2})}{10} = \dfrac{1 - \sqrt{2}}{2}$

d. $\sqrt{5} + \sqrt{2}$: $\dfrac{\sqrt{15} + \sqrt{6}}{\sqrt{3}} = \dfrac{\sqrt{3} \cdot \sqrt{5} + \sqrt{3} \cdot \sqrt{2}}{\sqrt{3}} = \dfrac{\sqrt{3}(\sqrt{5} + \sqrt{2})}{\sqrt{3}}$

---

13   To rationalize the denominator of a fraction that contains a binomial requires the use of a special product. Notice what happens when the two binomials $2 + \sqrt{3}$ and $2 - \sqrt{3}$ are multiplied.

$(2 + \sqrt{3})(2 - \sqrt{3})$
$2(2 - \sqrt{3}) + \sqrt{3}(2 - \sqrt{3})$
$4 - 2\sqrt{3} + 2\sqrt{3} - 3$
1

Since the two middle terms containing the radicals are alike except for their sign, their sum is zero. This always happens when two binomials that differ only in the sign of the second term are multiplied. The product then is a rational number.

A binomial that differs from another binomial only in the sign of the second term is called its *conjugate*. For example, the conjugate of $2 + \sqrt{3}$ is $2 - \sqrt{3}$. Their product is a rational number.

Example 1: Write the conjugate of $3 - \sqrt{5}$.

Solution

$3 + \sqrt{5}$

Example 2: Multiply $2 - \sqrt{7}$ by its conjugate.

Solution

$(2 - \sqrt{7})(2 + \sqrt{7}) = 4 + 2\sqrt{7} - 2\sqrt{7} - 7$
$= {}^-3$

Q22   Write the conjugates of the following irrational numbers:

a. $2 - \sqrt{5}$ _____   b. $3 + \sqrt{2}$ _____

c. ${}^-1 - \sqrt{2}$ _____   d. $\sqrt{3} - \sqrt{7}$ _____

STOP • STOP • STOP • STOP • STOP • STOP • STOP • STOP • STOP

A22   a. $2 + \sqrt{5}$ b. $3 - \sqrt{2}$

c. ${}^-1 + \sqrt{2}$ d. $\sqrt{3} + \sqrt{7}$

Q23   Multiply the following binomials times their conjugates:

a. $5 - \sqrt{2}$

b. $3 - \sqrt{5}$

c. $^-2 - \sqrt{3}$

d. $\sqrt{5} - \sqrt{2}$

STOP • STOP • STOP • STOP • STOP • STOP • STOP • STOP • STOP

A23   **a.** 23: $(5 - \sqrt{2})(5 + \sqrt{2}) = 25 - 2 = 23$

b. 4: $(3 - \sqrt{5})(3 + \sqrt{5}) = 9 - 5 = 4$

c. 1: $(^-2 - \sqrt{3})(^-2 + \sqrt{3}) = 4 - 3 = 1$

d. 3: $(\sqrt{5} - \sqrt{2})(\sqrt{5} + \sqrt{2}) = 5 + \sqrt{10} - \sqrt{10} - 2 = 3$

---

14   To rationalize the denominator of a fraction with a binomial denominator containing a square root, multiply both numerator and denominator times the conjugate of the denominator.

Example 1: Rationalize the denominator of $\dfrac{2 + \sqrt{3}}{1 - \sqrt{5}}$.

Solution

$$\frac{2 + \sqrt{3}}{1 - \sqrt{5}} = \frac{2 + \sqrt{3}}{1 - \sqrt{5}} \cdot \frac{1 + \sqrt{5}}{1 + \sqrt{5}}$$

$$= \frac{2 + 2\sqrt{5} + \sqrt{3} + \sqrt{15}}{1 + \sqrt{5} - \sqrt{5} - 5}$$

$$= \frac{2 + 2\sqrt{5} + \sqrt{3} + \sqrt{15}}{^-4}$$

The final fraction obtained above is considered to be in simplest radical form.

Example 2: Divide $4 + \sqrt{3}$ by $1 - \sqrt{3}$.

Solution

$$\frac{4 + \sqrt{3}}{1 - \sqrt{3}} = \frac{4 + \sqrt{3}}{1 - \sqrt{3}} \cdot \frac{1 + \sqrt{3}}{1 + \sqrt{3}}$$

$$= \frac{4 + 4\sqrt{3} + \sqrt{3} + 3}{1 - 3}$$

$$= \frac{7 + 5\sqrt{3}}{^-2}$$

Q24   Rationalize the denominators in the following fractions:

a. $\dfrac{2 + \sqrt{5}}{7 - \sqrt{5}}$

b. $\dfrac{3 - \sqrt{2}}{4 + \sqrt{3}}$

STOP • STOP • STOP • STOP • STOP • STOP • STOP • STOP • STOP

A24   **a.** $\dfrac{19 + 9\sqrt{5}}{44}$: $\dfrac{2 + \sqrt{5}}{7 - \sqrt{5}} \cdot \dfrac{7 + \sqrt{5}}{7 + \sqrt{5}}$

$$\frac{14 + 2\sqrt{5} + 7\sqrt{5} + 5}{49 - 5}$$

$$\frac{19 + 9\sqrt{5}}{44}$$

b. $\dfrac{12 - 3\sqrt{3} - 4\sqrt{2} + \sqrt{6}}{13}$: $\dfrac{3 - \sqrt{2}}{4 + \sqrt{3}} \cdot \dfrac{4 - \sqrt{3}}{4 - \sqrt{3}}$

$$\frac{12 - 3\sqrt{3} - 4\sqrt{2} + \sqrt{6}}{16 - 3}$$

$$\frac{12 - 3\sqrt{3} - 4\sqrt{2} + \sqrt{6}}{13}$$

Q25   Divide the following real numbers:
    a. 6 by $2 + \sqrt{7}$ **b.** $2 + \sqrt{3}$ by $\sqrt{3} + \sqrt{5}$

STOP • **STOP** • **STOP** • **STOP** • **STOP** • **STOP** • **STOP** • **STOP** • **STOP**

A25   **a.** $2\sqrt{7} - 4$: $\dfrac{6}{2 + \sqrt{7}} \cdot \dfrac{2 - \sqrt{7}}{2 - \sqrt{7}}$

$$\frac{12 - 6\sqrt{7}}{4 - 7}$$

$$\frac{12 - 6\sqrt{7}}{^-3}$$

$$\frac{^-3(^-4 + 2\sqrt{7})}{^-3}$$

$$^-4 + 2\sqrt{7}$$

b. $\dfrac{3 + 2\sqrt{3} - 2\sqrt{5} - \sqrt{15}}{^-2}$: $\dfrac{2 + \sqrt{3}}{\sqrt{3} + \sqrt{5}} \cdot \dfrac{\sqrt{3} - \sqrt{5}}{\sqrt{3} - \sqrt{5}}$

$$\frac{2\sqrt{3} - 2\sqrt{5} + 3 - \sqrt{15}}{3 - 5}$$

$$\frac{3 + 2\sqrt{3} - 2\sqrt{5} - \sqrt{15}}{^-2}$$

This completes the instruction for this section.

13.4   Exercises

1. Perform any indicated operations and write the answer in simplest radical form:
    a. $2\sqrt{3} + 5\sqrt{3}$ **b.** $^-\sqrt{5} + 4\sqrt{5}$

c.　$5\sqrt{6} - \sqrt{2} + \sqrt{6}$ 　　　　d.　$4\sqrt[3]{2} - 2\sqrt[3]{2} + \sqrt[3]{2}$

e.　$\sqrt{2} + \sqrt{3} - 7\sqrt{2} + \sqrt{3}$　f.　$4 + \sqrt{6} + 7 - \sqrt{6}$

g.　$\sqrt{12} + \sqrt{75}$ 　　　　　h.　$^-3\sqrt{3} + \sqrt{48}$

i.　$\sqrt{4x^3} - \sqrt{25x^3}$ 　　　　j.　$\sqrt{32} + 4\sqrt{\dfrac{1}{2}}$

k.　$\sqrt{\dfrac{5}{4}} + \sqrt{45}$ 　　　　l.　$\sqrt{8xy} + 2\sqrt{18xy}$

m.　$\sqrt{2} \cdot \sqrt{8}$ 　　　　　n.　$\sqrt{12} \cdot 3\sqrt{8}$

o.　$4\sqrt[3]{9} \cdot \sqrt[3]{24}$ 　　　　p.　$(2a\sqrt{a})(2\sqrt{a^2})(3\sqrt{a^3})$

2.　Simplify:

a.　$\sqrt{3}(2 + 4\sqrt{3})$ 　　　　b.　$2\sqrt{5}(\sqrt{5} + \sqrt{10})$

c.　$\sqrt{2}(\sqrt{3} + \sqrt{5})$ 　　　　d.　$\sqrt{2}(\sqrt{10} + \sqrt{40})$

e.　$(2 + \sqrt{3})(3 - \sqrt{3})$ 　　　f.　$(\sqrt{5} + 6)(2\sqrt{5} - 1)$

g.　$(1 + \sqrt{2})(1 - \sqrt{2})$ 　　　h.　$(\sqrt{2} + 3)^2$

i.　$\dfrac{\sqrt{15}}{\sqrt{3}}$ 　　　　　　j.　$\dfrac{\sqrt{6}}{\sqrt{2}}$

k.　$\dfrac{\sqrt{2}}{\sqrt{5}}$ 　　　　　　l.　$\dfrac{3\sqrt{8}}{\sqrt{3}}$

m.　$\dfrac{\sqrt{2xy}}{\sqrt{3x}}$ 　　　　　n.　$\dfrac{\sqrt[3]{4}}{\sqrt[3]{18}}$

o.　$\dfrac{2 + 4\sqrt{5}}{2}$ 　　　　　p.　$\dfrac{5 + \sqrt{50}}{5}$

q.　$\dfrac{6 - \sqrt{27}}{9}$ 　　　　　r.　$\dfrac{\sqrt{10} + \sqrt{20}}{\sqrt{2}}$

s.　$\dfrac{2 + \sqrt{3}}{1 - \sqrt{3}}$ 　　　　　t.　$\dfrac{5 + \sqrt{5}}{2 - \sqrt{5}}$

u.　$\dfrac{6 + \sqrt{10}}{1 - 2\sqrt{10}}$ 　　　　v.　$\dfrac{\sqrt{3} + \sqrt{7}}{2\sqrt{3} + \sqrt{7}}$

3.　Determine by substitution if the numbers given for the variable are solutions to the equations.

a.　$x = {}^-3 + \sqrt{10}$　in the equation　$x^2 + 6x - 1 = 0$

b.　$x = 2 + \sqrt{10}$　in the equation　$x^2 + 4x - 1 = 0$

13.4　Exercise Answers

1.　a.　$7\sqrt{3}$ 　　b.　$3\sqrt{5}$ 　　c.　$6\sqrt{6} - \sqrt{2}$ 　　d.　$3\sqrt[3]{2}$

　　e.　$2\sqrt{3} - 6\sqrt{2}$ 　　f.　11 　　g.　$7\sqrt{3}$ 　　h.　$\sqrt{3}$

　　i.　$^-3x\sqrt{x}$ 　　j.　$6\sqrt{2}$ 　　k.　$\dfrac{7\sqrt{5}}{2}$ 　　l.　$8\sqrt{2xy}$

　　m.　4 　　n.　$12\sqrt{6}$ 　　o.　24 　　p.　$12a^4$

2.　a.　$2\sqrt{3} + 12$ 　　b.　$10 + 10\sqrt{2}$ 　　c.　$\sqrt{6} + \sqrt{10}$ 　　d.　$6\sqrt{5}$

　　e.　$3 + \sqrt{3}$ 　　f.　$4 + 11\sqrt{5}$ 　　g.　$^-1$ 　　h.　$11 + 6\sqrt{2}$

　　i.　$\sqrt{5}$ 　　j.　$\sqrt{3}$ 　　k.　$\dfrac{\sqrt{10}}{5}$ 　　l.　$2\sqrt{6}$

m. $\dfrac{\sqrt{6y}}{3}$ **n.** $\dfrac{\sqrt[3]{6}}{3}$ **o.** $1 + 2\sqrt{5}$ **p.** $1 + \sqrt{2}$

q. $\dfrac{2 - \sqrt{3}}{3}$ **r.** $\sqrt{5} + \sqrt{10}$ **s.** $\dfrac{5 + 3\sqrt{3}}{-2}$ **t.** $^-15 - 7\sqrt{5}$

u. $\dfrac{2 + \sqrt{10}}{-3}$ **v.** $\dfrac{^-1 + \sqrt{21}}{5}$

3. a. yes   **b.** no

Chapter 13 Sample Test

At the completion of Chapter 13 it is expected that you will be able to work the following problems.

13.1   Introduction to Roots, Radicals, and Irrational Numbers

1. In the symbol $\sqrt[5]{20}$ match the following:
    a. 5   **1.** radicand
    b. 20   **2.** radical sign
    c. $\sqrt[5]{}$ **3.** index
2. Simplify if possible:
    a. the square roots of 36   **b.** $\sqrt{100}$ **c.** $^-\sqrt{49}$
    d. $^-\sqrt{0.09}$ **e.** $\sqrt{^-25}$ **f.** $\sqrt[3]{27}$
    g. $\sqrt[4]{^-1}$ **h.** $\sqrt{7}$
3. Indicate for each number below if it is rational, irrational, or not real:
    a. $0.3\overline{3}$ **b.** π **c.** $\sqrt{9}$ **d.** $\sqrt{7}$ **e.** $\sqrt[3]{^-1}$ **f.** $\sqrt{^-4}$

13.2   Use of Tables to Approximate Irrational Numbers

4. Use the table of powers and roots on page 600 to find the following roots:
    a. $\sqrt{484}$ **b.** $\sqrt[3]{2{,}744}$ **c.** $\sqrt{17}$ **d.** $\sqrt[3]{62}$
    e. $\sqrt{700}$ **f.** $\sqrt[3]{0.050}$ **g.** $\sqrt[3]{7{,}000}$ **h.** $\sqrt{0.0841}$

13.3   Simplest Radical Form

5. Write in simplest radical form:
    a. $\sqrt{150}$ **b.** $\sqrt[3]{24}$ **c.** $2\sqrt{9a^2b^4c^7}$
    d. $\sqrt[3]{^-x^6y^2}$ **e.** $\sqrt{\dfrac{32}{25}}$ **f.** $\sqrt[3]{\dfrac{^-27}{8}}$
    g. $\sqrt{\dfrac{7}{5}}$ **h.** $\sqrt{\dfrac{9y^2}{8x}}$

13.4   Operations on Radical Expressions

In problems 6–8 perform any indicated operations if possible and write in simplest radical form.

6. **a.** $3\sqrt{5} - \sqrt{5}$ **b.** $2\sqrt{3} - \sqrt{2} + 5\sqrt{3} + 2\sqrt{2}$

    c. $\sqrt{12} + 3\sqrt{3}$ **d.** $5\sqrt{2} - \sqrt{8} - \sqrt{18}$

    e. $\sqrt{8x^2y} + x\sqrt{50y}$ **f.** $\sqrt{24} + \sqrt{\dfrac{2}{3}}$

7. **a.** $\sqrt{3} \cdot \sqrt{24}$ **b.** $(3x\sqrt{x})(x\sqrt{x^2})(\sqrt{x})$

    c. $\sqrt{2}(2 + \sqrt{2})$ **d.** $\sqrt{3}(\sqrt{3} + \sqrt{5})$

    e. $(1 + \sqrt{3})(3 - \sqrt{3})$ **f.** $(2 + \sqrt{5})(2 - \sqrt{5})$

8. **a.** $\dfrac{\sqrt{6}}{\sqrt{2}}$ **b.** $\dfrac{\sqrt[3]{2}}{\sqrt[3]{4}}$

    c. $\dfrac{3 + \sqrt{27}}{3}$ **d.** $\dfrac{2 + \sqrt{3}}{1 - \sqrt{3}}$

Chapter 13 Sample Test Answers

1. **a.** 3   **b.** 1   **c.** 2
2. **a.** 6 and ⁻6   **b.** 10   **c.** ⁻7
    d. ⁻0.3   **e.** not a real number   **f.** 3
    g. not a real number   **h.** $\sqrt{7}$ (not possible to simplify further)
3. **a.** rational   **b.** irrational   **c.** rational
    d. irrational   **e.** rational   **f.** not real
4. **a.** 22   **b.** 14   **c.** 4.123
    d. 3.958   **e.** 26.46   **f.** 0.3684
    g. 19.13   **h.** 0.29
5. **a.** $5\sqrt{6}$ **b.** $2\sqrt[3]{3}$ **c.** $6ab^2c^3\sqrt{c}$

    d. $-x^2\sqrt[3]{y^2}$ **e.** $\dfrac{4\sqrt{2}}{5}$ **f.** $\dfrac{-3}{2}$

    g. $\dfrac{\sqrt{35}}{5}$ **h.** $\dfrac{3y\sqrt{2x}}{4x}$

6. **a.** $2\sqrt{5}$ **b.** $7\sqrt{3} + \sqrt{2}$ **c.** $5\sqrt{3}$

    d. 0   **e.** $7x\sqrt{2y}$ **f.** $\dfrac{7\sqrt{6}}{3}$

7. **a.** $6\sqrt{2}$ **b.** $3x^4$ **c.** $2\sqrt{2} + 2$
    d. $3 + \sqrt{15}$ **e.** $2\sqrt{3}$ **f.** ⁻1

8. **a.** $\sqrt{3}$ **b.** $\dfrac{\sqrt[3]{4}}{2}$

    c. $1 + \sqrt{3}$ **d.** $\dfrac{5 + 3\sqrt{3}}{-2}$

Chapter 14

Quadratic Equations

14.1   Solving Quadratic Equations by Factoring

---

1   The standard form for a *quadratic or second-degree equation* in x is $ax^2 + bx + c = 0$, where a, b, and c are real numbers and $a \neq 0$. The restriction $a \neq 0$ is necessary, because if $a = 0$, the equation $ax^2 + bx + c = 0$ becomes a first-degree or linear equation. Some examples of quadratic equations are:

$$4x^2 - 3x + 7 = 0$$

$$2x^2 = 6x - 5$$

$$3x = 8x^2$$

$$4 - x^2 = 0$$

Notice that some of these equations are not in standard form, but they can be converted to standard form. Some equations that are not quadratic equations are:

$$x^3 - 7x^2 + 5 = 0 \qquad \text{(highest exponent} > 2)$$

$$4x = 3 \qquad \text{(no second-degree term)}$$

$$4x - 7 = 9x^4 \qquad \text{(highest exponent} > 2)$$

Q1   Write "yes" if the equation is quadratic and "no" otherwise:

a. $8x^2 - x + 11 = 0$ _____   **b.** $15 - 3x = 0$ _____

c. $2x = x^2$ _____   **d.** $x^3 - 7x^2 + 8x - 3 = 0$ _____

STOP   •   STOP   •   STOP   •   STOP   •   STOP   •   STOP   •   STOP   •   STOP   •   STOP

A1   **a.** yes   **b.** no
      c. yes   **d.** no

---

2   When a quadratic equation in x is written in standard form, the terms are arranged on one side in descending order. If a quadratic equation is not already in standard form, it is often necessary to write it in this form.

Example 1: Write $2x^2 = 7x - 5$ in standard form.

Solution

To write $2x^2 = 7x - 5$ in standard form, subtract $7x$ from both sides of the equation and add 5 to both sides of the equation.

555

$$2x^2 = 7x - 5$$
$$2x^2 - 7x = 7x - 5 - 7x$$
$$2x^2 - 7x = {}^-5$$
$$2x^2 - 7x + 5 = {}^-5 + 5$$
$$2x^2 - 7x + 5 = 0$$

Example 2: Write $5x = 2x^2$ in standard form.

Solution

$$5x = 2x^2$$

(*Note:* It is desirable to have the coefficient on x^2 positive. Therefore, when placing a quadratic equation in standard form, the coefficient of the x^2 term is always made positive.)

$$5x - 5x = 2x^2 - 5x$$
$$0 = 2x^2 - 5x$$
$$2x^2 - 5x = 0$$

Example 3: Write $3x - 9x^2 = 2$ in standard form.

Solution

$$3x - 9x^2 = 2$$
$$3x - 9x^2 + 9x^2 = 2 + 9x^2$$
$$3x = 2 + 9x^2$$
$$3x - 3x = 2 + 9x^2 - 3x$$
$$0 = 9x^2 - 3x + 2$$
$$9x^2 - 3x + 2 = 0$$

Q2   Write in standard form:
   a. $7x^2 = 3x + 8$ **b.** $6 = 5x^2 - 3x$ **c.** $3x = 4x^2$

   d. $5x - 8x^2 = 4$

STOP   •   STOP   •   STOP   •   STOP   •   STOP   •   STOP   •   STOP   •   STOP   •   STOP

A2   **a.** $7x^2 - 3x - 8 = 0$ **b.** $5x^2 - 3x - 6 = 0$ **c.** $4x^2 - 3x = 0$
      d. $8x^2 - 5x + 4 = 0$

3   A solution to a quadratic equation is any value for the variable that converts the equation into a true statement. In the set of real numbers, a quadratic equation can have two solutions, one solution, or no solutions.

Example 1: Verify that $x = 6$ or $x = {}^-6$ are solutions to $x^2 - 36 = 0$.

Solution

Verification for $x = 6$:

$$x^2 - 36 = 0$$
$$(6)^2 - 36 \overset{?}{=} 0$$
$$36 - 36 \overset{?}{=} 0$$
$$0 = 0$$

Verification for $x = {}^-6$:

$$x^2 - 36 = 0$$
$$({}^-6)^2 - 36 \overset{?}{=} 0$$
$$36 - 36 \overset{?}{=} 0$$
$$0 = 0$$

Example 2. Find the solutions for $x^2 = 6 - 5x$ from the set $\{{}^-1, 1, 6, 3, {}^-6\}$.

Solution

$x = {}^-6, x = 1$

$x = {}^-1$:

$$x^2 = 6 - 5x$$
$$({}^-1)^2 \overset{?}{=} 6 - 5({}^-1)$$
$$1 \overset{?}{=} 6 + 5$$
$$1 = 11 \quad \text{(false)}$$

$x = 1$:

$$x^2 = 6 - 5x$$
$$(1)^2 \overset{?}{=} 6 - 5(1)$$
$$1 \overset{?}{=} 6 - 5$$
$$1 = 1 \quad \text{(true)}$$

$x = 6$:

$$x^2 = 6 - 5x$$
$$(6)^2 \overset{?}{=} 6 - 5(6)$$
$$36 \overset{?}{=} 6 - 30$$
$$36 = {}^-24 \quad \text{(false)}$$

$x = 3$:

$$x^2 = 6 - 5x$$
$$(3)^2 \overset{?}{=} 6 - 5(3)$$
$$9 \overset{?}{=} 6 - 15$$
$$9 = {}^-9 \quad \text{(false)}$$

$x = {}^-6$:

$$x^2 = 6 - 5x$$
$$({}^-6)^2 \overset{?}{=} 6 - 5({}^-6)$$
$$36 \overset{?}{=} 6 + 30$$
$$36 = 36 \quad \text{(true)}$$

Q3   Verify that $y = 2$ and $y = {}^-5$ are solutions for $y^2 + 3y - 10 = 0$.

STOP • **STOP** • **STOP** • **STOP** • **STOP** • **STOP** • **STOP** • **STOP** • **STOP**

A3 $y = 2$:

$$y^2 + 3y - 10 = 0$$
$$(2)^2 + 3(2) - 10 \overset{?}{=} 0$$
$$4 + 6 - 10 = 0 \quad \text{(true)}$$

$y = {}^-5$:

$$y^2 + 3y - 10 = 0$$
$$({}^-5)^2 + 3({}^-5) - 10 \overset{?}{=} 0$$
$$25 - 15 - 10 = 0 \quad \text{(true)}$$

Q4   Find the solution(s) for $x^2 - 6x = {}^-9$ from the set $\{2, {}^-1, 3, 4\}$.

STOP • **STOP** • **STOP** • **STOP** • **STOP** • **STOP** • **STOP** • **STOP** • **STOP**

A4 $x = 3$ (*Note:* This is a one solution quadratic.)

Q5   Verify that $x = 2 - \sqrt{3}$ and $x = 2 + \sqrt{3}$ are solutions for $x^2 - 4x + 1 = 0$.

STOP • **STOP** • **STOP** • **STOP** • **STOP** • **STOP** • **STOP** • **STOP** • **STOP**

A5

$$x = 2 - \sqrt{3}:$$
$$x^2 - 4x + 1 = 0$$
$$(2 - \sqrt{3})^2 - 4(2 - \sqrt{3}) + 1 \stackrel{?}{=} 0$$
$$4 - 4\sqrt{3} + 3 - 8 + 4\sqrt{3} + 1 = 0 \quad \text{(true)}$$

$$x = 2 + \sqrt{3}:$$
$$x^2 - 4x + 1 = 0$$
$$(2 + \sqrt{3})^2 - 4(2 + \sqrt{3}) + 1 \stackrel{?}{=} 0$$
$$4 + 4\sqrt{3} + 3 - 8 - 4\sqrt{3} + 1 = 0 \quad \text{(true)}$$

---

4   The solution of quadratic equations by factoring is based on the following property: The product of two real numbers is zero if and only if one of the factors is zero. That is, for all real numbers a and b, $ab = 0$ if and only if $a = 0$ or $b = 0$.

Examples:

1. $(x - 3)(x + 7) = 0$

The equation is expressed as the product of two factors equal to zero; hence, $(x - 3)(x + 7) = 0$ if and only if $x - 3 = 0$ or $x + 7 = 0$.

2. $5x(2x + 9) = 0$

The equation is expressed as the product of two factors equal to zero; hence, $5x(2x + 9) = 0$ if and only if $5x = 0$ or $2x + 9 = 0$.

Q6   **a.** If $(x - 5)(x + 5) = 0$, then _____ $= 0$ or _____ $= 0$.

         b. If $^-4x(3x + 7) = 0$, then _____ or _____ .

STOP • **STOP** • **STOP** • **STOP** • **STOP** • **STOP** • **STOP** • **STOP** • **STOP**

A6   **a.** $x - 5, x + 5$ **b.** $^-4x = 0, 3x + 7 = 0$

---

5   Many quadratic equations can be solved by the method of factoring using the following steps:

1. Write the equation in standard form.
2. Factor the left side of the equation.
3. Set each of the factors equal to zero.
4. Solve each of the resulting first-degree equations.
5. Check the solution(s) by substitution into the original equation.

Example 1: Solve $3x^2 = 12x$ by factoring.

Solution

$$3x^2 = 12x$$

Step 1: $3x^2 - 12x = 12x - 12x$

$$3x^2 - 12x = 0$$

Step 2: $3x(x - 4) = 0$

Step 3: $\qquad 3x = 0 \quad$ or $\quad x - 4 = 0$

Step 4: $\qquad \dfrac{3x}{3} = \dfrac{0}{3} \qquad\qquad x = 4$

$$x = 0$$

Step 5: Check: $\qquad 3x^2 = 12x \qquad\qquad 3x^2 = 12x$

$\qquad\qquad\qquad 3(0)^2 \overset{?}{=} 12(0) \quad 3(4)^2 \overset{?}{=} 12(4)$

$\qquad\qquad\qquad\qquad 0 = 0 \qquad\quad 3(16) \overset{?}{=} 48$

$$48 = 48$$

Example 2: Solve $^-12x = 15x^2$ by factoring.

Solution

$$^-12x = 15x^2$$

Step 1: $^-12x + 12x = 15x^2 + 12x$

$$0 = 15x^2 + 12x$$

$$15x^2 \mid 12x = 0$$

Step 2: $3x(5x + 4) = 0$

Step 3: $\qquad 3x = 0 \quad$ or $\quad 5x + 4 = 0$

Step 4: $\qquad x = 0 \qquad\qquad 5x = \ ^-4$

$$x = \dfrac{^-4}{5}$$

Step 5: Check: $\qquad ^-12x = 15x^2 \qquad\qquad ^-12x = 15x^2$

$\qquad\qquad 12(0) \overset{?}{=} 15(0)^2 \quad ^-12\left(\dfrac{^-4}{5}\right) \overset{?}{=} 15\left(\dfrac{^-4}{5}\right)^2$

$\qquad\qquad\qquad 0 \overset{?}{=} 15(0) \qquad\qquad \dfrac{48}{5} \overset{?}{=} 15\left(\dfrac{16}{25}\right)$

$\qquad\qquad\qquad 0 = 0 \qquad\qquad\qquad \dfrac{48}{5} = \dfrac{48}{5}$

Q7   Solve by factoring:

a. $x^2 + 5x = 0$ $\qquad\qquad\qquad\qquad$ **b.** $7x = 21x^2$

STOP • STOP • STOP • STOP • STOP • STOP • STOP • STOP • STOP

A7

a. $x = 0$ or $x = {}^-5$:

$$x^2 + 5x = 0$$
$$x(x + 5) = 0$$
$$x = 0 \quad \text{or} \quad x + 5 = 0$$
$$x = {}^-5$$

b. $x = 0$ or $x = \dfrac{1}{3}$:

$$7x = 21x^2$$
$$7x - 7x = 21x^2 - 7x$$
$$0 = 21x^2 - 7x$$
$$21x^2 - 7x = 0$$
$$7x(3x - 1) = 0$$
$$7x = 0 \quad \text{or} \quad 3x - 1 = 0$$
$$x = 0 \qquad\qquad 3x = 1$$
$$x = \dfrac{1}{3}$$

6

Many quadratic equations involve polynomials that are factorable into the product of two binomials.

Example 1: Solve $x^2 - 18x + 32 = 0$ by factoring.

Solution

$$x^2 - 18x + 32 = 0$$
$$(x - 16)(x - 2) = 0$$
$$x - 16 = 0 \quad \text{or} \quad x - 2 = 0$$
$$x = 16 \qquad\qquad x = 2$$

Check:

$$x^2 - 18x + 32 = 0 \qquad\qquad x^2 - 18x + 32 = 0$$
$$(16)^2 - 18(16) + 32 \overset{?}{=} 0 \qquad (2)^2 - 18(2) + 32 \overset{?}{=} 0$$
$$256 - 288 + 32 \overset{?}{=} 0 \qquad\qquad 4 - 36 + 32 \overset{?}{=} 0$$
$$0 = 0 \qquad\qquad\qquad 0 = 0$$

The solutions are $x = 16$ or $x = 2$.

Example 2: Solve $y^2 + 10 = 7y$ by factoring.

Solution

$$y^2 + 10 = 7y$$
$$y^2 + 10 - 7y = 7y - 7y$$
$$y^2 + 10 - 7y = 0$$
$$y^2 - 7y + 10 = 0$$
$$(y - 5)(y - 2) = 0$$
$$y - 5 = 0 \quad \text{or} \quad y - 2 = 0$$
$$y = 5 \qquad\qquad y = 2$$

Check:

$$y^2 + 10 = 7y \qquad\qquad y^2 + 10 = 7y$$
$$(5)^2 + 10 \overset{?}{=} 7(5) \qquad (2)^2 + 10 \overset{?}{=} 7(2)$$
$$35 = 35 \qquad\qquad 14 = 14$$

The solutions are $y = 5$ or $y = 2$.

Q8   Solve by factoring:
         a. $x^2 + 11x + 30 = 0$ **b.** $3x^2 = 2 - x$

STOP • STOP • STOP • STOP • STOP • STOP • STOP • STOP • STOP

A8   **a.** $x = {}^-5$ or $x = {}^-6$: $x^2 + 11x + 30 = 0$

$$(x + 5)(x + 6) = 0$$
$$x + 5 = 0 \quad \text{or} \quad x + 6 = 0$$
$$x = {}^-5 \qquad\qquad x = {}^-6$$

         b. $x = \dfrac{2}{3}$ or $x = {}^-1$: $3x^2 = 2 - x$

$$3x^2 + x = 2 - x + x$$
$$3x^2 + x = 2$$
$$3x^2 + x - 2 = 2 - 2$$
$$3x^2 + x - 2 = 0$$
$$(3x - 2)(x + 1) = 0$$
$$3x - 2 = 0 \quad \text{or} \quad x + 1 = 0$$
$$3x = 2 \qquad\qquad x = {}^-1$$
$$x = \frac{2}{3}$$

---

7   Quadratic equations are also often solved by factoring the difference of two squares.

Example 1: Solve $x^2 = 25$ by factoring.

Solution

$$x^2 = 25$$
$$x^2 - 25 = 0$$
$$(x - 5)(x + 5) = 0$$
$$x - 5 = 0 \quad \text{or} \quad x + 5 = 0$$
$$x = 5 \qquad\qquad x = {}^-5$$

Check: $x^2 = 25 \qquad x^2 = 25$

$$(5)^2 \overset{?}{=} 25 \qquad ({}^-5)^2 \overset{?}{=} 25$$
$$25 = 25 \qquad\quad 25 = 25$$

The solutions are $x = 5$ or $x = {}^-5$. The solutions may be written $x = \pm 5$. (Read: plus or minus 5.)

Example 2: Solve $36 = 9x^2$ by factoring.

Solution

$$36 = 9x^2$$
$$0 = 9x^2 - 36$$
$$9x^2 - 36 = 0$$
$$9(x^2 - 4) = 0$$
$$9(x - 2)(x + 2) = 0$$
$$x - 2 = 0 \quad \text{or} \quad x + 2 = 0$$
$$x = 2 \qquad\qquad x = {}^-2$$

Notice that the constant factor 9 can be ignored because it does not influence the values of x that will produce zero factors.

Check: $36 = 9x^2$ $36 = 9x^2$
$36 \stackrel{?}{=} 9(2)^2$ $36 \stackrel{?}{=} 9({}^-2)^2$
$36 = 9(4)$ $36 = 9(4)$

The solutions are $x = \pm 2$.

Q9   Solve by factoring:
        a. $x^2 - 81 = 0$ **b.** $16 = 49x^2$

STOP • **STOP** • **STOP** • **STOP** • **STOP** • **STOP** • **STOP** • **STOP** • **STOP**

A9   **a.** $x = \pm 9$: $x^2 - 81 = 0$
$$(x - 9)(x + 9) = 0$$
$$x - 9 = 0 \quad \text{or} \quad x + 9 = 0$$
$$x = 9 \qquad\qquad x = {}^-9$$

        b. $x = \dfrac{\pm 4}{7}$: $16 = 49x^2$
$$0 = 49x^2 - 16$$
$$49x^2 - 16 = 0$$
$$(7x - 4)(7x + 4) = 0$$
$$7x - 4 = 0 \quad \text{or} \quad 7x + 4 = 0$$
$$7x = 4 \qquad\qquad 7x = {}^-4$$
$$x = \frac{4}{7} \qquad\qquad x = \frac{{}^-4}{7}$$

Q10   Solve by factoring:
           a. $4x = 7x^2$ **b.** $3x^2 + x = 2$

           c. $1 = 9x^2$ **d.** $7x - 3 = 2x^2$

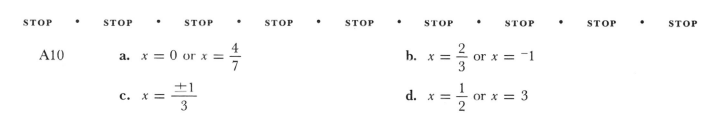

STOP • **STOP** • **STOP** • **STOP** • **STOP** • **STOP** • **STOP** • **STOP** • **STOP**

A10   **a.** $x = 0$ or $x = \dfrac{4}{7}$ **b.** $x = \dfrac{2}{3}$ or $x = {}^{-}1$

           c. $x = \dfrac{\pm 1}{3}$ **d.** $x = \dfrac{1}{2}$ or $x = 3$

---

8   When a quadratic equation is factorable as a perfect square trinomial, the equation has only one solution.

           Example 1: Solve $4x^2 - 12x + 9 = 0$ by factoring.

           Solution

$$4x^2 - 12x + 9 = 0$$
$$(2x - 3)(2x - 3) = 0$$

$$2x - 3 = 0 \qquad \text{or} \qquad 2x - 3 = 0$$
$$2x = 3 \qquad\qquad\qquad 2x = 3$$
$$x = \frac{3}{2} \qquad\qquad\qquad x = \frac{3}{2}$$

           The solution is $x = \dfrac{3}{2}$.

           Example 2: Solve $^{-}10x - 1 = 25x^2$ by factoring.

Solution

$$0 = 25x^2 + 10x + 1$$

$$25x^2 + 10x + 1 = 0$$

$$(5x + 1)^2 = 0$$

$$5x + 1 = 0 \qquad \text{or} \qquad 5x + 1 = 0$$

$$5x = {}^-1 \qquad \text{This solution is the same and need not be solved.}$$

$$x = \frac{{}^-1}{5}$$

The solution is $x = \dfrac{{}^-1}{5}$.

Q11   Solve by factoring:
   a. $9x^2 - 42x + 49 = 0$ **b.** $25x^2 = {}^-64 - 80x.$

STOP • **STOP** • **STOP** • **STOP** • **STOP** • **STOP** • **STOP** • **STOP** • **STOP**

A11   **a.** $x = \dfrac{7}{3}$ **b.** $x = \dfrac{{}^-8}{5}$

---

9   Not all quadratic equations can be solved by factoring. The following sections will develop solution procedures when factoring is not possible. For the purposes of this section, if the quadratic equation cannot be factored over the set of integers, it should be labeled "no solution by factoring."

Example 1: Solve $x^2 = {}^-16$ by factoring.

Solution

$x^2 = {}^-16$
$x^2 + 16 = 0$

No solution by factoring. This quadratic equation cannot be solved by factoring, because $x^2 + 16$ is prime (over the set of integers).

Example 2: Solve ${}^-4 = x^2 - 3x$ by factoring.

Solution

$$\begin{aligned}
{}^-4 &= x^2 - 3x \\
0 &= x^2 - 3x + 4
\end{aligned}$$

$x^2 - 3x + 4 = 0$

No solution by factoring. This quadratic equation cannot be solved by factoring, because $x^2 - 3x + 4$ is prime.

Q12   Solve by factoring:
       a. $5x^2 + 14x = 3$ **b.** $7 = x^2$

STOP • **STOP** • **STOP** • **STOP** • **STOP** • **STOP** • **STOP** • **STOP** • **STOP**

A12   **a.** $x = \dfrac{1}{5}$ or $x = {}^-3$:

$$5x^2 + 14x = 3$$
$$5x^2 + 14x - 3 = 0$$
$$(5x - 1)(x + 3) = 0$$
$$5x - 1 = 0 \quad \text{or} \quad x + 3 = 0$$
$$5x = 1 \qquad\qquad x = {}^-3$$
$$x = \frac{1}{5}$$

       b. no solution by factoring:

$$7 = x^2$$
$$0 = x^2 - 7$$
$$x^2 - 7 = 0$$

Q13   Solve by factoring:
       a. $x^2 - 13x + 12 = 0$ **b.** $64x^2 = 1$

       c. ${}^-5x = x^2$ **d.** $x^2 + 1 = x$

STOP • **STOP** • **STOP** • **STOP** • **STOP** • **STOP** • **STOP** • **STOP** • **STOP**

A13   **a.** $x = 12$ or $x = 1$ **b.** $x = \dfrac{\pm 1}{8}$

       c. $x = 0$ or $x = {}^-5$ **d.** no solution by factoring

10 Quadratic equations can be used to solve certain verbal problems. The steps to use are as follows:

Step 1: List the parts of the problem mathematically; use a variable for the unknown quantity. Give a pictorial representation if possible.

Step 2: Write an equation that involves the known and unknown quantities.

Step 3: Solve the resulting equation.

Step 4: List and check the solutions in the words of the problem.

Example: One positive number is five more than another positive number. If the product of the numbers is 84, find each of the numbers.

Solution

Step 1: Let $x =$ the first positive number, $x + 5 =$ the second positive number.

Step 2: The product of the numbers is 84; hence, $x(x + 5) = 84$.

Step 3:
$$x^2 + 5x = 84$$
$$x^2 + 5x - 84 = 0$$
$$(x - 7)(x + 12) = 0$$
$$x - 7 = 0 \quad \text{or} \quad x + 12 = 0$$
$$x = 7 \qquad\qquad x = {}^-12$$

Step 4: The solution $x = {}^-12$ is rejected, because the problem asks for positive numbers. Therefore, $x = 7$ is one number and $x + 5 = 7 + 5 = 12$ is the second number.

Check: 12 is five more than 7 and the product of the numbers, $(7)(12)$, is 84.

Q14 One positive number is one more than twice another positive number. If the product of the numbers is 78, find each of the numbers.

STOP • **STOP** • **STOP** • **STOP** • **STOP** • **STOP** • **STOP** • **STOP** • **STOP**

A14 6 and 13:
$$x = \text{one number}$$
$$2x + 1 = \text{second number}$$
$$x(2x + 1) = 78$$
$$2x^2 + x = 78$$
$$2x^2 + x - 78 = 0$$
$$(2x + 13)(x - 6) = 0$$
$$2x + 13 = 0 \quad\quad \text{or} \quad x - 6 = 0$$
$$2x = {}^-13 \qquad\qquad x = 6$$
$$x = \frac{{}^-13}{2}$$

The solution $x = \dfrac{-13}{2}$ is rejected, because the desired number is positive. Therefore, $x = 6$ is one number and $2x + 1 = 2(6) + 1 = 13$ is the second number. (Check not included.)

11

Example: The height of a right triangle is two more than twice its base. The area of the triangle is 30 square centimeters. Find the height and base of the triangle.

Solution

Step 1: $x = $ the base of the triangle.

            $2x + 2 = $ the height of the triangle.

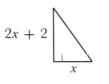

$2x + 2$

x

Step 2:   The area formula for a triangle is $A = \dfrac{1}{2} bh$. Therefore,

$$30 = \frac{1}{2} \cdot x(2x + 2)$$

$$30 = \frac{x}{2}(2x + 2)$$

$$2(30) = 2\left[\frac{x}{2}(2x + 2)\right]$$

$$60 = x(2x + 2)$$

$$60 = 2x^2 + 2x$$

$$0 = 2x^2 + 2x - 60$$

$$0 = x^2 + x - 30 \qquad \text{(dividing both sides by 2)}$$

$$x^2 + x - 30 = 0$$

$$(x + 6)(x - 5) = 0$$

$$x + 6 = 0 \qquad \text{or} \qquad x - 5 = 0$$

$$x = {}^-6 \qquad\qquad\qquad x = 5$$

The solution $x = {}^-6$ is rejected, because a triangle cannot have a negative dimension. Thus, the base of the triangle, x, is 5 centimeters and the height of the triangle $= 2x + 2 = 2(5) + 2 = 12$ centimeters.

Step 4: Check: Since 12 is $2(5) + 2$, the height is two more than twice its base. The area of the triangle is shown to be 30 square centimeters by using the formula $A = \dfrac{1}{2} bh$.

$$A = \frac{1}{2} bh$$

$$A = \frac{1}{2}(5)(12)$$

$$A = 30 \text{ square centimeters}$$

Q15   The length of a rectangle exceeds its width by six. If the area of the rectangle is 91 square meters, find the dimensions of the rectangle.

STOP • STOP • STOP • STOP • STOP • STOP • STOP • STOP • STOP

A15   7 by 13 m:

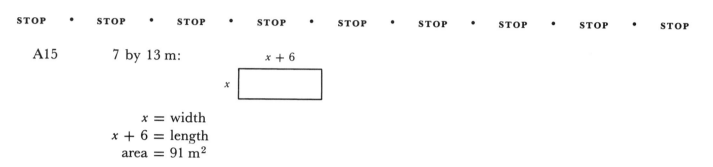

x = width
$x + 6$ = length
area = 91 m^2

Since the area formula for a rectangle is $A = lw$, the equation is

$$91 = x(x + 6)$$
$$91 = x^2 + 6x$$
$$0 = x^2 + 6x - 91$$
$$x^2 + 6x - 91 = 0$$
$$(x + 13)(x - 7) = 0$$
$$x + 13 = 0 \quad \text{or} \quad x - 7 = 0$$
$$x = {}^-13 \qquad\qquad x = 7$$

The solution $x = {}^-13$ is rejected. Thus, the width, x, is 7 meters and the length $= x + 6 = 7 + 6 = 13$ meters.

Check: The length, 13 meters, exceeds the width, 7 meters, by six. The area of the rectangle $= 7(13) = 91$ square meters.

This completes the instruction for this section.

14.1   Exercises

1. Solve by factoring:
   a. $5x^2 - 10x = 0$
   b. $x^2 + 1 = 0$
   c. $^-x^2 = {}^-x$
   d. $^-36x = 24x^2$
2. Solve by factoring:
   a. $x^2 - 12x + 35 = 0$
   b. $x^2 = {}^-5x + 36$
   c. $x = 42 - x^2$
   d. $24 - 14x + x^2 = 0$
   e. $x^2 + 21 = {}^-10x$
   f. $x^2 - 18x + 81 = 0$
3. Solve by factoring:
   a. $x^2 - 1 = 0$
   b. $16x^2 - 100 = 0$
   c. $6x^2 = 11$
   d. $4 = 49x^2$

4. Solve by factoring:
   a. $x^2 - 10x + 25 = 0$
   b. $4x - 3x^2 = 0$
   c. $36 = x^2$
   d. $5 = x^2$
   e. $2x^2 = 6x + 40$
   f. $^-12 + 2x^2 = 5x$
   g. $3x^2 - 23x + 14 = 0$
   h. $9x^2 - 45x = 0$

5. One positive number is one less than the square of another positive number. If their sum is 41, find the numbers.

6. The base of a right triangle is six less than twice the height. If the area of the triangle is 88 cm^2, find the dimensions of the triangle.

7. One natural number is three times the square of another. If the difference of the numbers is 44, find the numbers.

14.1   Exercise Answers

1. a. $x = 0$ or $x = 2$
   b. no solution by factoring

   c. $x = 0$ or $x = 1$
   d. $x = 0$ or $x = \dfrac{-3}{2}$

2. a. $x = 5$ or $x = 7$
   b. $x = ^-9$ or $x = 4$
   c. $x = ^-7$ or $x = 6$

   d. $x = 12$ or $x = 2$
   e. $x = ^-7$ or $x = ^-3$
   f. $x = 9$

3. a. $x = \pm 1$
   b. $x = \dfrac{\pm 5}{2}$

   c. no solution by factoring
   d. $x = \dfrac{\pm 2}{7}$

4. a. $x = 5$
   b. $x = 0$ or $x = \dfrac{4}{3}$
   c. $x = \pm 6$

   d. no solution by factoring
   e. no solution by factoring
   f. $x = \dfrac{-3}{2}$ or $x = 4$

   g. $x = \dfrac{2}{3}$ or $x = 7$
   h. $x = 0$ or $x = 5$

5. 6 and 35
6. height = 11 centimeters, base = 16 centimeters
7. 4 and 48

14.2   Solving Quadratic Equations by Completing the Square

1   The technique called "completing the square" can be used to solve quadratic equations. The perfect square trinomials discussed here will be similar to those studied in Section 11.5. However, in this section, trinomials will be converted into perfect squares only when the coefficient of x^2 is 1.

Some examples of perfect square trinomials are:

$(x - 4)^2 = x^2 - 8x + 16$
$(x + 3)^2 = x^2 + 6x + 9$
$(x + 7)^2 = x^2 + 14x + 49$

Recall that a trinomial $ax^2 + bx + c$ with $a = 1$ is a perfect square if the constant term is the square of one half the coefficient of x.

For the examples above, this relationship is shown as follows:

$$(x - 4)^2 = x^2 - 8x + 16$$

$$\left(\frac{1}{2} \cdot {}^-8\right)^2 = 16$$

$$(x + 3)^2 = x^2 + 6x + 9$$

$$\left(\frac{1}{2} \cdot 6\right)^2 = 9$$

$$(x + 7)^2 = x^2 + 14x + 49$$

$$\left(\frac{1}{2} \cdot 14\right)^2 = 49$$

In general, the polynomial $x^2 + bx + c$ is a perfect square if $\left(\frac{1}{2} \cdot b\right)^2 = c$.

Q1 The trinomial $x^2 - 10x + k$ is a perfect square if $\left(\frac{1}{2} \cdot \underline{\quad}\right)^2 = k$. Find k.

STOP • **STOP** • **STOP** • **STOP** • **STOP** • **STOP** • **STOP** • **STOP** • **STOP**

A1 $^-10, k = 25$: $\left(\frac{1}{2} \cdot {}^-10\right)^2 = (^-5)^2 = 25$

Q2 The trinomial $x^2 + 2x + k$ is a perfect square if $\left(\frac{1}{2} \cdot \underline{\quad}\right)^2 = k$. Find k.

STOP • **STOP** • **STOP** • **STOP** • **STOP** • **STOP** • **STOP** • **STOP** • **STOP**

A2 $2, k = 1$: $\left(\frac{1}{2} \cdot 2\right)^2 = 1^2 = 1$

2 Once the value of k is determined, the polynomial is factored as a perfect square trinomial.

Examples:

1. $x^2 + 18x + k = x^2 + 18x + 81 = (x + 9)^2$

    $$\left(\frac{1}{2} \cdot 18\right)^2 = 81$$

2. $x^2 - 12x + k = x^2 - 12x + 36 = (x - 6)^2$

    $$\left(\frac{1}{2} \cdot {}^-12\right)^2 = 36$$

3. $y^2 - 2y + k = y^2 - 2y + 1 = (y - 1)^2$

    $$\left(\frac{1}{2} \cdot {}^-2\right)^2 = 1$$

Q3 Find the value of k that makes the trinomial a perfect square and factor the result:

a. $x^2 + 4x + k = x^2 + 4x + \underline{\quad} = (\quad)^2$

b. $y^2 - 16y + k = y^2 - 16y + \underline{\quad} = (\quad)^2$

c. $x^2 - 20x + k = x^2 - 20x + \underline{\hspace{1cm}} = ($ $)^2$

d. $p^2 + 6p + k = p^2 + 6p + \underline{\hspace{1cm}} = ($ $)^2$

STOP • **STOP** • **STOP** • **STOP** • **STOP** • **STOP** • **STOP** • **STOP** • **STOP**

A3   **a.** $4, x + 2$ **b.** $64, y - 8$ **c.** $100, x - 10$ **d.** $9, p + 3$

3

Quadratic equations can also be solved by completing the square. The following steps are used to solve a quadratic equation by the method of completing the square.

1. Write the equation in the form

   $ax^2 + bx = c$

2. If $a \neq 1$, divide both sides of the equation by a.

3. Find $\left(\dfrac{1}{2} \cdot \dfrac{b}{a}\right)^2$ and *add it to* both *sides of the equation.*

4. Factor the left side as a perfect square and combine terms on the right side of the equation.

5. Take the square root of both sides of the equation.

6. Solve for x.

7. Write the two solutions.

Example 1: Solve $x^2 + 8x - 9 = 0$ by completing the square.

Solution

$x^2 + 8x - 9 = 0$

Step 1: $x^2 + 8x = 9$

Step 2: Not necessary, because $a = 1$

Step 3: $x^2 + 8x + 16 = 9 + 16$ notice that 16 must be added to *both* sides to form an equivalent equation

Step 4: $\quad (x + 4)^2 = 25$

Step 5: $\pm\sqrt{(x + 4)^2} = \pm\sqrt{25}$

$\qquad\qquad x + 4 = \pm 5$

Step 6: $\quad x + 4 - 4 = \pm 5 - 4$

$\qquad x = {}^+5 - 4 \qquad$ or $\qquad x = {}^-5 - 4$

$\qquad x = 1 \qquad\qquad\qquad\quad x = {}^-9$

Therefore, the solutions are $x = 1$ or $x = {}^-9$.

(*Note:* The $\pm$ on the left side of the equation in step 5 is usually not included, because using it leads to no new solutions to the equation.)

Example 2: Solve $2x^2 + 4x - 6 = 0$ by completing the square.

Solution

$2x^2 + 4x - 6 = 0$

Step 1: $\quad 2x^2 + 4x = 6$

Step 2: $\quad \dfrac{2x^2}{2} + \dfrac{4x}{2} = \dfrac{6}{2}$

$\qquad\qquad x^2 + 2x = 3$

Step 3: $x^2 + 2x + 1 = 3 + 1$

Step 4: $\quad (x + 1)^2 = 4$

Step 5: $\quad \sqrt{(x + 1)^2} = \pm\sqrt{4}$
$$x + 1 = \pm 2$$

Step 6: $\quad\quad x = {}^-1 \pm 2$

Step 7: $\quad\quad x = {}^-1 + 2 \quad$ or $\quad x = {}^-1 - 2$
$$x = 1 \quad\quad\quad\quad x = {}^-3$$

The solutions are $x = 1$ or $x = {}^-3$.

Q4　　　Solve $x^2 - 2x - 35 = 0$ by completing the square.

STOP • STOP • STOP • STOP • STOP • STOP • STOP • STOP • STOP

A4　　　$x = 7$ or $x = {}^-5$:　$x^2 - 2x - 35 = 0$
$$x^2 - 2x = 35$$
$$x^2 - 2x + 1 = 35 + 1$$
$$(x - 1)^2 = 36$$
$$\sqrt{(x - 1)^2} = \pm\sqrt{36}$$
$$x - 1 = \pm 6$$
$$x = 1 \pm 6$$
$$x = 1 + 6 \quad \text{or} \quad x = 1 - 6$$
$$x = 7 \quad\quad\quad\quad x = {}^-5$$

Q5　　　Solve $2x^2 - 12x + 10 = 0$ by completing the square.

A5 $\qquad x = 5$ or $x = 1$: $2x^2 - 12x + 10 = 0$

$$2x^2 - 12x = {}^-10$$

$$\frac{2x^2}{2} - \frac{12x}{2} = \frac{{}^-10}{2}$$

$$x^2 - 6x = {}^-5$$

$$x^2 - 6x + 9 = {}^-5 + 9$$

$$(x - 3)^2 = 4$$

$$\sqrt{(x - 3)^2} = \pm\sqrt{4}$$

$$x - 3 = \pm 2$$

$$x = 3 \pm 2$$

$$x = 3 + 2 \qquad \text{or} \qquad x = 3 - 2$$

$$x = 5 \qquad\qquad\qquad x = 1$$

---

4 $\qquad$ To complete the square on $2x^2 + 5x - 3 = 0$, proceed as follows:

$$2x^2 + 5x = 3$$

$$\frac{2x^2}{2} + \frac{5x}{2} = \frac{3}{2}$$

$$x^2 + \frac{5}{2}x = \frac{3}{2}$$

Since

$$\left(\frac{1}{2} \cdot \frac{5}{2}\right)^2 = \left(\frac{5}{4}\right)^2 = \frac{25}{16}, \text{ add } \frac{25}{16} \text{ to both sides.}$$

$$x^2 + \frac{5}{2}x + \frac{25}{16} = \frac{3}{2} + \frac{25}{16}$$

$$\left(x + \frac{5}{4}\right)^2 = \frac{24}{16} + \frac{25}{16}$$

$$\left(x + \frac{5}{4}\right)^2 = \frac{49}{16}$$

$$\sqrt{\left(x + \frac{5}{4}\right)^2} = \pm\sqrt{\frac{49}{16}}$$

$$x + \frac{5}{4} = \frac{\pm 7}{4}$$

$$x = \frac{{}^-5}{4} \pm \frac{7}{4}$$

$$x = \frac{{}^-5}{4} + \frac{7}{4} \qquad \text{or} \qquad x = \frac{{}^-5}{4} - \frac{7}{4}$$

$$= \frac{2}{4} \qquad\qquad\qquad\qquad = \frac{{}^-12}{4}$$

$$= \frac{1}{2} \qquad\qquad\qquad\qquad = {}^-3$$

The solutions are $x = \dfrac{1}{2}$ or $x = {}^-3$.

Q6 Solve $4x^2 - 7x + 3 = 0$ by completing the square.

STOP • STOP • STOP • STOP • STOP • STOP • STOP • STOP • STOP

A6 $x = 1$ or $x = \dfrac{3}{4}$: $\quad 4x^2 - 7x + 3 = 0$

$$4x^2 - 7x = {}^-3$$

$$\frac{4x^2}{4} - \frac{7x}{4} = \frac{{}^-3}{4}$$

$$x^2 - \frac{7}{4}x = \frac{{}^-3}{4}$$

$$x^2 - \frac{7}{4}x + \frac{49}{64} = \frac{{}^-3}{4} + \frac{49}{64}$$

$$\left(x - \frac{7}{8}\right)^2 = \frac{{}^-48}{64} + \frac{49}{64}$$

$$\left(x - \frac{7}{8}\right)^2 = \frac{1}{64}$$

$$\sqrt{\left(x - \frac{7}{8}\right)^2} = \pm\sqrt{\frac{1}{64}}$$

$$x - \frac{7}{8} = \frac{\pm 1}{8}$$

$$x = \frac{7}{8} \pm \frac{1}{8}$$

$$x = \frac{7}{8} + \frac{1}{8} \qquad \text{or} \qquad x = \frac{7}{8} - \frac{1}{8}$$

$$= \frac{8}{8} \qquad\qquad\qquad\qquad = \frac{6}{8}$$

$$= 1 \qquad\qquad\qquad\qquad = \frac{3}{4}$$

Q7   Solve $3x^2 + x - 10 = 0$ by completing the square.

STOP   •   **STOP**   •   **STOP**   •   **STOP**   •   **STOP**   •   **STOP**   •   **STOP**   •   **STOP**   •   **STOP**

A7 $x = \dfrac{5}{3}$ or $x = {}^{-}2$: $3x^2 + x - 10 = 0$

$$3x^2 + x = 10$$

$$\frac{3x^2}{3} + \frac{x}{3} = \frac{10}{3}$$

$$x^2 + \frac{1}{3}x = \frac{10}{3}$$

$$x^2 + \frac{1}{3}x + \frac{1}{36} = \frac{10}{3} + \frac{1}{36}$$

$$\left(x + \frac{1}{6}\right)^2 = \frac{121}{36}$$

$$\sqrt{\left(x + \frac{1}{6}\right)^2} = \pm\sqrt{\frac{121}{36}}$$

$$x + \frac{1}{6} = \frac{\pm 11}{6}$$

$$x = \frac{^{-}1}{6} \pm \frac{11}{6}$$

$$x = \frac{^{-}1}{6} + \frac{11}{6} \qquad \text{or} \qquad x = \frac{^{-}1}{6} - \frac{11}{6}$$

$$\qquad\quad = \frac{10}{6} \qquad\qquad\qquad\qquad\quad = \frac{^{-}12}{6}$$

$$\qquad\quad = \frac{5}{3} \qquad\qquad\qquad\qquad\quad\ = {}^{-}2$$

---

5   The technique of completing the square is most useful in solving quadratics that contain prime trinomials.

Example: Solve $x^2 + 4x - 1 = 0$.

Solution

$x^2 + 4x - 1$ is prime and cannot be factored using the techniques of Section 14.1. The equation can be solved by completing the square.

$$x^2 + 4x - 1 = 0$$
$$x^2 + 4x = 1$$
$$x^2 + 4x + 4 = 1 + 4$$
$$(x + 2)^2 = 5$$
$$\sqrt{(x + 2)^2} = \pm\sqrt{5}$$
$$x + 2 = \pm\sqrt{5}$$
$$x = {}^{-}2 \pm \sqrt{5}$$

The solutions are $x = {}^{-}2 + \sqrt{5}$ or $x = {}^{-}2 - \sqrt{5}$.

Q8   Solve $x^2 - 2x - 5 = 0$.

STOP • STOP • STOP • STOP • STOP • STOP • STOP • STOP • STOP

A8 $x = 1 + \sqrt{6}$ or $x = 1 - \sqrt{6}$: $x^2 - 2x - 5 = 0$
$$x^2 - 2x = 5$$
$$x^2 - 2x + 1 = 5 + 1$$
$$(x - 1)^2 = 6$$
$$\sqrt{(x - 1)^2} = \pm\sqrt{6}$$
$$x - 1 = \pm\sqrt{6}$$
$$x = 1 \pm \sqrt{6}$$
$$x = 1 + \sqrt{6} \quad \text{or} \quad x = 1 - \sqrt{6}$$

6   Many times it is necessary to simplify a radical that is part of the solution.

Example: Solve $3x^2 - 6x - 33 = 0$.

Solution

$$3x^2 - 6x - 33 = 0$$
$$3x^2 - 6x = 33$$
$$\frac{3x^2}{3} - \frac{6x}{3} = \frac{33}{3}$$
$$x^2 - 2x = 11$$
$$x^2 - 2x + 1 = 11 + 1$$
$$(x - 1)^2 = 12$$
$$\sqrt{(x - 1)^2} = \pm\sqrt{12}$$
$$x - 1 = \pm\sqrt{12}$$
$$x = 1 \pm \sqrt{12}$$

The radical $\sqrt{12}$ can be simplified.

$$x = 1 \pm \sqrt{4 \cdot 3}$$
$$x = 1 \pm 2\sqrt{3}$$

The solutions are $x = 1 + 2\sqrt{3}$ or $x = 1 - 2\sqrt{3}$.

Q9   Solve $2y^2 - 16y - 8 = 0$.

STOP • STOP • STOP • STOP • STOP • STOP • STOP • STOP • STOP

A9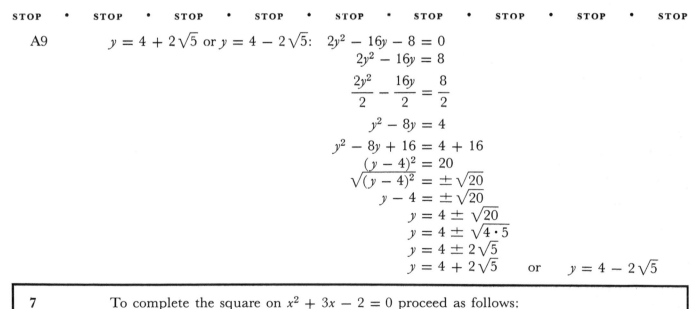

$y = 4 + 2\sqrt{5}$ or $y = 4 - 2\sqrt{5}$:

$$2y^2 - 16y - 8 = 0$$
$$2y^2 - 16y = 8$$
$$\frac{2y^2}{2} - \frac{16y}{2} = \frac{8}{2}$$
$$y^2 - 8y = 4$$
$$y^2 - 8y + 16 = 4 + 16$$
$$(y - 4)^2 = 20$$
$$\sqrt{(y - 4)^2} = \pm\sqrt{20}$$
$$y - 4 = \pm\sqrt{20}$$
$$y = 4 \pm \sqrt{20}$$
$$y = 4 \pm \sqrt{4 \cdot 5}$$
$$y = 4 \pm 2\sqrt{5}$$
$$y = 4 + 2\sqrt{5} \quad \text{or} \quad y = 4 - 2\sqrt{5}$$

7   To complete the square on $x^2 + 3x - 2 = 0$ proceed as follows:

$$x^2 + 3x - 2 = 0$$

$$x^2 + 3x = 2$$

Since $\left(\dfrac{1}{2} \cdot 3\right)^2 = \left(\dfrac{3}{2}\right)^2 = \dfrac{9}{4}$, add $\dfrac{9}{4}$ to both sides.

$$x^2 + 3x + \frac{9}{4} = 2 + \frac{9}{4}$$

$$\left(x + \frac{3}{2}\right)^2 = \frac{8}{4} + \frac{9}{4}$$

$$\left(x + \frac{3}{2}\right)^2 = \frac{17}{4}$$

$$\sqrt{\left(x + \frac{3}{2}\right)^2} = \pm\sqrt{\frac{17}{4}}$$

$$x + \frac{3}{2} = \frac{\pm\sqrt{17}}{2}$$

$$x = \frac{^-3}{2} \pm \frac{\sqrt{17}}{2}$$

$$x = \frac{^-3 \pm \sqrt{17}}{2}$$

The two solutions are $x = \dfrac{^-3 + \sqrt{17}}{2}$ or $x = \dfrac{^-3 - \sqrt{17}}{2}$.

Q10   Solve $y^2 + 5y + 5 = 0$

STOP • STOP • STOP • STOP • STOP • STOP • STOP • STOP • STOP

A10 $y = \dfrac{^-5 + \sqrt{5}}{2}$ or $y = \dfrac{^-5 - \sqrt{5}}{2}$:

$$y^2 + 5y + 5 = 0$$
$$y^2 + 5y = 5$$
$$y^2 + 5y + \frac{25}{4} = {^-5} + \frac{25}{4}$$
$$y^2 + 5y + \frac{25}{4} = \frac{^-20}{4} + \frac{25}{4}$$
$$\left(y + \frac{5}{2}\right)^2 = \frac{5}{4}$$
$$\sqrt{\left(y + \frac{5}{2}\right)^2} = \pm\sqrt{\frac{5}{4}}$$
$$y + \frac{5}{2} = \frac{\pm\sqrt{5}}{2}$$

$$y = \frac{^-5}{2} \pm \frac{\sqrt{5}}{2}$$

$$y = \frac{^-5 \pm \sqrt{5}}{2}$$

---

8 Some quadratic equations do not have real-number solutions.

Example: Solve $x^2 - 2x + 4 = 0$.

Solution

$$x^2 - 2x + 4 = 0$$
$$x^2 - 2x = {}^-4$$
$$x^2 - 2x + 1 = {}^-4 + 1$$
$$(x - 1)^2 = {}^-3$$
$$\sqrt{(x - 1)^2} = \pm\sqrt{^-3}$$
$$x - 1 = \pm\sqrt{^-3}$$
$$x = 1 \pm \sqrt{^-3}$$

Since $\sqrt{^-3}$ is not a real number, the solutions to the equation $x^2 - 2x + 4 = 0$ are not real numbers.

Solving quadratic equations of this sort is beyond the scope of this textbook. This topic will be studied, however, by the student who takes a course in intermediate algebra.

This completes the instruction for this section.

14.2 Exercises

1. What must k equal for each of the following to be a perfect square?
   a. $x^2 + 8x + k$ **b.** $x^2 + 7x + k$ **c.** $x^2 + 6x + k$
   d. $x^2 + 9x + k$

2. Solve by completing the square:
   a. $x^2 + 9x + 20 = 0$ **b.** $x^2 - 2x - 63 = 0$ **c.** $x^2 + 4x = 12$
   d. $3x^2 - x - 2 = 0$ **e.** $2x^2 + 5x - 2 = 0$ **f.** $3x^2 - 1 = {}^-6x$
   g. $x^2 - 2x = 1$ **h.** $3x^2 - 2x - 8 = 0$ **i.** $5x^2 - 6x + 1 = 0$
   j. $x^2 - 3x + 1 = 0$ **k.** $x^2 - 5x - 14 = 0$ **l.** $3x^2 - 18x + 6 = 0$

14.2 Exercise Answers

1. **a.** 16   **b.** $\dfrac{49}{4}$ **c.** 9

   d. $\dfrac{81}{4}$

2. **a.** $x = {}^-5$ or $x = {}^-4$ **b.** $x = {}^-7$ or $x = 9$ **c.** $x = 2$ or $x = {}^-6$

   d. $x = \dfrac{^-2}{3}$ or $x = 1$ **e.** $x = \dfrac{^-5 + \sqrt{41}}{4}$ or $\dfrac{^-5 - \sqrt{41}}{4}$

   f. $x = \dfrac{^-3 + 2\sqrt{3}}{3}$ or $x = \dfrac{^-3 - 2\sqrt{3}}{3}$

g. $x = 1 + \sqrt{2}$ or $x = 1 - \sqrt{2}$

h. $x = 2$ or $x = \dfrac{^-4}{3}$ **i.** $x = \dfrac{1}{5}$ or $x = 1$

j. $x = \dfrac{3 + \sqrt{5}}{2}$ or $x = \dfrac{3 - \sqrt{5}}{2}$

k. $x = {^-2}$ or $x = 7$ **l.** $x = 3 + \sqrt{7}$ or $x = 3 - \sqrt{7}$

14.3   Solving Quadratic Equations by the Quadratic Formula

1   In actual practice quadratic equations are usually solved by factoring or the quadratic formula. The method of solving quadratic equations by completing the square is often not practical, owing to its length. The *quadratic formula* can be derived by completing the square on the standard form for a quadratic equation, $ax^2 + bx + c = 0$, where a, b, and c are real numbers and $a \neq 0$.

$$ax^2 + bx + c = 0$$

Write the equation with c on the right.

$$ax^2 + bx + c - c = 0 - c$$
$$ax^2 + bx = {^-c}$$

Divide each term of the equation by a.

$$\frac{ax^2}{a} + \frac{bx}{a} = \frac{^-c}{a}$$

$$x^2 + \frac{b}{a}x = \frac{^-c}{a}$$

Since $\left(\dfrac{1}{2} \cdot \dfrac{b}{a}\right)^2 = \left(\dfrac{b}{2a}\right)^2 = \dfrac{b^2}{4a^2}$, add $\dfrac{b^2}{4a^2}$ to both sides of the equation.

$$x^2 + \frac{b}{a}x + \frac{b^2}{4a^2} = \frac{^-c}{a} + \frac{b^2}{4a^2}$$

Factor the left side and simplify the sum of the fractions on the right side.

$$\left(x + \frac{b}{2a}\right)^2 = \frac{^-4ac + b^2}{4a^2}$$

$$\left(x + \frac{b}{2a}\right)^2 = \frac{b^2 - 4ac}{4a^2}$$

Take the square root of both sides.

$$\sqrt{\left(x + \frac{b}{2a}\right)^2} = \pm\sqrt{\frac{b^2 - 4ac}{4a^2}}$$

$$x + \frac{b}{2a} = \frac{\pm\sqrt{b^2 - 4ac}}{2a}$$

Solve for x.

$$x + \frac{b}{2a} - \frac{b}{2a} = \frac{^-b}{2a} \pm \frac{\sqrt{b^2 - 4ac}}{2a}$$

$$x = \frac{^-b \pm \sqrt{b^2 - 4ac}}{2a}$$

The last equation is called the *quadratic formula* and gives two solutions,

$$x = \frac{^-b + \sqrt{b^2 - 4ac}}{2a} \qquad \text{or} \qquad x = \frac{^-b - \sqrt{b^2 - 4ac}}{2a}$$

for the quadratic equation $ax^2 + bx + c = 0$. (*Note:* At this time in your mathematical training, it is not essential for you to be able to derive the quadratic formula by completing the square. However, you will find it helpful to memorize the formula.)

To solve a quadratic equation using the quadratic formula:

1. Write the equation in standard form.
2. Identify the values for a, b, and c.
3. Substitute the values into the quadratic formula $x = \dfrac{^-b \pm \sqrt{b^2 - 4ac}}{2a}$ and simplify.
4. Write the two solutions.

Example: Solve $3x^2 - 8x + 5 = 0$.

Solution

Step 1: The equation is already in standard form.

Step 2: $a = 3$, $b = {}^-8$, $c = 5$

Step 3: $x = \dfrac{^-(^-8) \pm \sqrt{(^-8)^2 - 4(3)(5)}}{2(3)}$

$\quad = \dfrac{8 \pm \sqrt{64 - 60}}{6}$

$\quad = \dfrac{8 \pm \sqrt{4}}{6}$

$\quad = \dfrac{8 \pm 2}{6}$

Step 4: $x = \dfrac{8 + 2}{6} \qquad \text{or} \qquad x = \dfrac{8 - 2}{6}$

$\quad = \dfrac{10}{6} \qquad\qquad\qquad = \dfrac{6}{6}$

$\quad = \dfrac{5}{3} \qquad\qquad\qquad = 1$

The solutions are $x = \dfrac{5}{3}$ or $x = 1$.

Q1   **a.** For $2x^2 - 9x - 5 = 0$

$\quad a = \underline{\hspace{1cm}}, b = \underline{\hspace{1cm}}, c = \underline{\hspace{1cm}}.$

b. Solve $2x^2 - 9x - 5 = 0$ by the quadratic formula.

STOP • **STOP** • **STOP** • **STOP** • **STOP** • **STOP** • **STOP** • **STOP** • **STOP**

A1   **a.** $2, \, ^-9, \, ^-5$

      b. $x = 5$ or $x = \dfrac{^-1}{2}$: $x = \dfrac{^-b \pm \sqrt{b^2 - 4ac}}{2a}$

$$= \frac{^-(^-9) \pm \sqrt{(^-9)^2 - 4(2)(^-5)}}{2(2)}$$

$$= \frac{9 \pm \sqrt{81 + 40}}{4}$$

$$= \frac{9 \pm \sqrt{121}}{4}$$

$$= \frac{9 \pm 11}{4}$$

Therefore, $x = \dfrac{9 + 11}{4}$ or $x = \dfrac{9 - 11}{4}$

$$= \frac{20}{4} \qquad\qquad\qquad = \frac{^-2}{4}$$

$$= 5 \qquad\qquad\qquad\quad = \frac{^-1}{2}$$

Q2   **a.** For $2x^2 + 7x - 1 = 0$

      $a = \underline{\hspace{1cm}}, b = \underline{\hspace{1cm}}, c = \underline{\hspace{1cm}}.$

      b. Solve $2x^2 + 7x - 1 = 0$.

$$x = \frac{^-b \pm \sqrt{b^2 - 4ac}}{2a}$$

A2   **a.** $2, 7, {}^-1$ **b.** $x = \dfrac{{}^-7 + \sqrt{57}}{4}$ or $x = \dfrac{{}^-7 - \sqrt{57}}{4}$:

$$x = \frac{{}^-(7) \pm \sqrt{(7)^2 - 4(2)({}^-1)}}{2(2)}$$

$$= \frac{{}^-7 \pm \sqrt{49 + 8}}{4}$$

$$= \frac{{}^-7 \pm \sqrt{57}}{4}$$

---

2   To solve $2x^2 - 8x + 3 = 0$ by the quadratic formula, determine the values of a, b, and c and proceed as before.

$2x^2 - 8x + 3 = 0$

$a = 2, b = {}^-8, c = 3$

$$x = \frac{{}^-b \pm \sqrt{b^2 - 4ac}}{2a}$$

$$= \frac{{}^-({}^-8) \pm \sqrt{({}^-8)^2 - 4(2)(3)}}{2(2)}$$

$$= \frac{8 \pm \sqrt{64 - 24}}{4}$$

$$= \frac{8 \pm \sqrt{40}}{4}$$

To complete the solution, simplify the radical $\sqrt{40}$ using the procedures of Chapter 13.

$$= \frac{8 \pm \sqrt{4 \cdot 10}}{4}$$

$$= \frac{8 \pm 2\sqrt{10}}{4}$$

$$= \frac{\overset{1}{2}(4 \pm \sqrt{10})}{\underset{2}{4}} \qquad (\textit{Note:} \text{ It is helpful to factor the numerator before dividing out the common factor})$$

$$= \frac{4 \pm \sqrt{10}}{2}$$

The solutions are $x = \dfrac{4 + \sqrt{10}}{2}$ or $x = \dfrac{4 - \sqrt{10}}{2}$.

Q3   Solve $4x^2 - 6x + 1 = 0$ by the quadratic formula.

STOP • **STOP** • **STOP** • **STOP** • **STOP** • **STOP** • **STOP** • **STOP** • **STOP**

A3 $\qquad x = \dfrac{3 + \sqrt{5}}{4}$ or $x = \dfrac{3 - \sqrt{5}}{4}$: $\quad 4x^2 - 6x + 1 = 0$

$$a = 4,\ b = {}^-6,\ c = 1$$

$$x = \frac{{}^-({}^-6) \pm \sqrt{({}^-6)^2 - 4(4)(1)}}{2(4)}$$

$$= \frac{6 \pm \sqrt{36 - 16}}{8}$$

$$= \frac{6 \pm \sqrt{20}}{8}$$

$$= \frac{6 \pm \sqrt{4 \cdot 5}}{8}$$

$$= \frac{6 \pm 2\sqrt{5}}{8}$$

$$= \frac{2(3 \pm \sqrt{5})}{8}$$

$$= \frac{3 \pm \sqrt{5}}{4}$$

Q4 $\qquad$ Solve $2x^2 - 6x + 3 = 0$ by the quadratic formula.

STOP • **STOP** • **STOP** • **STOP** • **STOP** • **STOP** • **STOP** • **STOP** • **STOP**

A4 $\qquad x = \dfrac{3 + \sqrt{3}}{2}$ or $x = \dfrac{3 - \sqrt{3}}{2}$: $\quad 2x^2 - 6x + 3 = 0$

$$a = 2,\ b = {}^-6,\ c = 3$$

$$x = \frac{{}^-({}^-6) \pm \sqrt{({}^-6)^2 - 4(2)(3)}}{2(2)}$$

$$= \frac{6 \pm \sqrt{36 - 24}}{4}$$

$$= \frac{6 \pm \sqrt{12}}{4}$$

$$= \frac{6 \pm \sqrt{4 \cdot 3}}{4}$$

$$= \frac{6 \pm 2\sqrt{3}}{4}$$

$$= \frac{2(3 \pm \sqrt{3})}{4}$$

$$= \frac{3 \pm \sqrt{3}}{2}$$

3   To solve $x^2 = {}^-2 + 6x$ by the quadratic formula, the equation is first written in standard form.

$$x^2 = {}^-2 + 6x$$
$$x^2 - 6x + 2 = 0$$

The values for a, b, and c are now substituted into the quadratic formula. $a = 1$, $b = {}^-6$, $c = 2$. (*Note:* $a = 1$ because $x^2 = 1x^2$.)

$$x = \frac{{}^-b \pm \sqrt{b^2 - 4ac}}{2a}$$

$$= \frac{{}^-({}^-6) \pm \sqrt{({}^-6)^2 - 4(1)(2)}}{2(1)}$$

$$= \frac{6 \pm \sqrt{36 - 8}}{2}$$

$$= \frac{6 \pm \sqrt{28}}{2}$$

$$= \frac{6 \pm \sqrt{4 \cdot 7}}{2}$$

$$= \frac{6 \pm 2\sqrt{7}}{2}$$

$$= \frac{2(3 \pm \sqrt{7})}{2}$$

$$= 3 \pm \sqrt{7}$$

The solutions are $x = 3 + \sqrt{7}$ or $x = 3 - \sqrt{7}$.

Q5   Solve using the quadratic formula:

a. $2x^2 + 9x = 5$ **b.** $2x^2 = 1 - 3x$

STOP • STOP • STOP • STOP • STOP • STOP • STOP • STOP • STOP

A5   **a.** $x = \dfrac{1}{2}$ or $x = {}^-5$: $2x^2 + 9x = 5$

$$2x^2 + 9x - 5 = 0$$
$$a = 2, b = 9, c = {}^-5$$
$$x = \frac{{}^-(9) \pm \sqrt{(9)^2 - 4(2)({}^-5)}}{2(2)}$$
$$= \frac{{}^-9 \pm \sqrt{81 + 40}}{4}$$
$$= \frac{{}^-9 \pm \sqrt{121}}{4}$$
$$= \frac{{}^-9 \pm 11}{4}$$
$$x = \frac{{}^-9 + 11}{4} \qquad \text{or} \qquad x = \frac{{}^-9 - 11}{4}$$
$$= \frac{2}{4} \qquad\qquad\qquad = \frac{{}^-20}{4}$$
$$= \frac{1}{2} \qquad\qquad\qquad = {}^-5$$

b. $x = \dfrac{{}^-3 + \sqrt{17}}{4}$ or $x = \dfrac{{}^-3 - \sqrt{17}}{4}$: $2x^2 + 3x - 1 = 0$

$$a = 2, b = 3, c = {}^-1$$
$$x = \frac{{}^-(3) \pm \sqrt{(3)^2 - 4(2)({}^-1)}}{2(2)}$$
$$= \frac{{}^-3 \pm \sqrt{9 + 8}}{4}$$
$$= \frac{{}^-3 \pm \sqrt{17}}{4}$$

---

4   If in the equation $ax^2 + bx + c = 0$, the value of b or c is zero, the result is a quadratic equation with a missing first-degree term or a missing constant term. That is,

If $b = 0$:	If $c = 0$:
$ax^2 + bx + c = 0$	$ax^2 + bx + c = 0$
$ax^2 + 0x + c = 0$	$ax^2 + bx + 0 = 0$
$ax^2 + 0 + c = 0$	$ax^2 + bx = 0$
$ax^2 + c = 0$	The constant term is missing.

The first-degree term is missing.

Thus, when solving a quadratic equation that has a missing term, either the value of b or c is zero.

Example 1: Solve $3x^2 - 2 = 0$ by the quadratic formula.

Solution

$3x^2 - 2 = 0$
$a = 3, b = 0$ (because the first-degree term is missing)
$c = {}^-2$
$$x = \frac{{}^-(0) \pm \sqrt{(0)^2 - 4(3)({}^-2)}}{2(3)}$$

$$= \frac{0 \pm \sqrt{0 + 24}}{6}$$

$$= \frac{\pm \sqrt{24}}{6}$$

$$= \frac{\pm \sqrt{4 \cdot 6}}{6}$$

$$= \frac{\pm 2\sqrt{6}}{6}$$

$$= \frac{\pm \sqrt{6}}{3}$$

Thus, $x = \dfrac{\sqrt{6}}{3}$ or $x = \dfrac{^-\sqrt{6}}{3}$.

Example 2: Solve $7x^2 - 5x = 0$ by the quadratic formula.

Solution

$7x^2 - 5x = 0$

$a = 7, b = ^-5, c = 0$ \qquad (because the constant term is missing)

$$x = \frac{^-(^-5) \pm \sqrt{(^-5)^2 - 4(7)(0)}}{2(7)}$$

$$= \frac{5 \pm \sqrt{25 - 0}}{14}$$

$$= \frac{5 \pm \sqrt{25}}{14}$$

$$= \frac{5 \pm 5}{14}$$

$$x = \frac{5 + 5}{14} \qquad \text{or} \qquad x = \frac{5 - 5}{14}$$

$$= \frac{10}{14} \qquad\qquad\qquad = \frac{0}{14}$$

$$= \frac{5}{7} \qquad\qquad\qquad = 0$$

Thus, $x = \dfrac{5}{7}$ or $x = 0$.

Q6 Solve using the quadratic formula:

a. $2x = 3x^2$ \qquad\qquad\qquad **b.** $6x^2 - 5 = 0$

A6

a. $x = 0$ or $x = \frac{2}{3}$: $2x = 3x^2$

$$0 = 3x^2 - 2x$$

$$3x^2 - 2x = 0$$

$$a = 3, b = {}^-2, c = 0$$

$$x = \frac{{}^-({}^-2) \pm \sqrt{({}^-2)^2 - 4(3)(0)}}{2(3)}$$

$$= \frac{2 \pm \sqrt{4 - 0}}{6}$$

$$= \frac{2 \pm \sqrt{4}}{6}$$

$$= \frac{2 \pm 2}{6}$$

$$x = \frac{2 + 2}{6} \quad \text{or} \quad x = \frac{2 - 2}{6}$$

$$= \frac{4}{6} \qquad\qquad = \frac{0}{6}$$

$$= \frac{2}{3} \qquad\qquad = 0$$

b. $x = \frac{\sqrt{30}}{6}$ or $x = \frac{{}^-\sqrt{30}}{6}$: $6x^2 - 5 = 0$

$$a = 6, b = 0, c = {}^-5$$

$$x = \frac{{}^-(0) \pm \sqrt{(0)^2 - 4(6)({}^-5)}}{2(6)}$$

$$= \frac{0 \pm \sqrt{0 + 120}}{12}$$

$$= \frac{\pm \sqrt{120}}{12}$$

$$= \frac{\pm \sqrt{4 \cdot 30}}{12}$$

$$= \frac{\pm 2\sqrt{30}}{12}$$

$$= \frac{\pm \sqrt{30}}{6}$$

$$x = \frac{\sqrt{30}}{6} \quad \text{or} \quad x = \frac{{}^-\sqrt{30}}{6}$$

5

Notice that quadratic equations in which the constant term is zero are more easily solved by factoring:

Example 1: Solve $7x^2 - 5x = 0$ by factoring.

Solution

$$7x^2 - 5x = 0$$
$$x(7x - 5) = 0$$

$$x = 0 \quad \text{or} \quad 7x - 5 = 0$$
$$7x - 5 + 5 = 0 + 5$$
$$7x = 5$$
$$x = \frac{5}{7}$$

Thus, $x = 0$ or $x = \frac{5}{7}$.

Example 2: Solve $2x = 3x^2$ by factoring.

Solution

$$2x = 3x^2$$
$$0 = 3x^2 - 2x$$
$$3x^2 - 2x = 0$$
$$x(3x - 2) = 0$$
$$x = 0 \quad \text{or} \quad 3x - 2 = 0$$
$$3x - 2 + 2 = 0 + 2$$
$$3x = 2$$
$$x = \frac{2}{3}$$

Thus, $x = 0$ or $x = \frac{2}{3}$.

Q7   Solve:

     a. $4x^2 - 3 = 0$ **b.** $^{-}2x^2 + 9x = 0$

STOP • STOP • STOP • STOP • STOP • STOP • STOP • STOP • STOP

A7   **a.** $x = \dfrac{\sqrt{3}}{2}$ or $x = \dfrac{-\sqrt{3}}{2}$:

$$4x^2 - 3 = 0$$
$$a = 4, b = 0, c = ^{-}3$$
$$x = \frac{^{-}(0) \pm \sqrt{(0)^2 - 4(4)(^{-}3)}}{2(4)}$$
$$= \frac{0 \pm \sqrt{0 + 48}}{8}$$

b. $x = 0$ or $x = \dfrac{9}{2}$:

$$^{-}2x^2 + 9x = 0$$
$$x(^{-}2x + 9) = 0$$
$$x = 0 \quad \text{or} \quad ^{-}2x + 9 = 0$$
$$^{-}2x + 9 - 9 = 0 - 9$$
$$^{-}2x = ^{-}9$$
$$x = \frac{9}{2}$$

(solution continued on page 590)

$$= \frac{\pm\sqrt{48}}{8}$$

$$= \frac{\pm\sqrt{16 \cdot 3}}{8}$$

$$= \frac{\pm 4\sqrt{3}}{8}$$

$$= \frac{\pm\sqrt{3}}{2}$$

$$x = \frac{\sqrt{3}}{2} \quad \text{or} \quad x = \frac{^-\sqrt{3}}{2}$$

6 Some quadratic equations are solved more simply by factoring. If the polynomial $ax^2 + bx + c$ is prime, the quadratic equation $ax^2 + bx + c = 0$ must be solved by either completing the square or by the quadratic formula. Experience will enable you to choose the easiest method. For now, choose the method that works best for you.

Q8 Solve:

a. $2x^2 + 8x = 0$

b. $x^2 - 3x - 1 = 0$

c. $x^2 + 5x = {}^-4$

d. $8x^2 = 3$

STOP • STOP • STOP • STOP • STOP • STOP • STOP • STOP • STOP

A8
  a. $x = 0$ or $x = {}^-4$: $\quad 2x^2 + 8x = 0$
  $$2x^2 + 8x \text{ is factorable}$$
  $$2x(x + 4) = 0$$

  $2x = 0 \qquad$ or $\qquad x + 4 = 0$
  $\quad x = 0 \qquad\qquad\qquad x + 4 - 4 = 0 - 4$
  $$x = {}^-4$$

  b. $x = \dfrac{3 + \sqrt{13}}{2}$ or $x = \dfrac{3 - \sqrt{13}}{2}$: $\quad x^2 - 3x - 1 = 0$

  $$a = 1, b = {}^-3, c = {}^-1$$

  $$x = \frac{{}^-({}^-3) \pm \sqrt{({}^-3)^2 - 4(1)({}^-1)}}{2(1)}$$

  $$= \frac{3 \pm \sqrt{9 + 4}}{2}$$

  $$= \frac{3 \pm \sqrt{13}}{2}$$

  c. $x = {}^-1$ or $x = {}^-4$: $\quad x^2 + 5x = {}^-4$
  $$x^2 + 5x + 4 = 0$$
  $$x^2 + 5x + 4 \text{ is factorable}$$
  $$(x + 4)(x + 1) = 0$$

  $x + 4 = 0 \qquad$ or $\qquad x + 1 = 0$
  $\quad x = {}^-4 \qquad\qquad\qquad\quad x = {}^-1$

  d. $x = \dfrac{\sqrt{6}}{4}$ or $x = \dfrac{{}^-\sqrt{6}}{4}$: $\quad 8x^2 = 3$

  $$8x^2 - 3 = 0$$
  $$8x^2 - 3 \text{ is prime}$$
  $$a = 8, b = 0, c = {}^-3$$

  $$x = \frac{{}^-(0) \pm \sqrt{(0)^2 - 4(8)({}^-3)}}{2(8)}$$

  $$= \frac{0 \pm \sqrt{0 + 96}}{16}$$

  $$= \frac{\pm\sqrt{96}}{16}$$

  $$= \frac{\pm\sqrt{16 \cdot 6}}{16}$$

  $$= \frac{\pm 4\sqrt{6}}{16}$$

  $$= \frac{\pm\sqrt{6}}{4}$$

This completes the instruction for this section.

14.3 Exercises

1. Solve by factoring:
   a. $x^2 - x - 42 = 0$ $\qquad\qquad$ **b.** $9x^2 - 5x = 0$ $\qquad\qquad$ **c.** $16 = 9y^2$
   d. $6x^2 = 14 - 17x$

2. Solve by completing the square:
   a. $x^2 = {}^-2 + 6x$ b. $3x^2 - 8x + 5 = 0$
3. Solve by the quadratic formula:
   a. $x^2 - 3x + 2 = 0$ b. ${}^-x^2 + 6x - 1 = 0$ c. $x^2 + 8x = {}^-16$
   d. $4x^2 = 7$ e. ${}^-4x^2 = x$ f. $x^2 - x - 1 = 0$
4. Solve using the method of your choice:
   a. $x^2 + 5x + 5 = 0$ b. $x^2 - 25 = 0$ c. $x^2 = 3 - x$
   d. $3x - 10 = x^2$ e. $x = 4x^2$ f. $3x^2 = 12$

14.3   Exercise Answers

1. a. $x = 7$ or $x = {}^-6$ b. $x = 0$ or $x = \dfrac{5}{9}$ c. $y = \dfrac{4}{3}$ or $y = \dfrac{{}^-4}{3}$

   d. $x = \dfrac{2}{3}$ or $x = \dfrac{{}^-7}{2}$

2. a. $x = 3 + \sqrt{7}$ or $x = 3 - \sqrt{7}$ b. $x = \dfrac{5}{3}$ or $x = 1$

3. a. $x = 2$ or $x = 1$ b. $x = 3 + 2\sqrt{2}$ or $x = 3 - 2\sqrt{2}$

   c. $x = {}^-4$ d. $x = \dfrac{\sqrt{7}}{2}$ or $x = \dfrac{{}^-\sqrt{7}}{2}$

   e. $x = 0$ or $x = \dfrac{{}^-1}{4}$ f. $x = \dfrac{1 + \sqrt{5}}{2}$ or $x = \dfrac{1 - \sqrt{5}}{2}$

4. a. $x = \dfrac{{}^-5 + \sqrt{5}}{2}$ or $x = \dfrac{{}^-5 - \sqrt{5}}{2}$ b. $x = 5$ or $x = {}^-5$

   c. $x = \dfrac{{}^-1 + \sqrt{13}}{2}$ or $x = \dfrac{{}^-1 - \sqrt{13}}{2}$ d. $x = {}^-5$ or $x = 2$

   e. $x = 0$ or $x = \dfrac{1}{4}$ f. $x = \pm 2$

Chapter 14 Sample Test

At the completion of Chapter 14 it is expected that you will be able to work the following problems.

14.1   Solving Quadratic Equations by Factoring

1. Solve by factoring:
   a. $3x^2 - 27x = 0$ b. $x^2 + 2x - 15 = 0$ c. ${}^-x^2 = x$
   d. $5x = 36 - x^2$ e. $x^2 - 64 = 0$ f. $6x^2 - 11x + 4 = 0$
   g. $16x^2 = 49$ h. ${}^-12 + 2x^2 = 5x$

14.2   ## Solving Quadratic Equations by Completing the Square

2. Solve by completing the square:

   a. $x^2 + 4x - 12 = 0$ **b.** $x^2 - 11x + 28 = 0$ **c.** $x^2 - 6x = 1$
   d. $x^2 + 5x - 7 = 0$ **e.** $x^2 - 1 = 2x$ **f.** $x^2 + 6x = 0$
   g. $2x^2 - 3x - 2 = 0$ **h.** $3x^2 - 7x - 1 = 0$

14.3   ## Solving Quadratic Equations by the Quadratic Formula

3. Solve by the quadratic formula:

   a. $x^2 - x - 12 = 0$ **b.** $x^2 - 3 = 0$ **c.** $x^2 = 7x$
   d. $5x^2 - 49 = 0$ **e.** $2x^2 - x - 1 = 0$ **f.** $2x^2 + 10x + 5 = 0$
   g. $^-3x = 2x^2$ **h.** $5x^2 + 1 = 7x$

4. Solve by use of the method of your choice:

   a. $x^2 - 13x + 12 = 0$ **b.** $2x^2 = 11x$ **c.** $x^2 - 100 = 0$
   d. $2x^2 - x - 5 = 0$

Chapter 14 Sample Test Answers

1. **a.** $x = 0$ or $x = 9$ **b.** $x = 3$ or $x = ^-5$ **c.** $x = 0$ or $x = ^-1$

   d. $x = ^-9$ or $x = 4$ **e.** $x = \pm 8$ **f.** $x = \dfrac{1}{2}$ or $x = \dfrac{4}{3}$

   g. $x = \dfrac{\pm 7}{4}$ **h.** $x = \dfrac{^-3}{2}$ or $x = 4$

2. **a.** $x = ^-6$ or $x = 2$ **b.** $x = 4$ or $x = 7$

   c. $x = 3 + \sqrt{10}$ or $x = 3 - \sqrt{10}$ **d.** $x = \dfrac{^-5 + \sqrt{53}}{2}$ or $x = \dfrac{^-5 - \sqrt{53}}{2}$

   e. $x = 1 + \sqrt{2}$ or $x = 1 - \sqrt{2}$ **f.** $x = 0$ or $x = ^-6$

   g. $x = 2$ or $x = \dfrac{^-1}{2}$ **h.** $x = \dfrac{7 + \sqrt{61}}{6}$ or $x = \dfrac{7 - \sqrt{61}}{6}$

3. **a.** $x = 4$ or $x = ^-3$ **b.** $x = \pm \sqrt{3}$

   c. $x = 0$ or $x = 7$ **d.** $x = \dfrac{\pm 7\sqrt{5}}{5}$

   e. $x = 1$ or $x = \dfrac{^-1}{2}$ **f.** $x = \dfrac{^-5 + \sqrt{15}}{2}$ or $x = \dfrac{^-5 - \sqrt{15}}{2}$

   g. $x = 0$ or $x = \dfrac{^-3}{2}$ **h.** $x = \dfrac{7 + \sqrt{29}}{10}$ or $x = \dfrac{7 - \sqrt{29}}{10}$

4. **a.** $x = 12$ or $x = 1$ **b.** $x = 0$ or $x = \dfrac{11}{2}$ **c.** $x = \pm 10$

   d. $x = \dfrac{1 + \sqrt{41}}{4}$ or $x = \dfrac{1 - \sqrt{41}}{4}$

Appendix

Table I   English Weights and Measures

Units of Length

1 foot (ft or ′) = 12 inches (in or ″)
   1 yard (yd) = 3 feet
     1 rod (rd) = $5\frac{1}{2}$ yards = $16\frac{1}{2}$ feet
  1 mile (mi) = 1,760 yards = 5,280 feet

Units of Area

1 square foot = 144 square inches
1 square yard = 9 square feet
    1 acre = 43,560 square feet
1 square mile = 640 acres

Units of Weight

1 pound (lb) = 16 ounces (oz)
   1 ton = 2,000 pounds

Units of Capacity (Volume)
Liquid

1 pint (pt) = 16 fluid ounces (fl oz)
1 quart (qt) = 2 pints
1 gallon (gal) = 4 quarts = 8 pints

Dry

1 quart (qt) = 2 pints (pt)
1 peck (pk) = 8 quarts
1 bushel (bu) = 4 pecks = 32 quarts

Units of Capacity

1 cubic foot = 1,728 cubic inches
1 cubic yard = 27 cubic feet
1 gallon = 231 cubic inches
1 cubic foot = 7.48 gallons

Table II   Metric Weights and Measures

Units of Length

1 millimeter (mm) = 1000 micrometers (μm)
1 centimeter (cm) = 10 millimeters
1 decimeter (dm) = 10 centimeters
   1 meter (m) = 10 decimeters
1 dekameter (dam) = 10 meters
1 hectometer (hm) = 10 dekameters
1 kilometer (km) = 10 hectometers

Units of Weight

1 milligram (mg) = 1000 micrograms (μg)
1 centigram (cg) = 10 milligrams
1 decigram (dg) = 10 centigrams
   1 gram (g) = 10 decigrams
1 dekagram (dag) = 10 grams
1 hectogram (hg) = 10 dekagrams
1 kilogram (kg) = 10 hectograms
1 megagram (Mg) = 1000 kilograms
      (metric ton)

Units of Area

1 square centimeter = 100 square millimeters
1 square decimeter = 100 square centimeters
1 square meter = 100 square decimeters
1 square kilometer = 1 000 000 square meters
1 hectare (ha) = 10 000 square meters
1 are (a) = 100 square meters

Units of Capacity (Volume)

1 milliliter (ml) = 1 cubic
          centimeter (cc or cm^3)
1 centiliter (cl) = 10 milliliters
1 deciliter (dl) = 10 centiliters
1 liter (l) = 10 deciliters
     = 1000 cubic centimeters
1 deckaliter (dal) = 10 liters
1 hectoliter (hl) = 10 dekaliters
1 kiloliter (kl) = 10 hectoliters
1 megaliter (Ml) = 1000 kiloliters

Table III Formulas from Geometry **595**

Table III   Formulas from Geometry

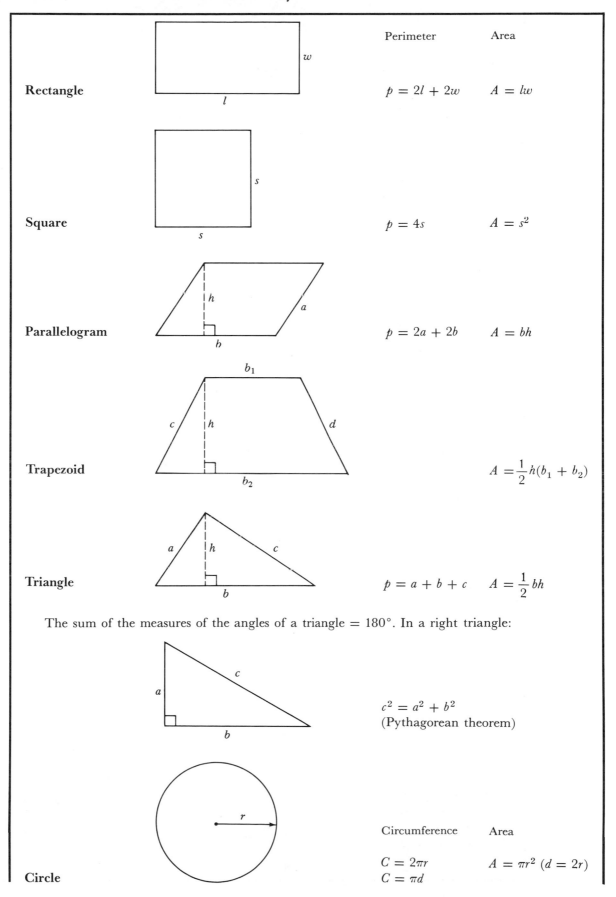

	Perimeter	Area
Rectangle	$p = 2l + 2w$	$A = lw$
Square	$p = 4s$	$A = s^2$
Parallelogram	$p = 2a + 2b$	$A = bh$
Trapezoid		$A = \dfrac{1}{2}h(b_1 + b_2)$
Triangle	$p = a + b + c$	$A = \dfrac{1}{2}bh$

The sum of the measures of the angles of a triangle $= 180°$. In a right triangle:

$$c^2 = a^2 + b^2$$
(Pythagorean theorem)

	Circumference	Area
Circle	$C = 2\pi r$ $C = \pi d$	$A = \pi r^2 \ (d = 2r)$

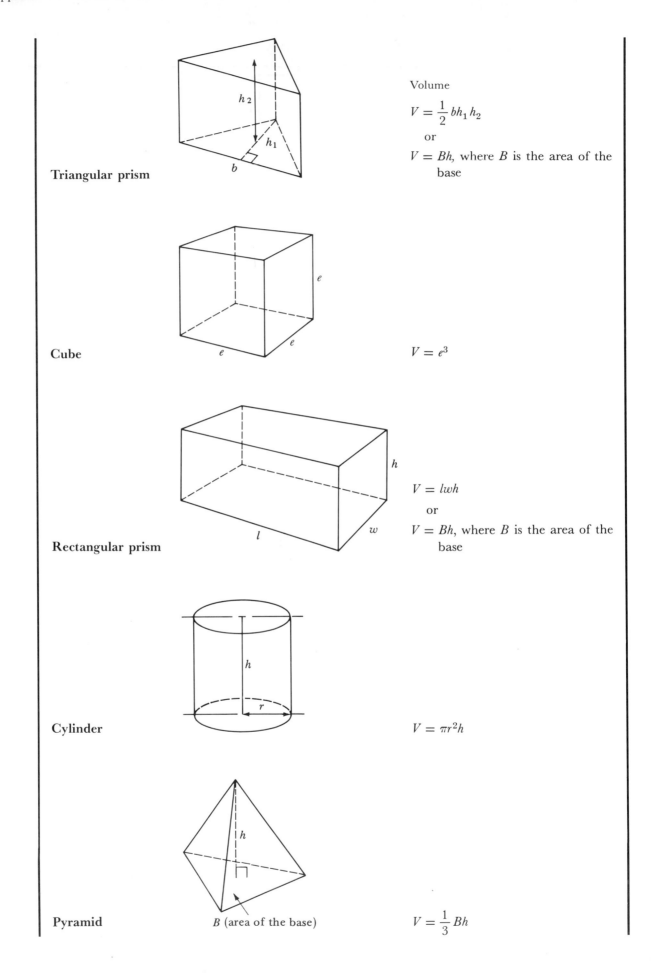

Triangular prism

Volume

$$V = \frac{1}{2} bh_1 h_2$$

or

$V = Bh$, where B is the area of the base

Cube

$$V = e^3$$

Rectangular prism

$$V = lwh$$

or

$V = Bh$, where B is the area of the base

Cylinder

$$V = \pi r^2 h$$

Pyramid

B (area of the base)

$$V = \frac{1}{3} Bh$$

Table III   Formulas from Geometry   **597**

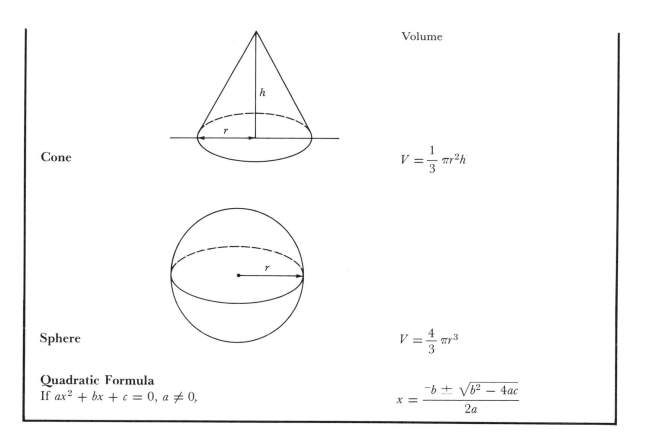

Volume

Cone

$$V = \frac{1}{3}\,\pi r^2 h$$

Sphere

$$V = \frac{4}{3}\,\pi r^3$$

Quadratic Formula

If $ax^2 + bx + c = 0$, $a \neq 0$,

$$x = \frac{-b \pm \sqrt{b^2 - 4ac}}{2a}$$

Table IV   Symbols

The section number indicates the first use of the symbol.

1.1	$\{a\}$	set containing the element a	Braces are used to denote a set
1.1	$\{\ \}$ or $\varnothing$	"empty set"	The set with no elements
1.1	$=$	"is equal to"	
1.1	$\in$	"is an element of"	
1.1	$\notin$	"is not an element of"	
1.1	$,\ldots$	"and so on"	Indicates that the list continues
1.2	$A \cap B$	"A intersect B" *or* "the intersection of A and B"	
1.2	$A \cup B$	"A union B" *or* "the union of A and B"	
1.3	$+$	"plus"	Indicates addition (sum) of two numbers
1.3	$-$	"minus"	Indicates subtraction (difference) of two numbers
1.3	$a \cdot b$	"a times b"	Raised dot to indicate the product of a and b
1.3	$(\ )$	"quantity"	Parentheses used as grouping symbols
1.3	$a(b) = (a)(b) = (a)b$		Product of a and b
1.3	$\div$	"divided by"	Indicates division (quotient between two numbers)
1.3	$.$	"point" *or* "and"	Decimal point used to separate the whole-number part from the decimal (fraction) part of the number
1.3	$\dfrac{a}{b}$	"a over b"	Fraction indicating a out of b equal parts
1.4	N		Denotes the set of natural numbers, $\{1, 2, 3, 4, \ldots\}$
1.4	W		Denotes the set of whole numbers, $\{0, 1, 2, 3, \ldots\}$
1.4	$\neq$	"is not equal"	
2.1	$^-$	"negative"	Raised minus sign to denote a negative number
2.1	$^+$	"positive"	Raised plus sign to denote a positive number
2.1	I		Denotes the set of integers, $\{\ldots, {}^-2, {}^-1, 0, 1, 2, \ldots\}$
2.2	^-a	"opposite of a"	Raised minus sign to indicate the opposite of a number
3.1	Q		Denotes the set of rational numbers
3.1	$0 \cdot \overline{ab}\ \ (a, b \in W)$		Indicates that the group of numbers under the bar repeat endlessly
3.2	LCD	"least common denominator"	
6.1	$a:b$	"a to b" *or* "a is to b"	Colon used in expressing a ratio of two numbers or quantities
7.1	$\%$	"percent"	Abbreviation for the word "percent"
7.4	$^\circ$	"degree"	Unit of measure of an angle

Table IV   Symbols   **599**

7.5	b_1 and b_2	"b sub-one and b sub-two"	Subscripts used to distinguish between the two bases of a trapezoid
7.5	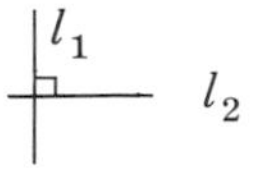 l_1, l_2		Indicates that l_1 and l_2 are perpendicular
7.5	π	"pi"	Constant whose value is generally given as approximately equal to 3.14 or $\dfrac{22}{7}$
7.5	$\doteq$	"is approximately equal to"	
8.1	$a > b$	"a is greater than b"	
8.1	$a < b$	"a is less than b"	
8.1	$a \leqslant b$	"a is less than or equal to b"	
8.1	$a \geqslant b$	"a is greater than or equal to b"	
8.1	$a \not< b$	"a is not less than b"	
8.1	$a \not\leqslant b$	"a is not less than or equal to b"	
8.1	$a \not> b$	"a is not greater than b"	
8.1	$a \not\geqslant b$	"a is not greater than or equal to b"	
8.1	$\{x \mid x < a\}$	"the set of all numbers x such that x is less than a"	Set builder notation
8.2	 a	open graph	Indicates that all points to the right of a are included in the graph (a is not included)
8.2	 a	closed graph	Indicates that a and all points to the right of a are included in the graph
9.1	(x, y)	"ordered pair x, y"	
11.1	a^n	"a to the nth"	Exponential notation
11.3	$[\ ]$	"quantity"	Brackets used as grouping symbols (used in the same manner as parentheses)
13.1	$\sqrt{a}$	"the principal (positive) square root of a"	
13.1	$-\sqrt{a}$	"the negative square root of a"	
13.1	$\sqrt[n]{a}$	"the principal nth root of a"	

Table V Powers and Roots

No.	Square	Cube	Square Root	Cube Root	No.	Square	Cube	Square Root	Cube Root
1	1	1	1.000	1.000	51	2,601	132,651	7.141	3.708
2	4	8	1.414	1.260	52	2,704	140,608	7.211	3.732
3	9	27	1.732	1.442	53	2,809	148,877	7.280	3.756
4	16	64	2.000	1.587	54	2,916	157,464	7.348	3.780
5	25	125	2.236	1.710	55	3,025	166,375	7.416	3.803
6	36	216	2.449	1.817	56	3,136	175,616	7.483	3.826
7	49	343	2.646	1.913	57	3,249	185,193	7.550	3.848
8	64	512	2.828	2.000	58	3,364	195,112	7.616	3.871
9	81	729	3.000	2.080	59	3,481	205,379	7.681	3.893
10	100	1,000	3.162	2.154	60	3,600	216,000	7.746	3.915
11	121	1,331	3.317	2.224	61	3,721	226,981	7.810	3.936
12	144	1,728	3.464	2.289	62	3,844	238,328	7.874	3.958
13	169	2,197	3.606	2.351	63	3,969	250,047	7.937	3.979
14	196	2,744	3.742	2.410	64	4,096	262,144	8.000	4.000
15	225	3,375	3.873	2.466	65	4,225	274,625	8.062	4.021
16	256	4,096	4.000	2.520	66	4,356	287,496	8.124	4.041
17	289	4,913	4.123	2.571	67	4,489	300,763	8.185	4.062
18	324	5,832	4.243	2.621	68	4,624	314,432	8.246	4.082
19	361	6,859	4.359	2.668	69	4,761	328,509	8.307	4.102
20	400	8,000	4.472	2.714	70	4,900	343,000	8.367	4.121
21	441	9,261	4.583	2.759	71	5,041	357,911	8.426	4.141
22	484	10,648	4.690	2.802	72	5,184	373,248	8.485	4.160
23	529	12,167	4.796	2.844	73	5,329	389,017	8.544	4.179
24	576	13,824	4.899	2.884	74	5,476	405,224	8.602	4.198
25	625	15,625	5.000	2.924	75	5,625	421,875	8.660	4.217
26	676	17,576	5.099	2.962	76	5,776	438,976	8.718	4.236
27	729	19,683	5.196	3.000	77	5,929	456,533	8.775	4.254
28	784	21,952	5.292	3.037	78	6,084	474,552	8.832	4.273
29	841	24,389	5.385	3.072	79	6,241	493,039	8.888	4.291
30	900	27,000	5.477	3.107	80	6,400	512,000	8.944	4.309
31	961	29,791	5.568	3.141	81	6,561	531,441	9.000	4.327
32	1,024	32,768	5.657	3.175	82	6,724	551,368	9.055	4.344
33	1,089	35,937	5.745	3.208	83	6,889	571,787	9.110	4.362
34	1,156	39,304	5.831	3.240	84	7,056	592,704	9.165	4.380
35	1,225	42,875	5.916	3.271	85	7,225	614,125	9.220	4.397
36	1,296	46,656	6.000	3.302	86	7,396	636,056	9.274	4.414
37	1,369	50,653	6.083	3.332	87	7,569	658,503	9.327	4.431
38	1,444	54,872	6.164	3.362	88	7,744	681,472	9.381	4.448
39	1,521	59,319	6.245	3.391	89	7,921	704,969	9.434	4.465
40	1,600	64,000	6.325	3.420	90	8,100	729,000	9.487	4.481
41	1,681	68,921	6.403	3.448	91	8,281	753,571	9.539	4.498
42	1,764	74,088	6.481	3.476	92	8,464	778,688	9.592	4.514
43	1,849	79,507	6.557	3.503	93	8,649	804,357	9.644	4.531
44	1,936	85,184	6.633	3.530	94	8.836	830,584	9.695	4.547
45	2,025	91,125	6.708	3.577	95	9,025	857,375	9.747	4.563
46	2,116	97,336	6.782	3.583	96	9,216	884,736	9.798	4.579
47	2,209	103,823	6.856	3.609	97	9,409	912,673	9.849	4.595
48	2,304	110,592	6.928	3.634	98	9.604	941,192	9.899	4.610
49	2,401	117,649	7.000	3.659	99	9,801	970,299	9.950	4.626
50	2,500	125,000	7.071	3.684	100	10,000	1,000,000	10.000	4.642

Glossary

1. When a bar is placed over a vowel, the vowel says its own name.
2. A curved mark over a vowel indicates the following sounds:
   a. ă as in at
   b. ě as in bed
   c. ĭ as in it
   d. ŏ as in ox
   e. ŭ as in rug

Acute (ŭ kūt′)
  (angle) Angle whose measure is less than 90°.
  (triangle) Triangle with three angles less than 90°.

Add (ăd) To combine into one sum or quantity.

Addend (ad′ ĕnd) Number or quantity to be added to another.

Addition (ă dĭ′ shŭn) Act of adding.

Additive (ăd′ ĭ tĭv) **(inverses)** (*See* Opposites.)

Algebraic (ăl′ jŭ brā′ ĭk) **(expression)** Open expressions containing variables: such as $a + 3$, $5y - 2x$, $9(b + 7)$, $(x - 2)(x + 3)$, $\dfrac{x}{y} - z$, etc.

Altitude (ăl′ tĭ tōōd)
  (of a parallelogram) Perpendicular distance between two parallel sides.
  (of a trapezoid) Perpendicular distance between the two parallel sides.
  (of a triangle) Line segment from the vertex of a triangle perpendicular to the base.

Angle (ăng′ g′l) Two rays with a common endpoint.

Area (air′ ē ŭ) Measure (in square units) of the region within a closed curve (including polygons) in a plane.

Ascending (ă sĕnd′ ĭng) **(order of the terms of a polynomial with respect to a particular variable)** Terms of a polynomial arranged so that the degree of each term increases from left to right with respect to a particular variable.

Associative (ŭ sō′ shŭ tĭv) Indicates that the grouping of three numbers in an addition or multiplication can be changed without affecting the sum or product.

Base (bās)
  (of a cylinder) Two circular regions in a right-circular cylinder.
  (of a power) Number being raised to a power, such as 3 in 3^5.
  (of a triangle) Side opposite a vertex.

Bases (bās′ ĭz) **(of a trapezoid)** Parallel sides.

Binomial (bī nō′ mē ŭl) Polynomial that has exactly two terms.

Braces (brās′ ĭz) Symbols, { }, used for grouping expression or to indicate a set, such as $\{0, 1, 2, \ldots\}$.

Circle (sur′ kŭl) Set of all points in a plane whose distance from a given point (center) is equal to a positive number, r (radius).

Circumference (sur kŭm′ fur ĕns) **(of a circle)** Distance around.

Coefficient (kō′ ĕ fĭsh′ ĕnt) (*See* Numerical coefficient.)

Commutative (kŭ mū′ tŭ tĭv) Indicates that the order of two numbers in an addition or multiplication can be changed without affecting the sum or product.

Complementary (kŏm plŭ mĕn′ tŭ rē) **(angles)** Two or more angles whose measures added together result in a sum of 90°.

Complex (kŏm′ plĕks) **(fraction)** Rational expression that has a fraction in its numerator or its denominator or both.

Composite (kŏm pŏz′ ĭt) **(number)** Natural number that has more than two natural-number factors.

Cone (kōn) **(right-circular)** Circular region (base) and the surface made up of line segments that connect the circle with a point (vertex) located on a line through the center of the circle and perpendicular to the plane of the circle.

Conjugates (kŏn′ jōō gŭts) Two binomials that differ only in the sign of the second term, such as $a + b$ and $a - b$.

Consecutive (kŏn sĕk′ ū tĭv) **(integers)** Integers that differ by 1.

Consistent (kŏn sĭs′ tĕnt) **(equations)** Two equations in two unknowns (variables) which have only one solution.

Constant (kŏn′ stănt) Numbers without a literal coefficient, such as $5, -7, \frac{7}{8}, 0$.

Coordinate axes (kō or′ dĭ nĭt) (ăk′ sēz) x axis and y axis in a rectangular-coordinate system.

Coordinates (kō or′ dĭ nĭts) **(of a point)** Two numbers that locate a point in a rectangular-coordinate system, such as $(3, 2)$, which indicates that the x coordinate is 3 and the y coordinate is 2.

Cube (kūb) Polyhedron (solid where all faces are polygons) of six equal faces.
(of a number) Indication that a number is being used three times as a factor, such as 2^3 $(2 \cdot 2 \cdot 2)$.
(root) The value b such that $\sqrt[3]{a} = b$ if $b^3 = a$.

Cylinder (sĭl′ ĭn dur) **(right-circular)** Two circular regions with the same radius in parallel planes (bases) connected by line segments perpendicular to the planes of the two circles.

Cylindrical (sĭ lĭn′ drĭ kŭl) Relating to, or having the properties of, a cylinder.

Decimal (dĕs′ ĭ mŭl) **(fraction)** Fraction whose denominator is a power (multiple) of 10, such as $0.5 = \frac{5}{10}$ or $2.67 = \frac{267}{100}$.

Degree (dŭ grē′)

(measurement of an angle) $\frac{1}{360}$ of a complete revolution of a ray around its endpoint.

(of a polynomial) Greatest degree of any of its terms.
(of a term) The degree of a constant is zero. The number zero has no degree. The degree of a term of one variable agrees with the exponent on the variable (the degree of a term with more than one variable is equal to the sum of all exponents on individual variables).

Denominator (dŭ nŏm′ ĭ nā tur) Bottom number in a fraction, such as 3 in $\frac{2}{3}$.

Dependent (dē pĕn′ dĕnt) **(equations)** Two equations in two unknowns (variables) which have an infinite number of solutions.

Descending (dē sĕnd′ ĭng) **(order of the terms of a polynomial with respect to a particular variable)** Terms of a polynomial arranged so that the degree of each term decreases from left to right with respect to a particular variable.

Diameter (dī ăm′ ŭ tur) **(of a circle)** Twice the radius (of the circle).

Difference (dĭf′ ur ĕns) Result of a subtraction of two numbers, such as 5 in $7 - 2 = 5$.

Digit (dĭj′ ĭt) Any one of the symbols 0, 1, 2, 3, 4, 5, 6, 7, 8, 9.

Disjoint (dĭs joint′) **(sets)** Two or more sets that have no common elements (the intersection of disjoint sets is empty).

Distributive (dĭs trĭb′ ū tĭv) **(property of multiplication over addition or subtraction)** Property that permits changing sums or differences to products (or vice versa); that is,

$$a(b + c) = ab + ac$$
$$a(b - c) = ab - ac$$
$$(a + b)c = ac + bc$$
$$(a - b)c = ac - bc$$

Divide (dĭ vīd′) Process of determining the number of equal parts in a number or quantity.

Dividend (dĭv′ ĭ dĕnd) Number in a division problem which is being divided, such as 15 in $15 \div 3$.

Division (dĭ vĭ′ zhŭn) Act of dividing.

Divisor (dĭ vī′ zur) Number by which another number (dividend) is divided, such as 3 in $15 \div 3$.

Elements (ĕl′ ŭ ments) Things that make up a set.

Empty (ĕmp′ tē) **(set)** Set with no elements.

Equation (ē kwā′ zhŭn) Statement that the expressions on opposite sides of an equal sign represent the same number, such as $2 + 3 = \dfrac{12 - 2}{2}$ (mathematical statement of equality).

Equilateral (ē′ kwĭ lāt′ ur ŭl) **(triangle)** Triangle with three equal sides.

Equivalent (ŭ kwĭv′ ŭ lent)
  (equations) Two or more equations that have the same solution set.
  (expressions) Two or more expressions that have the same evaluation for all replacements of the variable(s).

Exponent (ĕks′ pō nŭnt) Indicates the number of times the base is used as a factor. For example, in 2^3, 3 is the exponent and indicates that 2 (base) is being used three times as a factor ($2 \cdot 2 \cdot 2$).

Extremes (ĕks trēmz′) **(of a proportion)** First and fourth terms of a proportion, such as 2 and 35 in $2:5 = 14:35$, or $\dfrac{2}{5} = \dfrac{14}{35}$.

Factor (făk′ tur) One of the values in a multiplication expression, such as 3 in $(3)(5)$.

Finite (fī′ nīt) **(set)** Set in which the elements can be counted and the count has a last number.

Fraction (frăk′ shŭn) Number which indicates that some whole has been divided into a number of equal parts and that a portion of the equal parts is represented.

Hexagon (hĕk′ sŭ gŏn) Polygon with six sides.

Horizontal (hŏr′ ĭ zŏn′ tŭl) **(line)** Line parallel to the plane of the horizon.

Hypotenuse (hī pŏt′ ŭ nōōs) **(of a right triangle)** Side opposite the right angle.

Improper (ĭm prŏp′ ur) **(fraction)** Fraction that has a value greater than or equal to 1.

Inconsistent (ĭn kŏn sĭs′ tĕnt) **(equations)** Two equations in two unknowns (variables) which have no common solution.

Index (ĭn′ dĕks) **(of a radical)** Number that indicates the desired root, such as 3 in $\sqrt[3]{8}$ (the index 2 is understood when the radical sign is being used to indicate square root).

Inequality (ĭn′ ē kwăl′ ĭ tē) Expression consisting of two unequal quantities, with the sign of inequality ($<$, $>$, $\neq$) between them.

Infinite (ĭn′ fĭ nĭt) **(set)** Set whose count is unending.

Integer (ĭn′ tŭ jur) Any number that is in the following list: ..., −2, −1, 0, 1, 2, ...

Intersection (ĭn′ tur sĕk′ shŭn) **(of two sets)** Third set that contains those elements, and only those elements, that belong to both (one and the other) of the original sets.

Inverse (ĭn vurs′) **(operations)** Addition and subtraction, multiplication and division are said to be inverse operations, since one undoes the effect of the other; that is, $5 + 2 − 2 = 5$ or $5 − 2 + 2 = 5$ and $7(2) ÷ 2 = 7$ or $7 ÷ 2(2) = 7$.

Irrational (ĭr răsh′ ŭn ŭl) **(numbers)** Infinite nonrepeating decimals (square roots of positive nonperfect squares are but one class of irrational numbers).

Isosceles (ī sŏs′ ĕ lēz) **(triangle)** Triangle with at least two equal sides.

Kilogram (kĭl′ ŭ grăm) Base (standard) unit of weight in the metric system (1 kilogram = 1000 grams).

Lateral surface (lăt′ ur ŭl) (sur′ fĭs) **(of a cylinder)** Curved surface connecting the bases.

Least common denominator (LCD) Smallest number that is exactly divisible by each of the original denominators of two or more fractions.

Like terms Terms of an algebraic expression which have exactly the same literal coefficients (including exponents).

Line (līn) Set of points represented with a picture, such as ⟵——————⟶ (A line is always straight, has no thickness, and extends forever in both directions.)

Linear (lĭn′ ē ur)
(equation) Equation that can be written in the form $ax + by + c = 0$, where a, b, and c are real numbers, a and b both not zero.
(measurement) Measurement along a (straight) line.

(line) segment (sĕg′ mĕnt) Portion of a line between two points, including its endpoints.

Liter (lē′ tur) Base (standard) unit of volume or capacity in the metric system.

Literal coefficient (lĭt′ ur ŭl) (kō′ ĕ fĭsh′ ŭnt) Letter factor of an indicated product of a number and one or more variables.

Means (mēnz) **(or a proportion)** Second and third terms of a proportion, such as 5 and 14 in $2:5 = 14:35$ or $\frac{2}{5} = \frac{14}{35}$.

Meter (mē′ tur) Base (standard) unit of length in the metric system.

Minuend (mĭn′ ū ĕnd) Number or quantity from which another (subtrahend) is to be subtracted, such as 17 in $17 − 3$.

Mixed number Understood sum of a whole number and a proper fraction.

Monomial (mō nō′ mē ŭl) Polynomial that contains only one term.

Multiple (mŭl′ tĭ p'l) Product of a quantity by an integer.

Multiplication (mŭl′ tĭ plĭ kā′ shŭn) Act of multiplying.

Multiplicative (mŭl′ tĭ plĭk′ ŭ tĭv) **(inverses)** (*See* Reciprocals.)

Multiply (mŭl′ tĭ plī) To take by addition a certain number of times.

Natural (năt′ jŭ rŭl) **(or counting number)** Any number that is in the following list: 1, 2, 3, 4, ...

Negative (nĕg′ ŭ tĭv) **(number)** Any number that is located to the left of zero on a horizontal number line.

Numeral (nōō′ mur ŭl) A symbol that represents a number, such as V, ⫻, and 5, which are all numerals naming the number five.

Numerator (noo′ mur ā′ tŭr) Top number in a fraction, such as 2 in $\frac{2}{3}$.

Numerical (noo mĕr′ ĭ kŭl) **(coefficient)** Number factor of an indicated product of a number and a variable.

Obtuse (ŏb toos′)
  (angle) Angle whose measure is between 90° and 180°.
  (triangle) Triangle that has one angle greater than 90°.

Octagon (ŏk′ tŭ gŏn) Polygon with eight sides.

Open expression Expression in which the position of an unknown number is held by a letter (variable). (*See* Algebraic expression.)

Open sentence Equation that does not contain enough information to be judged as either true or false, such as $x + 2 = 7$.

Opposite (ŏp′ ŭ zĭt) **(sides of a quadrilateral)** Pairs of sides that do not intersect.

Opposites (ŏp′ ŭ zĭtz) **(additive inverses)** Two numbers on a number line which are the same distance from zero, such as ⁻5 and 5, $6\frac{2}{3}$ and $⁻6\frac{2}{3}$. (The sum of opposites is zero; that is, $a + ⁻a = 0$.)

Ordered pair (of numbers) Expression (x, y) which locates a point on a rectangular-coordinate system (x indicates the distance and direction in a horizontal direction from the origin, y indicates the distance and direction in a vertical direction).

Origin (or′ ĭ jĭn) Point of intersection of the coordinate axes in a rectangular-coordinate system; identified by the ordered pair $(0, 0)$.

Parallel (păr′ ŭ lĕl) **(lines)** Two or more lines in the same plane that have no points in common.

Parallelogram (păr′ ŭ lĕl′ ŭ gram) Quadrilateral with opposite sides parallel.

Parentheses (pŭ rĕn′ thŭ sēz) Symbols, ( ), used for grouping expressions, such as $5 + (a + 3)$ or $2(x − 1)$; or to indicate an ordered pair of numbers, such as $(2, 3)$.

Pentagon (pĕn′ tŭ gŏn) Polygon with five sides.

Percent (pur sĕnt′) Fraction with a denominator of 100, such as $\frac{7}{100} = 0.07 = 7$ percent.

Perfect square
  (integer) Integer obtained by squaring an integer.
  (monomial) Square of a monomial.
  (rational number) Square of a rational number.
  (trinomial) Square of a binomial (trinomial of the form $a^2 \pm 2ab + b^2$).

Perimeter (pŭ rĭm′ ŭ tur) **(of a polygon)** Total length of all the sides of the polygon.

Perpendicular (pur′ pĕn dĭk′ ū lur) **(lines)** Two intersecting lines that form a right angle.

Plane (plān) Flat surface (such as a tabletop) that extends infinitely in every direction.

Point Location in space that has no thickness (represented on paper by a dot).

Polygon (pŏl′ ĭ gŏn) Plane closed figure of three or more angles.

Polynomial (pŏl′ ĭ nō′ mē ŭl) Algebraic expression made up of sums, differences, and products of variables and numbers.

Positive (pŏz′ ĭ tĭv) **(number)** Any number that is located to the right of zero on a horizontal number line.

Power (pow′ ur) Expression used when referring to numbers with exponents, such as 3^5, "the fifth power of three."

Prime (prīm)
    (number) Natural number other than 1 divisible by exactly two natural-number factors, itself and 1.
    (polynomial over the set of integers) Polynomial that does not contain an integer factor other than 1 or −1 and cannot be factored into two other polynomials with integer coefficients.

Principal (prĭn′ sĭ p'l) Amount of money invested or borrowed.

Product (prŏd′ ŭkt) Result of a multiplication problem.

Proper (prŏp′ ur) **(fraction)** Fraction that has a value less than 1.

Proportion (prŭ pōr′ shŭn) Statement of equality between two ratios.

Quadrant (kwŏd′ rănt) Any of the four parts into which a plane is divided by rectangular coordinate axes lying in that plane.

Quadratic (kwŏd răt′ ĭk) **(equation)** Equation that can be written in the form $ax^2 + bx + c = 0$, where a, b, and c are real numbers and $a \neq 0$.

Quadrilateral (kwŏd′ rĭ lăt′ ur ŭl) Polygon with four sides.

Quantity (kwŏn′ tĭ tē) Indicates an expression that has been placed within parentheses (can represent an entire expression, such as the quantity $x + 7$).

Quotient (kwō′ shŭnt) Number or quantity resulting from the division of one number or quantity (dividend) by another (divisor).

Radical sign (răd′ ĭ kŭl) (sīn) Symbol $\sqrt{}$ ($\sqrt[n]{a}$ is the nth root of a).

Radicand (răd′ ĭ kănd) Number under the radical sign, such as 2 in $\sqrt[3]{2}$.

Radii (rā′ dē ī) Plural of radius.

Radius (rā′ dē ŭs) **(of a circle)** Line segment connecting the center of a circle to any point on the circle (measure of this distance).

Ratio (rā′ shō) Quotient of two quantities or numbers.

Rational (răsh′ ŭn ŭl)
    (expression) Expression in which both numerator and denominator are polynomials, excluding all possible values of the variables which would produce a zero denominator.
    (number) Any number that can be written in the form $\dfrac{p}{q}$, where p and q are integers and $q \neq 0$ (terminating or infinitely repeating decimals).

Ray (rā) Part of a line on one side of a point which includes the point (endpoint of the ray).

Real (rē ŭl) **(number)** Either a rational or irrational number (the union of the set of rational numbers and the set of irrational numbers is the set of real numbers).

Reciprocals (rŭ sĭp′ rŭ kălz) **(multiplicative inverses)** Two numbers whose product is 1, such as 5 and $\dfrac{1}{5}$, $\dfrac{-5}{8}$ and $\dfrac{-8}{5}$, x and $\dfrac{1}{x}$ ($x \neq 0$).

Rectangle (rĕk′ tăng g'l) Parallelogram whose sides meet at right angles.

Rectangular (rĕk tăng′ gū lur) **(coordinate system)** Formed by crossing (intersecting) vertical and horizontal number lines.

Reduce (rŭ doōs′) Process of dividing both numerator and denominator of a fraction by some common factor.

Relation (rŭ lā′ shŭn) Set of ordered pairs of numbers.

Replacement set Set of permissible values of a variable in an open (algebraic) expression or equation.

Right
> **(angle)** Any of the angles formed by two intersecting lines, where all four angles are equal (the measure of a right angle is 90°).
> **(triangle)** Triangle with one angle of 90°.

Scalene (skā lēn′) **(triangle)** Triangle that has no pair of sides equal.

Set (sĕt) Well-defined (membership in the set is clear) collection of things.

Set notation The use of braces to indicate a set, such as {2, 3, 7} indicates the set containing the numbers 2, 3, and 7.

Solution (sŭ loo′ shŭn) **(truth set)** Set of all values from the replacement set which converts the open sentence into a true statement.

Solutions (sŭ loo′ shŭnz) **(to an equation)** All values of the variable from the replacement set that convert the open sentence (equation) into a true statement.

Square (skwair) Rectangle whose sides are of equal length.
> **(of a number)** Indication that a number is being used twice as a factor, such as 3^2 $(3 \cdot 3)$.
> **(roots)** The square roots of a number are the values that when squared will produce the number, such as the square roots of 16 are 4 and ⁻4.

Subset (sŭb′ sĕt) Set in which each element is also an element of another set.

Subtract (sŭb trăkt′) To withdraw or take away, as one number from another.

Subtraction (sŭb trăk′ shŭn) Act of subtracting.

Subtrahend (sŭb′ trŭ hĕnd′) Number or quantity to be subtracted from another (minuend), such as 3 in $17 - 3$.

Sum (sŭm) Result of an addition of two or more numbers.

Supplementary (sŭp′ lĭ mĕn′ tŭ rē) **(angles)** Two or more angles whose measures added together result in a sum of 180°.

Terms (turmz) Parts of an expression separated by an addition symbol.

Trapezoid (trăp′ ŭ zoid) Quadrilateral in which only one pair of opposite sides are parallel.

Triangle (trī′ ăng g′l) Polygon with three sides.

Trinomial (trī nō′ mē ŭl) Polynomial that has exactly three terms.

Union (ūn′ yŭn) **(of two sets)** Third set that contains all those elements that belong to either (one or the other) of the original sets.

Variable (vair′ ĭ ŭ b′l) Letter in an algebraic expression that can be replaced by any one of a set of many numbers.

Vertex (vur′ tĕks) **(of an angle)** Common endpoint of two rays that form an angle.

Vertical (vur′ tĭ kŭl) **(line)** Line perpendicular to the plane of the horizon.

Vertices (vur′ tĭ sēz) **(of a polygon)** Plural of vertex.

Volume (vŏl′ yŭm) Measure (in cubic units) of the space within a closed solid figure.

Whole (hōl) **(number)** Any number that is in the following list: 0, 1, 2, 3, . . . (the next whole number is formed by adding 1 to the previous whole number).

***x* axis** (x ăk′ sĭs) Horizontal number line in a rectangular coordinate system.

***y* axis** (y ăk′ sĭs) Vertical number line in a rectangular coordinate system.

Index